picta

platensis

In California the salamander *Ensatina eschscholtzi* is found in the Coast Range and in the Sierra Nevada mountains, distributed in a circular pattern around the Great Valley. Down the west coast, from *picta* to *eschscholtzi,* these salamanders present a reddish, unblotched appearance; on the eastern side of the state, an increasingly blotchy pattern is seen, culminating in the black-and-yellow *klauberi.*

With one exception, the various populations of this species interbreed where their ranges overlap. Where *klauberi* and *eschscholtzi* overlap, however, little or no interbreeding occurs. These "end" populations of a classic "circle of races" are of considerable biological interest. The lack of interbreeding has led to questions about the degree of kinship between these terminal races. Biologists are asking whether evolution has not resulted in the two becoming separate species, instead of remaining members of the same.

oregonensis

croceater

klauberi

xanthoptica

eschscholtzi

Fundamental Concepts of Biology

SECOND EDITION

GIDEON E. NELSON
University of South Florida

GERALD G. ROBINSON
University of South Florida

RICHARD A. BOOLOOTIAN
Institute of Visual Medicine
Los Angeles, California

John Wiley & Sons, Inc. *New York · London · Sydney · Toronto*

Illustrations for the Second Edition by MISS LANEY HICKS
with illustrations included from the first edition by
Mrs. Geraldine Beye Robinson

Library of Congress Catalogue Card Number: 75-100329

SBN 471 63151 5

Printed in the United States of America

10 9 8 7 6 5 4 3 2 1

Preface

In preparing the revised edition of *Fundamental Concepts of Biology*, we maintained the same objectives that guided our writing of the first edition.

The same order of chapters has been retained, except for the placement of Development after Post-Mendelian Genetics, but many topics have been revised and rewritten. This was necessary, in our opinion, for three reasons: to update certain materials, to expand some topics with additional experimental evidence and applications, and to improve the clarity or readability of certain sections.

Topics that have been expanded include protein and enzyme structure, role of the environment in photosynthesis, structure and function of cells, hormones and growth substances, mitosis and meiosis, operon theory, speciation, and trends in evolution.

New topics include photosynthetic pigments, experimental materials involving DNA and RNA, receptors, invertebrate nervous systems, human sex chromosomes and linkage, Hans Spemann's work on organizers, and a consideration of a concept of the chemical origin of life.

The illustrations for the second edition have been significantly revised for clarification and to effect better coordination with the text. Many new illustrations appear for the first time. The color yellow has been added to augment the visual presentation. It has been used in a functional manner to call attention to pathways, cycles, and so forth—wherever the additional color can help the interpretation of an illustration. The captions have been expanded to allow fuller understanding of visual information.

The artwork for the second edition was designed and executed by Miss Laney Hicks, with illustrations included from the first edition by Mrs. Geraldine Beye Robinson. The assistance of Dr. Wendel Lim, Miss Jane Larson, and Miss Margery Koerner to Miss Hicks is gratefully acknowledged. In addition, Mrs. Margaret Schmidt was helpful in preparing some of the illustrations for reproduction.

New photographs have been included in this edition and old ones were replaced. Many of the photomicrographs were made from Turtox slides, which were of outstanding quality.

We were pleased with the acceptance of the first edition by teachers and students across the country. Especially helpful were the comments, suggestions, and criticisms made by some users. We hope that this revised edition incorporates the best of these.

Gideon E. Nelson
Gerald G. Robinson
Richard A. Boolootian

August, 1969

Preface to the First Edition

This textbook was written with two primary objectives. The first is to help students develop an understanding of the operation of biological systems through acquaintance with selected basic biological concepts and principles. The topics chosen are, in our experience and judgment, those that are most important to the understanding of biological systems as a whole. Some of the material is presented in a fairly traditional manner, but more of it, we hope, breaks away from conventional treatments which present surveys of the field of biology.

The second objective, which also influenced the selection of topics included in this book, is to provide a foundation of subject matter that enables students to achieve a better interpretation and evaluation of the types of biological information they encounter in newspapers, magazines, and semiscientific journals—in short, in everyday life. The value of this type of activity to an informed citizenry seems obvious in this age of rapid scientific advance.

The scheme of presentation we have followed starts with the smallest fundamental units of living material and gradually progresses toward the largest functioning biological system, the world of living things. Chapters II and III deal with the structural features of living matter, commencing with its basic chemical and physical composition and then examining in considerable detail the unit called the cell. These chapters are followed by three devoted to photosynthesis, respiration, and transport, the key metabolic processes which are so characteristic of life.

The subject of regulation dominates Chapters VII through XI. Here we treat the coordination of cellular activities, the coordination and control among cells by hormones and the nervous system, the maintenance of the entire organism, and finally, the integration of behavior made possible by communication among organisms. We then proceed to the subjects of reproduction and development, which are discussed in Chapters XII–XIV. This leads logically to a consideration of the genetics of individuals and populations (Chapters XV–XVIII). We devote four chapters to this important topic because we believe that every educated person should have a broad knowledge of heredity and its many applications to human society.

Darwin's contribution to the principles of evolution has been termed the major unifying principle in biology. The significant subject of evolution together with its evidence and mechanisms occupy the next two chapters, XIX and XX. The final five chapters of the book examine organisms, their environments, and the relationships between the two. The last of these chapters selects a major habitat type, aquatic environments, and applies to it the major principles presented in the preceding four.

A book such as this presents a number of opportunities for the introduction of outside readings. Many teachers of biology prefer two types of readings: those of a historical and classical nature, and material such as *Scientific American* offprints which present recent information on various topics. Many references of this sort are available at reasonable cost to be utilized as individual instructors see fit.

We would like to acknowledge the assistance of those who have contributed to the clarity and accuracy of this text by reading all or part of the manuscript and suggesting possible modifications. These include Mr. James Campbell, Dr. Marvin Cantor, Dr. Elof Carlson, Dr. Jack E. Fernandez, Dr. John W. Hall, Dr. Joe R. Linton, Dr. Knut Norstog, Dr. Eugene Odum, and Dr. James D. Ray, Jr. In addition we would like to recognize the assistance of Mrs. Janice Duncan, Mr. Arthur Hepner, Miss Joyce McKee, and Mr. Thomas L. Sears in the preparation of the manuscript.

We wish to thank especially the General Biological Supply House, Inc., for the use of prepared microscope slides, from which we made many of the photomicrographs appearing in this book.

Gideon E. Nelson
Gerald G. Robinson
Richard A. Boolootian

June, 1966

Contents

CHAPTER I

Life: Its Characteristics and Study

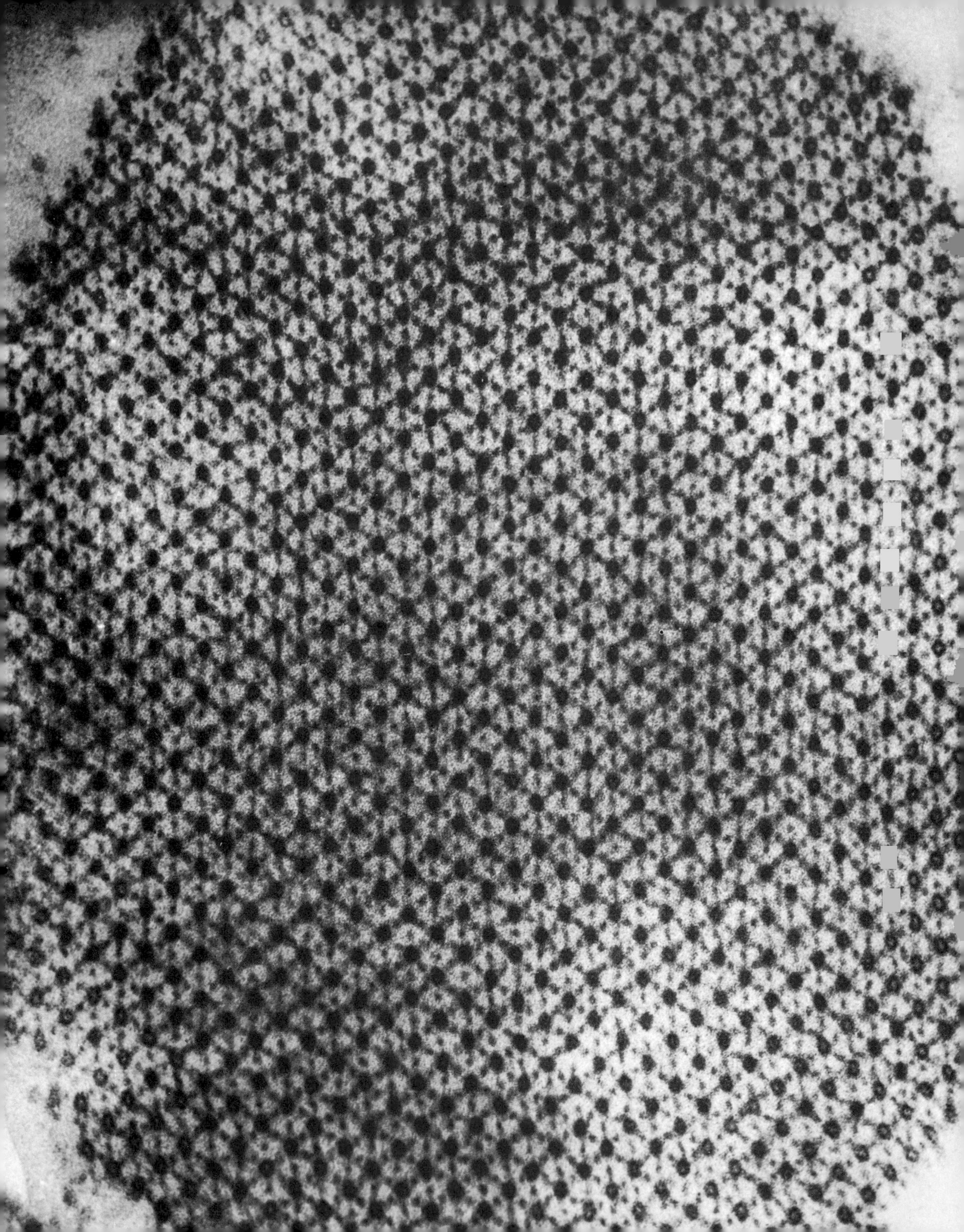

An electron photomicrograph of muscle tissue in cross section (×215,000) (Dr. H. E. Huxley).

CHAPTER I

Life: Its Characteristics and Study

Man turns to science in his efforts to understand the mysteries of the universe. For knowledge of the living world he relies on biology, the science of life itself. A tiny acorn grows into a massive oak tree; a bougainvillea plant that lends rich coloring to a tropical isle cannot survive the northern Minnesota climate; a child cannot easily be told apart from his identical twin; a quarterback throws a touchdown pass three weeks after bone surgery. What actually makes each of these dramatic events possible becomes clear from an understanding of the fundamentals of modern biology.

To many students, biology suggests dissecting animals in a laboratory, collecting leaves and flowers, and memorizing obscure Latinized names for the myriad plants and animals. These activities no more characterize the whole field of biology than heating chemicals in a beaker typifies the science of chemistry. The true scope of biology is all of life.

The science of biology treats life as a series of processes whose natures may be discovered through observation and experimentation. These processes cover a vast range of phenomena—from how cells develop into tissues and organisms, to the changes that occur in organisms over a period of time. They include such matters as how living organisms obtain their energy from chemical sources, how certain mechanisms control and regulate the functioning of plant and animal parts, and how organisms

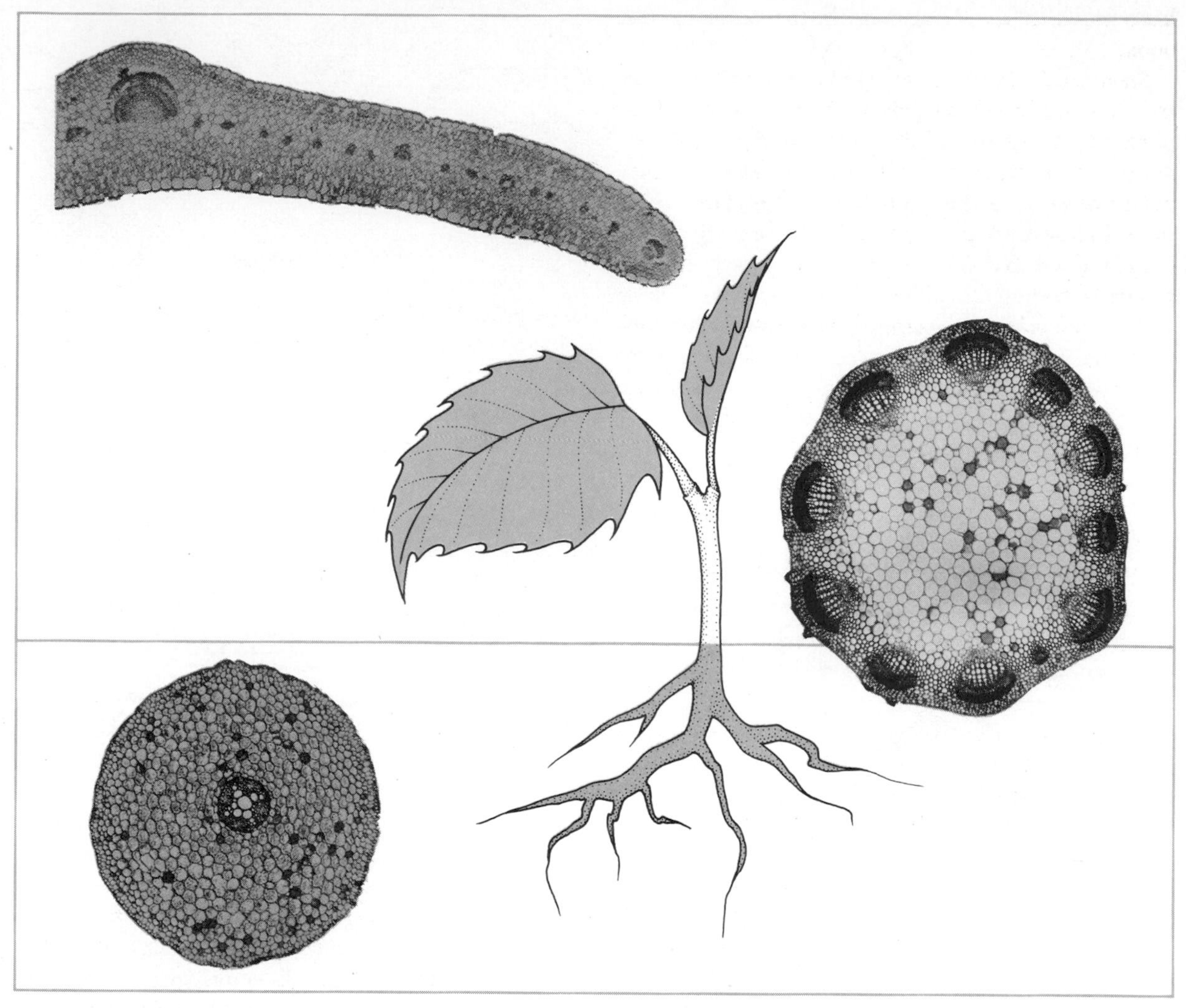

Figure 1.1. A generalized flowering plant consists of several principal organs including the root (yellow), leaf (blue), and stem. The structural components of these three organs are shown in cross section.

reproduce and pass on their traits from one generation to the next. Our knowledge of the living world gains perspective from a description of its features and behavior, and from extensive study of biological principles.

In a broad sense, biology deals with all living things, including both plants and animals. The plant and animal kingdoms contain millions of organisms in an almost limitless variety of sizes and shapes. These differ as widely as the single-celled bacterium and the million-celled poinsettia plant, or the microscopic amoeba and the six-foot, one-inch man. Despite the profusion of different organisms, all share certain features to some degree. The study of biology seeks to explain the basis of these similarities and disparities.

Characteristics of Living Things

All living things have structure, experience metabolism, engage in regulation and control, reproduce and develop, inherit from their forebears, and adapt to their surroundings ultimately participating in the process of evolution. These many areas point to some of the major characteristics of life that have occupied biologists for years. Since we believe that these topics will continue to be important, we shall have much to say about them in the course of this book.

Structure. Most flowering plants, like the cherry tree and the rosebush, consist of several principal parts called organs. All of us are familiar with their names—root, stem, leaf, flower. Each of these organs performs an important function in the plant. If we take a close look at them under laboratory conditions, we see that each organ is composed of many groups of tissues, all having jobs of their own (Figure 1.1). In the stem, for example, one tissue may conduct the passage of water and salts, another may transport food, another may supply storage facilities, and there are also supporting tissues and an outer protective tissue. If we place any of these tissues under a microscope for a still closer look, we observe that it consists of a mass of cells, which are the basic units of biological organization.

When we turn this hierarchy around, in the realm of the so-called higher plants, we note that a group of similar cells with a similar function forms a tissue, a group of tissues forms an organ, and a group of organs composes the plant itself. The location and arrangement of the tissues and organs are generally similar in all plants even though their structures may differ. For example, a cactus plant and a pear tree may share many internal features while having quite different external appearances.

The so-called lower plants, like mosses, algae, or fungi, are simpler in structure than the higher plants. Some, like liverworts, may consist only of cells and tissues, and many of the algae are single celled (Figure 1.2).

The structural—or *morphological*—plan of animals follows the same course as in plants, except that the range of complexity is much greater. Thus, the cat is composed of many *systems* of organs, such as the nervous, glandular, and muscular systems. Each of these systems is made up of many organs. For example, the upper bone of the leg is an organ of the skeletal system (Figure 1.3). As in plants, the organs of animals are groups of tissues which, in turn, are clusters of cells.

As in the plant kingdom, lower animals have less complex structural formations than higher forms of animal life. *Hydra,* for example, is barely beyond the cell-tissue level of organization, and most of the minuscule animals called protozoans consist only of single cells (Figure 1.4). It should be noted, however, that some protozoans contain numerous tiny parts

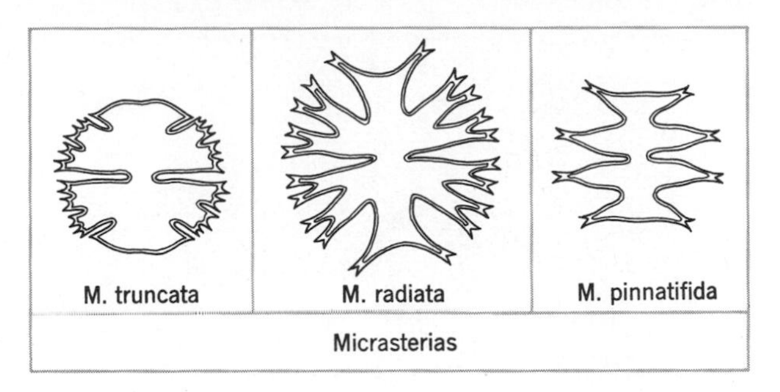

Figure 1.2. These three species of the freshwater alga Micrasterias *vary widely in shape despite the fact that each is a single cell.*

(organelles) and are far from simple in their organization.

All living things, from the smallest to the largest, thus share structure as a common characteristic. In reality, an organism's structure is inseparable from its function since each influences and modifies the other. Nevertheless, we separate them for convenience in discussion and description.

Metabolism. Living organisms acquire and utilize the energy they need through intricate chemical changes that take place mostly within the cells. These important processes constitute metabolism, a second significant characteristic of living things. Although quite varied, the chemical reactions involved in metabolism show marked similarities among a wide range of organisms. In their own unique ways, two organisms as unlike as lotus flowers and porcupines obtain nourishment, strength, hardiness, and other requisites for survival through similar chemical reactions.

Every biological activity—whether growth, reproduction, movement, or production of chemical products—demands an expenditure of energy. For maintenance and repair of its biological structure alone, every form of life continually consumes an enormous amount of energy. The chemical process that captures this energy occurs mostly in the cells and is much the same in most instances. For all living organisms, the ultimate energy supply, in the form of carbon compounds, is derived from *photosynthesis,* a metabolic process dependent on sunlight and the green coloring matter of plants (Figure 1.5).

Research into metabolism falls, for the most part, into the field of biochemistry. Biologists have identified many of the enzymes that are crucial to metabolism, and have attempted to find out exactly where

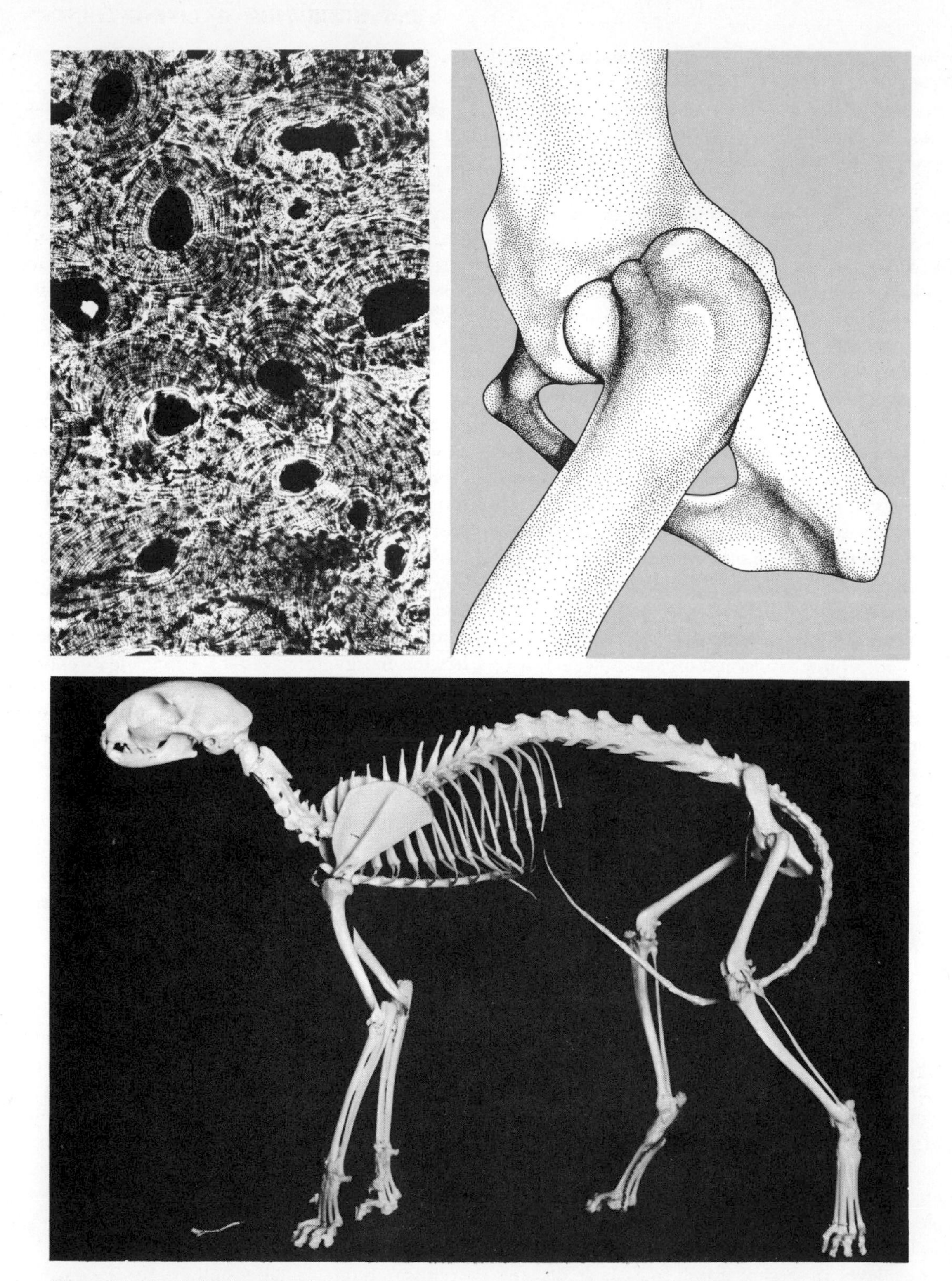

Figure 1.3. The skeletal system of the cat consists of many organs, including the upper leg and pelvic bones shown at the right above. Bones consist of the specialized cells and tissues shown at the upper left ($\times 140$).

they function in various biochemical reactions in the cell. Researchers have also sought to determine the structure of proteins—in particular, the sequences of amino acids, the basic chemical units from which proteins are built.

Photosynthesis has been subjected to extensive study. Use of new research tools like the electron microscope has acquainted man with the structure of the *chloroplasts* (Figure 1.5), the site where photosynthesis occurs in the cell. In addition, biochemical studies have revealed many details about the sequence of chemical reactions, the nature of the substances involved, and their relationship to the visible structure of the cells.

Regulation. Just as engineers have devised intricate systems of automatic controls for the operation of huge assembly lines, organisms have developed their own regulatory mechanisms. Such activity, through many self-regulatory control systems, is essential if all the biological processes are to function effectively. The control mechanisms must execute their roles not only within cells but also among the cells composing an organism, and in the integration of an organism with its surroundings, which are continually changing. The study of two well-known control mechanisms, the nervous and hormonal systems, has been part of biology for a long time. In fact, the characteristic of response to stimuli (irritability)—the consequence of these systems—is often listed as a basic feature of life. The study of other areas, like behavior and intracellular chemical control, is relatively new.

Reproduction and Development. Every living thing reproduces and grows. In every organism, the first step in reproduction occurs at the cellular level. This step is *cell division*. It may lead directly to the formation of a new organism—as in the single-celled plant or animal—to the formation of specialized sex cells and hence new organisms—as in the ovary of a woman—or to an increase in the size of an individual—as in a growing pine tree. Cell division is a complex process for which the inanimate world has no counterpart.

Growth is unique. It results from cell division and chemical activities within cells. Nonliving or inanimate objects like crystals may increase in size, but their "growth" follows from an addition of new atoms to the outside of the object rather than from internal events.

The study of developmental biology, called *embryology*, is one of the most fascinating areas of biological investigation. New techniques have taught scientists to grow plant and animal embryos outside of their normal environment and to apply biochemical methods to problems of development. These advances have inspired new research into how the fertilized egg gives rise to an adult organism (Figure 1.6).

One urgent aspect of studies on reproduction in today's heavily populated world is the question of

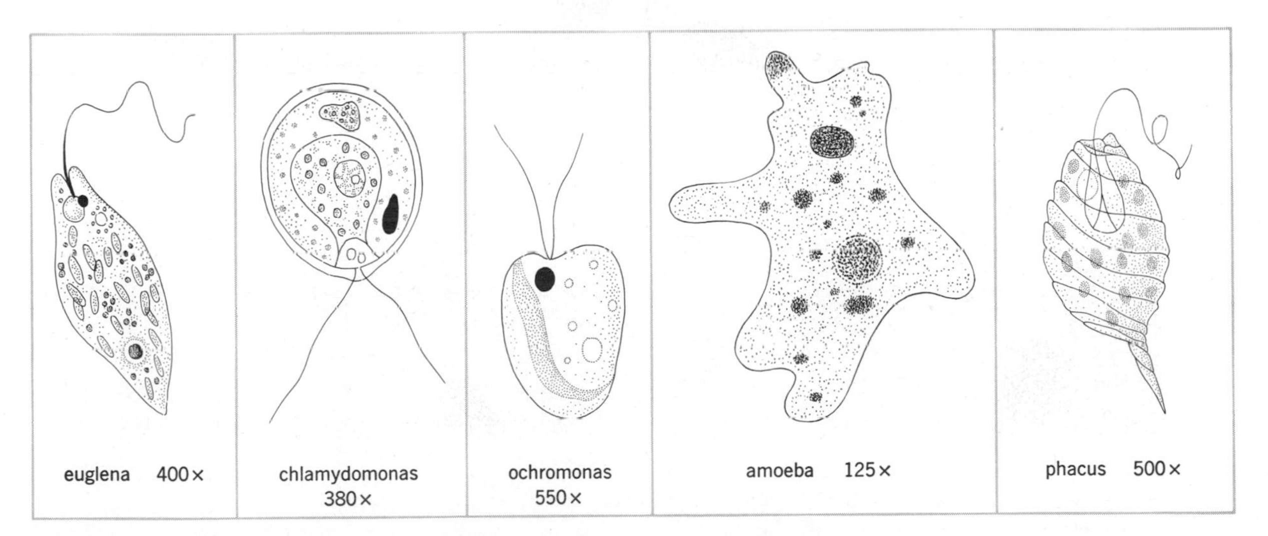

Figure 1.4. A variety of unicellular organisms. Note the diversity in form and structure.

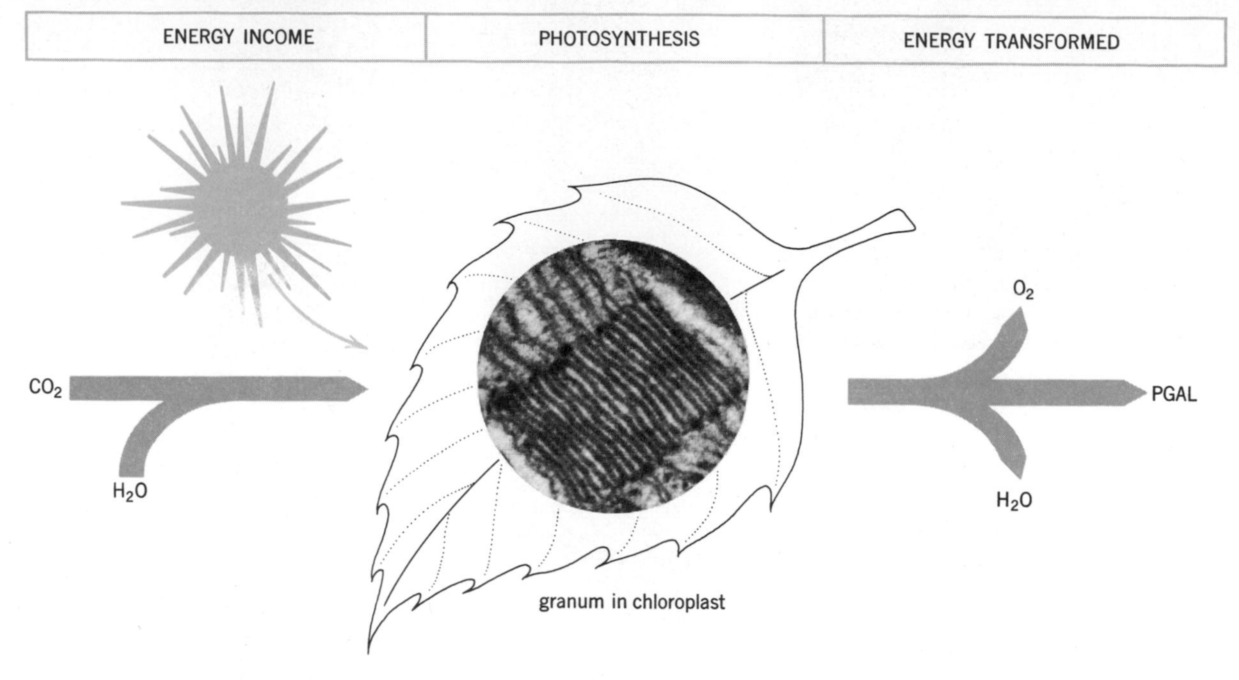

Figure 1.5. A generalized scheme of energy capture in photosynthesis and energy expenditure in respiration. The photograph, taken through an electron microscope, shows a granum, the site of photosynthesis—PGAL is the carbohydrate end product of photosynthesis.

birth control among human beings. New drugs have demonstrated a capacity to alter the normal hormonal system that controls reproduction, and thus prevent conception. Such drugs may eventually be used everywhere to slow down the population increase which at present threatens the world's food resources.

Heredity. Any mention of biological reproduction necessarily leads to the subject of heredity. Nearly all of our knowledge of heredity—or *genetics,* as this area of biology is usually called by biologists—has been acquired since the beginning of this century. Thus, genetics is one of the newer areas of biology.

The general mechanisms of transfer from parents to offspring are well known. But the center of greatest interest at this time is a giant molecule called DNA (deoxyribonucleic acid). All hereditary information is evidently coded into this basic hereditary material, which is found in all the cells of living things. DNA not only serves as an hereditary mechanism but also controls all biochemical events in the cell. Studies on metabolism, for example, often become involved with functions related to DNA. We shall have much to say about DNA in the course of this book. Some biologists predict that the discovery of its significance may eventually rank as the foremost biological advance of the century.

Adaptation and Evolution. The last of the major characteristics of life to engage our attention are adaptation and evolution. Anyone who travels from north to south immediately becomes aware of a marked change in plant and animal life. Fir trees may be seen against the snow in one clime and live oak among the marshes in another. In both places, plants and animals are adjusted to their surroundings. This adaptation is accomplished through the interaction of hereditary materials with the environment. As the adaptation continues over a long period of time, it becomes part of a gradual process of evolution (Figure 1.7).

Life has doubtless existed on earth for a longer time than we can readily conceive, apparently for at least two billion years. The origin and early evolution of life remains one of the baffling puzzles of biological science. Recent discoveries of primitive microscopic organisms indicate that they may be nearly two billion years old. The knowledge ob-

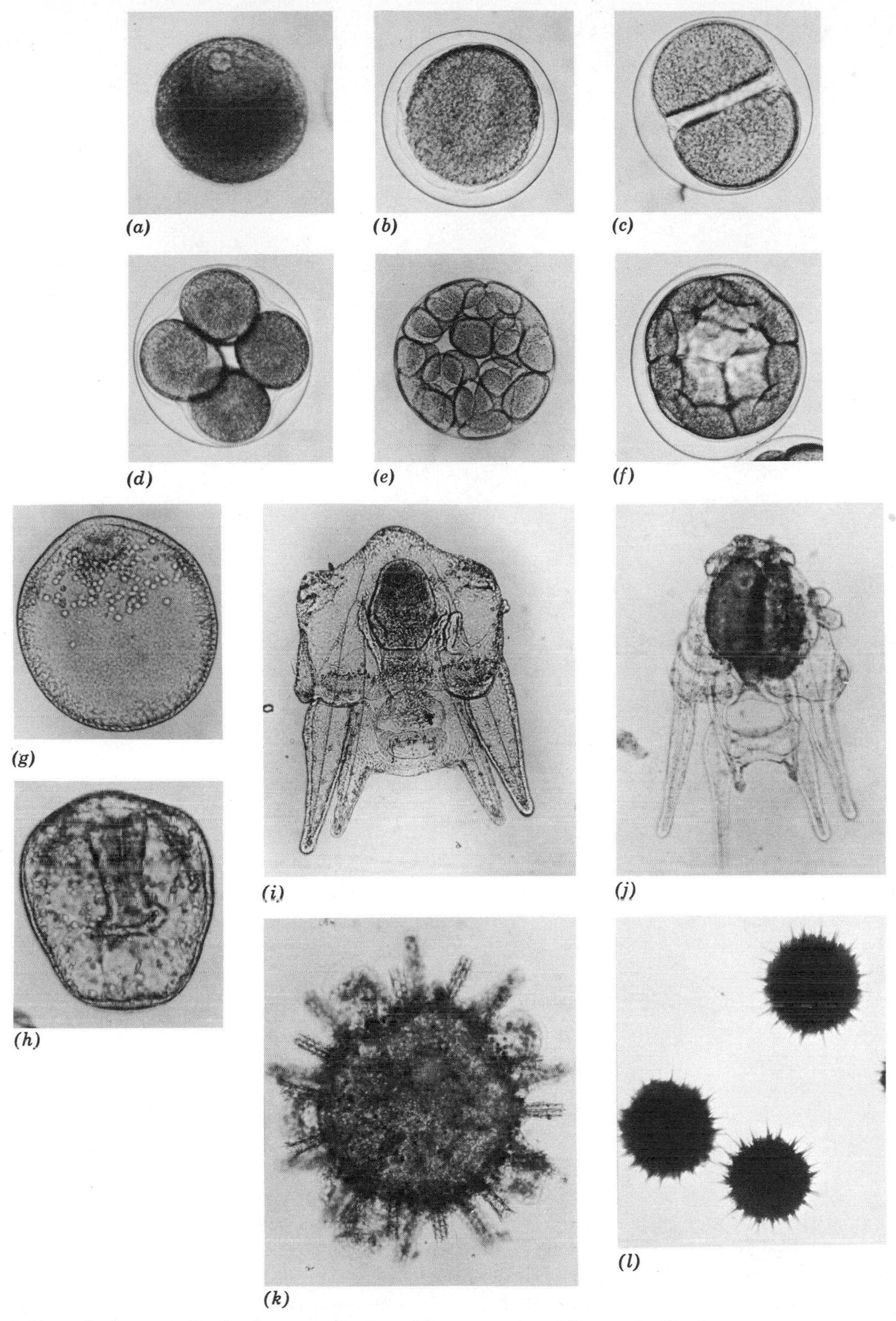

***Figure 1.6.** Successive stages in the development of a sea urchin commencing with an unfertilized egg* (a) *and proceeding through larval stages* (i *and* j) *to a small adult form* (l). *The unfertilized egg measures approximately 80 microns in diameter; the juvenile form* (k) *measures approximately 1000 microns. Sea urchin embryos are widely used for experimental studies in developmental biology.*

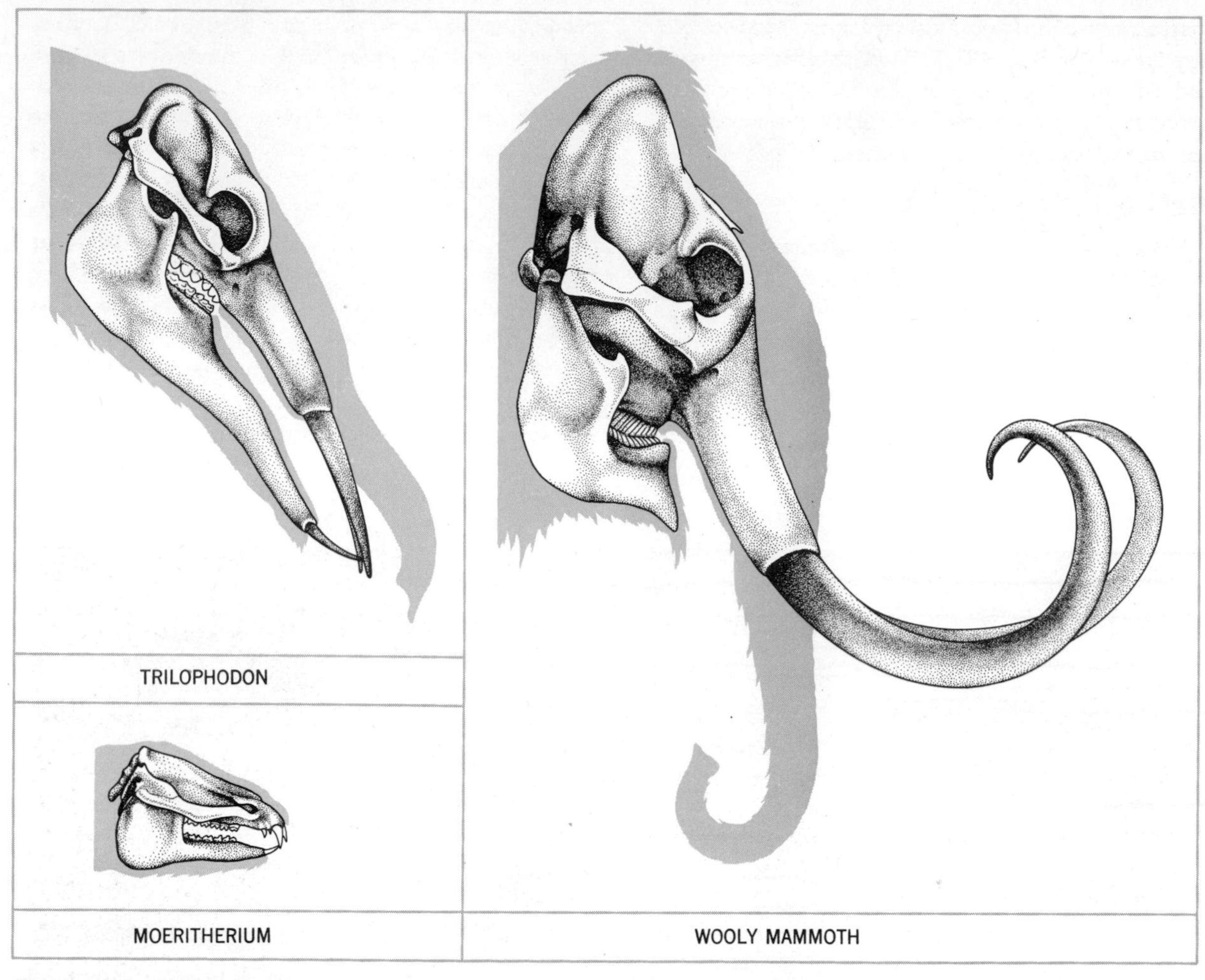

Figure 1.7. Some unusual adaptations have occurred during the evolution of elephants. The ancestral form (lower left) was relatively small and unspecialized—note the features of its skull and teeth. By contrast, fossils of moderate age (upper left) and of recent times (right) show the development of conspicuous adaptations as indicated by the tusks, long proboscis, and shortening of the skull.

tained from studying them provides a basis for making inferences about the environment in which these plants thrived. These inferences, in turn, shed light on the kinds of events that could have occurred in earlier epochs.

Many important factors bear on the adaptation and evolution of living organisms. Some of the major ones are the flow of energy from individual producers (plants) to individual consumers (animals), the interaction of populations of organisms with their environment, and the distribution of principal communities of living things over the surface of the earth (Figure 1.8). Studies on these subjects belong to the branch of biology called *ecology*. Ecological research is particularly intense in the areas of energy flow and population dynamics. The findings may someday enable man to achieve a better adjustment of his growing numbers to the available resources on the earth.

From all this, it is evident that the numerous areas of biology are not disjoined, but are parts of one whole. Through their linkage, the smallest development in the cell, for example, may directly affect the study of the organism and its habitat or, for that matter, any other biological endeavor. Modern biology is not standing still. Indeed, this highly diversified though fully integrated science continues to probe further for greater knowledge and more exact answers to the profound question, "What is life?"

Methods of Study

Like other scientists, biologists approach their investigations within a broad framework of what is termed scientific methodology, but it would be erroneous to assume that all biologists, or all scientists, utilize this approach in the same manner.

Before an individual can really begin an investigation he must be familiar with previous studies regarding his topic. He can then avoid needless duplication of effort and utilize the data from other studies in his own work. This is generally accomplished by a thorough review of the appropriate scientific literature, especially journals that report research studies in his field. With this knowledge, the investigator is ready to proceed with his own work. From this point on, the sequence of steps varies considerably, depending on the investigator and the nature of the material with which he is working.

Let us say that a researcher has decided approximately what he wants to study and knows how he is going to do it. He may have a tentative hypothesis

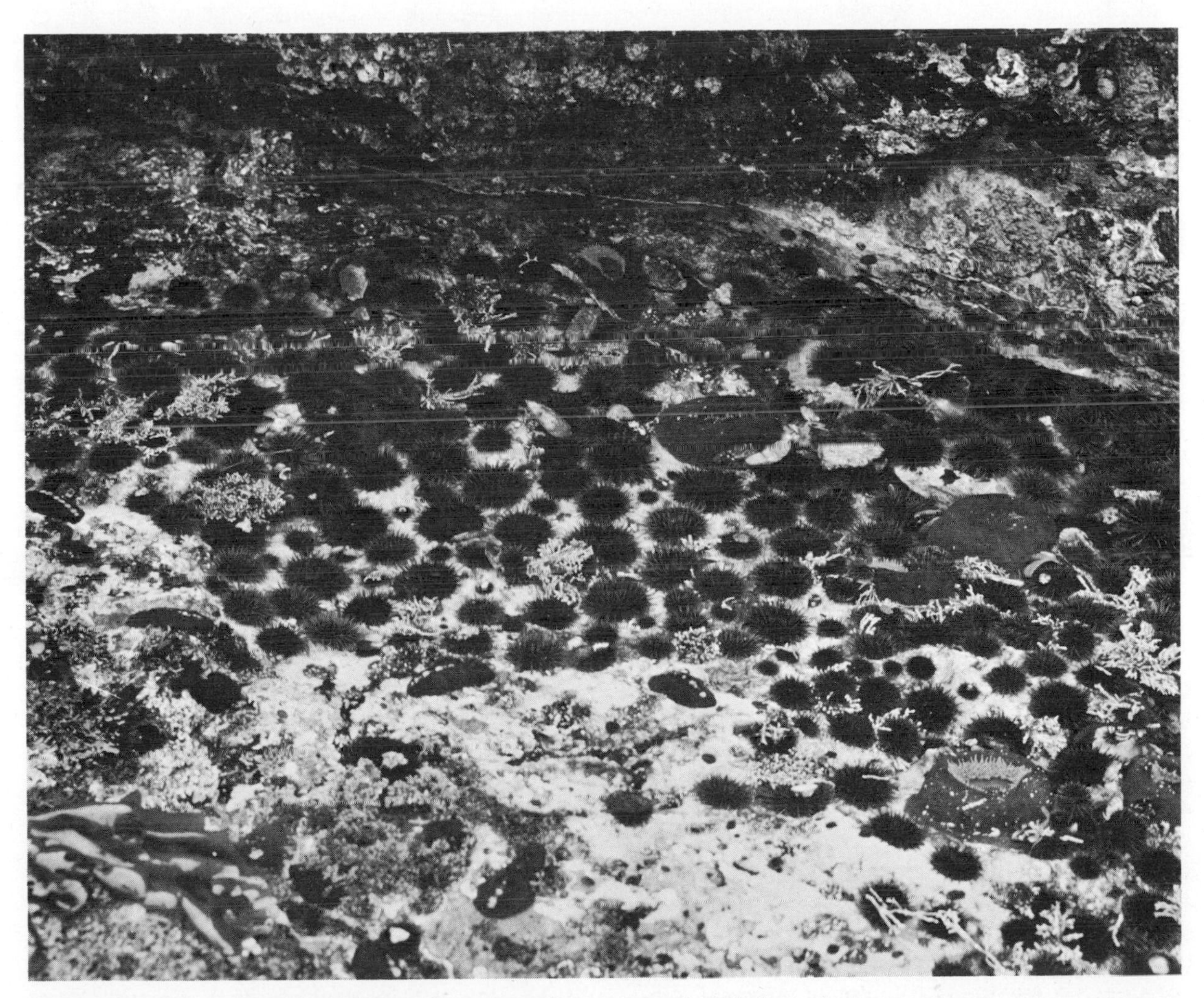

Figure 1.8. A marine community in shallow water. Sea urchins, sea anemones, chitins, and algae form an interacting community of organisms.

and perhaps some general ideas about how the investigation will turn out. He proceeds to gather data by making observations; these may originate in nature or in the laboratory. In addition, his observations are usually aided by scientific equipment: camera, microscope, chemical apparatus, etc. Eventually he considers his observations (facts) and proposes a statement that relates them or summarizes their significance. This is an hypothesis, a working statement. If possible, additional observations or experiments are performed to test the validity of the hypothesis.

Scientists rely heavily on experimentation to verify an hypothesis when the material is suitable to this technique. Experiments are designed to discover which of the variables in a situation are significant. In other words, experimentation is a method for determining the most probable cause-and-effect relationship. The experimenter is bound by only two rules. His experiments must be conducted in such a way that one set of tests differs from another set by only one factor: that is, they must have proper controls. Furthermore, they must be of such a nature that they may be repeated by other individuals. *Repeatability* is one of the most crucial aspects of the scientific method, for it is the major basis for the acceptance or rejection of many hypotheses.

If the hypothesis is supported (verified) by whatever tests the experimenter applies, then it reaches a level of greater certainty. There is now a higher probability that it correctly interprets a set of facts. In a way, this is the essence of science—not to "prove" or "disprove," but to indicate levels of certainty or probability.

At this point, the investigator has contributed two kinds of knowledge: the facts he has gathered, and the hypotheses or theories he has derived from the facts. These are important additions to the field of biology.

Theories are useful not only for synthesizing data but also because predictions can frequently be made from them, thus leading to entirely new lines of investigation. For example, from Darwin's observation on the way in which young plants grew toward light, he constructed the hypothesis that the influence of light on the stem tip was transmitted to the rest of the plant by a chemical factor. This hypothesis led to experiments that proved him correct.

A theory that has been repeatedly verified and appears to have wide application in biology may become a biological principle. These are sometimes called biological laws, although this does not change their status as statements that apply with a high degree of probability to a wide range of biological events. They are still man-made and subject to change if additional facts emerge in the future. Throughout this book, the term *principle* is used in this sense.

An important aspect of scientific investigation is the imagination, the creativity of the scientist. The proper interpretation of data, the design of experiments, and the formulation of useful new theories frequently involve a certain amount of intuition—a hunch about how to proceed. This may sound quite unscientific, yet many scientists admit that it is important in their investigatory work.

In addition to gathering facts, hypothesizing, and testing hypotheses, someone must, at intervals, attempt to summarize the research efforts of other biologists in order to present the current status or progress being made in a biological field. Frequently, this type of material appears as a monograph or an extensive review article in a scientific journal. The knowledge presented in the chapters to follow was acquired through the techniques described.

Biology and Public Affairs

People seldom realize the important role of biology in their everyday life. Every community contains individuals who use biological concepts as a basic part of their professions: agricultural and soil specialists, conservationists, public health personnel, sanitation engineers, water works supervisors, farmers, medical people, and many others. In a general sense, all of these individuals are applied biologists, and they are usually highly knowledgeable in the areas of biology relevant to their professions.

Many of the problems a citizen encounters in human communities are biological in nature. Air and water pollution, the maintenance of the health standards, the use of pesticides, the fluoridation of drinking water, and the disposal of garbage all fall into this category. Moreover, many of these biological problems plus some new ones are causing concern at the national and international levels. The biological consequences of radioactive fallout, the conservation of natural resources, methods of increasing agricultural yields, and a rapidly multiplying world population are a few of the problems of mankind that are directly based on biological principles.

In biology as in other sciences one often encounters a distinction between basic and applied research. Scientific investigations that are not directed at immediate practical applications are classed as pure or basic research. For example, a biochemist might invest many years studying the effects of certain chemicals on the energy-releasing reactions of insects. His interest may lie entirely in insect physiology rather than in practical uses of his findings. The value of this kind of activity lies primarily in its contribution to man's understanding of the world around him. Most research of this type is conducted by individuals associated with academic institutions. In fact, one of the major attractions of a college teaching career is the opportunity to engage in research or scholarly activities of one's own choosing.

Applied research attempts to solve a problem of immediate concern or, in some cases, tries to find utilitarian uses for a new scientific discovery. The basic materials for these endeavors mainly arise from pure research, and in this way the two activities are intimately related. As an example, a scientist employed by an insecticide manufacturer may devise a new bug killer based on the biochemist's discoveries in insect physiology.

Obviously, applied science, or technology, is extremely important in modern life. Controversy sometimes arises over the relative importance of basic and applied research since both are expensive to maintain, and basic research does not seem as "useful" as applied science. But no one can predict with certainty the ultimate value or utilitarian nature of a seemingly unimportant effort in basic research. This judgment can be made only in retrospect. Consider the enormous significance of our present knowledge about the structure of the atom. Yet, in the early 1900's, study of the atom was a field of basic research with no forseeable practical use. It thus seems absolutely necessary to continue to support and encourage the process of basic research in order to obtain, occasionally, an advance in man's understanding of himself and nature.

Principles

1. Life is a series of processes that can be studied scientifically. Living things are characterized by a combination of structure, metabolism, regulatory devices, reproduction, heredity, and evolutionary history.

2. Scientists employ a variety of techniques and approaches in acquiring new knowledge. Such knowledge must be capable of verification by others.

3. An understanding of fundamental biological concepts often facilitates intelligent decisions in problems encountered in daily life.

Suggested Readings

Beveridge, W. I. B., *The Art of Scientific Investigation.* Revised edition. W. W. Norton and Co., New York, 1957.

"Careers in Biology," *American Biology Teacher,* Vol. 30, No. 4 (April, 1968).

Loeb, Jacques, *The Mechanistic Conception of Life.* Harvard University Press, Cambridge, Mass., 1965.

Moment, Gairdner, "Biological Science Today," in *Frontiers of Modern Biology,* Houghton Mifflin Company, Boston, 1962.

Polanyi, Michael, "Life's Irreducible Structure," *Science,* Vol. 160, No. 3834 (June 21, 1968).

Schrodinger, Erwin, *What Is Life?* Cambridge University Press, Cambridge, England, 1963.

Simpson, George Gaylord, "Biology and the Public Good," *American Scientist,* Vol. 55, No. 2 (June, 1967).

Wald, George, "Innovation in Biology," *Scientific American,* Vol. 199 (September, 1958). Offprint No. 48, W. H. Freeman and Co., San Francisco.

Questions

1. How is life defined in this chapter?

2. Would you say that this is a definition that emphasizes a structural or a functional approach? Why?

3. In your own experience, how do you distinguish between living and nonliving entities?

4. List the differences between living and nonliving objects using the characteristics of living things from this chapter. How does this compare with your ideas from Question 3?

5. Why is there not a single scientific method, that is, a list of steps, which all scientists follow in their investigations?

6. List one possible sequence of steps one might follow in a scientific investigation. In what ways could you change the sequence of steps and still accomplish the same investigation?

7. Why is it erroneous to state that science can prove or disprove a hypothesis?

8. Can science prove or disprove a fact? Why?

9. Explain how basic research provides the materials for applied research.

10. What are the materials for *basic* research?

CHAPTER
II

Life: Its Chemical Basis

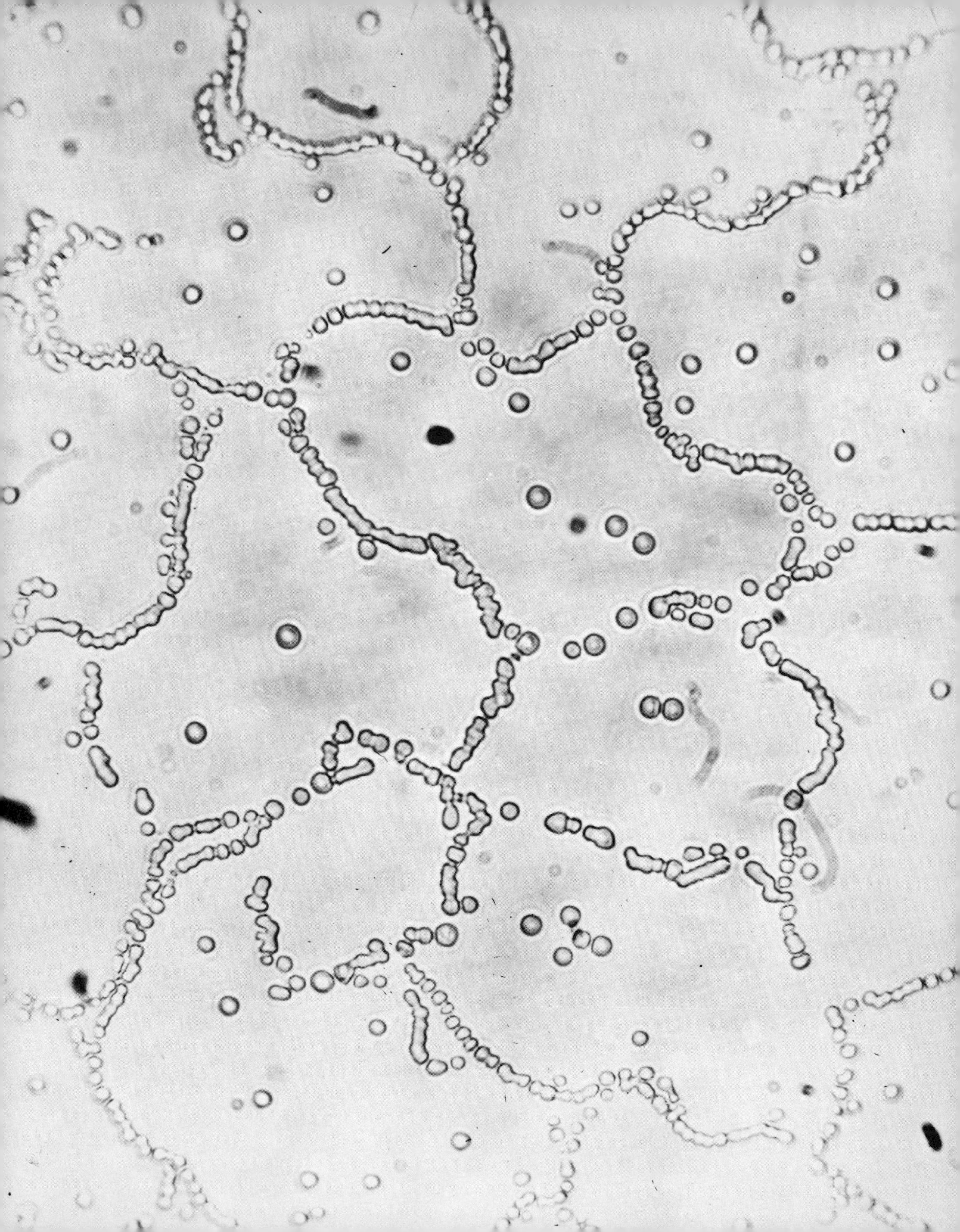

These microspheres consisting of proteinlike material were produced during experimental studies on the origin of life (Dr. Sidney Fox).

CHAPTER II

Life: Its Chemical Basis

As we have seen, every organism has a number of characteristics that help us to identify it as living. The more important of these include the chemical activities that occur in the organism. If we look at the general structure of the atom and then turn to some of the types of atoms that are common to living things, we can acquire a general understanding of these vital chemical processes.

Atomic Structure. Atoms are the smallest particles of elements that enter into chemical reactions. Nevertheless, atoms are not indivisible. Each atom consists of a relatively heavy, compact central nucleus, and lighter particles called electrons that orbit the nucleus at some distance from its center (Figure 2.1). Electrons are virtually weightless and each one carries a negative electrical charge. Atomic nuclei are composed of protons and neutrons, except for the hydrogen nucleus, which contains only one proton. Each proton and neutron has one unit of atomic weight (an arbitrary unit) and is about 1800 times heavier than the electron. An atom's weight results almost entirely from its protons and neutrons. A proton has a positive charge, whereas a neutron is neutral.

Looking at the structure of atoms of different elements, we see that each element has a distinctive number of protons (Figure 2.1). An element is a substance whose atoms all contain the same number of protons and the same number of electrons.

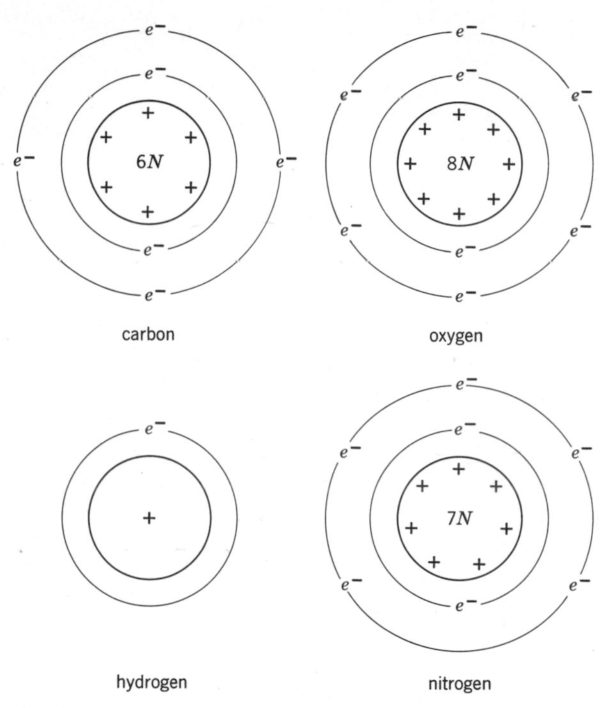

Figure 2.1. A schematic representation of four atoms important in biological systems. The inner circle represents the nucleus of the atom containing protons (+) and neutrons (N). The electrons are represented by e⁻. Carbon, for example, has six neutrons and six protons in the nucleus and six electrons in rapid motion around the nucleus.

Furthermore, since the number of protons equals the number of electrons, an atom is electrically neutral.

We may have given the impression that the number of neutrons present is the same in every atom of an element. This is not always true. In the atoms of some elements the number of neutrons varies. Atoms of carbon, for example, may have one of three different atomic weights, 12, 13, or 14, depending on the number of their neutrons. These different kinds of atoms of the same element are called *isotopes*, and in this instance are designated C^{12}, C^{13}, and C^{14} (Figure 2.2). Each of these contains six protons, but has six, seven, or eight neutrons, respectively.

Bonds and Energy. Atoms combine chemically with one another, that is, form bonds, by gaining, losing, or sharing electrons. When the atoms of two or more different elements combine in this way, a *compound* (such as water, H_2O) is created. The symbol H_2O also represents a *molecule*, the smallest combination or particle retaining all the properties of the compound itself.

Atoms that gain electrons become negatively charged, whereas those that lose them become positively charged, having originally been electrically neutral. (See Figure 2.3.) These charged particles are called *ions*. Negatively charged ions (Cl^-, for example) are attracted to postively charged ions (Na^+) because the opposite charges attract each other. The resulting force that binds these ions together is an *ionic bond*.

When immersed in water, compounds held together by ionic bonds tend to separate, or dissociate,

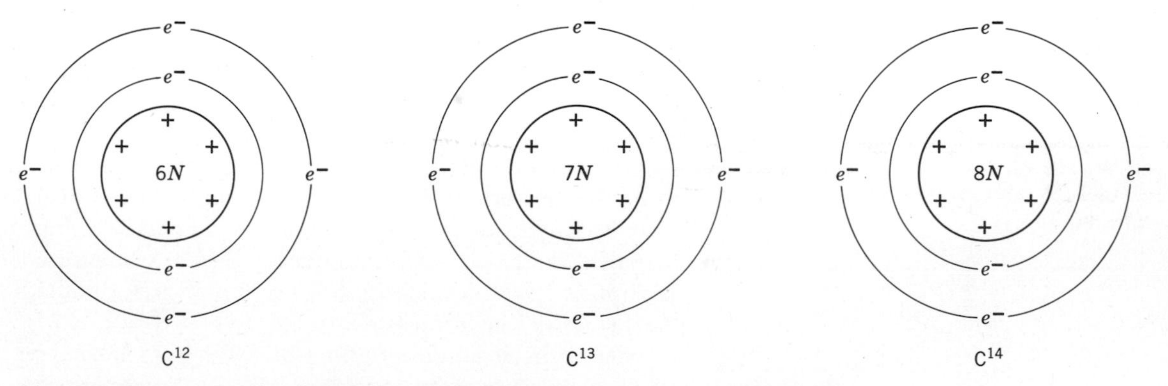

Figure 2.2. The isotopes of carbon differ only in the number of neutrons contained in the nucleus. The numbers of protons and electrons in these three isotopes are the same. C^{14} is synthetically prepared and has been extensively used in clarifying metabolic pathways.

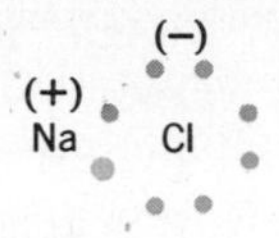

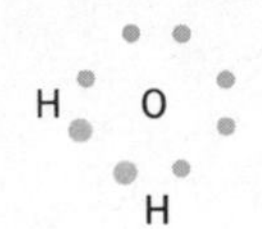

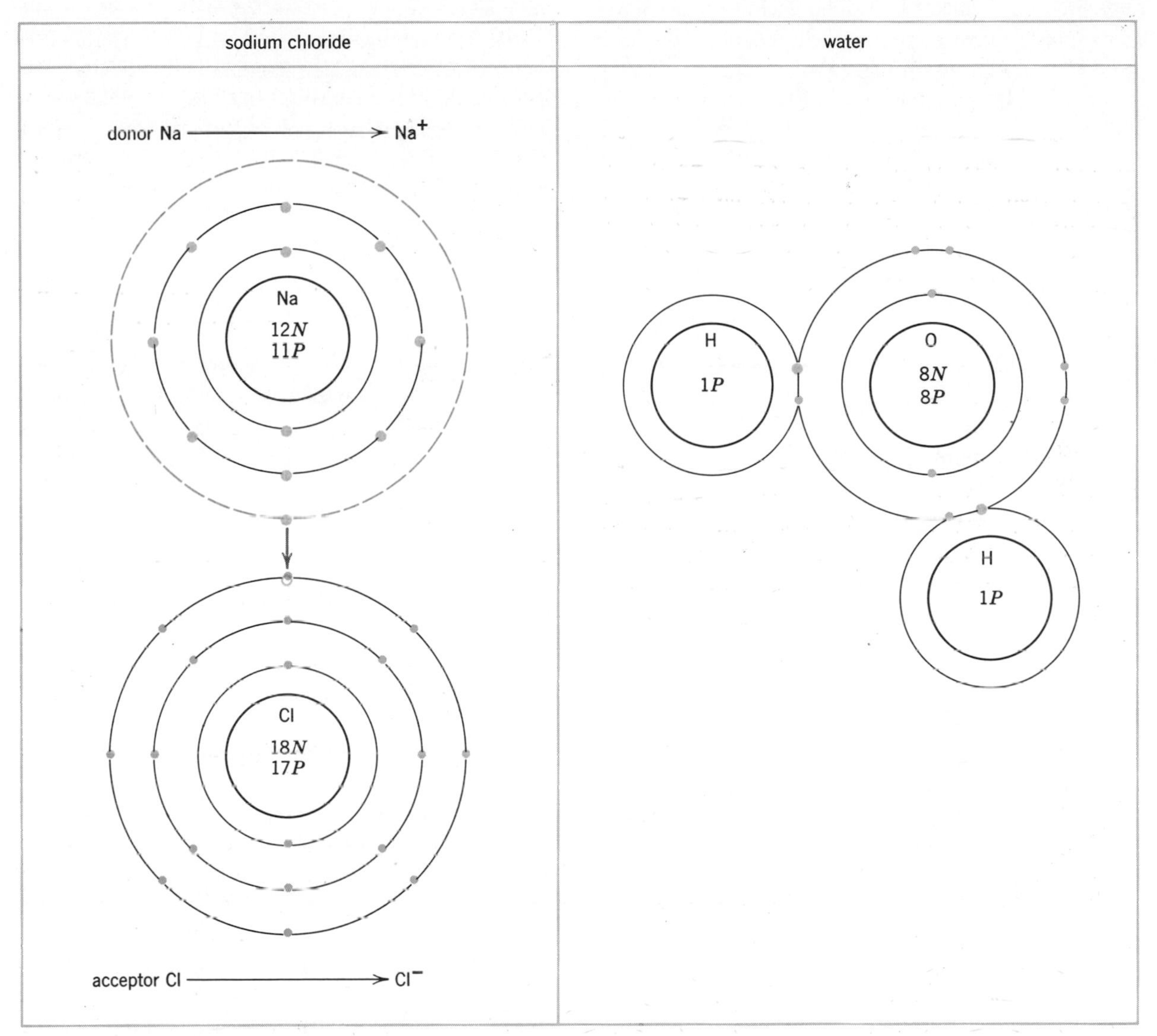

Figure 2.3. Representation of two types of bonding. In sodium chloride (NaCl), common table salt, the chloride ion and the sodium ion are held together by an ionic bond. This bond was formed by the transfer of an electron (arrow). The water molecule (H_2O) contains covalent bonds.

into their constituent ions because of the action of the water. Many substances required by biological systems exist in nature in ionic form. Such mineral salts as sodium (Na^+), chloride (Cl^-), potassium (K^+), calcium (Ca^{++}), and phosphate ($PO_4^{\equiv}$) are good examples. The minus and plus signs indicate the kind and size of charge on each ion.

Another type of bond found in many molecules is the *covalent bond* (Figure 2.3). Here the atoms share electrons. Molecules containing covalent bonds do not dissociate when placed in water, but remain intact. Carbon (C), oxygen (O), hydrogen (H), and nitrogen (N), which constitute about 95 percent of the material in cells, engage in this type of bonding. All of the cell's larger molecules and many of its smaller ones contain such bonds.

Chemical bonds and reactions—the changing of bonds—are so important that it is convenient to classify molecules according to their role in these bonds and changes. Molecules furnishing electrons during a reaction are called electron *donors*. Those that gain electrons during the process are called electron *acceptors*. Some molecules gain electrons only to lose them to some other molecule in a very short time; these are designated electron *carriers*.

Bonds contain energy, the ability to do work. This results from the interaction of the electrons and nuclei of the bonded atoms. If we measure the amount of energy present between two atoms, we find that this amount varies as the distance between the atoms changes (Figure 2.4). When the atoms are close to each other, the paths of their electrons overlap. The repulsion of these negatively charged particles tends to drive the two atoms apart, and the amount of energy necessary to keep them together is high. As the distance between the atoms grows, the amount of energy needed decreases markedly, and then increases somewhat to a level that is more nearly constant. Point *A* in Figure 2.4 represents the distance separating the atoms in a molecule; this is the point at which the least energy is required to hold the atoms together.

In order to break a bond, we must move its atoms farther apart. To do so we must add energy. Generally, when one bond is broken, another forms. If the new bond contains more energy than the first, we assume energy must have been added. On the other hand, if the new bond contains less energy, we assume energy has been released, which may be used to change bonds elsewhere or which may escape as heat. In either case, a certain amount of energy, the *activation energy* (Figure 2.5), must be initially added to move the atoms apart so that the reaction may proceed. If the new bond contains less energy, the reaction will continue because the energy released serves to activate changes in other molecules. This kind of reaction supplies energy to be utilized in the cell's varied activities.

Common Substances in Living Systems

Small Molecules. Water is the most abundant substance in living cells. It consists of small, simple molecules composed of two hydrogen atoms attached to one oxygen atom (H_2O). The three distinct roles that water plays in cells underline its importance. It takes part in some reactions (see Chapters IV and V), it serves as a medium (or solvent) for other reactions, and it serves as a basis for the transport of materials. In reality, the chemistry of life is dominated by the chemistry of water.

Chemical reactions occur between individual atoms, ions, or molecules, and not between large aggregations of these particles. As these particles move about in the water, they come into contact with other particles and a reaction occurs.

The small carbon dioxide (CO_2) molecule contains one carbon atom and two oxygen atoms. The role of carbon dioxide in living systems is not nearly as diverse as that of water, but it plays a part in both respiration and photosynthesis, two processes of great importance. All the carbon in the larger carbon-containing (organic) compounds found in the living systems comes directly or indirectly from carbon dioxide. As we shall see later, the concentration of CO_2 in an organism may be important in the control of other activities. In mammals the rates of heartbeat and respiration are outstanding examples.

Molecular oxygen (O_2) is required by almost all organisms. It is utilized in respiration, the releasing of energy in the cell (see Chapter V). Since oxygen is a product of photosynthesis (see Chapter IV), the supply in the atmosphere is maintained at a nearly constant level. The small size of the oxygen molecule lessens the problems involved in its transport be-

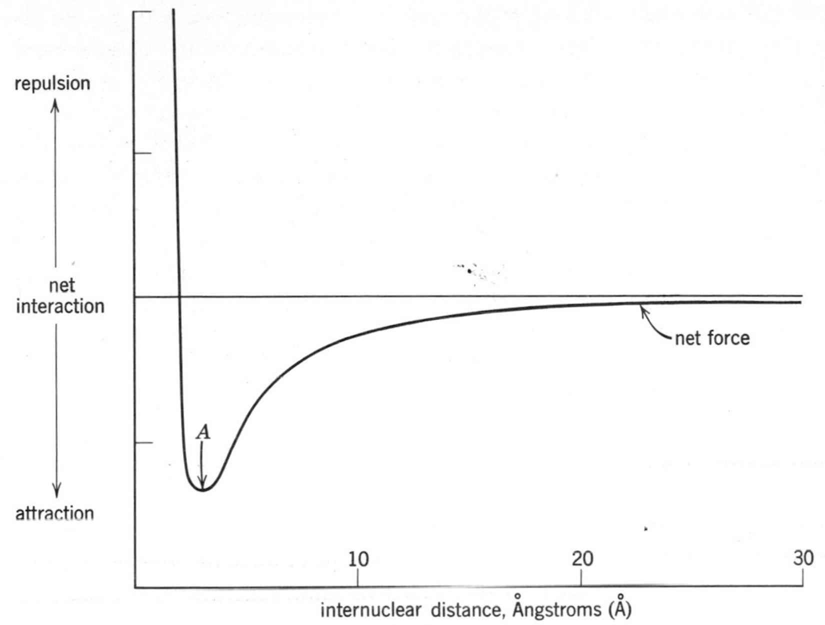

Figure 2.4. Interaction of two charged ions such as K^+ and Cl^-. Because of electrostatic attraction, the two ions will approach one another. When the two ions get very close together their electron orbits will begin to overlap, and this causes a strong repulsion. The distance at which the attractive and repulsive forces are equal is indicated by the intersection of the curve with the horizontal denoting zero net interaction. The least amount of energy required to hold atoms together is shown at point A.

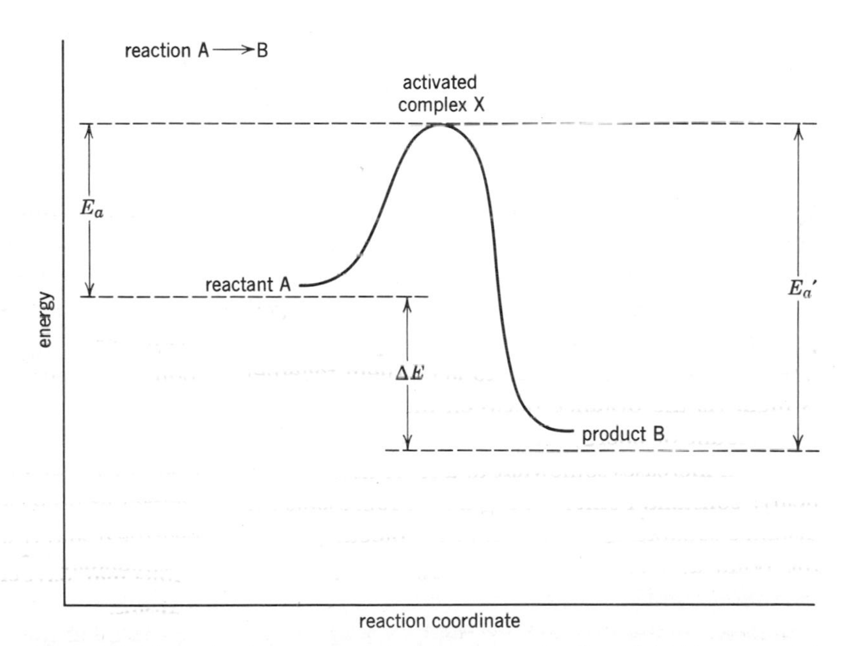

Figure 2.5. The activation energy required for the reaction from A to B is represented by E_a; for the reverse reaction it is indicated by E_a'. In both directions, the reaction proceeds through the activated complex X. ΔE signifies the difference in energy between compounds A and B, and therefore the net energy yield of the reaction from A to B.

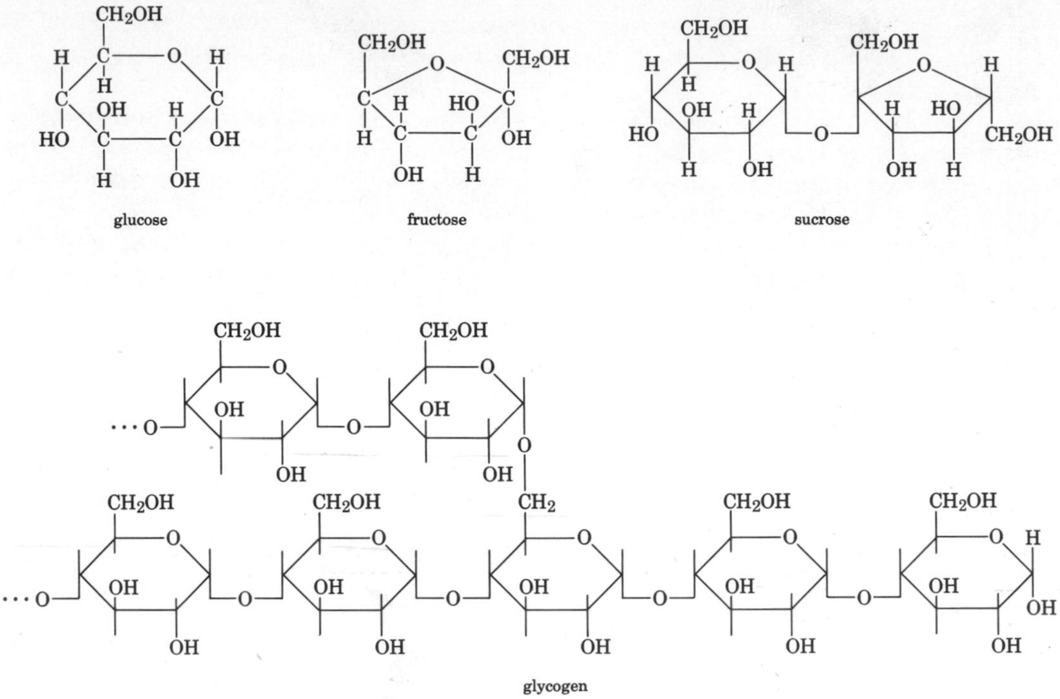

Figure 2.6. One molecule of glucose and one molecule of fructose compose the table sugar sucrose. Glycogen is comprised of many glucose units.

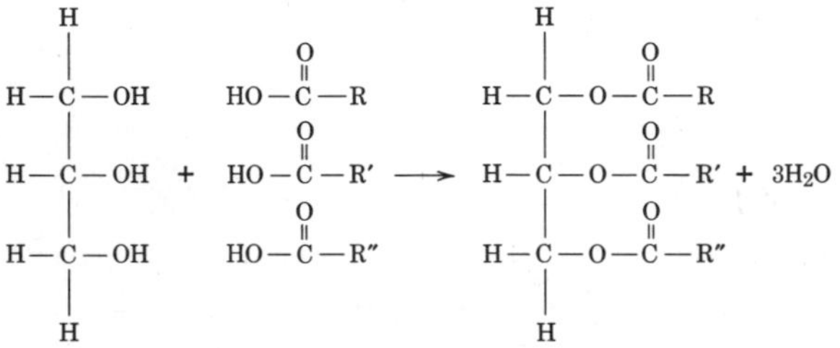

Figure 2.7. The synthesis of one molecule of fat (right) from one molecule of glycerol (left) and three molecules of fatty acids (center).

tween the environment and the cells (Chapter VI).

The last small molecule we shall mention is the ammonia molecule (NH_3). A common source of ammonia is the decomposition of proteins. The organism must have a method of disposing of ammonia because even a moderate concentration is injurious to cells. The methods for the removal of ammonia often include the incorporation of ammonia into a larger molecule and then disposal of the portion of the molecule containing the ammonia. A higher concentration of these segments may be tolerated by the organism because they are less toxic. Because many plants are able to use NH_3 or the products of bacterial action on NH_3 as a nitrogen source for protein synthesis, NH_3 is a common constituent of fertilizers.

The mineral salts are composed of small ions. They function in numerous ways as parts of enzymes, or as portions of the cellular environment necessary for enzyme or protein action. Ca^{++}, $PO_4^{\equiv}$, Na^+, and K^+ are a few examples.

Larger Molecular Substances. The other classes of compounds that we shall survey—carbohydrates, lipids, proteins, and nucleic acids—are characterized by larger molecules. Each of these kinds of molecules includes many carbon atoms bonded to each other. Sometimes, the chains of carbon atoms are very long. In many plants these molecules are produced inside the cell from simple molecules such as CO_2, H_2O, and NH_3, whereas in other organisms simpler carbon-containing compounds brought into the cell serve as a raw material for synthesis of the larger molecules.

Carbohydrates are composed of carbon, hydrogen, and oxygen. All molecules of carbohydrates contain the atoms of hydrogen and oxygen in a two-to-one ratio. Each molecule is composed of a chain of carbon atoms with hydrogen and oxygen bonded to them. The smallest carbohydrates, called *simple sugars,* are those that cannot be made to react with water to produce a simpler form. These vary in size; the smallest have a chain of three carbon atoms, whereas the largest have seven. More complex carbohydrates are formed by the joining of two or more of the simple sugars.

The commonest of the simple sugars is *glucose* ($C_6H_{12}O_6$). Its structure is

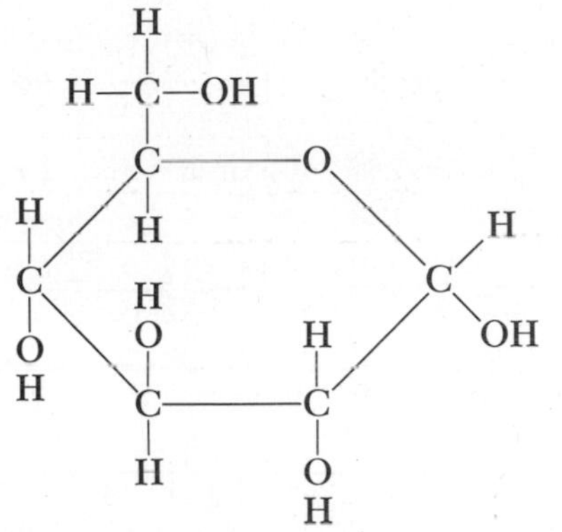

The connecting lines represent molecular bonds. Note the repetitions of the H—C—OH unit. This is typical of sugars. *Starch, glycogen,* and *cellulose,* as well as many other complex carbohydrates, are formed by bonding a number of glucose molecules. Besides glucose there are other six-carbon sugars. Combinations of these with glucose result in another series of sugars. Common table sugar, sucrose, is one (Figure 2.6).

Carbohydrates have two functions, energy storage and strengthening of the cell. Energy storage is the more common of these. Glycogen and starch, respectively, are animal and plant examples of carbohydrates used mainly for this purpose. Cellulose is a carbohydrate used to strengthen the structure of the cell. In most plants it constitutes the cell wall. Animals do not commonly use carbohydrates in this way.

Lipids, the second large group of compounds, includes fats and phospholipids. One molecule of a fat is formed from one molecule of *glycerol,* a three-carbon relative of the sugars (note the H—C—OH in Figure 2.7) combined with three molecules of *fatty acids.* The letters R, R′, and R″ in Figure 2.7 represent chains of different lengths. These chains consist of carbon atoms and their attached hydrogen atoms, but the letters R, R′, and R″ do not specify the lengths of the chains or their exact structures. One fat, for example, might contain three fatty-acid molecules of the same structure. Two of the commonest fatty acids are palmitic and stearic acids (Figure 2.8). For these, the letter R would represent chains having lengths of 15 and 17 carbon atoms, respectively.

Fatty acids contain the *carboxyl group,* $-\overset{\overset{\Large O}{\|}}{C}-OH$. This group is characteristic of the most important

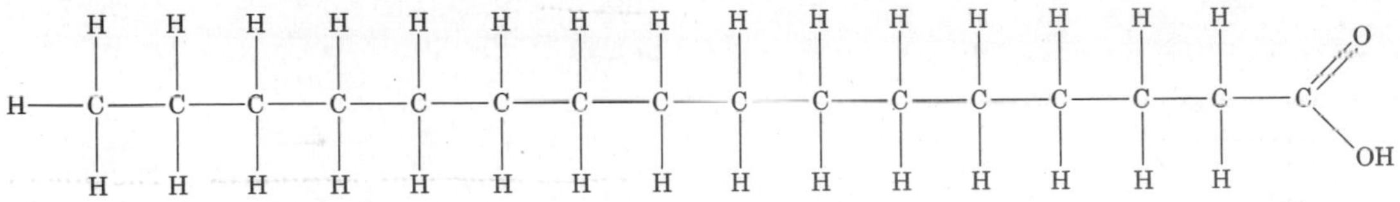

palmitic acid

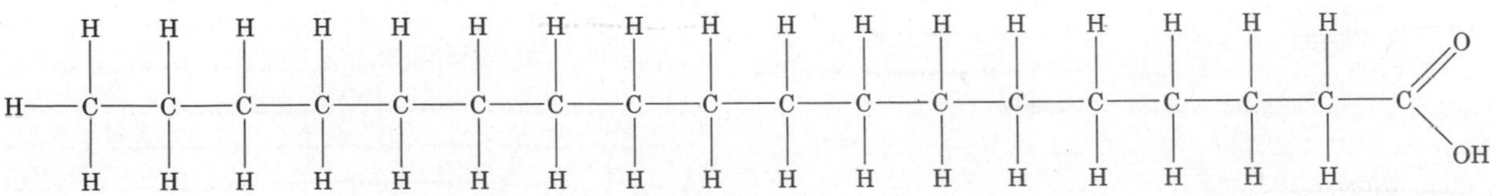

stearic acid

***Figure 2.8** Two common fatty acids. Note the repeating* $-\overset{\overset{H}{|}}{\underset{\underset{H}{|}}{C}}-$ *unit. This long hydrocarbon (hydrogen-carbon) chain imparts the characteristic properties of the lipids.*

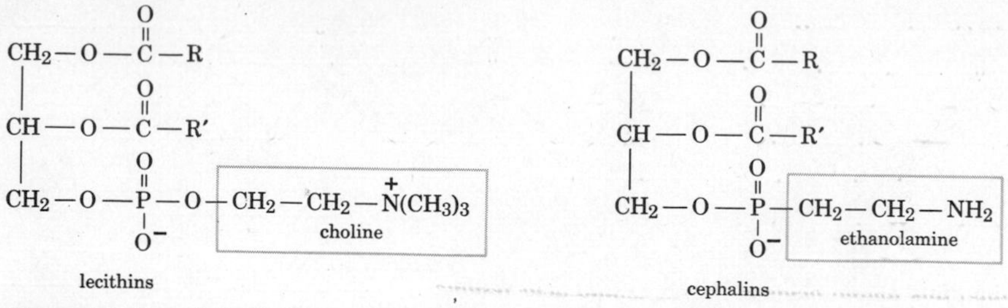

$$\begin{array}{l} CH_2-O-\overset{O}{\overset{\|}{C}}-R \\ | \\ CH-O-\overset{O}{\overset{\|}{C}}-R' \\ | \\ CH_2-O-\overset{O}{\overset{\|}{P}}-O-\boxed{CH_2-CH_2-\overset{+}{N}(CH_3)_3} \\ \quad\quad\quad\ \ O^- \quad\quad\quad\quad \text{choline} \end{array}$$

lecithins

$$\begin{array}{l} CH_2-O-\overset{O}{\overset{\|}{C}}-R \\ | \\ CH-O-\overset{O}{\overset{\|}{C}}-R' \\ | \\ CH_2-O-\overset{O}{\overset{\|}{P}}-\boxed{CH_2-CH_2-NH_2} \\ \quad\quad\quad\ \ O^- \quad\quad\quad \text{ethanolamine} \end{array}$$

cephalins

Figure 2.9. Phospholipids represent a large variety of natural lipids which contain not only carbon, hydrogen, and oxygen, but also phosphorus and nitrogen. Best known of the phospholipids are the lecithins and the cephalins.

kind of organic acids occurring in living systems. Many such simple acids play an important role in respiration in cells. We shall be referring frequently to *acetic acid*, a two-carbon acid, and to *pyruvic acid* and *lactic acid*, three-carbon acids. These three do not actually exist as acids in the cell but as the dissociated salts—acetate, pyruvate, and lactate—which are the products of reactions between acids and bases.

In structure, phospholipids are similar to fats except that one of the fatty acids has been replaced by a phosphate group and a nitrogen-containing base (Figure 2.9). Most phospholipids can be identified by the nature of the nitrogenous base.

The portion of the phospholipid molecule that contains the phosphate and base is attracted to water, whereas the remainder of the molecule has an affinity for oils, chloroform, and ether rather than water. The two ends of the molecule are arranged in characteristic fashion in any mixture of water and fats.

The lipids serve four functions. First, phospholipids are important in the structure of the cell, as we shall see in discussing the makeup of membranes (Chapter III). Second, lipids are important energy reserves; any intake of food in excess of the amount needed at the time may result in an increase in the body's fat reserves, as many people realize from firsthand experience. Third, the waxes, another group of lipids, serve a protective role on the leaves of plants and the skin or fur of animals. Fourth, in the form of steroids, they serve as chemical coordinating agents.

Whether we examine the carbohydrates or the lipids, we find similar molecules in all organisms. Some variations exist, but they are minor compared with the variation in proteins and nucleic acids, the other two groups of carbon-containing compounds common in living things. These latter compounds show tremendous diversity partly because of their large size and also because the smaller molecules from which they are synthesized may be arranged in any order. Remember that although there is a

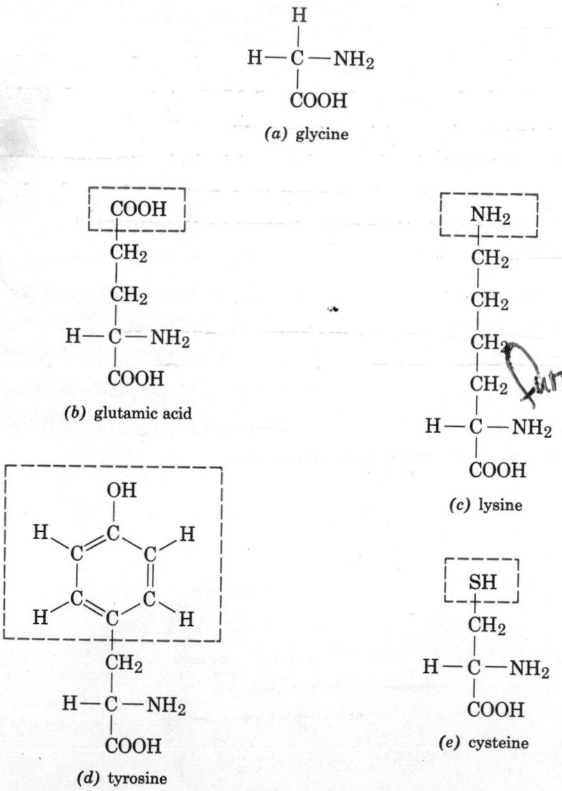

$$\begin{array}{c} H \\ | \\ H-C-NH_2 \\ | \\ COOH \end{array}$$

(*a*) glycine

$$\begin{array}{c} \boxed{COOH} \\ | \\ CH_2 \\ | \\ CH_2 \\ | \\ H-C-NH_2 \\ | \\ COOH \end{array}$$

(*b*) glutamic acid

$$\begin{array}{c} \boxed{NH_2} \\ | \\ CH_2 \\ | \\ CH_2 \\ | \\ CH_2 \\ | \\ CH_2 \\ | \\ H-C-NH_2 \\ | \\ COOH \end{array}$$

(*c*) lysine

$$\begin{array}{c} \boxed{\text{OH-substituted benzene ring } (C_6H_4OH)} \\ | \\ CH_2 \\ | \\ H-C-NH_2 \\ | \\ COOH \end{array}$$

(*d*) tyrosine

$$\begin{array}{c} \boxed{SH} \\ | \\ CH_2 \\ | \\ H-C-NH_2 \\ | \\ COOH \end{array}$$

(*e*) cysteine

Figure 2.10. Five amino acids. (a) *The simplest amino acid, glycine.* (b) *The acidic amino acid glutamic acid.* (c) *The basic amino acid lysine.* (d) *The amino acid tyrosine containing the benzene ring.* (e) *Cysteine, an amino acid containing a sulfhydryl group, a sulfur with a hydrogen attached. The characteristic part of each molecule is indicated.*

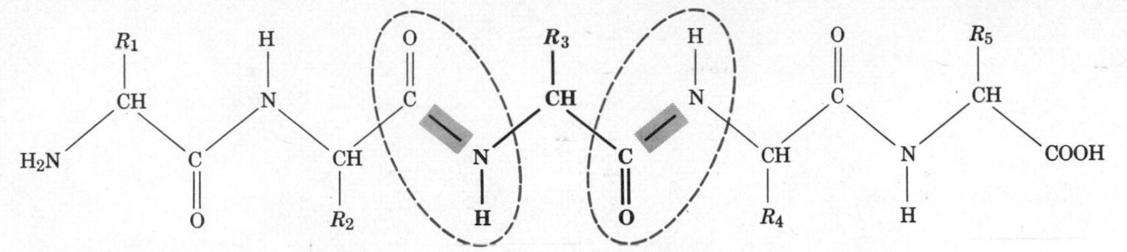

Figure 2.11. Peptide bonds (blue shading) link amino acids. (A single amino acid residue is indicated by the boldface type.) Chains of amino acids are called polypeptides. Large polypeptide chains are called proteins.

wide variety of fatty acids, a maximum of three can be combined with glycerol to form any particular fat. This limits the number of possible combinations.

The third group of large molecules, the *proteins*, are built of units called amino acids. The word acid in the name indicates the presence of the carboxyl group. The word amino tells us that an *amino group* $-\mathrm{N}\begin{smallmatrix}\diagup \mathrm{H}\\ \diagdown \mathrm{H}\end{smallmatrix}$ is present. Each amino acid has at least one amino group and one carboxyl group. We represent a generalized amino acid as

$$\begin{array}{c} \mathrm{R} \\ | \\ \mathrm{H{-}C{-}N}\begin{smallmatrix}\diagup \mathrm{H}\\ \diagdown \mathrm{H}\end{smallmatrix} \\ | \\ \mathrm{C} \\ \mathrm{HO}\diagup \;\; \diagdown\!\!\diagdown \mathrm{O} \end{array}$$

The part of the molecule symbolized by the letter R varies from one hydrogen atom to a group of carbon atoms with hydrogen and other atoms attached. Each amino acid has a distinctive R-group. Figure 2.10 shows some amino acids. Many amino acids are interconvertible with fatty acids through exchange of the amino group. Removal of this group from an amino acid results in a fatty acid and ammonia.

The reaction between the amino group of one amino acid and the carboxyl group of another amino acid results in the formation of a *peptide* bond between them. This is a carbon-to-nitrogen bond (Figure 2.11). A large number of amino acids held together by peptide bonds forms a protein. There are about two dozen different amino acids found in various proteins. Since the amino acids may be linked together in any order, and each different order of amino acids is a different protein, the variety of possible proteins approaches infinity. This variety is so great, in fact, that each organism has its unique protein combination.

In addition to the peptide bonds holding the amino acids together in a long chain, other bonds give each protein a distinctive shape. The protein takes on a more or less twisted or contorted configuration when sulfur-to-sulfur bonds link portions of the amino acid chain. These bonds are sparsely distributed in a protein. They form between the sulfhydryl groups (Figure 2.10*e*) of two molecules of the amino acid cysteine. These bonds may cause sharp bends in the amino acid chain (Figure 2.12).

Proteins are often held in a helical shape by hydrogen bonds that form between the remaining oxygen of a carboxyl group and the remaining hydrogen of an amino group. This weak bond is solely a result of the attraction of the small negative charge on the oxygen and the small positive charge on the hydrogen. Because these remnants are available from every amino acid in the chain, hydrogen bonds are common in proteins (Figure 2.13).

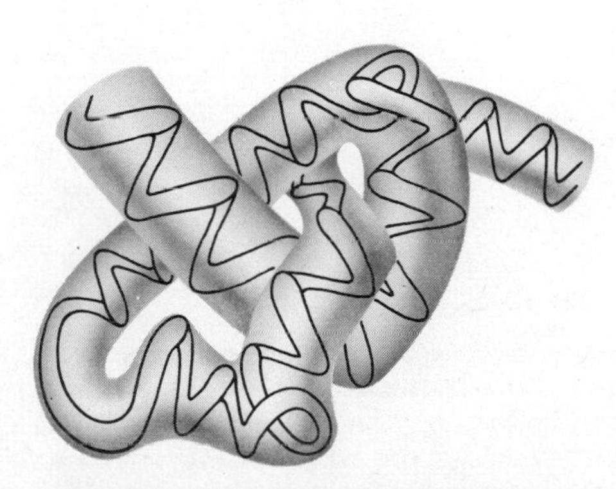

Figure 2.12. Configuration of a portion of a protein.

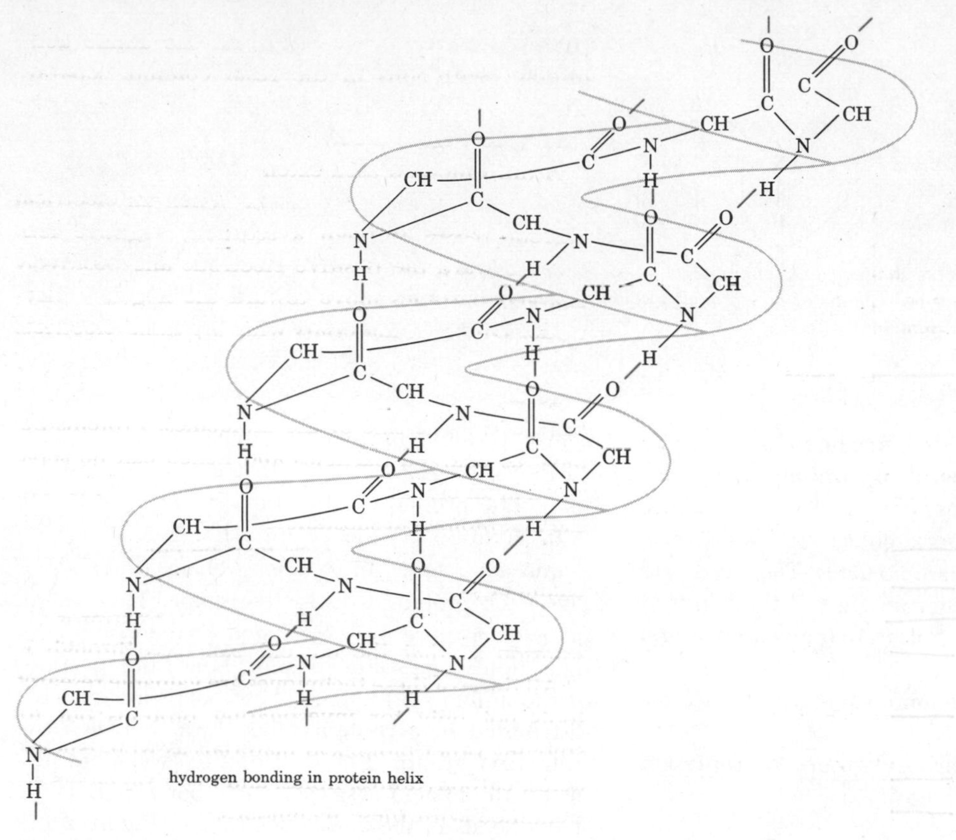

Figure 2.13. The helical configuration of a protein is maintained by hydrogen bonds (blue lines).

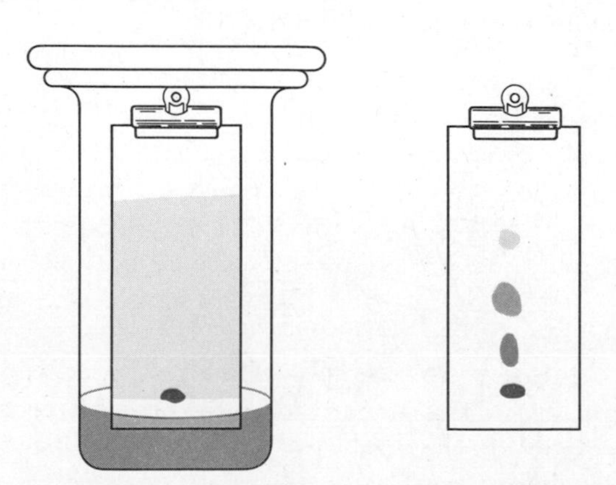

Figure 2.14. One form of paper chromatography. After the paper strip is removed and dried, it is sprayed with a chemical that makes the amino acid spots visible.

An analysis of the kinds of amino acids in a protein usually involves *chromatography,* a name given to several widely used laboratory techniques for separating mixtures of similar materials. Of course, the protein must be degraded chemically into separate amino acids before these methods are applied.

In *paper* chromatography, a drop of the amino acid mixture is placed near one end of a strip of filter paper and allowed to dry. This end of the strip is arranged to touch an appropriate solvent (Figure 2.14 left). The solvent rises in the paper by capillary action and dissolves the amino acids. These in turn move at different rates to become, eventually, separate spots along the paper (Figure 2.14 right). The amino acids on the chromatogram can be identified by comparison with chromatograms of known amino acids. Many refinements and variations of this method are in use today.

Ion exchange chromatography is also frequently used in amino acid studies. This method involves passing the amino acid mixture through a long tube packed with synthetic resins. As the amino acids interact with ions in the resin column, separate bands form. Each band contains a different amino acid. (See Figure 2.15.)

A third method used extensively in studying proteins is that of *electrophoresis.* When an electrical current passes through a solution, negative ions move toward the positive electrode and positively charged particles move toward the negative electrode. Different materials with the same electrical charge are also separated because they migrate at different rates. The rate depends on the amount of charge and the size of the molecules. Proteins behave as charged particles and hence can be separated in an electrical field. In fact, they were among the first biological substances to be separated by this technique. Electrophoresis may be conducted in a chamber similar to that shown in Figure 2.16, or along a strip of filter paper moistened with a salt solution so that the current can pass through it.

All three of these techniques are valuable research tools not only for investigating proteins but for studying other biological materials as well. Nucleic acids, carbohydrates, lipids, and plant pigments are examined with these techniques.

Antibodies are proteins built up by the individual that help to inactivate foreign proteins and render them harmless. Although this mechanism is usually helpful, it may pose a drastic problem in certain artificially induced cases. For example, blood transfusions, tissue transplants, and organ transplants often fail or prove damaging because of antibody reactions.

In living systems many proteins, functioning differently from antibodies, may form an integral part of the structure of cells. All membranes, for example, have a protein component; in Chapter III we shall see how proteins in membranes contribute to many structures within the cell. In muscle cells, two proteins form the specialized contractile structure. Other proteins contribute to the structure of supporting and protective elements such as skin, hair, nails, bone, and cartilage.

Enzymes are catalysts; in biological systems they are the substances that change the rate of a reaction without themselves being changed. Remember that reactions often require activation energy. In common with many other types of catalysts, enzymes reduce the amount of activation energy required (Figure 2.17). As a result, many reactions that occur

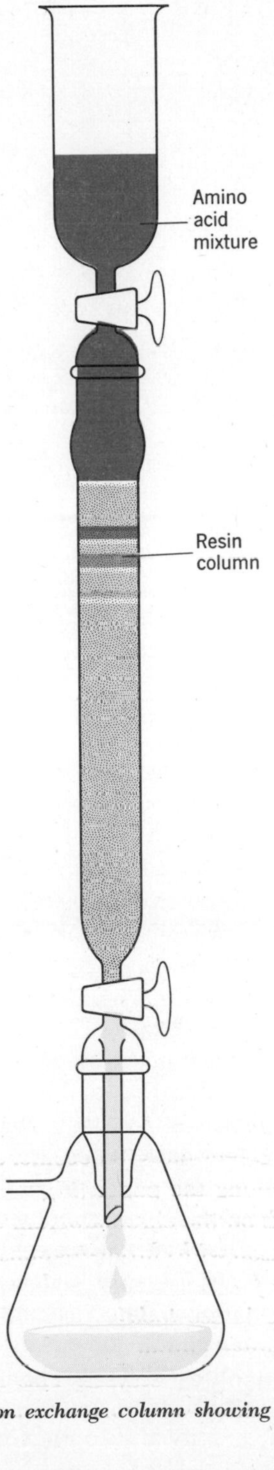

Amino acid mixture

Resin column

***Figure 2.15.** An ion exchange column showing several bands of amino acids.*

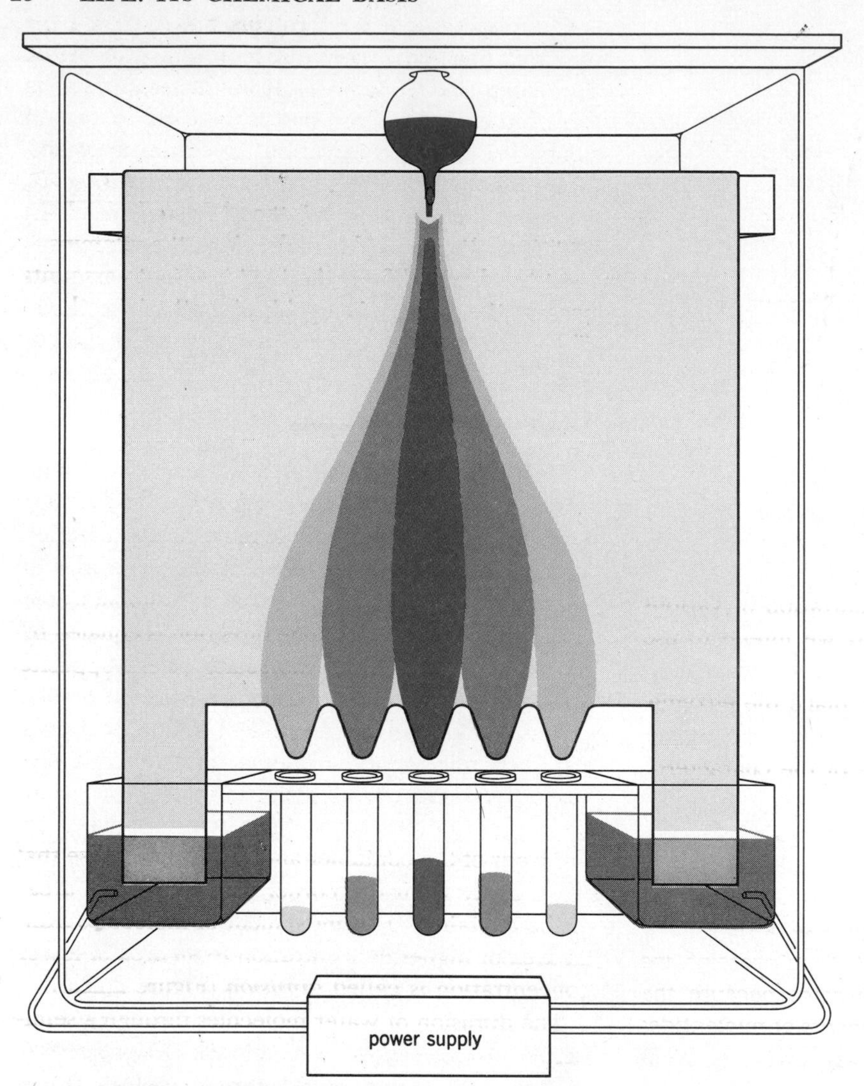

Figure 2.16. Electrophoresis separates proteins as indicated by the different shades of blue. A mixture of proteins is fed to the paper at the top and the various proteins are collected at the base of the paper. Each component migrates differently in the electric field.

rapidly in cells proceed slowly in the absence of the proper enzymes. Virtually all reactions in biological systems are controlled by enzymes.

Enzymes have complex and varied surface shapes. If they are to affect any reaction of a molecule, the surfaces of the molecule and the enzyme must fit together (Figure 2.18). Since the surface of a given enzyme will fit with only a few different molecules, enzymes are specific in their action. Often an enzyme will act on only one kind of molecule.

All known enzymes are wholly or partly protein. This gives them many of their characteristics. For example, they are sensitive to heat and to the relative amounts of acids and bases in the environment. Any marked change in temperature or acidity will change the activity of the enzymes and may even deactivate some, for such changes will alter the pattern of hydrogen bonding and, therefore, the shape of the enzyme. Consequently, all enzymes, whether inside the cell like the majority or secreted outside the cell like many digestive enzymes, show their typical activity at its optimum rate only if the

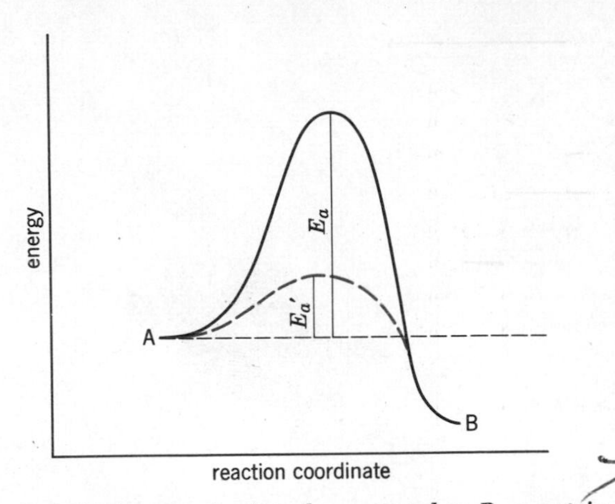

Figure 2.17. In order for A to be converted to B, a certain amount of activation energy (E_a) is required. When the proper enzyme is present, less energy is required for activation (E_a').

temperature, acidity, and concentration of various ions fall within narrow limits. If we intend to use an enzyme for any process, we must carefully control the conditions in which we place the enzyme. For example, the aging of meat takes place best at moderate temperatures because of the characteristics of the enzymes involved.

Nucleic acids are the fourth group of carbon-containing molecules found in all cells. These are composed of long chains of *nucleotides*. Each nucleotide has three parts: an organic base, a sugar, and a phosphate (Figures 2.19–2.21). As in proteins, we find a large variety of nucleic acids because the number of the possible arrangements of nucleotides approaches infinity. Nucleic acids have two functions: the control of cell activity and information storage. We shall discuss these topics in Chapters VII and XVII.

Carbohydrates, lipids, proteins, and nucleic acids are similar enough in structure to be interconvertible in the cell (see Chapter V). It might seem, then, as though the intake of any one of these groups should suffice for complete nutrition. However, most animals cannot synthesize all fatty acids or amino acids, but must obtain some of them through food. In addition, other complex molecules—the vitamins—must be obtained from the food in minute amounts because the cells are incapable of synthesizing them. These substances are often incorporated in the enzymes synthesized by the cells.

Some Physical Phenomena

Molecular Movement. All molecules are in constant, and usually random, motion. On the average, a given molecule is as likely to move in one direction as it is to move in any other. If the molecules of a substance are evenly distributed throughout a container, the movement in one direction is equaled by the movement of other molecules in the opposite direction, and no *net* movement of material results.

If we remove all the molecules from the container and then reintroduce a number of them into one corner, a random movement causes them to disperse. Molecules that move toward the corner bounce off the walls of the container and each other; those that move away from the corner pass into other areas of the container. This movement of molecules from an area of higher concentration to an area of lower concentration is called *diffusion* (Figure 2.22*a*).

The diffusion of water molecules through a semipermeable membrane (a membrane through which only the solvent may pass) is termed *osmosis*. It can be demonstrated with nonliving membranes such as cellophane. This is a passive process since energy

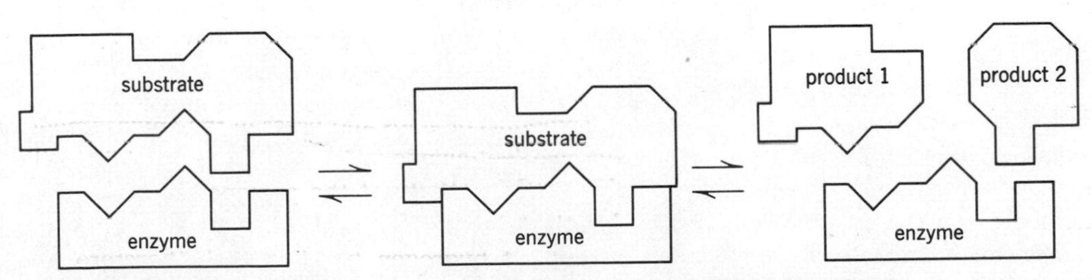

Figure 2.18. Enzymes are highly specific, as is indicated by this schematic "lock and key" model. The reaction shown is reversible. The enzyme itself remains the same and is unchanged during the course of the reaction. The word substrate *is used to denote a compound that is acted upon by an enzyme.*

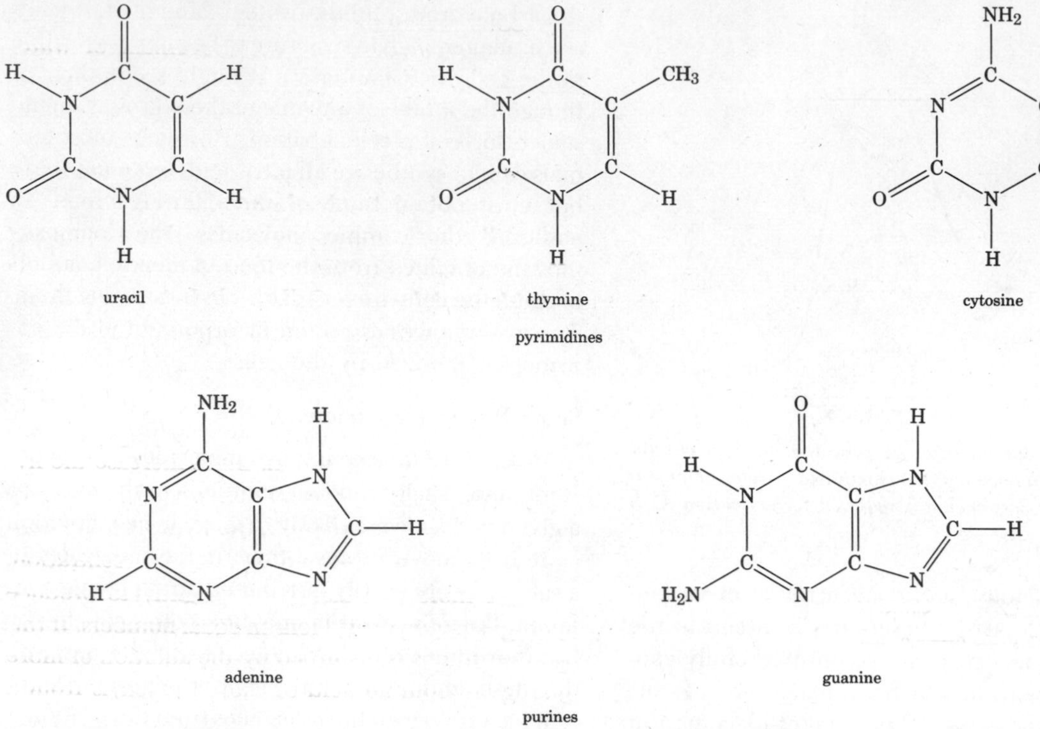

uracil thymine cytosine

pyrimidines

adenine guanine

purines

Figure 2.19. The organic bases of the nucleic acids are the pyrimidines and purines. The three common pyrimidines are uracil, thymine, and cytosine. The two common purines are adenine and guanine.

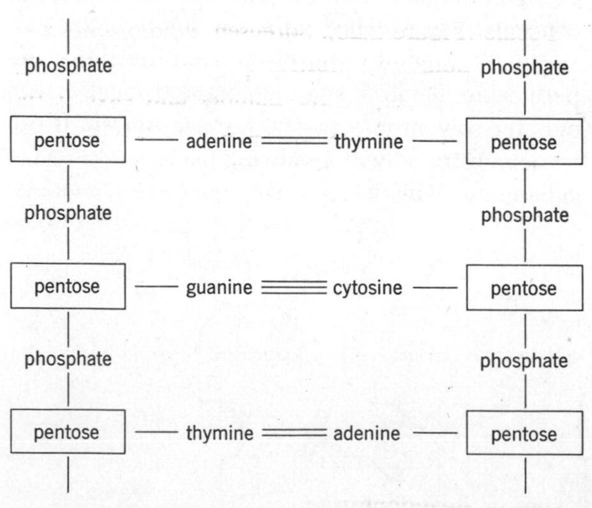

Figure 2.20. Nucleic acids are made up of nucleotides linked together between the phosphate and the pentose (sugar) in repeating units. The bases protrude from the chain and serve to link two complementary chains.

is not expended in moving the molecules through the membrane.

Water also diffuses through living cell membranes. It has become customary to call this event osmosis also; however, cell membranes are *differentially* permeable rather than semipermeable because they allow certain kinds of particles other than the solvent to pass through. Small uncharged particles like water, carbon dioxide, and ammonia move through easily. Larger molecules, such as sugar, and charged particles pass through more slowly, if at all.

These latter particles do indeed move in and out of cells, but to achieve their transport the cell must expend energy. For example, energy is required to move glucose through a cell membrane. This is also the case when molecules or ions are moved from an area of lower concentration to one of higher concentration (Figure 2.22*c*). Potassium ions, for example, are usually found in a greater concentra-

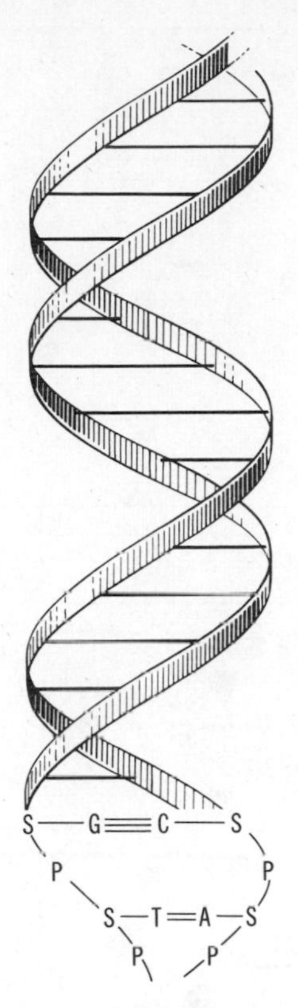

Figure 2.21. This schematic diagram illustrates the spatial configuration of the double-stranded DNA molecule. The bars represent paired purine and pyrimidine bases while the spiral backbones are composed of alternating sugar and phosphate units.

tion inside cells than in the cell's immediate environment. The cell must expend energy in order to maintain this uneven relationship. The term *active transport* is applied to situations where a cell expends energy in transporting materials through its boundaries.

The importance of osmosis in cells can be demonstrated by considering the effects of different concentrations of sugar solutions on sea urchin eggs. If the eggs are placed in a solution that has a lower concentration of dissolved particles than do the eggs, water molecules will move from the area of their higher concentration to the area of their lower concentration, namely, into the eggs (Figure 2.22*b*). Consequently, the eggs will swell and eventually burst. If the sugar solution has a higher concentration of dissolved particles than the eggs, water will move out of the eggs. Through this loss of water, the eggs will shrink.

This movement is the *net* movement of water molecules. In either case, some of the water molecules will move in each direction, but many more will move in one direction than in the other. If we place the eggs in a sugar solution that has the same concentration of dissolved particles as the eggs, the eggs will remain unchanged in size because no *net* movement of water occurs.

Acids and bases. Cells and their components function in a liquid medium. The acidity, alkalinity, or neutrality of this medium is an important factor in chemical activities in cells.

Acids are substances that contribute hydrogen ions (H^+) in an aqueous medium. Bases increase the hydroxyl (OH^-) ion concentration when dissolved in water. In a neutral solution, the hydrogen ion concentration equals the hydroxyl ion concentration.

Water molecules also ionize slightly to produce hydrogen and hydroxyl ions in equal numbers. If this balance of ions is disturbed by the addition of more hydrogens from an acid or more hydroxyls from a base, the solution changes from neutral to an acidic or alkaline condition.

It is impractical to describe the actual numbers of hydrogen and hydroxyl ions in water because of the small number present. We would be discussing ratios of 10^{-7} (1/10,000,000) hydrogen ions for every 56 molecules of water. To cope with this problem, a Swedish chemist, Jönen Sørensen, developed the pH scale (Figure 2.23). Sørensen, who did not like negative numbers, arbitrarily converted the exponent −7 to a positive number and gave it the designation *pH.* A pH of 7 is a description of the concentration of hydrogen ions in pure water. Since the numbers of hydroxyl and hydrogen ions are equal, this is a neutral solution. As the pH becomes larger (8, 9, 10, etc.), the concentration of hydrogen ions decreases, that is, the solution becomes more basic. As the pH becomes smaller (6, 5, 4, etc.), the concentration of hydrogen ions increases, that is, the solution becomes more acidic. This scale is a convenient description of the conditions in and around cells.

Types of Reactions

The chemical reactions in cells are varied and frequently complex. They may become easier to understand if we classify them into five general

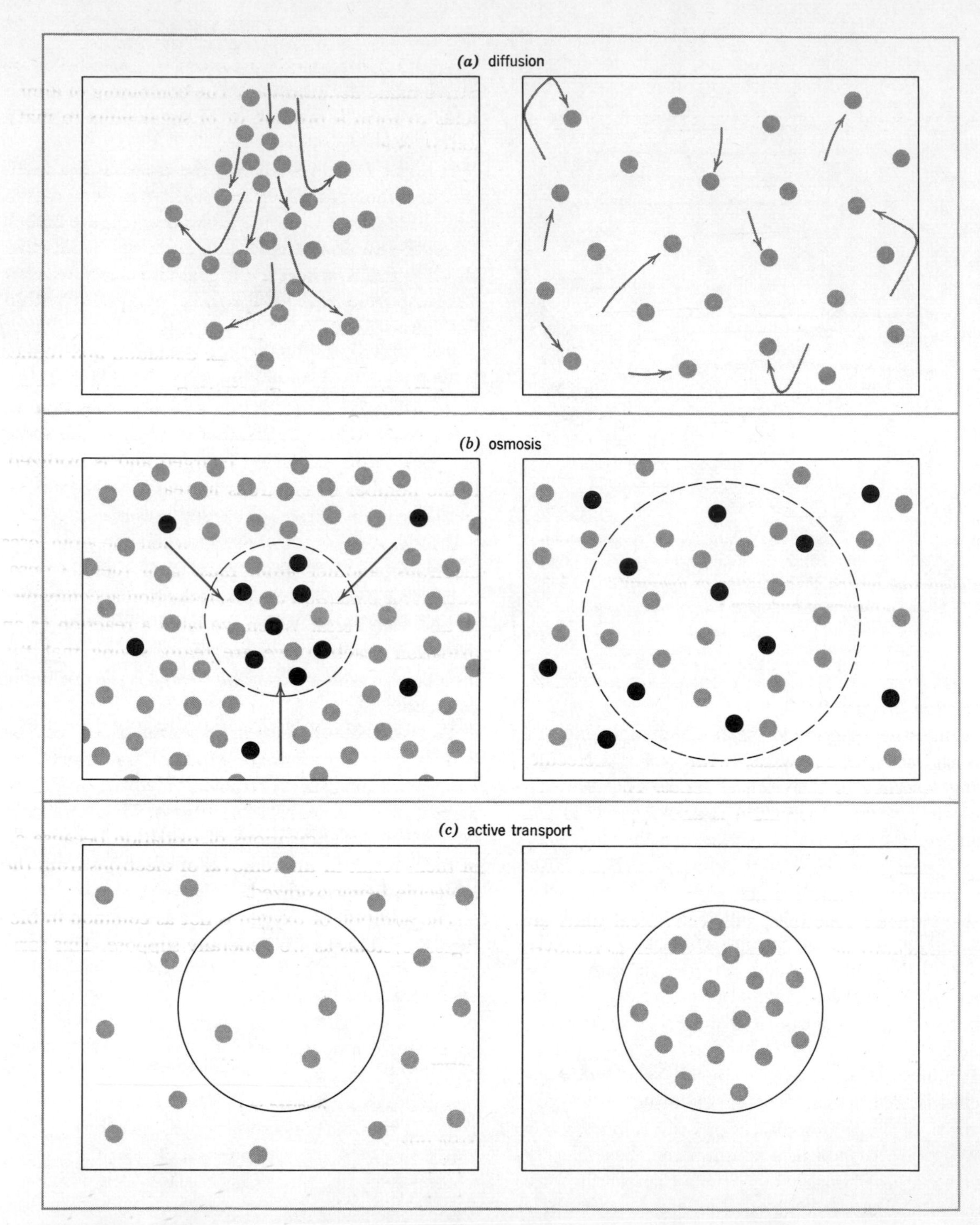

Figure 2.22. (a) ***Diffusion,*** (b) ***osmosis,*** (c) ***active transport.*** (a) ***In diffusion, molecules introduced into a container*** (***left***) ***become evenly distributed after a period of time*** (***right***). (b) ***In osmosis, the diffusion of water may cause a cell to swell, as shown. The blue dots represent water molecules; the black dots represent dissolved particles. In the left-hand drawing the arrows indicate the direction of water movement into the original cell; the right-hand drawing shows the cell's final size. Under different conditions, water may move out of the cell, causing the cell to shrink.*** (c) ***Active transport enables a cell to accumulate molecules even when there are more inside the cell than outside. Cells may also expel other molecules by the same process.***

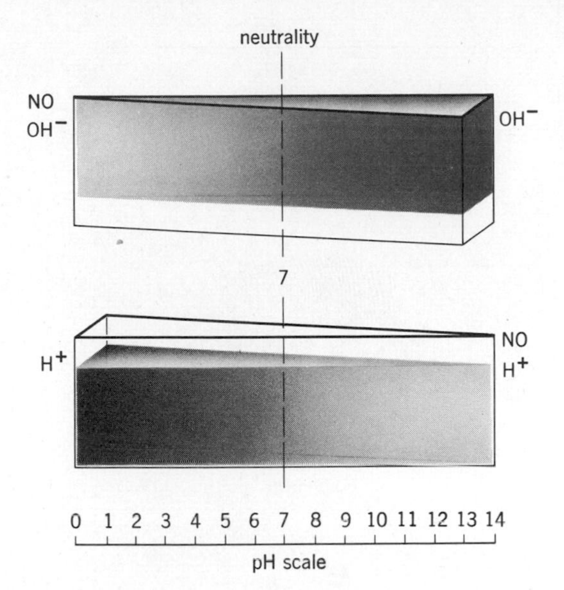

Figure 2.23. The pH scale. The hydrogen ion concentration is highest at low pH and decreases with increasing pH. The opposite holds true for the concentration of hydroxyl ions. At pH 7.0, the concentrations of hydrogen ions and hydroxyl ions are equal.

groups: *digestion, synthesis, transfer, oxidation,* and *reduction* (Figure 2.24).

A digestive reaction is one in which molecules are broken into smaller units, with water molecules being attached to the broken bonds. The term hydrolysis ("water dissolving") is also applied to this reaction. Examples of digestion are the degrading of proteins (Figure 2.25*a*), lipids, and carbohydrates into simpler chemical units.

In synthetic reactions, small chemical units are combined into larger ones, and water is removed (Figure 2.25*b*). (Hence, synthesis is given the alternative name *dehydration.*) The combining of amino acids to form a protein, or of sugar units to make starch, is an example.

Transfer reactions involve the transfer of a small part of a molecule, such as a hydrogen or an amino group, from one molecule to another (Figure 2.25*c*). Transfer reactions are vital to chemical events like photosynthesis, where hydrogens and electrons must be transported from compound to compound within the chloroplast.

The final pair of reactions, oxidation and reduction, always occur together (Figure 2.26). Both involve a change in the number of electrons that an atom controls. If the number decreases, the atom becomes more positively charged and is oxidized. If the number of electrons increases, the atom becomes more negatively charged and is reduced.

It is important to notice that when one atom loses electrons, another atom must gain them. Consequently, if oxidation occurs, reduction accompanies it, and vice versa. When we label a reaction as an oxidation reaction, we are really saying that the molecule in which we are interested is the one being oxidized.

The oxidation of molecules within a cell may be detected by any of three kinds of treatments: (1) addition of oxygen, (2) removal of hydrogen (Figure 2.25*d*), or (3) removal of electrons. These are only three different indications of oxidation because all of them result in the removal of electrons from the molecule being oxidized.

The addition of oxygen is not as common in biological systems as we generally suppose. This form

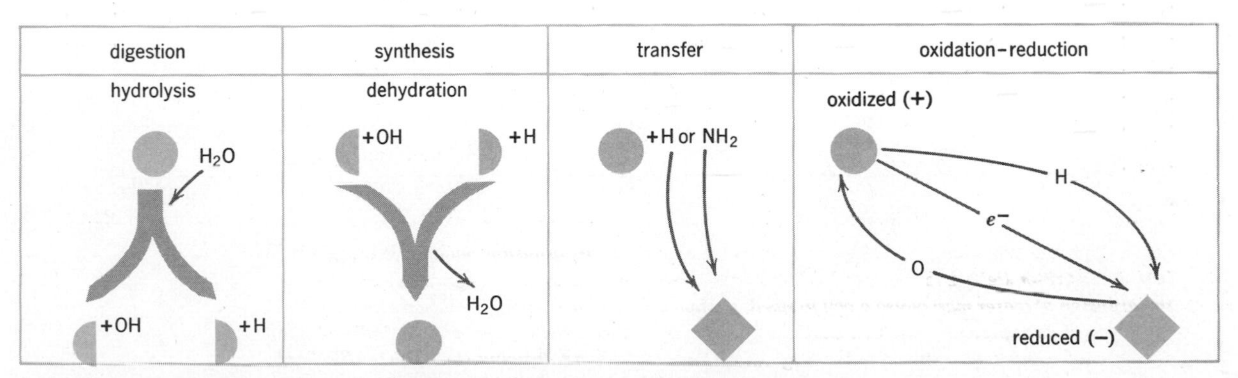

Figure 2.24. A summary of the basic reactions that occur in the cell. The molecules are represented in yellow. Examples of these reactions are given in Figure 2.25.

$$NH_2-CH(R)-C(=O)-NH-CH(R)-C(=O)-NH-CH(R)-COOH + 2H_2O \rightleftharpoons 3NH_2-CH(R)-COOH$$

tripeptide water amino acids

(a)

glycine + alanine $\rightleftharpoons$ glycyl-alanine + H_2O

(b)

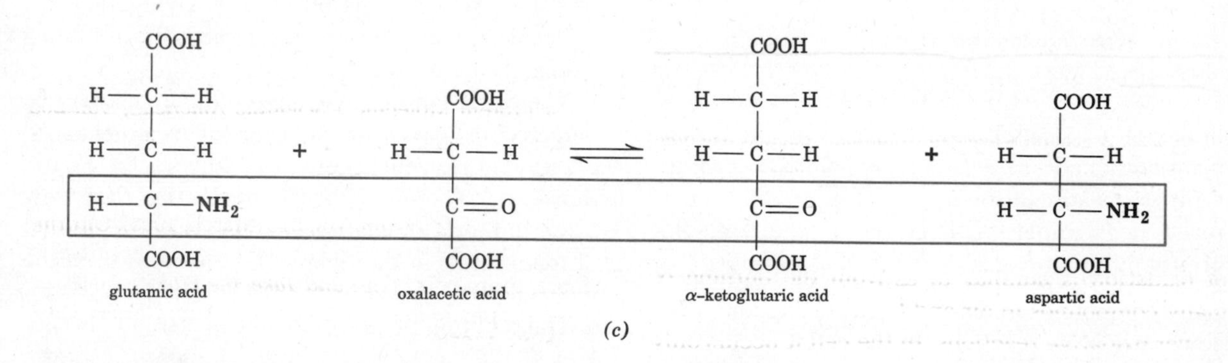

(c)

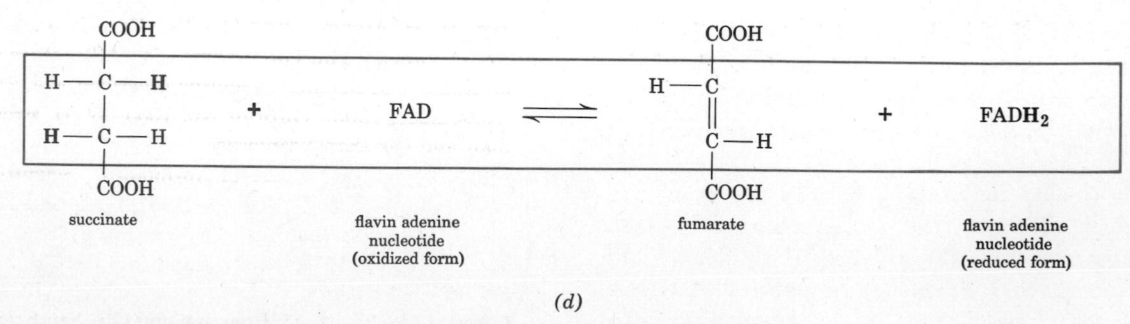

(d)

Figure 2.25. (a) *The hydrolysis of proteins or peptides into amino acids is a typical digestive reaction.* (b) *The condensation of two amino acids into a dipeptide is an important synthetic reaction.* (c) *The transfer of an amino group from one compound to another illustrates a transfer reaction.* (d) *The oxidation of succinate by the oxidized form of flavin adenine nucleotide (FAD) is an important reaction in respiration. During the course of the reaction, succinate is oxidized to fumarate and FAD is reduced to $FADH_2$.*

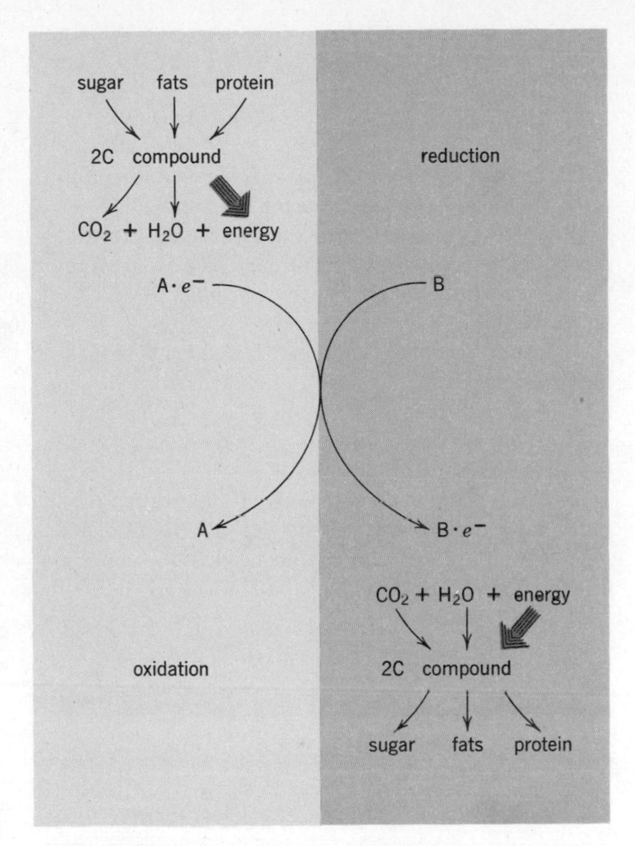

Figure 2.26. A general scheme of oxidation-reduction reactions that indicates the relation of three classes of food to each type of reaction.

of oxidation is familiar to us from the burning of many compounds in air and from rusting and other slower oxidative reactions. In the cell it occurs only in the final step of cellular respiration (see Chapter V).

The removal of hydrogen (*dehydrogenation*) and the concurrent removal of electrons from the molecule are processes commonly found in cells. The importance of these processes in energy release will be noted in Chapter V.

Life can be described in part by describing the chemical makeup of living organisms. But much more is involved. These chemical substances are highly organized into structural entities which we call cells. Life is, at least in this sense, more than the sum of its parts.

Principles

1. The laws governing the chemical and physical events in living matter are the same as those acting on inanimate matter.
2. Living things are characterized by four types of complex molecules.
3. The method of transport of molecules in living matter depends on their size and on their charge.

Suggested Readings

Allfrey, Vincent G. and Alfred E. Mirsky, "How Cells Make Molecules," *Scientific American*, Vol. 205 (September, 1961). Offprint No. 92, W. H. Freeman and Co., San Francisco.

Asimov, Isaac, *The Chemicals of Life*. New American Library of World Literature, New York, 1954.

Baker, Jeffrey J. W. and Garland E. Allen, *Matter, Energy, and Life*, Addison-Wesley Publishing Co., Reading, Mass., July, 1965.

Holter, Heinz, "How Things Get into Cells," *Scientific American*, Vol. 205 (September, 1961). Offprint No. 96, W. H. Freeman and Co., San Francisco.

Kendrew, John C., "The Three-Dimensional Structure of a Protein Molecule," *Scientific American*, Vol. 205 (December, 1961). Offprint No. 121, W. H. Freeman and Co., San Francisco.

Merrifield, R. B., "The Automatic Synthesis of Proteins," *Scientific American*, Vol. 218 (March, 1968). Offprint No. 320, W. H. Freeman and Co., San Francisco.

Moore, Frances D., *Give and Take, the Biology of Tissue Transplantation*. Doubleday and Co., Garden City, N.Y., 1965.

Neurath, Hans, "Protein-Digesting Enzymes," *Scientific American*, Vol. 211 (December, 1964). Offprint No. 198, W. H. Freeman and Co., San Francisco.

Phillips, David C., "The Three-Dimensional Structure of an Enzyme Molecule," *Scientific American*, Vol. 215 (November, 1966). Offprint No. 1055, W. H. Freeman and Co., San Francisco.

Porter, R. R., "The Structure of Antibodies," *Scientific American*, Vol. 217 (October, 1967). Offprint No. 1083, W. H. Freeman and Co., San Francisco.

Roberts, John D., "Organic Chemical Reactions," *Scientific American*, Vol. 197 (November, 1957). Offprint No. 85, W. H. Freeman and Co., San Francisco.

Questions

1. What is the simplest atom from a structural viewpoint? Is this atom of importance in living matter?
2. Does the symbol H_2O represent a compound, a molecule, or both? Explain.
3. How would you test whether the atoms in a substance were held together by covalent bonds or ionic bonds?
4. Which type of bonding is found in table salt? Which type is found in table sugar?
5. What is the relation between the bonds that hold atoms together and the energy that is essential to maintaining life?
6. What three features make water vital to life?
7. Do CO_2 and NH_3 have any importance to cells except as waste products? Explain.
8. What are the chief carbon-chain compounds in living tissues? List their major functions.
9. What characteristic group of atoms always designates that an organic compound is an acid? Which of the following contain this group: amino acid, fatty acid, glycerol, acetic acid?
10. How does a phospholipid differ structurally from a fat?
11. Review the structural arrangements found in proteins, then deduce what is meant by *polypeptide.*
12. What evidence is presented in the chapter to indicate that every living being probably has some proteins that are unique to it? How is this possible? (*Hint.* Check on the building units of proteins.)
13. Describe some techniques by which mixtures of biological materials, such as amino acids, may be separated and analyzed.
14. What are the (*a*) advantages, and (*b*) disadvantages to cells of diffusion and osmosis?
15. What is the relation of active transport to osmosis?
16. What does a pH of 7.8 signify? A pH of 5.9?
17. Name five types of chemical reactions important in cells and explain each one.

CHAPTER III

Life: Its Structural Basis

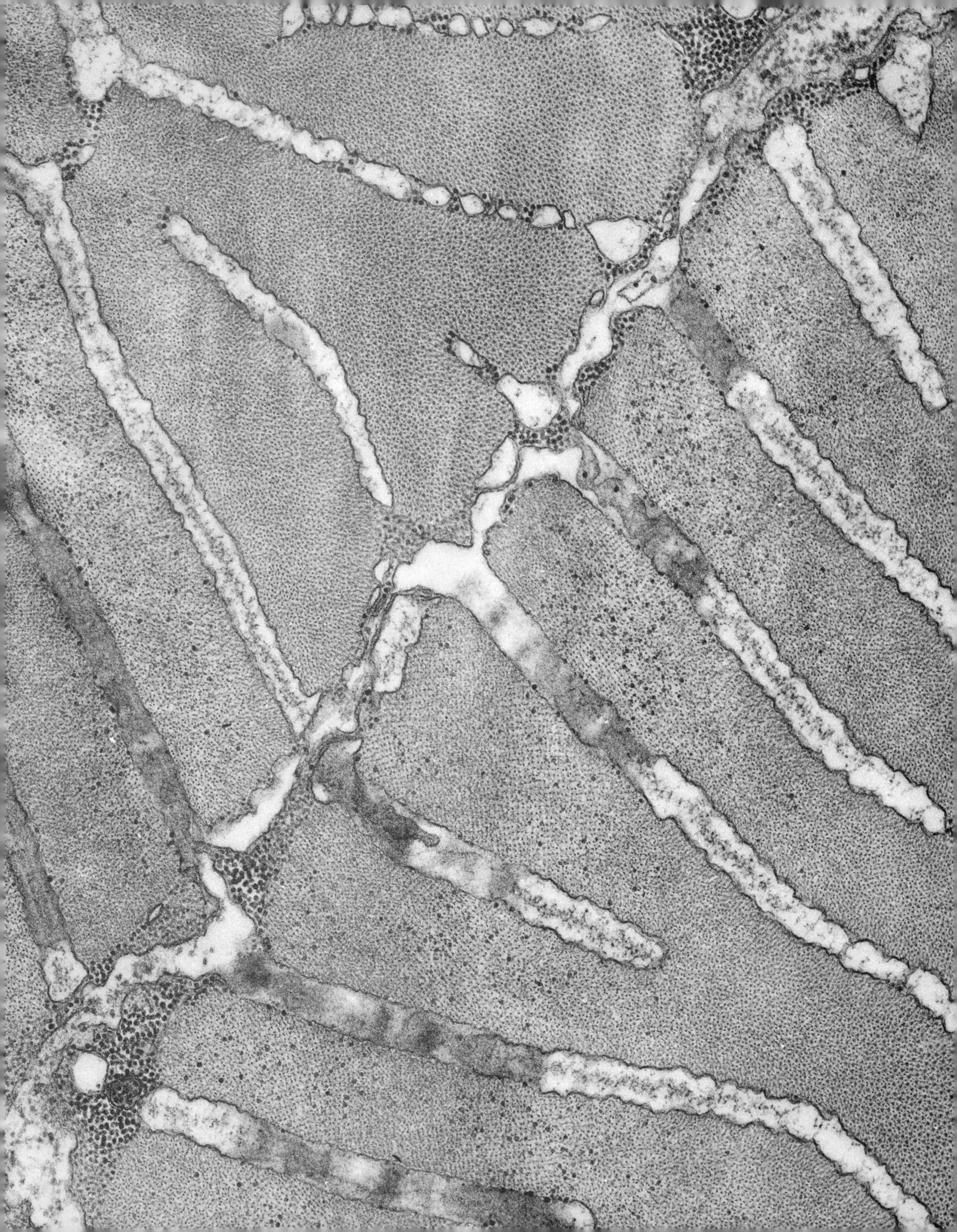

Electron photomicrograph of two fish muscle fibers showing the distribution of some cytoplasmic elements (×46,000) (Dr. Keith R. Porter).

CHAPTER III

Life: Its Structural Basis

The study of cells, *cytology*, is a major field of biology. Much significant biological knowledge has been obtained by studying the various parts of the cell, by examining it as a functional entity, and by considering its role in relation to other cells. Usually the cell is the basic unit of reference for understanding the structure or function of any living entity. In this chapter we review briefly the history of the cell concept, examine the traditional or "classical" idea of cell structure, and then relate it to newer ideas concerning cells that have been derived from studies made with modern cytological "tools."

The discovery of cells and the subsequent realization of their basic importance in biology could not take place until lenses and microscopes were developed. By the beginning of the seventeenth century, various types of magnifying devices had begun to disclose the microscopic structure of many materials. A number of significant biological events took place during that century. An Italian, Marcello Malpighi, performed the first microscopic studies on the embryos of plants and animals and also discovered capillaries. Robert Hooke, an Englishman, using a 30-power "microscope" noted the honeycomblike nature of a thin slice of cork and he termed the tiny spaces "cells." A Dutchman, Anton Van Leeuwenhoek, used an improved 270-power microscope and was able to describe blood cells, spermatozoa, bacteria, protozoans, and many other previously unknown objects.

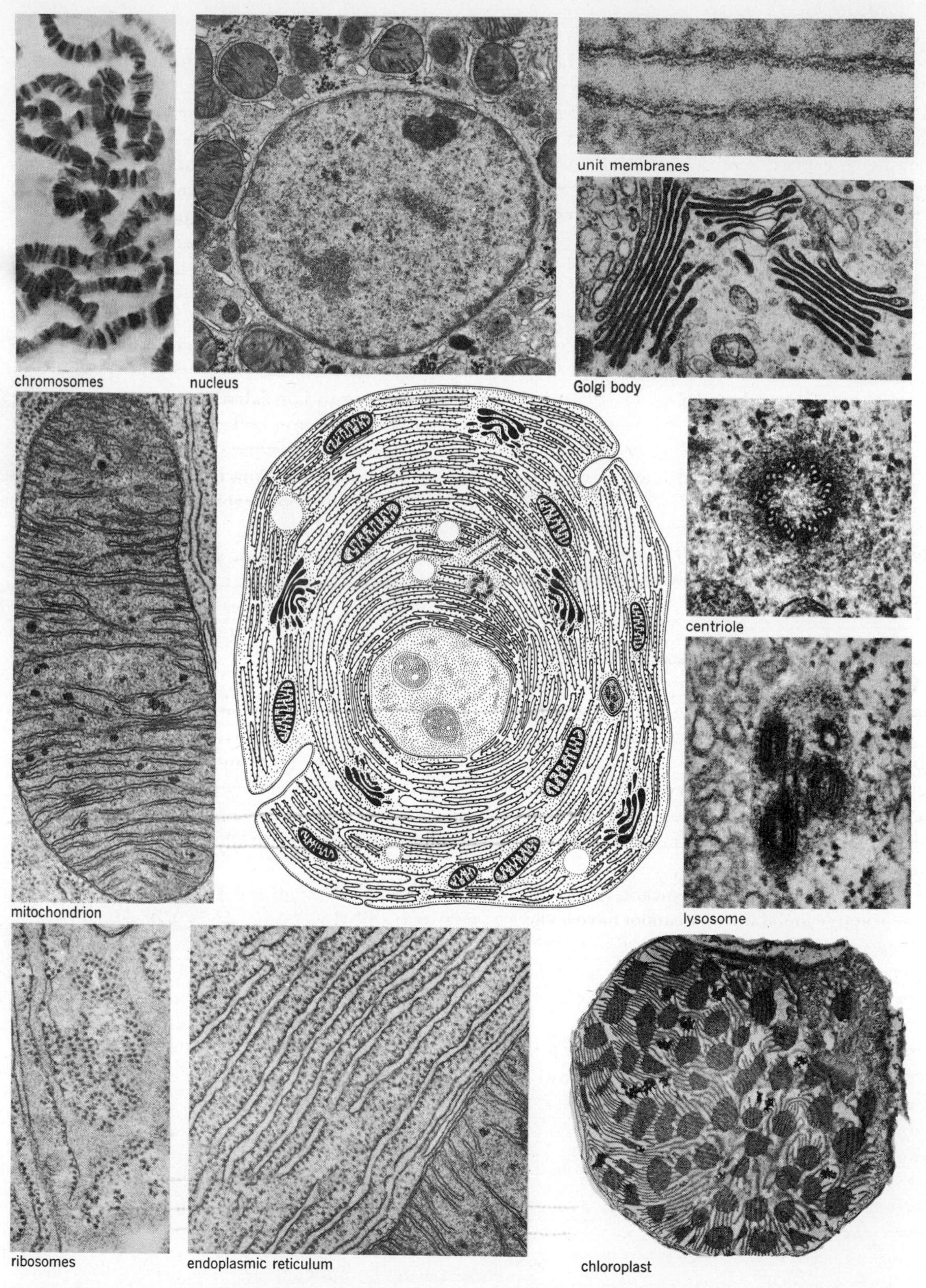

Figure 3.1. A contemporary idea of a generalized cell as shown by the electron microscope. The chloroplast is not shown in the generalized cell.

Over two hundred years passed, however, before biologists began to suspect that most living matter was composed of cells. Early in the 1800s, a number of zoologists and botanists began, as a result of their observations, to suggest a principle that was eventually to become one of the most significant principles in biology, the *cell theory*. This concept and the theories to which it gave rise proposed that all living things were composed of cells or of cells and their products, that all cells arose from preexisting cells, that all were basically alike in chemical composition, and that the activity of an organism was the outcome of the activities of its constituent cells.

The cell theory cannot be credited to any particular individual; rather, it evolved from the researches and writings of many biologists. To mention a few, Mirbel in 1802 concluded that plants were made up of cells, Lamarck in 1809 stated that cellular tissue was the general matrix of all organization, and Schwann in 1838 emphasized that entire animals and plants were aggregates of cells.

Some General Features of Cells

A majority of cells are microscopic in size and only a few, such as birds' eggs and some algae, are macroscopic. The size of a cell is limited by what might be termed a surface-to-volume dilemma. Small bodies have a greater ratio of surface area to volume than larger ones. Cells continually obtain materials necessary for their metabolism from the surrounding environment. Since this acquisition occurs through their surface membranes, the surface-volume relationship is vitally important. A cell that is extremely active metabolically cannot have a very large volume. If a cell must be both large and active, its surface area must, in some way, be increased disproportionately. The long thin processes of some nerve cells are examples.

Other problems are also encountered in relation to cell size: how much cytoplasm is to be controlled by the nucleus? What will give the cell sufficient physical support as it increases in size? These and other chemical and physical factors usually limit cell size to the microscopic level.

The size of cells has obviously affected the methods used to investigate their structure and function. Coincident with the improvement of microscopes came other developments: many techniques for staining specific parts of the cell, new ways to cut exceedingly thin slices of tissues so that various cell aggregates could be studied, and microsurgical instruments to manipulate and remove parts within the cell. Today, cells and tissues may be cultured or grown in laboratory glassware and thus studied in great detail. The development of extremely high-speed centrifuges, known as *ultracentrifuges*, has made it much easier to separate parts of cells and obtain concentrations of these parts for study.

The use of radioactive isotopes or "tracers" is also important in the study of cells and tissues. For example, carbon 14 (C^{14}) is an isotope of ordinary carbon 12 (C^{12}) and can be used in place of it in chemical reactions. Since C^{14} is also energy-emitting (radioactive), it can be detected and followed by means of appropriate laboratory instruments as it is incorporated into a cell or passes from one cell to another. By following an atom or molecule through chemical reactions in the cell, we can often acquire formerly unobtainable information about the reactions.

The *electron microscope* has also provided new knowledge and greater detail of the structure of cells (Figure 3.1). This instrument uses beams of electrons rather than light rays. As a result, an observer can see far more detail than is possible with conventional microscopes. By using an electron microscope, a cytologist can study the minute structure of cell parts rather than just their overall shapes and sizes.

These and other tools and techniques have contributed much new knowledge to the field of cytology since 1950. Concepts of cell structure, for example, changed drastically as the electron microscope revealed the ultrastructure, that is, submicroscopic morphology, of the cell's interior. In fact, a new area called *molecular biology* developed out of the studies on the structure and functioning of living matter at the molecular level. Molecular biology is an exciting frontier in contemporary biology since new discoveries and concepts are emerging from research laboratories almost daily.

The shapes of cells vary greatly, as illustrated in Figure 3.2. Cells like those forming tissues and organs demonstrate the existence of a close relationship between a cell's form and function. Free-living cells such as protozoans (Figure 3.3) and algae show an even greater range of forms or shapes, from simple spheres to the bizarre and complex.

Structures in the Cell

If one asked a cytologist to describe the structure of a cell, he would be likely to respond, "What *kind* of cell?" There are many structural differences be-

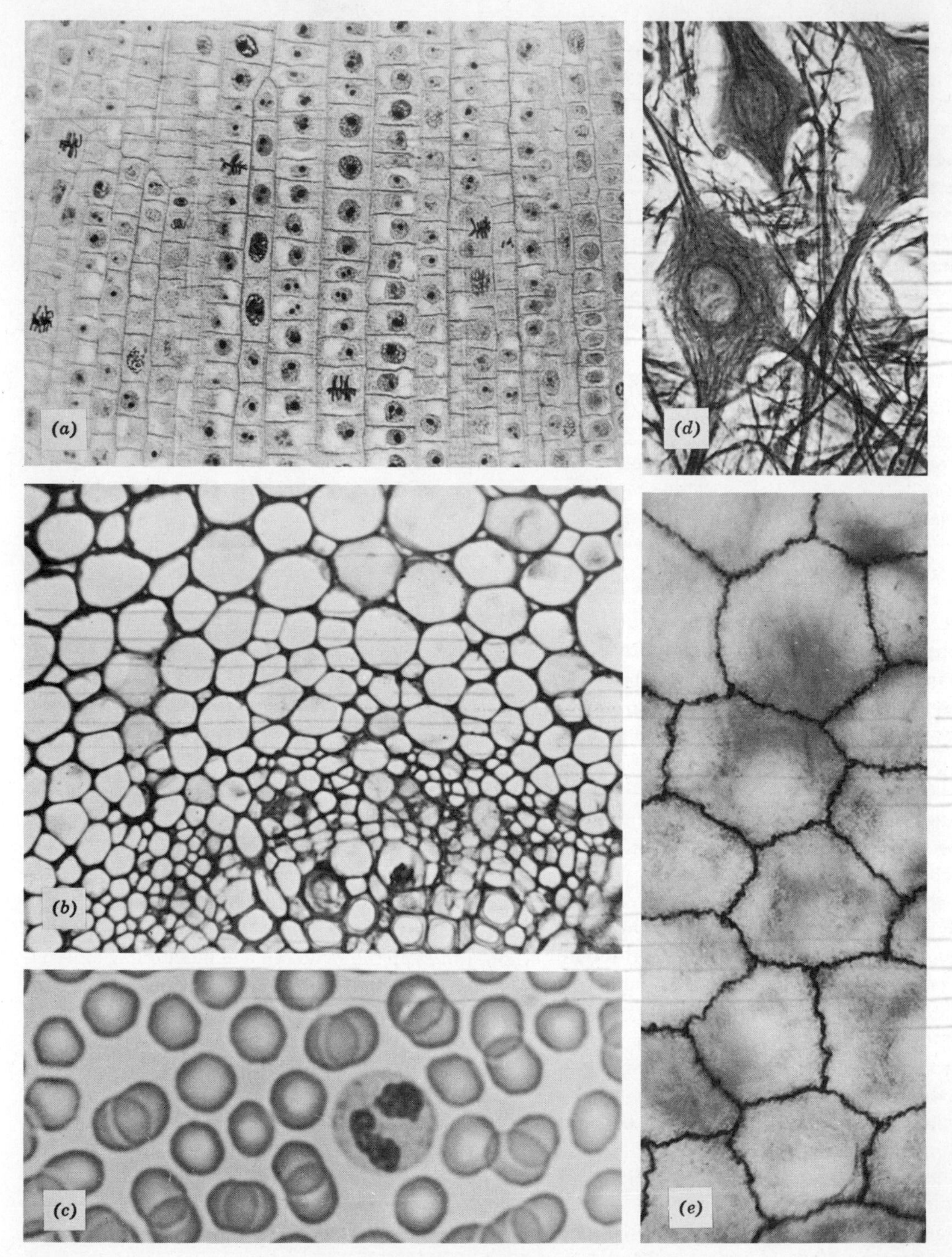

Figure 3.2. Cells of various shapes. (a) *Onion root tip;* (b) *section of a leaf showing part of a vein;* (c) *human blood cells;* (d) *human nerve cells;* (e) *the type of cell that lines internal body cavities of animals.*

tween a nerve cell and a blood cell, for example. What we describe here is a composite cell that contains the structures this chapter emphasizes. A cell of exactly this type probably does not exist. It is only a useful model.

In describing a cell, it is convenient to consider two major regions, the *cytoplasm* and the *nucleus*. Until recently, the cytoplasm was often described as a viscous, fluidlike material without definite structure, containing a number of functional bodies such as mitochondria, the Golgi complex, chloroplasts, centrioles, and a miscellaneous assortment of granules, droplets, and pigments. Electron microscopy has disclosed additional structures, in particular the endoplasmic reticulum. This indicates that the cytoplasm can no longer be considered "structureless." As these cytoplasmic parts are described, we should note that many of them consist of membranes of essentially similar structure.

Cell Membrane. All cells are bounded by an extremely thin membrane, often called a plasma membrane. Its existence was first revealed by its biological activities rather than its appearance under the microscope. In fact, the membrane is not even visible with an ordinary microscope. Biologists noted long ago, however, that cells must be surrounded by a permeability barrier in order to account for the differential passage of materials in and out of the cell. Also, the surface of a cell could be punctured with a microdissection needle or simply pushed inward as though a covering were present.

For some time it was thought that the plasma membrane was analogous to a sack made of thin cellophane, or other semipermeable materials, through which small molecules like water diffused freely but larger molecules like sugar could not. Osmosis in this cell membrane model supposedly imitated osmosis in real cell membranes. The analogy is correct only in a crude sense, because cellular membrane structure and function have almost no resemblance to the structure and function of artificial membranes.

In the 1930s, studies on cell membranes involving their permeability, surface tension, and electrical conductivity led J. F. Danielli and H. Davson to propose a theory about the chemical composition of plasma membranes. Their suggestion was that a cell membrane consists of two thin, adjacent layers of lipid molecules enveloped between two sheets or films of protein molecules (Figure 3.4). The lipids

Figure 3.3. Variation in cell shape in two groups of protozoans: foraminifera (top) and radiolaria (General Biological Supply House, Inc.).

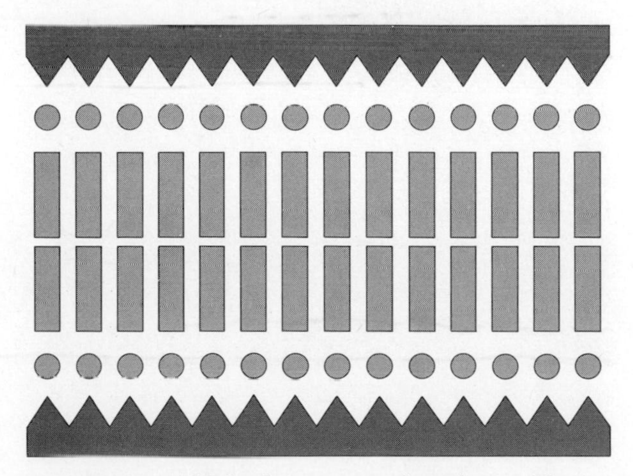

Figure 3.4. A model for the structure of membranes. The two outer layers are protein and the two medial layers are lipid.

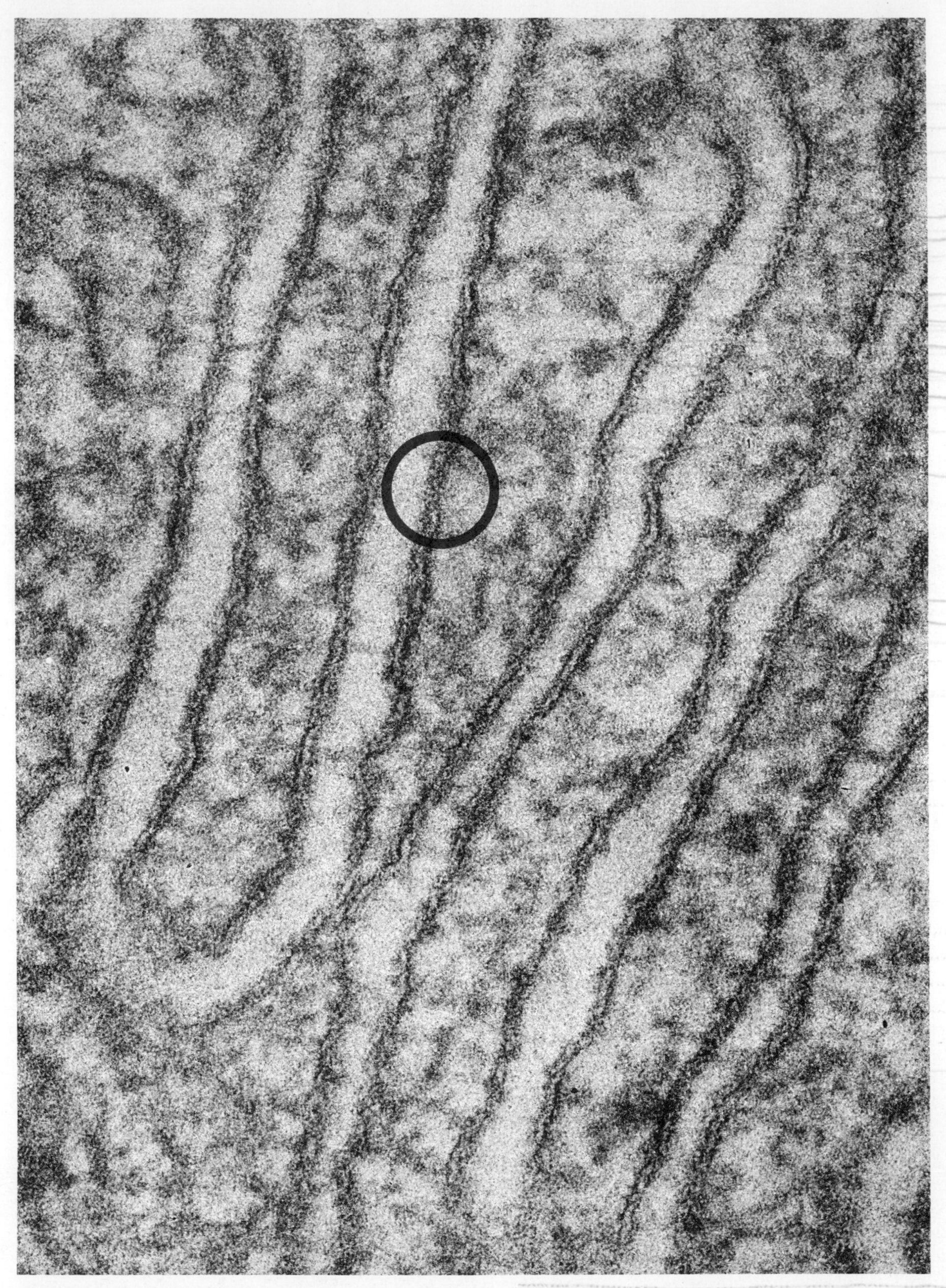

Figure 3.5. An electron photomicrograph of a cell membrane that demonstrates the "unit membrane" structure. The blue circle encloses one unit membrane. This membrane is greatly folded (×*400,000*) (*Dr. F. Sjostrand*).

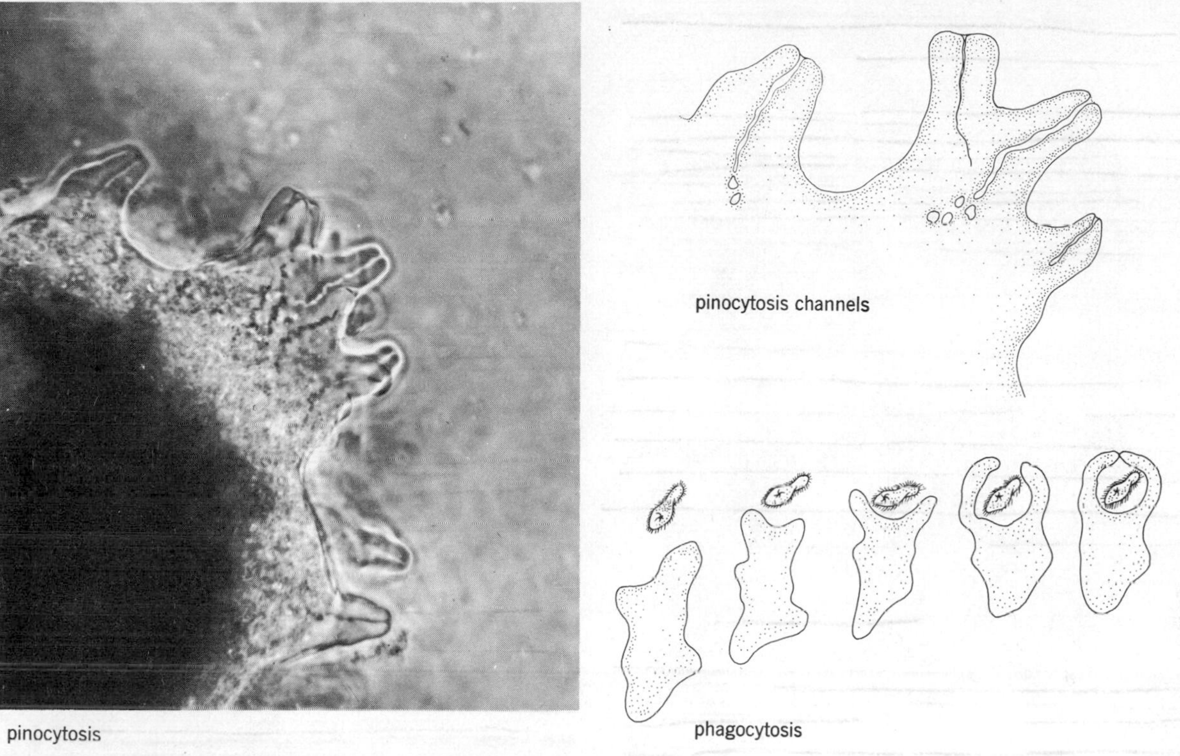

Figure 3.6. Amoeba showing pinocytosis (Dr. David Prescott) and phagocytosis.

are arranged with the fatty acid portions pointed toward the center of the membrane and the water-soluble portion (phosphate and nitrogen base) near the protein layers. This structure has been compared to a butter sandwich in which the lipid material constitutes the "butter." This theory about the structure of the cell membrane, proposed in 1935, is known as the Danielli-Davson model. Additional supporting evidence since 1935 has led to a general acceptance of the idea.

The lipids in membranes are now known to be mostly phospholipids and cholesterol. Less is known about the protein portion except that it is loosely bound to the lipid, is a structural protein, and comprises 60 to 80 percent of some membranes.

In recent years, plasma membranes have been examined with the electron microscope. Under high magnification the membrane appears as a triple-layered structure consisting of two opaque lines separated by a clear zone as shown in Figure 3.5. What relation does this have to the Danielli-Davson model? Dr. J. D. Robertson considered this question and, on the basis of his own studies on nerve cell membranes, proposed what is now termed the *unit membrane theory.* In the unit membrane, the two opaque lines correspond to the two protein sheets of the Danielli model, and the clear zone between them represents the lipid portion of the membrane. Robertson's concept is now generally accepted so that the term unit membrane appears in place of cell membrane in many scientific journals.

Critics of the Danielli model and unit membrane idea point out that these studies were based mostly on red cell corpuscle membranes (red cell "ghosts") and on myelin sheath material from nerve cells; the chemical composition of membranes from different sources varies greatly in lipid-to-protein ratio; the only similarity among cell membranes studied so far is their general resemblance in electron micrographs. These are serious objections and must be explained eventually if the unit membrane and lipoprotein "sandwich" concepts are to become biological principles. In the meantime, we must watch the scientific literature for additional evidence about the nature of the membrane.

Whatever its structure, the unit membrane is im-

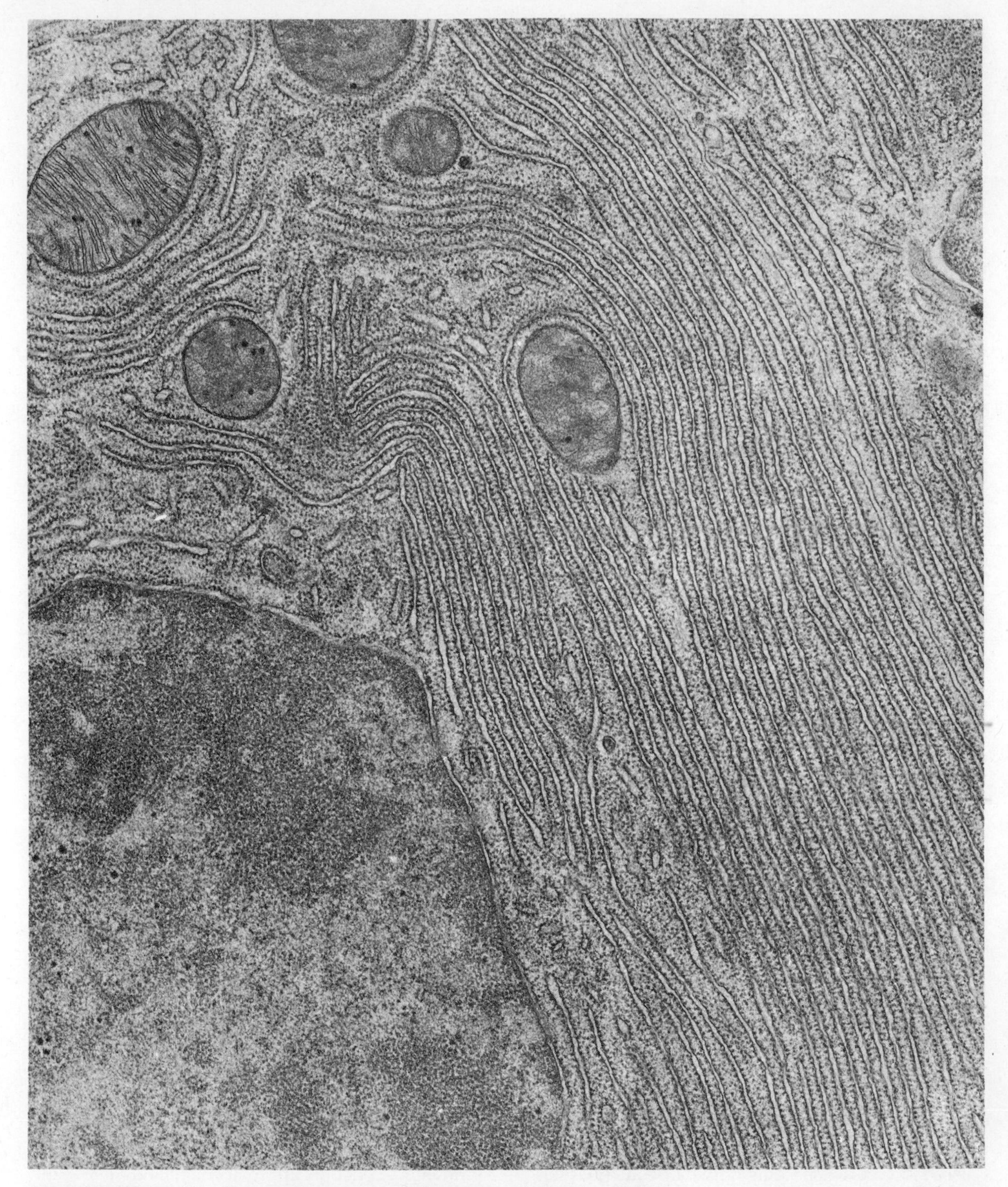

Figure 3.7. (Above) This electron photomicrograph shows the endoplasmic reticulum as numerous diagonal lines (×30,000) (Dr. Keith R. Porter). (Right) The small black dots lining the ER in this more highly magnified view are ribosomes (×184,000) (Dr. F. Sjostrand).

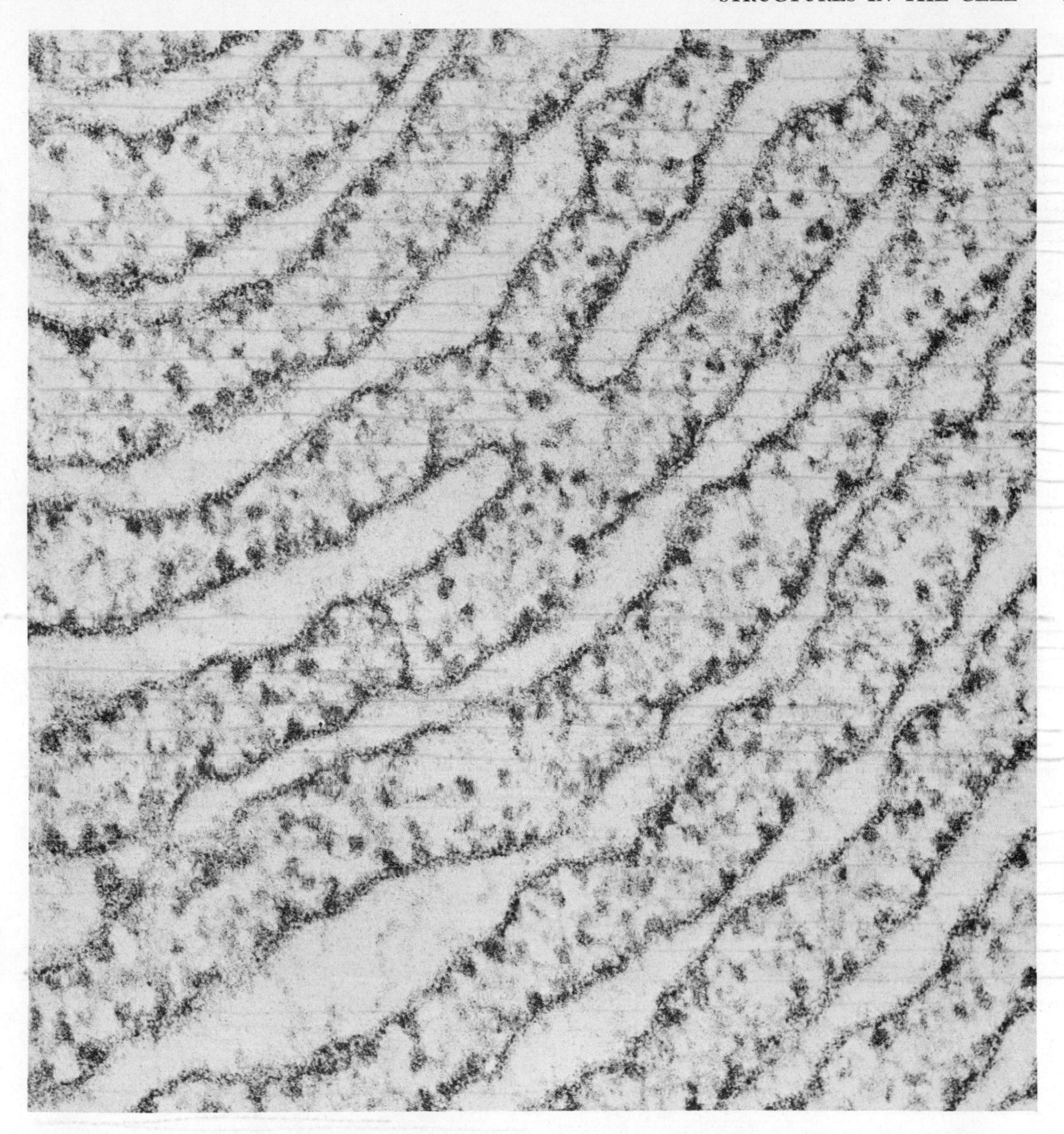

portant to the cell because everything passing into or out of a cell must go through it. This passage is not a simple movement like the diffusion of water molecules through an artificial membrane. Instead, it is highly regulated by the membrane itself. The membrane is differentially permeable as discussed in Chapter II in relation to osmosis and active transport. There is little conclusive evidence yet as to *how* materials pass through the latticeworklike layers of molecules that form the membrane; this remains an important and challenging problem in biology.

The cell membrane shows many specializations and functions. For example, it frequently forms tiny canals leading into the cell. Fluids may flow into these canals and then be pinched off in the cytoplasm, a process termed *pinocytosis* (Figure 3.6). Or the membrane may surround and engulf particles of material much too large to pass through by diffusion—this is termed *phagocytosis.* Where adjacent cells meet, their membranes are often interlocked by fingerlike convolutions. In cells where absorption is a major function, the free surfaces consist of thousands of microvilli (minute projections of the cell membrane) which result in a greatly increased surface area for the cell.

Endoplasmic Reticulum. One of the most startling structures revealed by electron microscope photographs was the network of fine membranes filling the supposedly structureless portion of the cytoplasm. Further studies showed that this *endoplasmic reticulum* (hereafter called ER) consists of a complex system of membrane-bound tubules, swollen areas, and flattened sacs, all with many interconnections. Figure 3.7 shows the extensive ramifications of the system. The ER may also be continuous with the plasma membrane, as it shows a unit membrane structure.

Some cells contain relatively large fluid-filled spaces termed *vacuoles.* It appears likely that most of these membrane-enclosed spaces are expanded portions of the ER. Vacuoles function in various ways. Thus, in many plant cells vacuoles nearly fill the cell to give it support or turgidity. They may also act as sites where unusable chemical products can be dumped or stored. Some animal cells, like *Paramecium,* use vacuoles for regulating water content. Many cells hold food particles in food vacuoles for intracellular digestion.

In most cells, the ER is associated with the making of proteins; ribosomes, the sites of protein synthesis, are located adjacent to it. Such a network is called rough ER because the ribosomes give it a granular appearance. Certain liver cells, some gland cells, and a few other cell types have a smooth ER with no attached ribosomes. Chemical evidence suggests that smooth ER functions in the transport of glucose and glycogen, and synthesis of steroids, and the absorption of lipid molecules.

The tubular organization of the ER suggests an intracellular transport system for metabolic products as well as a widespread distribution of enzyme-rich membranes. In addition, the ER may provide a path by which materials might enter a cell from the external environment. Pinocytosis, for example, could represent a function of the ER.

In summary, ER is a versatile ultrastructure with many shapes and adaptations. It provides a large surface area for chemical activities, localizes specific syntheses, and serves as a transport device.

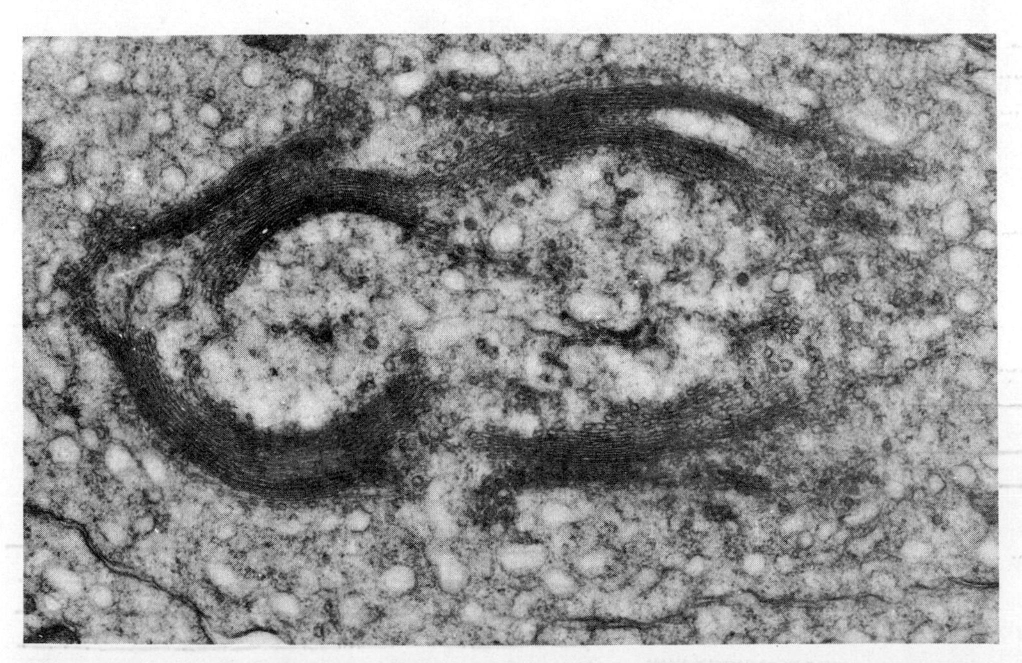

Figure 3.8. The Golgi body, made of a platelike series of membranes. Notice the association of numerous tiny droplets with these contoured membranes* (×*11,500*) *(Dr. L. Evans Roth).

Ribosomes. As seen with the electron microscope, ribosomes are tiny granular bodies. They are associated with ER in most cells (see Figure 3.7), but also occur free in the cytoplasmic matrix of cells such as yeast, growing plant cells, and embryonic nerve cells.

Ribosomes are receiving considerable attention from researchers because they are the sites of protein assembly. Since they are such tiny units, an ultracentrifuge is used to obtain a concentrated mass. From these analyses we know that ribosomes are uniform in size, structure, and chemical composition within the tissues that have been examined. Chemically, a ribosome is about one half ribonucleic acid (RNA) and about one half protein, although the proportion varies somewhat in different organisms. It has been determined that ribosomes are composed of two subunits of different size and shape, but the significance of this fact is not known.

When engaged in protein synthesis, ribosomes associate in clusters called polyribosomes. This function, which is described in detail in Chapter VII, is of interest to medical researchers, since any malfunctioning of the ribosomal system seriously affects the entire cell. Some antibiotics, for example, inhibit bacterial growth by interfering with their protein synthesis. Perhaps certain human diseases have a similar basis.

Golgi Apparatus. The Golgi apparatus or complex consists of tubules, tiny flattened sacs, and small vacuoles formed within unit membranes. (See Figure 3.8.) This complex is often localized in certain areas of the cytoplasm, but in plants and invertebrates it is scattered about.

Many cytologists view the Golgi complex as a specialization or derivation of the ER since they are structurally similar.

Numerous experiments have related the Golgi system to cell secretion. Substances like hormones, enzymes, and lipids are packaged in portions of the Golgi complex to be secreted later. The role of Golgi membranes in forming these substances is uncertain. The fact that certain substances are formed in the ER and later move into Golgi tubules has been shown by the use of radioautography. In this experimental technique, a radioisotope of an element commonly used by a cell is placed in a cell culture. The culture is placed on sensitive photographic film at successive time intervals, so that the isotope takes its own picture as a tiny exposure spot on the film.

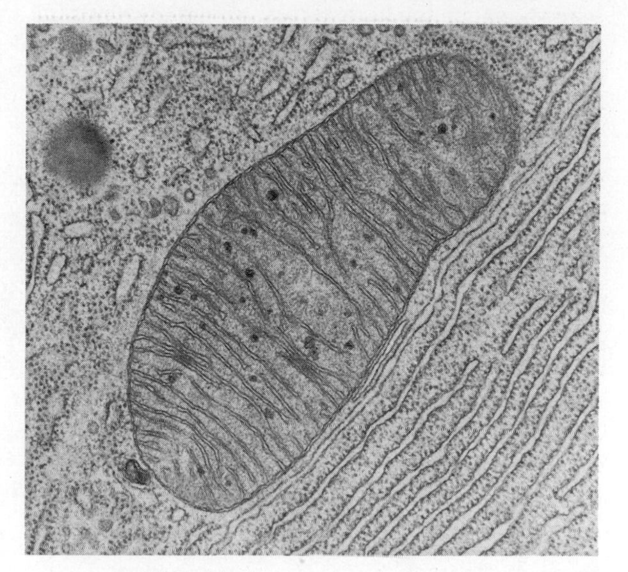

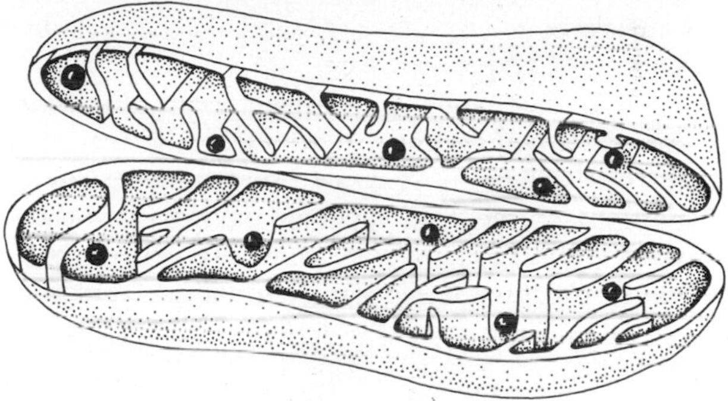

Figure 3.9. (Top) An electron photomicrograph of a mitochondrion in a bat pancreas cell (×53,000) (Dr. Keith R. Porter) (Bottom) A three-dimensional diagrammatic representation of a mitochondrion.

In this way, molecules of material can be traced as they pass from the ER to Golgi sacs or elsewhere in the cell.

Mitochondria. Another membranous structure, and one of great importance in cellular functions, is the mitochondrion. Mitochondria are relatively large bodies that appear as rods and spheres when viewed with a conventional microscope. The form and size of these bodies vary with different conditions within the cell, mostly in relation to nutrition. There may be only a few thousand mitochondria in lower cells, while there are 150,000 in each egg of one species of sea urchin.

A mitochondrion consists of two layers of unit membrane. The outer one is smooth and simply envelopes the body. The inner membrane is folded to form a series of shelflike plates within the mitochondrion as shown in Figure 3.9. Small spheres

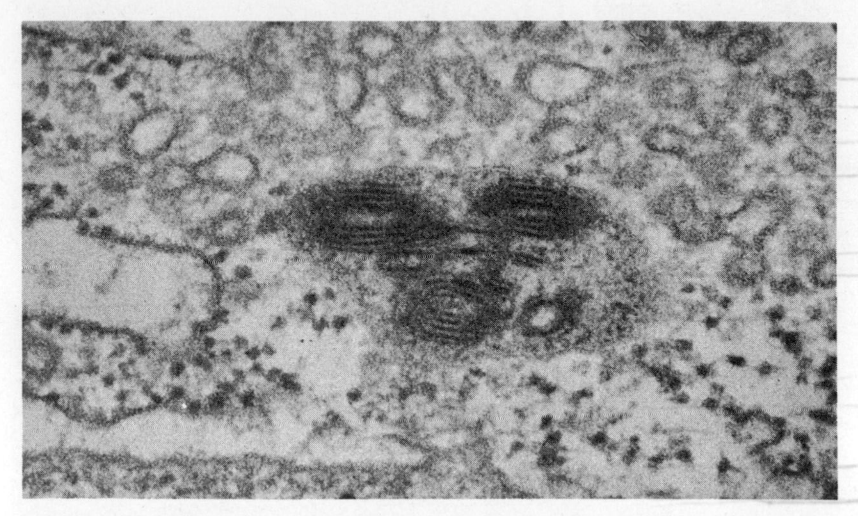

Figure 3.10. An electron photomicrograph of a lysosome. The appearance of this cell organelle varies with its functional state (×120,000) (Dr. F. Sjostrand).

termed intramitochondrial granules frequently occur within the matrix. Their function is not known.

Mitochondria play their most important role in the complex of chemical reactions known as cellular respiration, which provides the cell's energy. The numerous enzymes involved in this fundamental event, which is described further in Chapter V, are localized in the inner membrane of the mitochondrion. The shelflike plates inside the mitochondrion could reasonably be considered an adaptation that increases the enzyme-bearing surface area for cellular respiration.

Chemical analyses show that mitochondria from a wide variety of organisms contain deoxyribonucleic acid (DNA), a nucleic acid supposedly confined to the nucleus. Current evidence indicates that mitochondrial DNA may control the production of proteins peculiar to these organelles.

Mitochondria are sensitive indicators of injury to cells. They may fragment, degenerate, or swell into large vacuoles. Abnormal changes in mitochondria are typical of scurvy and possibly other diseases. Disintegration of mitochondria leads to death of the cell.

Lysosomes. Lysosomes are small membranous sacs distributed through the cytoplasm (see Figure 3.10). They were originally thought to be small mitochondria until cytochemical tests showed that they contain groups of hydrolytic enzymes.

The enzymes within lysosomes are known to hydrolyze (digest) nucleic acids, proteins, and carbohydrates; hence, these are special intracellular digestive bodies. In addition to digesting organic materials within food vacuoles, lysosomal enzymes also have intracellular functions. For example, the breakdown of cells and tissues that occurs when a tadpole changes into a frog is caused by lysosomal enzymes. When cells are starved, these enzymes digest various cell parts as a source of energy for the cell. Lysosomes are particularly large in phagocytic cells, such as the white blood cells that engulf bacteria.

Lysosomes are also involved in a number of pathogenic conditions. An excess of vitamin A causes the enzymes to be released and thus destroy the cell. Cell aging may have a relation to lysosomal action. One investigator applied the term "suicide bags" to lysosomes because of their potentially destructive qualities. Dead cells quickly disintegrate due to the release of enzymes as lysosomes rupture. This may explain the tenderizing effect of letting meat age in cold storage.

Plastids. Plastids constitute still another cellular structure made of membranes. These large cytoplasmic bodies occur in all plants and algae with few exceptions.

One group of plastids are colorless and serve functions such as starch or oil storage. Other specialized

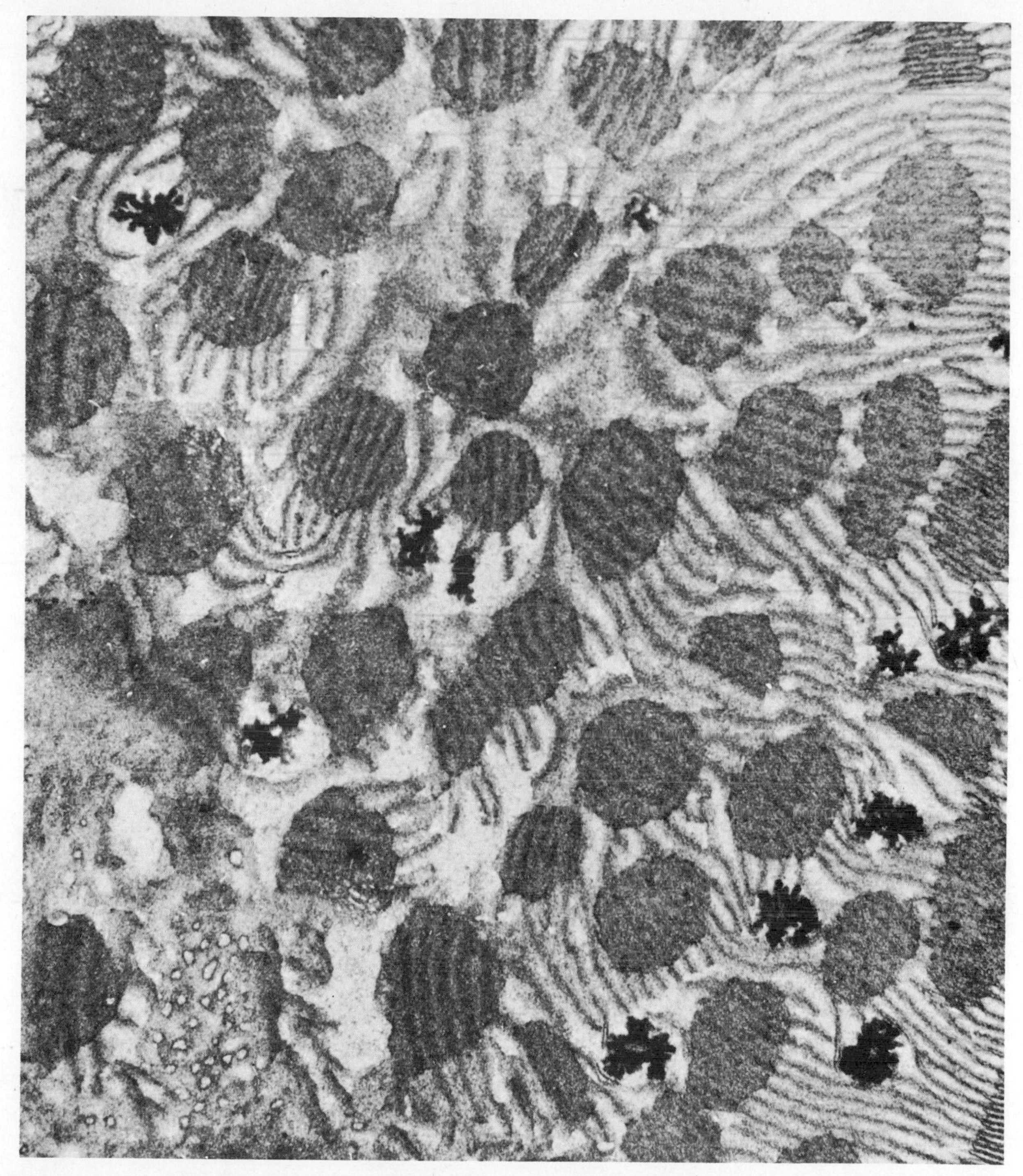

Figure 3.11. A chloroplast. This is an oblique section through a single plastid showing profiles of grana, the dark spherical areas (×25,000) *(Dr. L. K. Shumway and Dr. T. E. Weier).*

plastids are responsible for most of the color of plant parts. The red color of a ripe tomato is an example. The best known of the colored plastids is, of course, the *chloroplast;* it has been studied thoroughly because it is the site of photosynthesis.

The chloroplast is enclosed by a membrane and, in most plants, is divided internally by numerous additional membranes that contain the chlorophyll (Figure 3.11). Areas of closely packed membranes within the chloroplast are termed *grana.* In the electron microscope, a granum has the appearance of a stack of thin disks connected to adjacent grana by intergrana membranes (Figure 3.12). A fluid fills the remainder of the chloroplast. Many variations in chloroplast structure exist among groups of plants. Some, for example, lack grana, but in virtually all cases the chlorophyll is associated with some membranous structure.

DNA occurs in chloroplasts and may be related to the self-duplicating ability of these bodies.

Centrioles. Centrioles have been known in animal cells for a long time, and appear as two dark granules located adjacent to the nucleus. Electron microscopy indicates that each of these granules is actually a tiny rod lying at a right angle to the other. Each rod consists of nine tubules, made up of one to three units, arranged in a circle (Figure 3.13). This precise arrangement has been found in all centrioles that have been studied, but its functional significance is not known at present. During cell division in animals and at least a few algae, centrioles control the formation of the spindle (see Chapter XII). Under the microscope the spindle appears as a body of tiny tubules formed during cell division and involved in the movement of chromosomes.

Centrioles are also capable of reproducing daughter centrioles and motile bodies called cilia and flagella. These motile processes are found on cells throughout the animal kingdom and in many plants. They function in locomotion, create currents in fluids and, in some cases, act as sensory receptors. The ultrastructure of cilia and flagella is like that of the centriole but with two additional tubules running down the center.

Cytoplasmic Adaptations. In relation to the cytoplasm, several points should be emphasized. First, the cytoplasm is that part of the cell which *differentiates* (specializes) for a particular function.

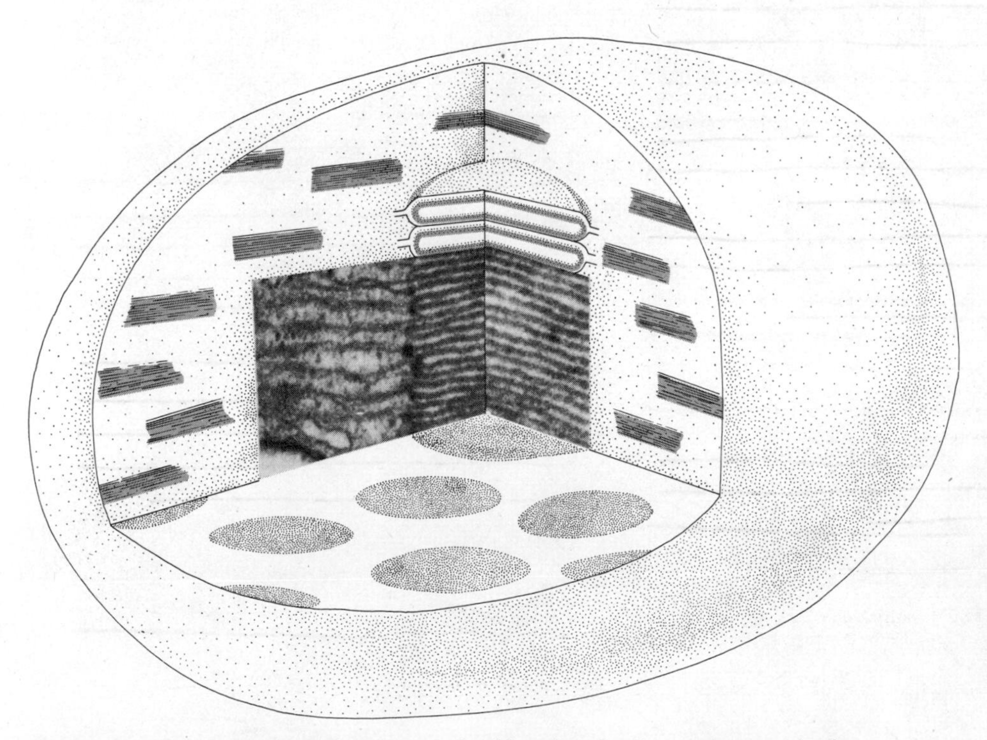

Figure 3.12. An interpretation of the structure of a chloroplast showing granum organization and intergrana membranes.

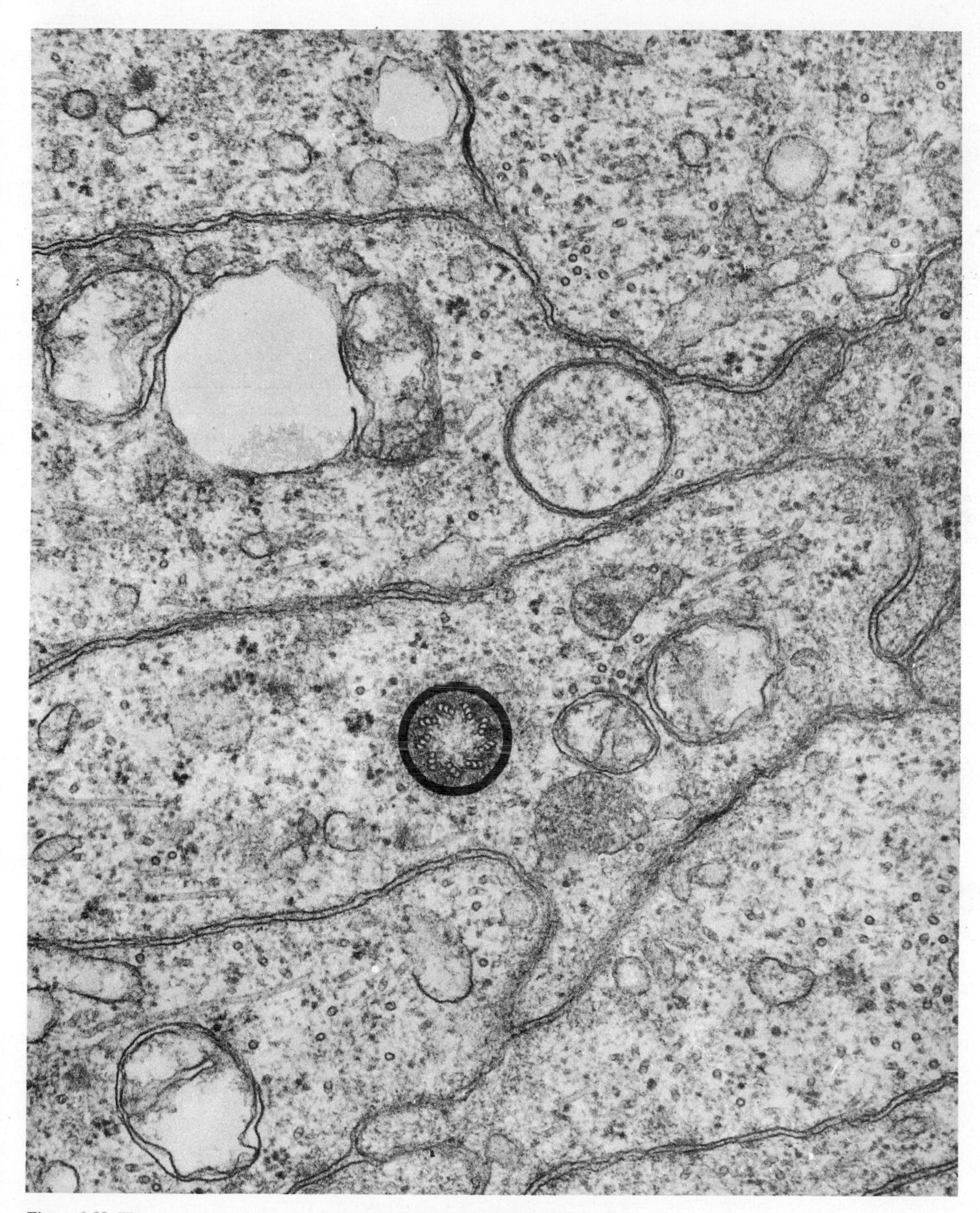

Figure 3.13. The centriole appears as a ring of nine tubules (×66,000) (Dr. Keith R. Porter).

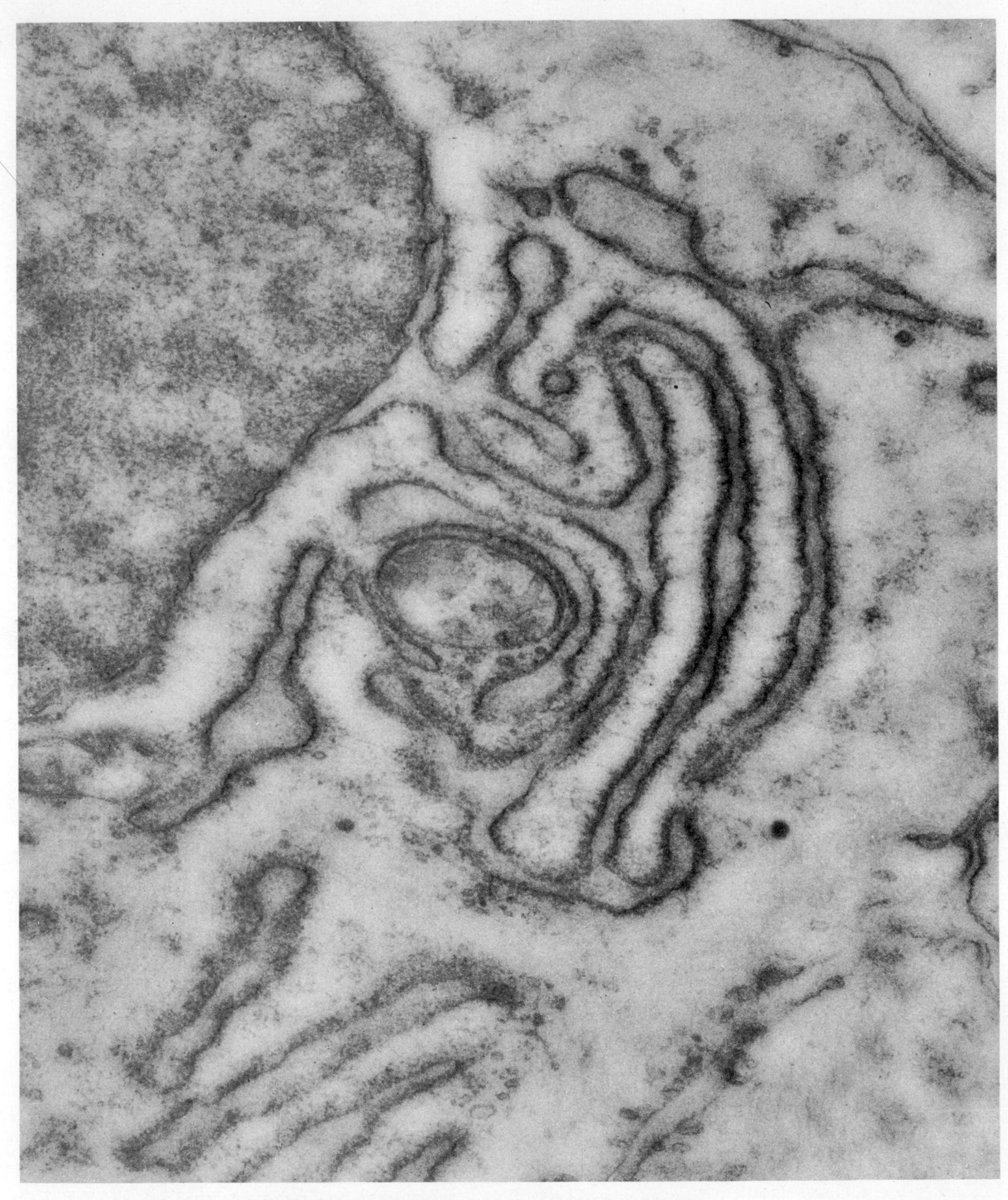

Figure 3.14. The membrane surrounding the dark spherical nucleus (upper left) is continuous with the endoplasmic reticulum (approx. ×75,000) (Dr. Keith R. Porter).

Thus, the functional element of a nerve cell is its cytoplasm. The same statement can be made for muscle cells, secretory cells, or specialized plant cells. Another way of stating this is to say that genetic information in the nucleus is always expressed by some type of activity or modification of structure in the cytoplasm.

The second point to remember is that the adaptations, or specialization, shown by the cytoplasm are a reflection of the functions performed by the cell. In other words, there is an intimate relation between the morphology of a cell and its function. This may seem too obvious to need emphasis; nevertheless the form-function concept is a basic one in biology and is found not only at the molecular and cellular level but also in considering tissues, organs, and organisms.

If one wonders why the form-function relationship appears universally in biological materials, one should keep in mind that everything from molecules to organisms must exist in an environment. Later, we shall consider a process, called *natural selection*, which matches up the most efficient "fit" between materials and their environment (see Chapter XX). This fit we call adaptation which, after all, is no more than form following function.

In the past few years, DNA has been detected by chemical tests in the cytoplasm of a variety of cells chick embryo cells, frog and sea urchin eggs, fern eggs, and growing root cells. In most cases, the DNA appears to be localized in mitochondria and chloroplastids. The presence of this basic hereditary material in cytoplasmic structures raises many questions. It may be that it controls the activities and reproduction of these parts. However, the following experimental evidence suggests another explanation. When labeled DNA precursor material was first incorporated into mitochondrial DNA, it later appeared in DNA in the nucleus. This was interpreted as meaning that DNA was synthesized in the mitochondria and then transferred to the nucleus. Whether this event occurs in the life cycle of most cells is not known at present.

Nuclear Structures. Electron photomicrographs show that the nucleus is surrounded by two membranes, the outer one continuous with the endoplasmic reticulum (Figure 3.14). This two-layered nuclear envelope appears to be porous. A number of permeability experiments indicate, however, that the pores are not open gaps: only certain materials pass between cytoplasm and nucleus. Thus, nucleic acids, sugars, and polypeptides pass through, but certain proteins cannot. Puncture of the nuclear membrane causes nucleoplasm to flow out and the nucleus dies.

Inside the nucleus we find darkly staining, threadlike material, dense bodies of granular material termed *nucleoli*, and nuclear sap. None of this material shows up well under the electron microscope; hence other techniques have been used to study its ultrastructure. The threadlike material is known to consist of DNA, RNA, and certain protein substances. Chromosomes, rod-shaped bodies found in the cell during cell division, are composed of these same materials and appear to be the carriers of heredity. Hence, it is assumed that the threadlike material represents uncoiled chromosomes containing the hereditary substance.

The nucleolus appears granular in electron photomicrographs, and there is evidence that the granules are ribosomes that later pass into the cytoplasm. Nucleoli and ribosomes contain many of the same sized subunits, and both contain proteins made up of the same amino acids. All evidence strongly indicates that the nucleolus is the site where ribosomal RNA is either synthesized or at least assembled into ribosomes. Some investigators consider the nucleolus as reserve chromosome material that is utilized in nuclear division, since it disappears during this process. Nuclear sap, as the term indicates, is a fluid that probably contains a complex assortment of chemical substances employed in the nucleus.

Many experiments have indicated that the nucleus is the control center for the cell. Specifically, it appears to direct the synthesis of enzymes in the cytoplasm, which in turn regulate cellular functions. It is also the hereditary reservoir, a feature closely related to the function of control. The nucleus, unlike the cytoplasm, seldom shows any morphological specializations. It is usually spherical or disklike, although some cells, like certain types of white blood cells, have odd-shaped nuclei.

Cells: Basic Units. The basic molecular configurations and chemical compounds that compose the cell form an exceedingly plastic, adaptable unit of structure and function in the living world. Perhaps, as some biologists contend, all of the important problems concerning life processes can be solved by studying cells. Certainly the current studies of cell biology are revealing many new facts about cell structure, which in turn will provide many new hypotheses concerning cell functions. Probably the most startling and significant consequences of studies in this new field are in relation to the DNA molecule, a topic to be taken up in detail in Chapter VII.

The idea that cells are the basic structural units of life remains valid. Some groups of microorganisms, such as bacteria and blue-green algae (groups sometimes termed the "lower protists"), do not have a definitive nucleus and therefore differ in their structure from the conventional notion of a cell. Nevertheless, they are bounded by functional cell membranes and contain proteins, nucleic acids, lipids, and carbohydrates of the same molecular structure as that found in the remainder of the living world. These organisms violate the cell principle only by lacking a highly organized nucleus. Otherwise, they are cells in a functional sense.

A considerable amount of current research is concerned with the mechanics of cell specialization. Why do cells with identical hereditary codes or messages in their nuclei differentiate into unlike bodies—for example, nerve cells and muscle cells? What keeps the process so orderly that exactly the right amount of each tissue and organ is formed? What is the role of cytoplasmic DNA? Dozens of related questions are also unanswered.

Principles

1. The cell is a convenient basic unit of biology.
2. The form of a cell is an adaptation to its performance.
3. Many of the cell's structures are composed of membranes.
4. Structural specialization entails primarily amount and placement of cytoplasmic structures.

Suggested Readings

Allison, Anthony, "Lysosomes and Disease," *Scientific American,* Vol. 217 (November, 1967). Offprint No. 1085, W. H. Freeman and Co., San Francisco.

Brachet, Jean, "The Living Cell," *Scientific American,* Vol. 205 (September, 1961). Offprint No. 90, W. H. Freeman and Co., San Francisco.

de Duve, Christian, "The Lysosome," *Scientific American,* Vol. 208 (May, 1963). Offprint No. 156, W. H. Freeman and Co., San Francisco.

Dippell, Ruth V., "Ultrastructure of Cells in Relation to Function," in *This Is Life,* edited by Willis H. Johnson and William C. Steere. Holt, Rinehart and Winston, New York, 1962.

Gray, George W., "The Ultracentrifuge," *Scientific American,* Vol. 184 (June, 1951). Offprint No. 82, W. H. Freeman and Co., San Francisco.

Hokin, Lowell and Mabel, "The Chemistry of Cell Membranes," *Scientific American,* Vol. 213 (November, 1965). Offprint No. 1022, W. H. Freeman and Co., San Francisco.

Hooke, Robert, "Of the Schematisme or Texture of Cork, and of the Cells and Porès of Some Other Frothy Bodies," in *Great Experiments in Biology,* edited by M. L. Gabriel and S. Fogel. Prentice-Hall, Englewood Cliffs, N.J., 1955, pp. 3–5.

Robertson, J. David, "The Membrane of the Living Cell," *Scientific American,* Vol. 206 (April, 1962). Offprint No. 151, W. H. Freeman and Co., San Francisco.

Rustad, Ronald C., "Pinocytosis," *Scientific American,* Vol. 204 (April, 1961).

Solomon, Arthur K., "Pores in the Cell Membrane," *Scientific American,* Vol. 203 (December, 1960). Offprint No. 76, W. H. Freeman and Co., San Francisco.

Swanson, Carl P., *The Cell.* Second edition. Prentice-Hall, Englewood Cliffs, N.J., 1964, pp. 1–61.

Questions

1. Give several reasons why the cell is considered a fundamental unit in biology.
2. What are some of the fundamental units in other sciences?
3. Describe how additions to our knowledge about cells have depended on improvements in tools and technology. Is this true of other sciences?
4. List the parts of cells that are made of membranes and the parts that are not. Which list is longer?
5. Assign a function to each of the parts you listed.
6. Does the membranous structure of a cell part have any relation to the functioning of that part? Explain with examples.
7. What is meant by the phrase "cytoplasmic adaptations"? Give several examples.
8. In what way does this idea express the form-function concept in biology?
9. In what ways do the form and structure of the nucleus reflect its function?
10. Why does the nucleus, unlike the cytoplasm, seldom show any morphological specializations?

CHAPTER
IV

Photosynthesis: Energy Fixation

An electron photomicrograph of grana in a chloroplast (×158,000) (Dr. L. K. Shumway and Dr. T. E. Weier).

CHAPTER IV

Photosynthesis: Energy Fixation

High-school discussions of photosynthesis are often limited to the statement, "All the food energy on earth comes from the sun and is trapped by green plants," plus the chemical equation

$$6CO_2 + 6H_2O + \text{Energy} \xrightarrow{\text{Chlorophyll}} C_6H_{12}O_6 + 6O_2$$

For the most part, both expressions are true summaries of photosynthesis, for this process does transform energy from light into a universally usable form: carbon-to-carbon bonds. These summaries, however, hide much of what really occurs. Yet even these were the result of a long series of experiments and were not known until the late nineteenth century.

Historical Background

One of the earliest experiments relating to photosynthesis was carried out by Jean-Baptiste Van Helmont in the 1740s. Van Helmont weighed dried earth, placed it in a large pot, and then planted a weighed willow shoot in it. For a period of five years, only water was added to the soil. Then the willow tree and the soil were weighed. The weight of the soil had changed only slightly, but the tree had gained over 160 pounds. Van Helmont concluded that the tree had arisen from water alone. He did not realize that carbon dioxide from the air had contributed to its gain.

In the late eighteenth century Joseph Priestly, an English

Figure 4.1. Ingenhousz found that the green portions of plants such as leaves, green stem, and green seeds produced oxygen when kept in light (below) but not when kept in darkness (above).

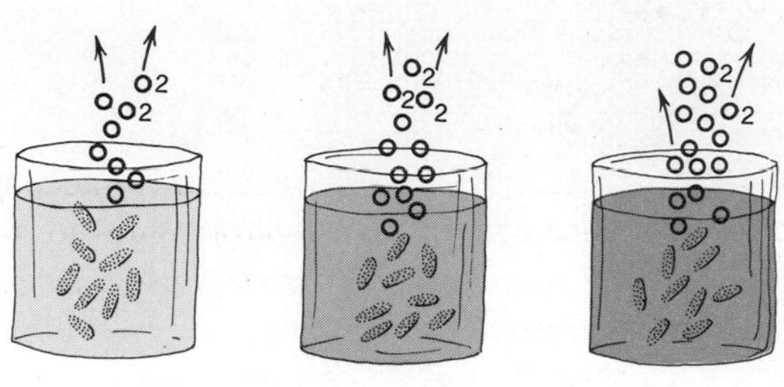

water with increased amounts of heavy oxygen

bicarbonate with increased amounts of heavy oxygen

Figure 4.2. The oxygen released during photosynthesis comes from water, not carbon dioxide. Compare the amount of heavy oxygen released as its concentration in the water is increased (above) with that released as its concentration in bicarbonate is increased (below).

pastor and chemist, showed that plants could utilize carbon dioxide and produce oxygen but he did not call these substances by these names. Not long after the Priestly experiments, Jan Ingenhousz, a Dutch scientist, demonstrated that only leaves and green stems could carry out such activity—and only when illuminated. Ingenhousz' experiments were simple. Placing different parts of plants such as leaves, green stems, older stems, wood, and seeds in separate sealed glass vessels, he left them in the dark for several hours and then tested the air by introducing a burning candle. The candle did not burn in any of the trials. He then placed the vessels in the light and found that after a few hours the candle burned in any vessel that contained green parts of the plant but would not burn in the other vessels (Figure 4.1). Ingenhousz also showed that the brighter the light, the more rapid was the evolution of oxygen.

Just after 1800, Nicholas de Saussure, a Swiss scientist, carried out the first quantitative studies of photosynthesis and showed that the amount of oxygen produced was the same as the amount of carbon dioxide utilized by the plant. He carefully measured the amounts of oxygen, nitrogen, and carbon dioxide in the vessels before and after illuminating the plants in them. He also weighed the plants. By comparing the amounts of gases before and after the experiment, he was able to show the changes that were brought about by photosynthesis.

It was possible at this time to write the following general equation for photosynthesis:

$$\text{Carbon dioxide} + \text{Water} \xrightarrow[\text{Light}]{\text{Green plants}} \text{Living material} + \text{Oxygen}$$

Notice that this does not indicate why the light was needed, what component of the green parts of plants functioned in this reaction, or what was the organic product.

By 1875, the equation

$$6CO_2 + 6H_2O + \text{Energy from light} \xrightarrow{\text{Chlorophyll}} C_6H_{12}O_6 + 6O_2$$

had been experimentally determined. At that time it was thought to be a simple one-step reaction. Over the next sixty years evidence slowly accumulated to the contrary. Investigators showed that many organisms could incorporate carbon dioxide into carbohydrates without light as a source of energy. This incorporation seemed to occur in the same way as in photosynthesis, except that a different source of energy was used to drive the reaction. Shortly after 1900, F. F. Blackman, an English plant physiologist, studied the rates at which photosynthesis occurred under various conditions and showed that the process had to be a composite of several different reactions. Today, it is necessary to talk of the light and dark reactions of photosynthesis in order to distinguish those reactions for which light is essential from those that take place by utilizing the products of the light reactions, but do not directly depend on light.

Until modern biochemical techniques became available, it was thought that the oxygen released in photosynthesis came from the carbon dioxide. It is now known, however, that water is the source of the oxygen. Although several studies had given indirect evidence, it was not until 1941 when Samuel Ruben, an American biochemist, and his coworkers confirmed this by using isotopes as tracers in the study of photosynthesis. They grew suspensions of single-celled plants in two solutions. In one of these the water molecules contained heavy oxygen (O^{18}), while in the other, bicarbonate, a source of carbon dioxide, contained the heavy oxygen. The gases released by the plants were then analyzed for heavy oxygen. If the amount of heavy oxygen in the water molecules was increased, the amount of heavy oxygen released by the plant increased. However, increased amounts of heavy oxygen in the bicarbonate did not change the amount released by the plant (Figure 4.2).

More detailed information awaited the development in 1954 of a technique for carrying out the complete process of photosynthesis in chloroplasts that were freed from the cell. This technique makes it possible to examine photosynthesis without interference from other reactions.

It has been learned that chlorophyll and the enzymes involved in photosynthesis are integral parts of the structure of the chloroplasts. The multiple layers of lipoprotein membrane composing the grana of the chloroplasts (Figure 3.11) present a large amount of chlorophyll to light while maintaining the necessary close relationship between chlorophyll and the enzymes.

Nearly all the reactions of photosynthesis have been experimentally demonstrated often enough to inspire confidence in our present understanding of the general processes. However, the mechanisms by which these reactions occur are still relatively unknown.

The Photosynthetic Reactions

ADP and ATP. Before we can discuss the reactions that constitute photosynthesis, we must consider the ADP-ATP* cycle, which serves as the basic system for energy transfer in the cell. The high-energy bond in ATP is the immediate source of energy for all reactions in the cell (Figure 4.3). When this bond is broken, leaving ADP plus a simple inorganic phosphate group (P_i), the released energy can be used in reactions. When the right amount of energy becomes available, the high-energy bond between ADP and P_i is reconstituted, resulting in ATP. The usual sources of energy are sunlight or the oxidation of some complex molecule.

Even though photosynthesis is a series of many steps, it is convenient to divide the process into light and dark reactions as a starting point for our discussion. The reactions that require light and are therefore called the light reactions can be summarized by the equation

$$ADP + P_i + H_2O + \text{Energy (light)} \xrightarrow{\text{Chlorophyll}} ATP + H + O_2$$

Notice that ADP, P_i, and water are used during these reactions and that ATP, hydrogen (attached

* ADP (adenosine diphosphate) and ATP (adenosine triphosphate) are substances that differ by one phosphate group. This group is attached to the rest of the molecule by a bond (high-energy bond) containing a large amount of energy relative to most bonds in biological systems. Adenosine is an organic base plus a five-carbon sugar. (See Chapter VII.)

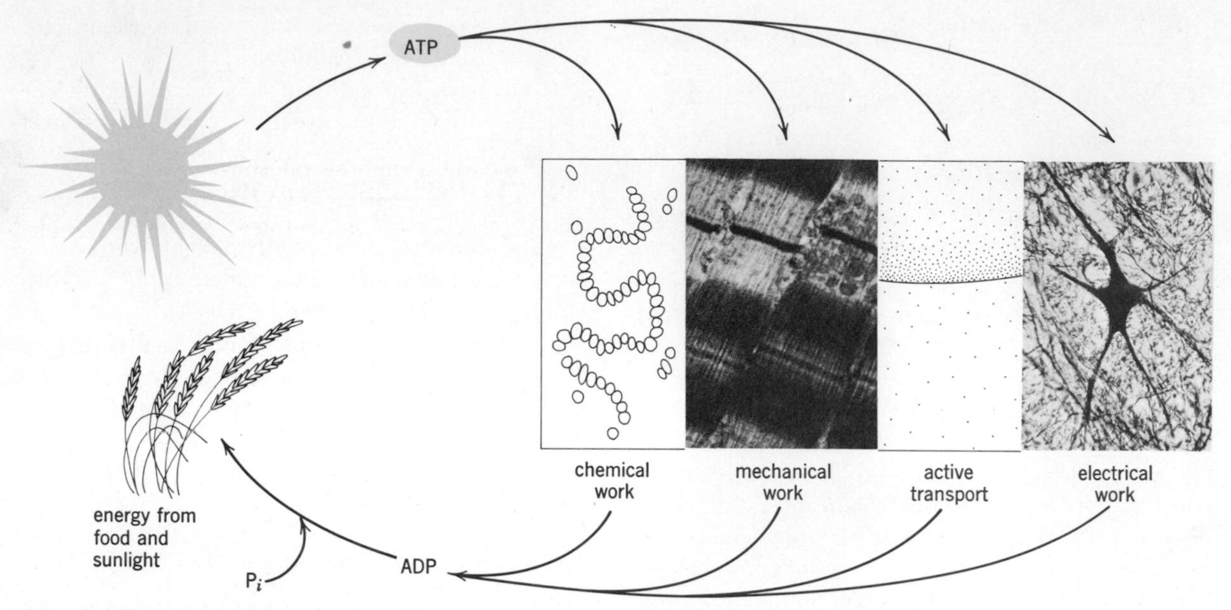

Figure 4.3. The ADP-ATP cycle. Energy from ATP is used by the cell in such work as synthesis, muscle contraction, active transport, and neural transmission. (*Redrawn from "How Cells Transform Energy," A. L. Lehninger. Copyright © 1961 by* Scientific American, Inc. *All rights reserved.*)

to a carrier molecule), and oxygen are the products. Of these, the equation used to summarize photosynthesis shows only water as a reactant and oxygen as a product. We may then expect that the dark reactions of photosynthesis will use up the remaining products of the light reactions (ATP and hydrogen) and will produce ADP and P_i, which were shown as reactants. Thus, the net changes that result from photosynthesis do not show any overall changes in these materials.

The dark reactions may be summarized by

$$ATP + CO_2 + H \rightarrow ADP + P_i + \text{Carbohydrate (PGAL)}$$

They occur in plants any time that the reactants—ATP, carbon dioxide, and hydrogen—are all present. ADP, inorganic phosphate, and PGAL (phosphoglyceraldehyde, a three-carbon relative of sugars) result. Notice that this equation, like the one used to summarize the light reactions, shows only what the reactants and the products are; it is not a balanced equation.

Light Reactions. Let us now look at the light and dark reactions in a little more detail. Chlorophyll absorbs energy from sunlight. This extra energy makes the chlorophyll molecule unstable. As a result an electron, which carries the excess energy, is released. This electron is picked up by a hydrogen carrier (NADP) which is then able to remove hydrogen from water, that is, split the water molecule. The other product of this reaction is OH^-. This ion donates an electron to the cytochrome system (see p. 77) leaving OH. Several of these combine to form water and release O_2 (Figure 4.4, left).

The electrons that are transferred to the *cytochrome system* pass from one of the complex substances that make up this system to the next. As the transfers occur, energy is obtained that is used to form ATP from ADP and P_i. The last acceptor for these electrons is chlorophyll. The return of the electrons to the chlorophyll restores its original condition. This allows it again to absorb energy from light.

Again, the summary for the light reactions is:

1. Energy from light is absorbed by chlorophyll.
2. This energy is used to split water and to build ATP from ADP and P_i.
3. The oxygen from the water is released but the hydrogen is picked up by hydrogen carrier molecules.

Dark Reactions. In the dark reactions (Figure 4.4, right) a five-carbon sugar (ribulose diphosphate) unites with a carbon dioxide molecule to form, briefly, a six-carbon chain. This immediately splits into two three-carbon units. Energy is transferred to these three-carbon compounds by splitting ATP, leaving ADP + P_i. The energy is used to rearrange some of the bonds within the compounds.

Hydrogen is now transferred to the three-carbon units from hydrogen-carrier molecules ($NADPH_2$ which obtained the hydrogen from water during light reactions), resulting in the formation of PGAL molecules. Much of the PGAL is used to reconstitute the five-carbon ribulose phosphate with which we started this description. The fate of the remaining PGAL is discussed shortly. The carrier molecules

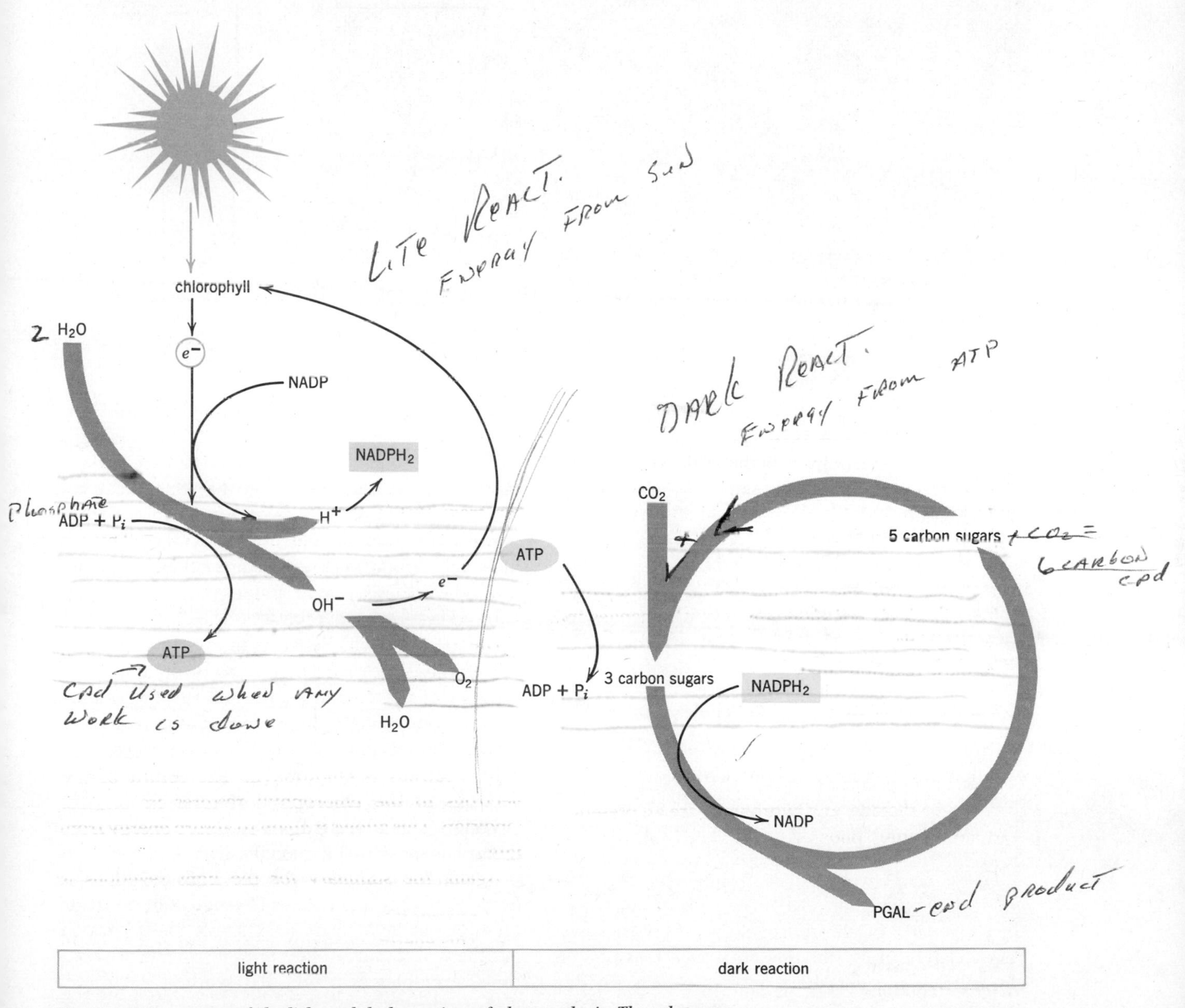

Figure 4.4. A summary of the light and dark reactions of photosynthesis. The substances shown in yellow contain relatively large amounts of energy.

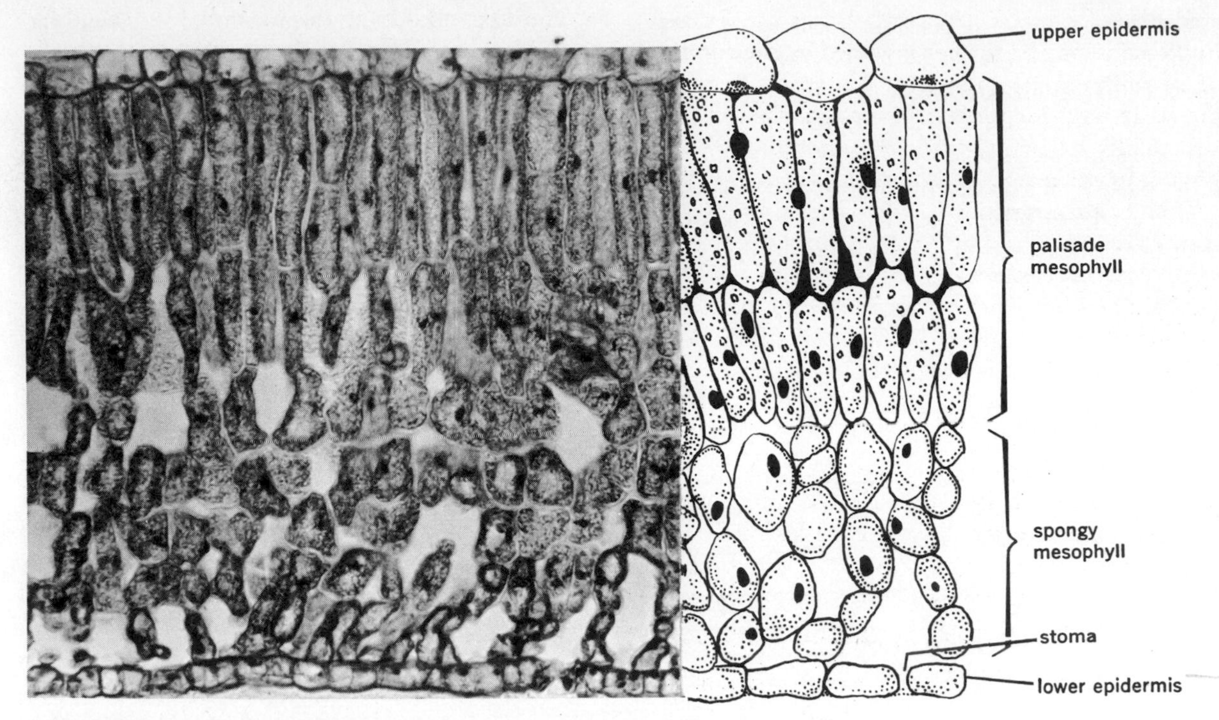

Figure 4.5. Cross section of a leaf showing the typical organization. The cells comprising the mesophyll contain numerous chloroplasts, which make this layer appear darker than the upper and lower epidermis.

also continue to participate in the cycle by obtaining more hydrogen from water molecules during the light reactions.

In summary, the dark reactions show:

1. The incorporation of carbon dioxide by its addition to a five-carbon sugar.
2. The splitting of the six-carbon chain into two three-carbon molecules.
3. The utilization of ATP and hydrogen carriers to change the three-carbon molecules into PGAL.

Fate of PGAL. In the above reactions, we indicate that the end product of photosynthesis is PGAL rather than glucose because, strictly speaking, all of the reactions of photosynthesis are completed upon the formation of PGAL.

This concept was verified through the research conducted by Melvin Calvin and his associates at the University of California over the past twenty-five years. Calvin grew illuminated suspensions of single-celled plants for short periods of time in the presence of radioactive carbon dioxide (CO_2 with C^{14}). This isotope enabled the researchers to identify which compounds incorporated the CO_2. From such experiments it was concluded that the direct end product of photosynthesis was not glucose. Rather, six-carbon sugars like glucose are synthesized from the three-carbon PGAL units after the termination of photosynthesis.

In addition to providing the basic food (fuel) for the plant cell itself, PGAL is the starting point for the synthesis of all major groups of organic molecules in cells. For example, the union of two of these three-carbon units will produce six-carbon sugars such as glucose and fructose. These, in turn, are used as basic units for synthesizing complex carbohydrates such as cellulose, starch, and glycogen.

Long ago, man recognized that certain plants, like the grains, provided a good, easily stored food supply. This food, of course, is concentrated carbohydrate, in the form of starch. Through domestication, these plants have become the staple foods throughout the world. Sugar beets and sugar cane are additional examples of plants that synthesize carbohydrates valuable to man.

PGAL may also be modified to glycerol and fatty

acids, the substances from which fats are formed. Fatty acids may be further modified by the addition of an amino group to form amino acids. These are the basic units for protein formation. Thus, we see that the light energy originally captured by chlorophyll is found in many different kinds of compounds.

If all of these organic groups have the same precursor in PGAL, it would make a beautifully efficient system if the compounds of one group could be digested and the products rearranged to form compounds of another group. In fact, this occurs regularly in both plant and animal cells.

Site of Photosynthesis

The leaves of green plants are the most familiar sites of photosynthesis. Because the processes of photosynthesis are the same in each leaf, all leaves have some structural similarities. The cross section of a "typical" leaf is shown in Figure 4.5. Most of the photosynthesis occurs in the *mesophyll* of the leaf.

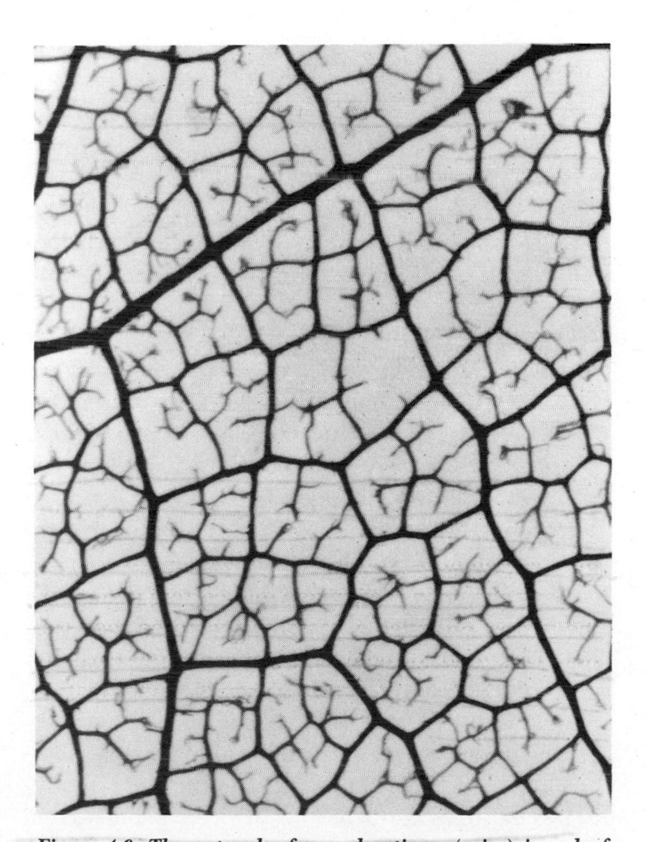

Figure 4.6. The network of vascular tissue (veins) in a leaf.

The upper layer of the mesophyll is called the *palisade* because of its closely packed, vertically oriented cells. The concentration of chlorophyll is higher here than anywhere else in the leaf. The rest of the mesophyll, the *spongy mesophyll*, consists of cells that have considerable space between them. Atmospheric gases circulate rather freely through these spaces.

Within the mesophyll, branching cylinders of *vascular* tissue (veins) traverse the leaf (Figure 4.6). These carry water and minerals to the leaves and organic products away from the leaves.

Water loss from the upper and lower surfaces of the leaf is reduced by *epidermis*, a tissue with an outer layer of waxy material called cutin. The lower surface differs from the upper because it has small openings called *stomata* which allow gases to move between the spaces in the mesophyll and the atmosphere. (The operation of the stomata is described in Chapter X.)

Adaptive Variations. The leaves of many plants show marked differences from our "typical" leaf, because the environments in which plants live vary widely. The alfalfa leaf is representative of leaves found in plants that require moderate or large amounts of moisture. Both the upper and lower epidermal layers contain stomata (Figure 4.7). A maximum supply of carbon dioxide is made available in this way but water loss is also maximal. If the water supply is inadequate, wilting occurs.

Plants like the Russian thistle, creosote bush, and oleander (Figure 4.8) show many adaptations to the dry conditions under which they live. These include an increase in the thickness of the palisade layers, a reduction (in some cases, absence) of the spongy mesophyll, thickened cutin, water-storage cells, and stomata located in pits in the leaves.

The thicker palisade layer and smaller spongy mesophyll lessen the amount of cell surface exposed to the air and to evaporation inside the leaf. The thick cutin reduces evaporation from the outer surface of the leaf, and the protected stomata also inhibit the loss of water.

Even if we compare leaves from the same plant, those taken from parts of the plant that are in the shade differ considerably from those that were in the sun (Figure 4.9). Sun leaves typically show more palisade layers and a thicker cutin on the epidermis. This indicates that the palisade may be an adaptation for exposed situations.

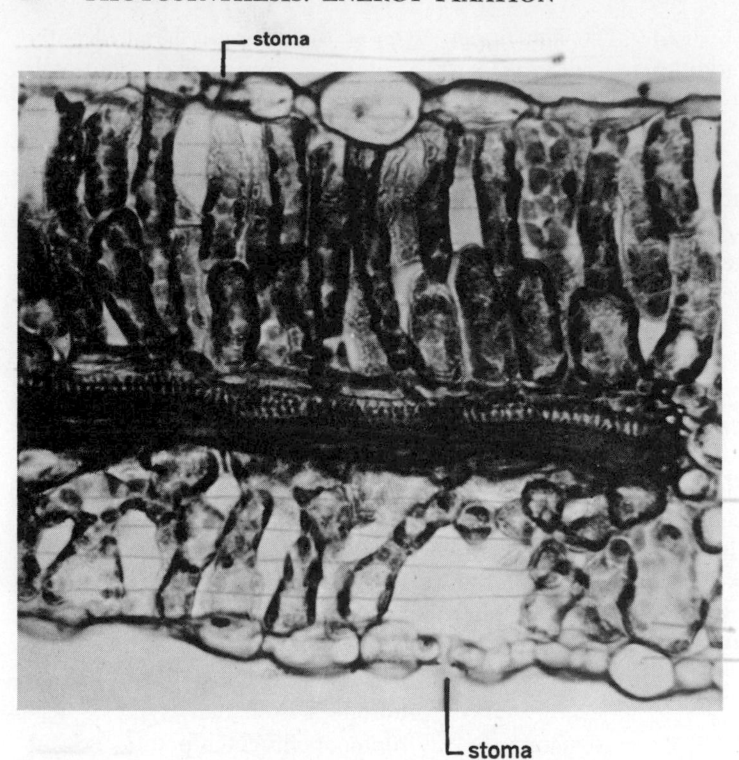

Figure 4.7. As this cross section of an alfalfa leaf shows, stomata are present in both the upper and lower epidermal layers.

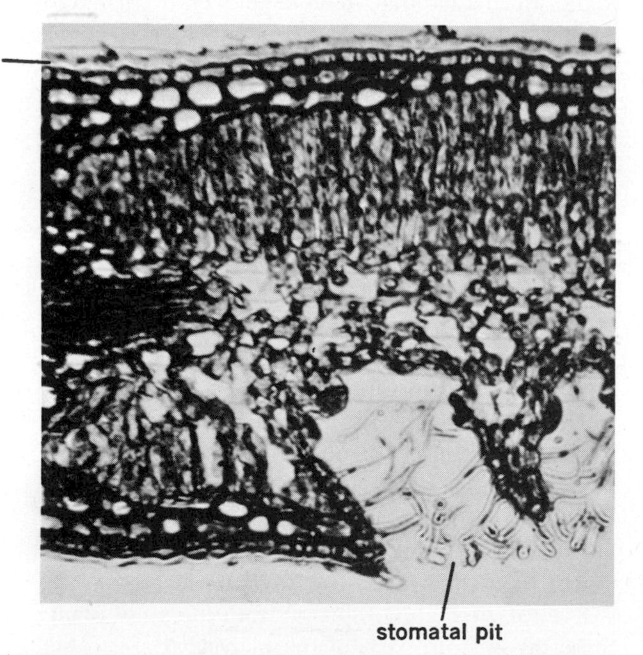

Figure 4.8. Cross section of an oleander leaf showing the thickened cutin layer, the thickened palisade, and stomata located in pits.

Photosynthetic Pigments. As described in Chapter III, chlorophyll is localized in membranes within the chloroplast. All the chemical events discussed under light and dark reactions occur within these specialized bodies.

The green material in leaves can be extracted with a solvent and analyzed by the chromatographic technique described in Chapter II. Such studies indicate that the extracted material is a mixture of pigments rather than a single compound. In higher plants, these are chlorophyll *a* and chlorophyll *b*. Several additional varieties of chlorophyll are known.

Chlorophyll *a* seems to be the specific pigment required for photosynthesis, with the other pigments complementing its function. This is illustrated by the phenomenon called *enhancement:* that is, chlorophylls *a* and *b* together produce a higher photosynthetic rate than chlorophyll *a* alone. The accessory pigment in some manner enables chlorophyll *a* to make a more efficient conversion of light energy into chemical energy. The functional relationships of the various pigments in the photochemical process are currently the subject of much research.

The chlorophylls in a leaf are always accompanied by yellow and orange pigments classed as carotenoids. These pigments are usually masked by the green chlorophylls but can appear as bright leaf coloration in the fall of the year. It is not presently known how the carotenoids interact with chlorophyll in photosynthesis.

Studies of purified chlorophyll extracts show that certain wavelengths (colors) of light are absorbed to a greater degree than others. In addition, this *absorption spectrum* pattern is unique for each type of chlorophyll and can be used as a technique to separate or identify them. Figure 4.10 shows absorption spectra for chlorophylls *a* and *b*. It is notable that the red-orange and the blue wavelengths of light are absorbed and used to a much greater extent than the other wavelengths. Can you deduce from the graph why chlorophyll is *green*?

Knowledge about the absorption spectrum for chlorophyll has a potential practical application for

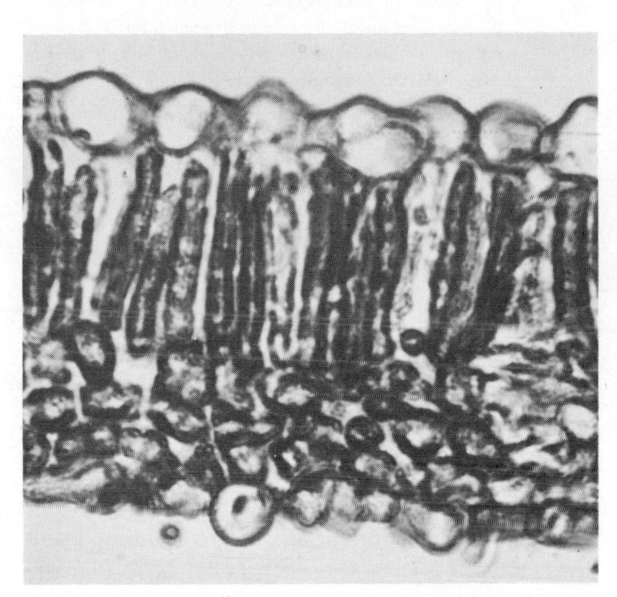

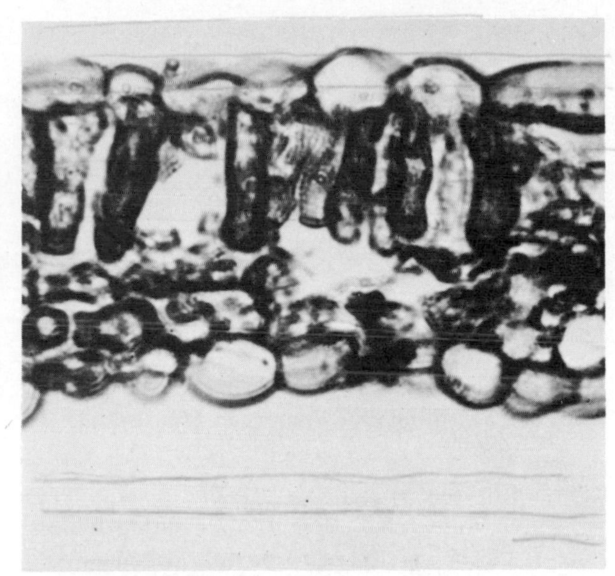

Figure 4.9. Cross section of maple leaves grown in the sun (left) and in the shade (right).

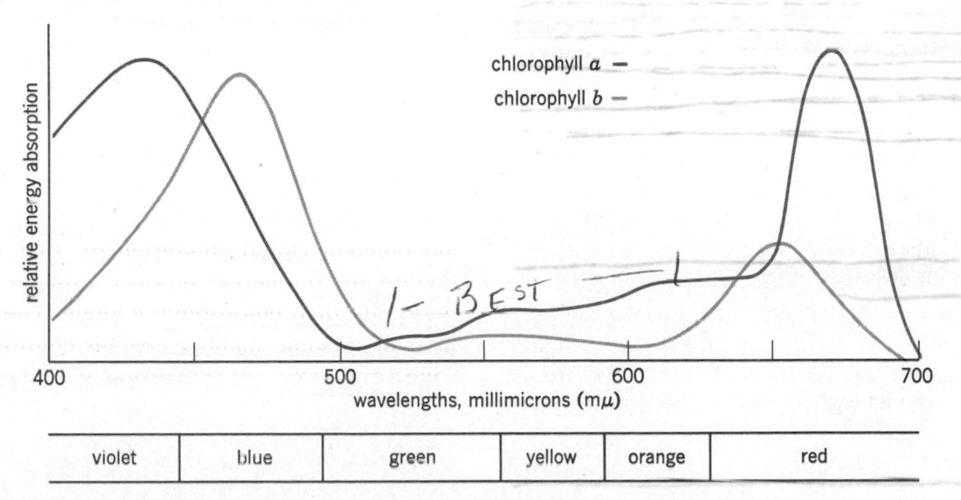

Figure 4.10. The absorption spectra for chlorophylls a *and* b*. The spectra are similar but not identical.*

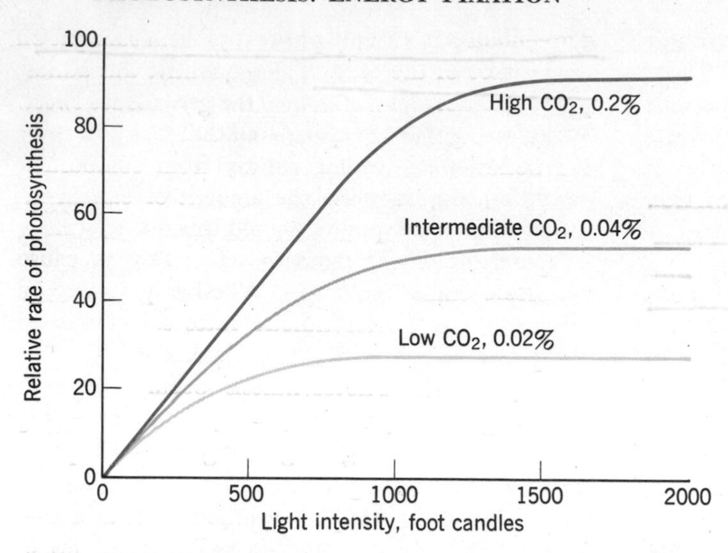

Figure 4.11. Both light intensity and carbon dioxide concentration influence the rate of photosynthesis. In weak light, the rate is directly proportional to intensity. In very bright light, the rate is not influenced by small changes in intensity but is dependent on the carbon dioxide concentration and temperature. (*W. H. Johnson and W. C. Steere,* This Is Life, *Holt, Rinehart and Winston, New York, 1962.*)

fields like horticulture. By controlling the wavelengths of light used in a greenhouse, an increased growth rate is achieved.

Role of Environmental Factors in Photosynthesis

The efficiency or rate of photosynthesis is determined primarily by light, carbon dioxide, and temperature. Water and soil are contributing factors only as they influence plant growth in general.

The photosynthetic rate appears to be determined by the interplay of light, carbon dioxide, and temperature rather than by any single one of these factors. For example, carbon dioxide concentration and light intensity are related as shown in Figure 4.11. Obviously light must always be present for photosynthesis to occur, but temperature and carbon dioxide concentration may modify its influence.

Light. As indicated in Figure 4.11, the photosynthetic rate increases, in weak light, in proportion to light intensity. Beyond a maximum intensity, however, light does not increase the rate and may even become inhibitory. Thus, direct sunlight in the temperate zone may considerably exceed the optimum intensity needed by many plants. Shade-adapted species, such as those that inhabit a forest floor, attain maximum photosynthetic rates at very low light intensities.

If the light available to a plant becomes too scanty, its energy production from photosynthesis may be equaled by its energy output in cellular respiration. This is called the *compensation point* and is illustrated in Figure 4.12. At this point, light becomes a limiting factor in plant growth. This occurs commonly in plants growing in the shade.

Light also differs in quality as well as intensity under certain conditions. Thus, light on cloudy days is richer in blue and green wavelengths. Light filtered through the foliage of trees contains a lot of green wavelengths. Aquatic plants are especially subject to differences in light quality because water quickly filters out the longer wavelengths (red-orange). At increasing depths, the light consists mostly of the blue-green wavelengths. It is for this

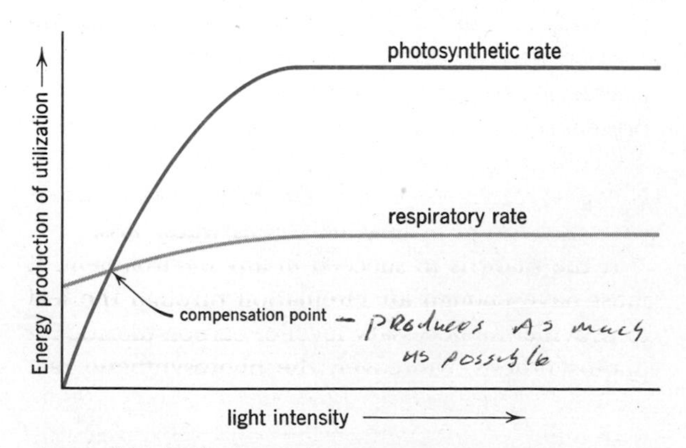

Figure 4.12. Under low light intensities, the two processes of photosynthesis and respiration may balance each other in energy. This balance is indicated on the graph as the compensation point.

reason that underwater objects take on a bluish appearance.

Carbon Dioxide. Carbon dioxide comprises only 0.03 percent of the atmosphere by volume. Yet it is the sole source of carbon atoms for photosynthesis. Carbon dioxide is being continually withdrawn from the atmosphere by photosynthesis, and returned to the air by respiration, volcanic activity, and combustion. Oceans contain even more CO_2 (in a number of dissolved forms) than the atmosphere. Here it becomes involved in a variety of activities, like forming shells and coral, as well as photosynthesis. In fact, the large-scale equilibrium of gases between oceans and atmosphere probably accounts for the constancy of carbon dioxide concentration in the atmosphere.

Carbon dioxide enters a plant via the stomata and must dissolve in moisture on the surface of palisade and spongy mesophyll cells in order to pass into these cells. Hence, the carbon dioxide is delivered to chloroplasts in the dissolved state, or as carbonic acid. Because CO_2 is very soluble in water, aquatic plants live in an environment considerably richer in carbon dioxide than the atmosphere.

Increasing the carbon dioxide concentration in air increases the rate of photosynthesis for a while, but a concentration of about 10 percent is toxic to plants. As we saw previously, light intensity also influences the photosynthetic rate. In nature, foggy (and smoggy) weather leads to an increase in atmospheric carbon dioxide, followed by an increase in photosynthesis until light intensity becomes a limiting factor.

A dilemma becomes apparent when the interplay of sunlight and carbon dioxide is considered. Increased light intensities do not result in increased photosynthesis unless plenty of carbon dioxide is present (Figure 4.11). Free circulation of air between the cells is necessary to supply this carbon dioxide. However, an increase in the circulation of air along with absorption of heat heightens water loss.

If the plant is to succeed in any environment, it must have enough air circulation through the leaf to provide the necessary level of carbon dioxide for photosynthesis. Moreover, the photosynthetic rate must be high enough to allow the plant to compete with other plants. On the other hand, the circulation of air cannot be so great that the loss of water becomes critical.

Carbon dioxide acts somewhat like glass panes in a greenhouse in trapping heat rays from sunlight on the surface of the earth. Consequently, this atmospheric phenomenon is termed the *greenhouse effect*. Many biologists have speculated that man's increasing pollution of the air, mostly from combustion activities, may increase the amount of carbon dioxide in the atmosphere. Should this occur, surface temperatures could increase sufficiently to cause widescale climatic changes. Whether plants could increase their photosynthetic rates sufficiently to compensate for this situation is unknown.

Temperature. Photosynthesis occurs within a wide range of temperatures, from just above freezing to over 100°F. During the growing season it is probably never a limiting factor.

In bright light the photosynthetic rate increases in proportion to increases in temperature up to about 90°F. However, in low light intensities, photosynthesis increases only slightly with temperature increases.

Water. Although water is one of the basic raw materials used in the photochemical reaction, plant tissues nearly always contain an adequate amount for this purpose. It is estimated that less than one percent of the water absorbed by a land plant is used in photosynthesis.

Photosynthesis is indispensable to the existence of life for two major reasons. First, the release of oxygen serves to restore the atmospheric supply and to maintain it at a level consistent with the requirements of organisms needing oxygen. Second, the production of the energy-containing molecules utilized by all organisms originates from photosynthesis in green plants.

Principles

1. Photosynthesis is the method by which light energy, not usually utilizable by most organisms as an energy source, is transformed into a readily utilizable form in chemical bonds.
2. Photosynthesis is essentially reduction of carbon dioxide with hydrogen obtained by splitting water molecules.
3. The energy obtained from this, or any other source, is transferred to high-energy phosphate bonds in ATP before being used for cellular activity.
4. The various structural arrangements of leaves are adaptations to facilitate photosynthesis in their own environments.

Suggested Readings

Arnon, Daniel I., "Photosynthesis as an Energy Conversion Process," in *Frontiers of Modern Biology*, coordinated by Gairdner B. Moment. Houghton Mifflin Company, Boston, 1962.

Arnon, Daniel I., "The Role of Light in Photosynthesis," *Scientific American*, Vol. 203 (November, 1960). Offprint No. 75, W. H. Freeman and Co., San Francisco.

Butler, W. L. and Robert J. Downs, "Light and Plant Development," *Scientific American*, Vol. 203 (December, 1960). Offprint No. 107, W. H. Freeman and Co., San Francisco.

French, C. S., "Photosynthesis," in *This Is Life*, edited by W. H. Johnson and W. C. Steere. Holt, Rinehart and Winston, New York, 1962, pp. 3–38.

Lehninger, Albert L., "How Cells Transform Energy," *Scientific American*, Vol. 205 (September, 1961). Offprint No. 91, W. H. Freeman and Co., San Francisco.

Priestley, Joseph, "Observations on Different Kinds of Air," in *Great Experiments in Biology*, edited by M. L. Gabriel and S. Fogel. Prentice-Hall, Englewood Cliffs, N.J., 1955, pp. 155–157.

Rabinowitch, Eugene I., and Govindjee, "The Role of Chlorophyll in Photosynthesis," *Scientific American*, Vol. 213 (July, 1965). Offprint No. 1016, W. H. Freeman and Co., San Francisco.

Questions

1. What relationships, if any, existed between historical advances in the knowledge of photosynthesis and improvements in scientific technique? Can you give a specific example?

2. Explain why it is necessary to use isotopes to prove that water is split during photosynthesis.

3. Summarize, in your own words, what occurs during the light reactions of photosynthesis. Do the same thing for the dark reactions. (The flow sheet diagram should help you do this.)

4. What major uses may be made of PGAL in cells?

5. What is the direct (immediate) source of energy for all cells? Does this mean that cells do not use PGAL or other organic compounds for energy directly? Explain.

6. Why is most of photosynthesis described as a *reduction reaction*?

7. Explain why the formula $6CO_2 + 6H_2O + \text{Energy} \xrightarrow{\text{Chlorophyll}} C_6H_{12}O_6 + 6O_2$ is adequate as a *summary* of photosynthesis, but is inadequate to describe the chemical reactions that take place.

8. What external features of leaf shapes and forms adapt them for photosynthesis?

9. Is the internal anatomy of a leaf adapted to facilitate photosynthesis? In what ways?

10. Tell what effect, if any, each of the following environmental factors has on photosynthesis: light intensity, air temperature, relative hours of daylight and darkness, humidity, amount of oxygen in the atmosphere, type of soil, amount of water in the soil.

11. Does there appear to be any correlation between the use of visible portions of the spectrum and the early evolutionary development of green plants in water?

CHAPTER
V

Respiration: Energy Harvest

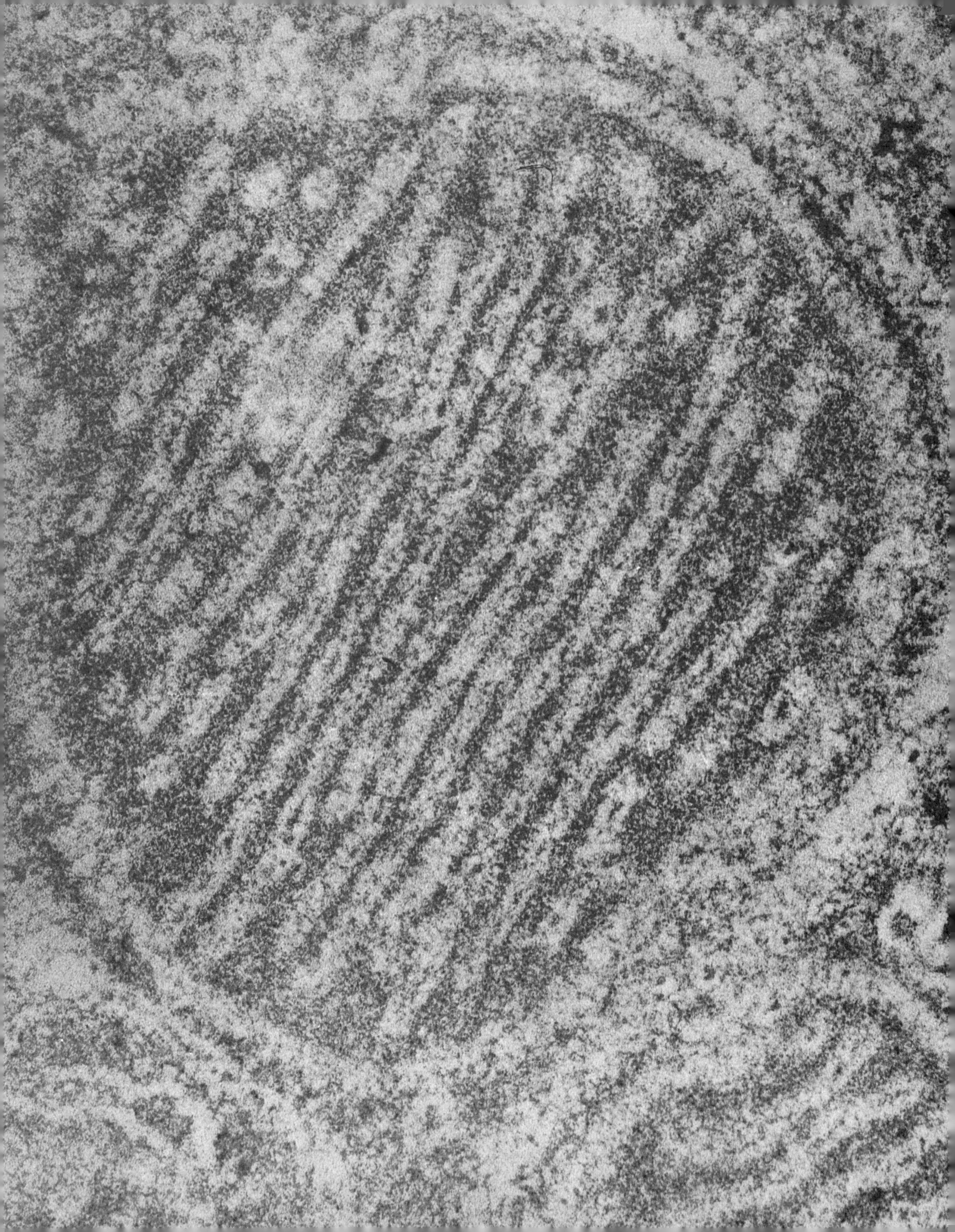

Electron photomicrograph of a mitochondrion (×400,000) (Dr. F. Sjostrand).

CHAPTER V

Respiration: Energy Harvest

Photosynthesis is the mechanism that transforms the energy for most of the living world into bonds in carbon-containing compounds. However, many organisms have no ability to carry out photosynthesis and are dependent for energy upon the intake of complex carbon energy sources. The energy in these sources comes either directly or indirectly from green plants. Regardless of how organisms get the energy-containing compounds to the cell, all of them must break down these compounds in a way that will allow the cell to harvest a maximum of utilizable energy. The sum of these energy-yielding processes is called *respiration.*

The term respiration is used in two ways. The processes we are referring to here occur within each cell and are called *cellular respiration.* More familiarly, respiration is used to mean the exchange of respiratory gases (carbon dioxide and oxygen) between the organism and the environment. This is *external respiration.* (See pp. 83–85.)

Organisms generally store energy within the cell in the form of either complex carbohydrates or fats. Proteins may also be used as a source of energy during periods of starvation but do not primarily serve an energy storage function. To obtain energy from any of these materials, the cell must first break them down into simpler molecules. This digestion always consists of hydrolytic reactions. As we have seen, most carbohydrates yield glu-

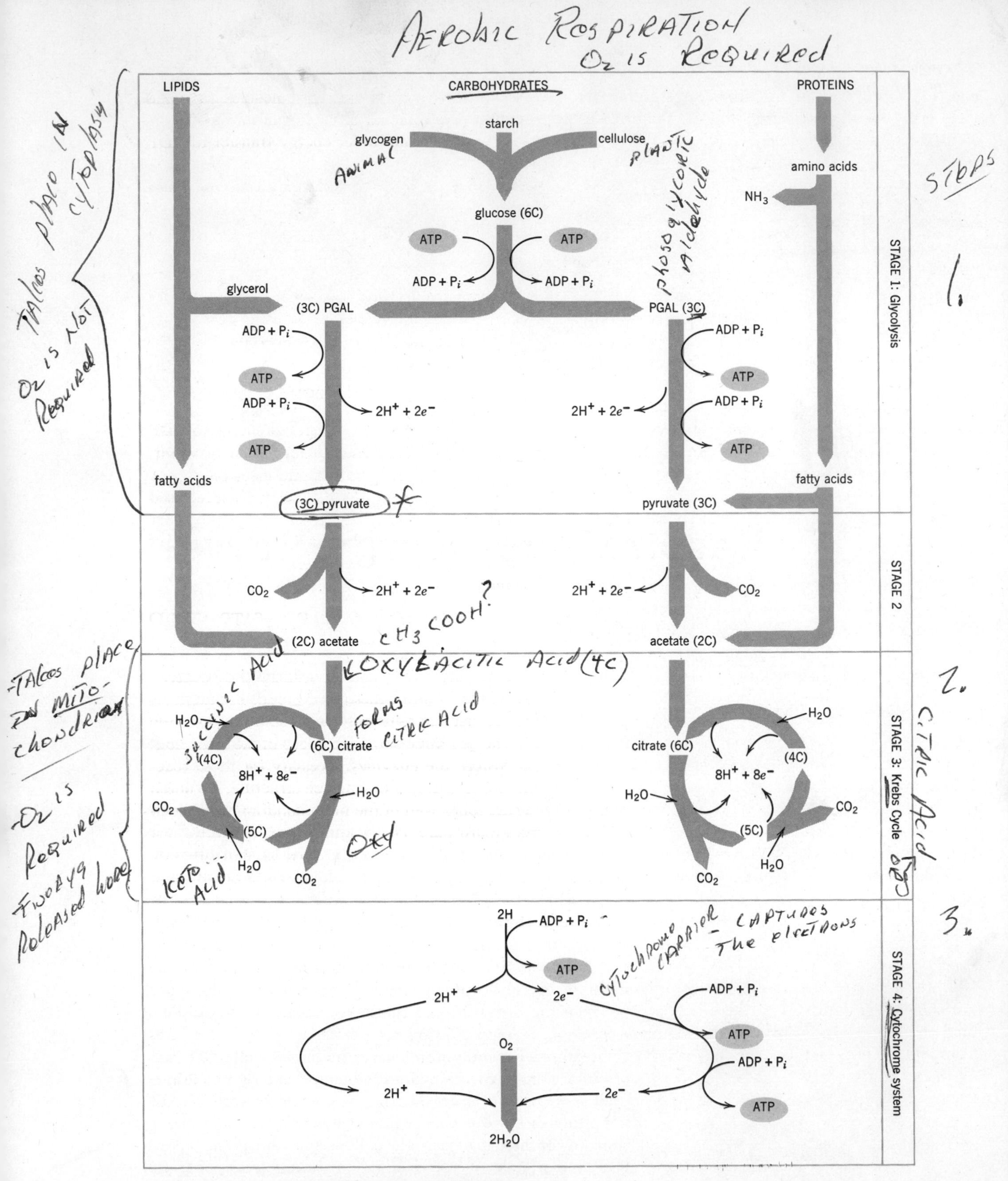

Figure 5.1. Summary of the major events of respiration.

cose or closely related six-carbon sugars; the fats yield fatty acids and glycerol, whereas the proteins yield amino acids. First let us look in detail at the reactions that glucose undergoes. Later we shall see how the other types of molecules fit this scheme.

Reactions of Respiration

The energy found in glucose molecules is released by their oxidation, that is, cell respiration. In this oxidative process, dehydrogenation is one of the major events. Dehydrogenation causes energy changes in the molecules that will donate hydrogen, resulting in the release of energy. A step-by-step repetition of this reaction produces an orderly, controllable supply of energy. Each hydrogen removal is accompanied by the removal of an electron from the donor molecule. Eventually, these electrons and the hydrogens are captured by oxygen molecules to form water, one of the end products of cellular respiration. Perhaps it surprises you to learn that oxygen functions in respiration as an ion acceptor and *not* as a substance used in "burning" carbohydrates. By following the electrons as they are transported from the donor molecules to oxygen, the final electron acceptor, we can gain insight into the methods of energy release in the cell.

The equation for respiration is usually given as

$$C_6H_{12}O_6 + 6O_2 \rightarrow 6H_2O + 6CO_2 + \text{Energy}$$

Although this equation contains the reactants and the products of the overall reactions, it neither describes anything about the processes nor tells what other substances are necessary. Rather than being a simple, one-step reaction as we might infer from the equation, respiration is an intricate series of reactions. It should be remembered, in the descriptions that follow, that each of the many reaction steps is controlled by a specific enzyme. It is estimated that nearly 100 enzymes function in the cellular respiration of glucose.

For convenience, we may divide respiration into four stages (Figure 5.1). It is important to keep in mind that we are summarizing the reactions that occur and are not treating them in detail or balancing them chemically.

Stage 1: Glycolysis

$$2ADP + 2P_i + \text{glucose (6C)}^* \rightarrow 2 \text{ pyruvate (3C)} + 4H^+ + 4e^- + 2ATP$$

The major events are: (*a*) cleavage of a six-carbon molecule to two three-carbon molecules; (*b*) some oxidation (the removal of hydrogen ions and electrons); and (*c*) some direct energy transfer to ADP and P_i to form ATP.

Stage 2: Bridge Between Glycolysis and the Krebs Cycle

$$\text{Pyruvate (3C)} \rightarrow \text{acetate (2C)} + CO_2 + 2H^+ + 2e^-$$

The major events are: (*a*) some oxidation; and (*b*) decarboxylation (removal of CO_2) of a three-carbon compound to form a two-carbon molecule.

Stage 3: The Krebs Cycle

$$3H_2O + \text{acetate (2C)} \rightarrow 2CO_2 + 8H^+ + 8e^-$$

The major events are: (*a*) transfer of energy to ADP and P_i to form ATP; and (*b*) formation of water. the removal of hydrogen ions and electrons.

In stages 1, 2, and 3, $H^+ + e^-$ have been released to carrier molecules. These must now pass to other molecules and finally to oxygen to form water. This occurs in the cytochrome system.

Stage 4: Cytochrome System

$$6ADP + 6P_i + 4H^+ + 4e^- + O_2 \rightarrow 6ATP + 2H_2O$$

The major events are: (*a*) transfer of energy to ADP and P_i to form ATP; and (*b*) formation of water.

As for the sites of these four groups of reactions, only the first occurs through most of the cytoplasmic sap. The last three stages all occur in the mitochondria, where the enzymes necessary for these reactions are found. In studying cell structure, we noted that the inner wall of the mitochondrion was folded (see Figure 5.2). This folding enlarges the surface area on which the enzymes are located, thereby increasing the amount of the enzymes present and the speed at which respiration may occur. Let us consider each of these four stages in more detail.

Glycolysis. For glycolysis ("glucose splitting") to occur, the energy level of glucose must be increased. Activation energy is transferred from ATP to glucose by the shift of a high-energy phosphate group. This transfer has its paradoxical side; in order to obtain ATP from respiration, we must first expend ATP (see Figure 5.1). The resulting molecule is then split into two three-carbon molecules, which we call PGAL.

*6C and 3C indicate that glucose and pyruvate are, respectively, a six-carbon and a three-carbon compound. H^+ and e^- are the symbols for a hydrogen ion and an electron, respectively.

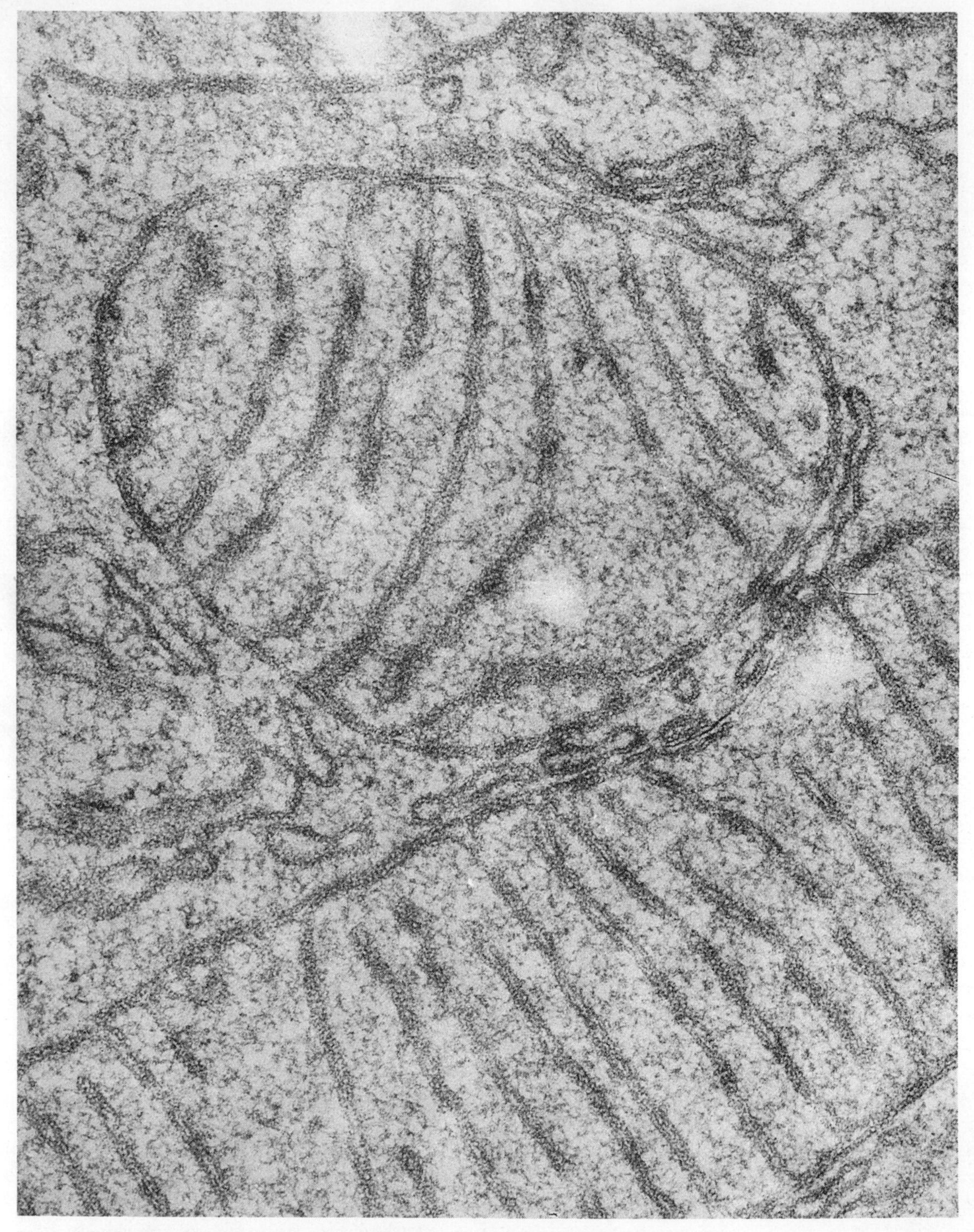

Figure 5.2. Mitochondria showing the folds of the inner wall (×*40,000*) (*Dr. F. Sjostrand*).

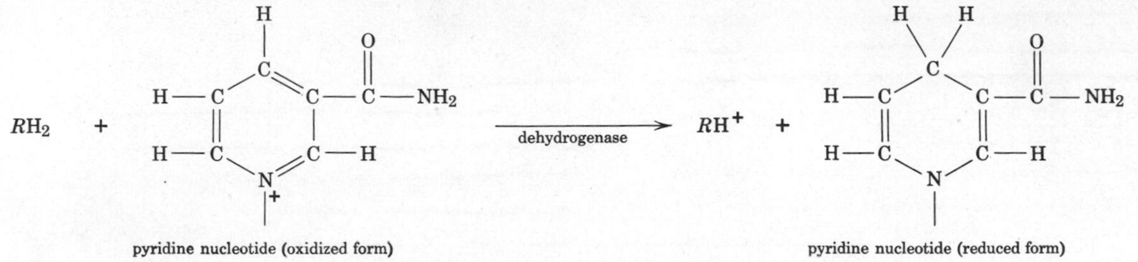

$$RH_2 + \text{pyridine nucleotide (oxidized form)} \xrightarrow{\text{dehydrogenase}} RH^+ + \text{pyridine nucleotide (reduced form)}$$

Figure 5.3. A pyridine nucleotide carrier receives a hydrogen from the donor, RH_2, and transports it as shown. Dehydrogenase is an enzyme that facilitates the transfer to the pyridine nucleotide.

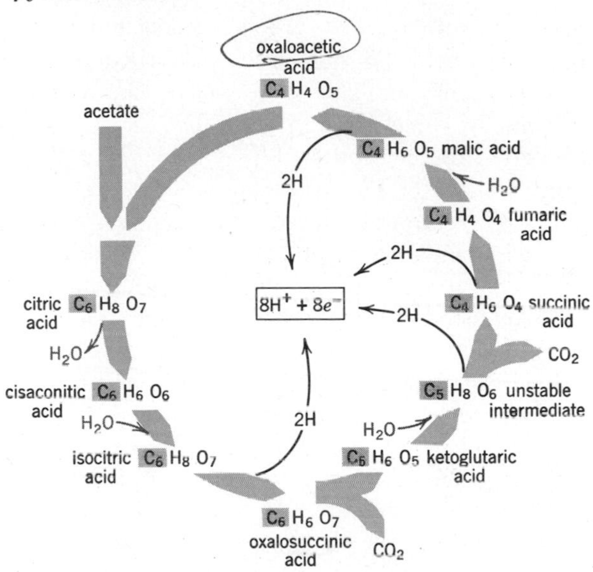

Figure 5.4. The Krebs cycle. The major portion of the oxidation, indicated by the released hydrogens and electrons, occurs in this elaborate cycle.

PGAL is an important crossroad in the metabolism of the cell because it is common to the pathways of photosynthesis, other syntheses, and respiration (see pp. 64–65).

Three types of reactions occur in the transformation of PGAL to pyruvate: the rearrangement of bonds in the molecule, the direct transfer of energy to ADP and P_i to form ATP, and the removal of hydrogen with electrons. The amount of ATP produced directly in glycolysis is twice as much as was used in supplying the activation energy. A net gain of energy is evident.

The hydrogen atoms released here and elsewhere in respiration do not appear in the form of free hydrogen. Rather, they are donated to molecules that serve as hydrogen and electron carriers. These carriers are complex organic molecules, called pyridine nucleotides (Figure 5.3), which are necessary for respiration. After the electrons are transferred from the pyridine nucleotides to the cytochrome system, the hydrogen is released into the cytoplasmic sap as H^+.

Many cells cannot synthesize some portions of the pyridine nucleotide molecules but must bring them in already made in the form of some of the B vitamins. In fact, vitamins B_2, K, and E as well as iron form essential parts of various carrier molecules. These must be provided in the diet of the organism; a deficiency will result in metabolic disorders.

Bridge. In the second stage of respiration, the three-carbon molecule pyruvate enters into the mitochondria from the cytoplasmic sap. Here, one of its carbons is removed, in the form of carbon dioxide, leaving a two-carbon molecule, acetate. In addition, more hydrogen atoms are transferred to a hydrogen carrier.

The Krebs Cycle. The acetate now enters a series of reactions called the Krebs cycle (Figure 5.4) after its discoverer, the English biochemist Sir Hans Krebs. Acetate is combined with a four-carbon molecule to form the six-carbon citrate from which the cycle gets another of its names, the *citric acid cycle.* By the successive removal of carbon dioxide, the citrate is first changed to a five-carbon molecule and then to a four-carbon molecule. This decarboxylation, along with other reactions such as the addition of water and the transfer of hydrogen atoms to a carrier, results in the reconstitution of the original four-carbon molecule. This completes the cycle. The four-carbon molecule may now combine with another acetate molecule and the extensive series of reactions occurs again (Figure 5.4).

Cytochrome System. Thus far, we have talked of several reactions in which carbon dioxide is pro-

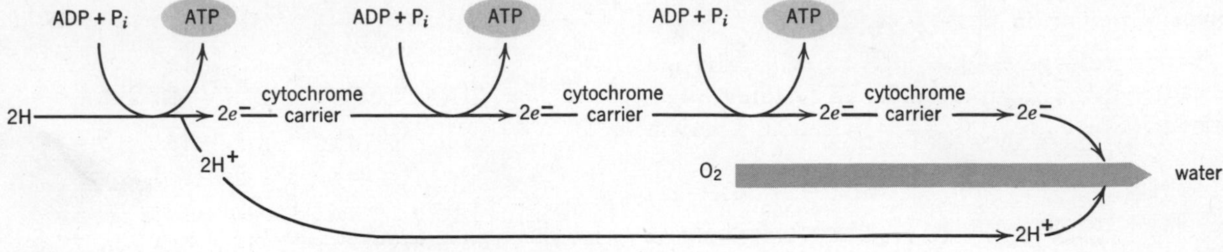

Figure 5.5. The cytochrome system accounts for most of the ATP formation.

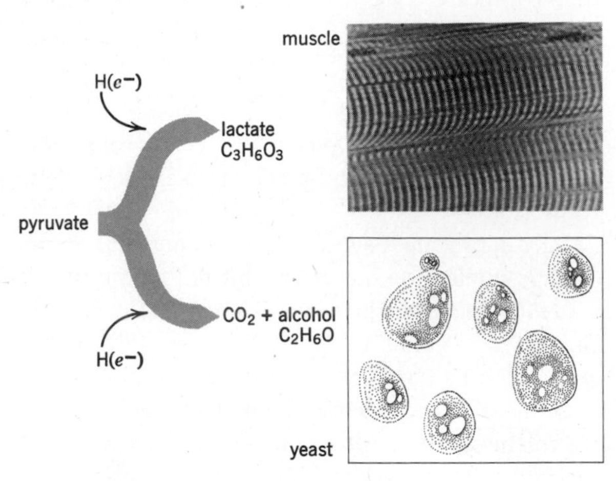

Figure 5.6. Alternate pathways of anaerobic respiration in muscle and yeast.

duced or hydrogen atoms are transferred to a carrier. What happens to these substances? Carbon dioxide is an end product of respiration and diffuses from the cell as a waste product. The electrons from the hydrogen atoms, however, are transferred successively through a series of complex organic molecules (the cytochromes) to oxygen. This series of transfers is the cytochrome system (Figure 5.5). *Most of the energy available to the cell from respiration is obtained by the transfer of electrons in the cytochrome system.* At this time hydrogen ions from the cytoplasmic sap are also picked up by the oxygen and water is formed.

An important consideration is that the cytochrome system releases energy in small amounts rather than in one large amount. If this latter event occurred, as in many noncellular oxidations, much of the energy would be lost as heat. Instead, the cell is able to harvest a considerable amount of the available energy in the form of ATP. For example, one molecule of glucose yields 38 ATP units in the course of cellular respiration. This represents an impressive efficiency of about 44%.

Several familiar poisons function by disrupting normal respiratory pathways. Cyanide blocks the transfer of electrons from the cytochrome system to oxygen by combining with one of the cytochromes and inactivating it. Fluoride also acts as a respiratory poison by reacting with the magnesium ions required by some of the enzymes that function in glycolysis. It should be noted that the levels of fluoride used for preventing cavities in teeth are far less than the amount that inhibits enzyme action.

Respiration of Proteins and Lipids. As indicated in Figure 5.1, lipids may also participate in cellular respiration. In order for proteins to do this, they must first be digested into amino acids. These in turn are stripped of their amino groups and converted into fatty acids. Proteins are not often used in this way; lipids, on the other hand, are quite important in cellular respiration, representing a material of high energy content that is easily stored and quickly available for use in the respiratory cycle.

Most fatty acids are long chains of carbon atoms with some oxygen and a great number of hydrogen atoms attached. These chains are broken down by the successive removal of two-carbon fragments (acetate). The last fragment may contain either two- or three-carbon atoms (may be either acetate or pyruvate), depending on the number of carbons in the original fatty acid. The subsequent oxidation reactions are thus exactly the same as those involved in the oxidation of carbohydrates.

Glycerol is another product of the digestion of fats. It is a three-carbon compound similar to PGAL and may enter the glycolytic reactions at the same point.

Some Variations in Respiration

So far we have assumed that all cells carry out oxidation in the same way under all conditions. This is not true. We find all kinds of variations from this basic route.

Fermentation is a familiar variation carried out by yeast and many other microorganisms. Fermentation follows the glycolytic pathway as far as pyruvate. Then, however, pyruvate is changed into ethyl alcohol (a two-carbon molecule) and carbon dioxide. These reactions require the transfer of hydrogen atoms from the carrier to pyruvate to complete the transition to alcohol (Figure 5.6).

Two aspects of this pathway should be emphasized. First, it requires no oxygen consumption. This makes it useful to many organisms that live in environments lacking oxygen (*anaerobic* environments). Second, the addition of hydrogen atoms to pyruvate provides a means of disposing of the hydrogen obtained during glycolysis. Pyruvate substitutes for oxygen as a final hydrogen and electron acceptor. If there were no final acceptor, reduced carrier molecules would accumulate, no oxidized carrier molecules would be available to accept hydrogen atoms from PGAL, and all the reactions of glycolysis would cease.

Anaerobic Respiration in Muscle. Another variation in the pathway for respiration is found in muscle cells under anaerobic conditions. This variation of respiration is similar to fermentation in that it undergoes the same reactions from glucose to pyruvate. However, the pyruvate is now transformed to lactate, another three-carbon molecule (Figure 5.6). During this transformation, pyruvate substitutes for oxygen as in fermentation so that no oxygen is required.

Why is it that all respiration in muscle cells does not occur by this route? Two characteristics of the pathway make this undesirable. Only about five percent of the energy obtained by the normal aerobic pathway is acquired by anaerobic glycolysis. Second, lactate tends to build up in the cell. If the concentration rises too high, muscular fatigue and soreness result. This pathway is used only when a need for energy arises during a shortage of oxygen, as in extreme muscular exertion.

Sulfur bacteria. Certain bacteria show a third respiratory variation. In sulfur bacteria, wide variety exists in the patterns of respiration. One group of bacteria may use sulfur rather than oxygen as the final hydrogen acceptor.

$$4H_2 + \underset{\text{sulfuric acid}}{H_2SO_4} \rightarrow H_2S + 4H_2O$$

The first hydrogen in this equation comes from several different sources. These bacteria use hydrogen gas if nearby bacteria are producing it. If not, they may use a number of organic molecules like pyruvate as hydrogen sources. A source is essential; which source does not matter. Students of chemistry know the similarity of the reactions of oxygen and sulfur. The final product, H_2S or hydrogen sulfide, is famous for its "rotten egg" odor because the bacteria that produce it are found in decaying eggs.

Oxidation involving the removal of electrons is an important feature that all forms of respiration share. However, great diversity exists in the details of the processes of respiration found in different kinds of cells. Both this variation in detail and the similarity in the bases of phenomena are in themselves characteristic of living systems.

Principles

1. Respiration is the process by which useful energy is made available to cells by the breakdown of fuel (carbon-containing) molecules within each cell.
2. The three major events of respiration are: (*a*) Oxidation of the fuel molecules by dehydrogenation. (*b*) The energy is released in small quantities by the stepwise transfer of hydrogen to oxygen to form water. (*c*) This energy is utilized in the synthesis of ATP.

Suggested Readings

Green, David E., "The Mitochondrion," *Scientific American,* Vol. 210 (January, 1964). Offprint No. 175, W. H. Freeman and Co., San Francisco.

Lehninger, Albert L., "Energy Transformation in the Cell," *Scientific American,* Vol. 202 (May, 1960). Offprint No. 69, W. H. Freeman and Co., San Francisco.

Lehninger, Albert L., "How Cells Transform Energy," *Scientific American,* Vol. 205 (September, 1961). Offprint No. 91, W. H. Freeman and Co., San Francisco.

McElroy, W. D., "Energy and Life," in *Frontiers of Modern Biology,* coordinated by Gairdner B. Moment. Houghton Mifflin Company, Boston, 1962.

Racker, Efraim, "The Membrane of the Mitochondrion," *Scientific American,* Vol. 218 (February, 1968). Offprint No. 1101, W. H. Freeman and Co., San Francisco.

Questions

1. Follow a molecule of glucose through the events of cellular respiration, using the flow sheet as a guide.
2. What is erroneous about the common statement that cells burn foodstuffs for energy?
3. Explain the statement that the primary function of cellular respiration is to transfer energy into phosphate bonds.
4. Review the five general types of reactions that occur in cells (Chapter II), then see how many of them take place during cellular respiration.
5. During which stage of respiration would you expect to find the largest number of ATP molecules formed?
6. During the respiratory process, H^+ are released into the cell. What does this do to the pH of the cell? How does the cytochrome system eventually utilize these hydrogens?
7. Why is glycolysis an *anaerobic* reaction even in aerobic organisms?
8. Anaerobic respiration produces far less useful energy for a cell than does aerobic respiration. Why?
9. Yeasts can live in either oxygen-poor or oxygen-rich environments. Do you think they produce ethyl alcohol in the presence of oxygen? Why?
10. Muscles that are used regularly seldom become sore even with vigorous exercise. Can you think of some reasons why this should be true?
11. Sulfur bacteria illustrate what basic concept about respiration?
12. In what specific parts of cells does respiration take place? In what structural way might one cell be better adapted for respiration than another, e.g., an active cell compared to a relatively inactive one?
13. A deficiency of vitamin B_2 is often associated with a general inability to utilize foods as an energy source. What relation does this have to cellular respiration?

CHAPTER VI

Systems for Treatment and Transport of Materials

Barnacles feeding
(© *Walt Disney Productions*).

CHAPTER
VI

Systems for Treatment and Transport of Materials

In order to maintain life and to grow, organisms must obtain a variety of substances or the materials from which to synthesize them. In Chapter II we outlined the general types of materials found in living systems. The inorganic materials comprise small molecules which, upon reaching the cells, pass readily through membranes. Animals generally obtain the complex organic materials in forms that must be digested before they can be transported through cell membranes. In larger animals, the digested materials must be moved through the cells of the wall of the digestive tract, carried through the organism, and then transported into their cells. Even though most plants manufacture their own organic materials, these materials must still be transported from the site of production to the site of utilization. This involves problems similar to those met in animals.

Three different processes are involved: intake of materials, digestion, and transport to cells. Transport through membranes, which plays an important part in these processes, was discussed in Chapter II.

Intake of Materials: External Respiration

Generally, external respiratory structures are areas specialized for the exchange of oxygen and carbon dioxide with the surrounding environment. An examination of a series of these structures in different organisms shows how they are adapted to an organism's needs and environment.

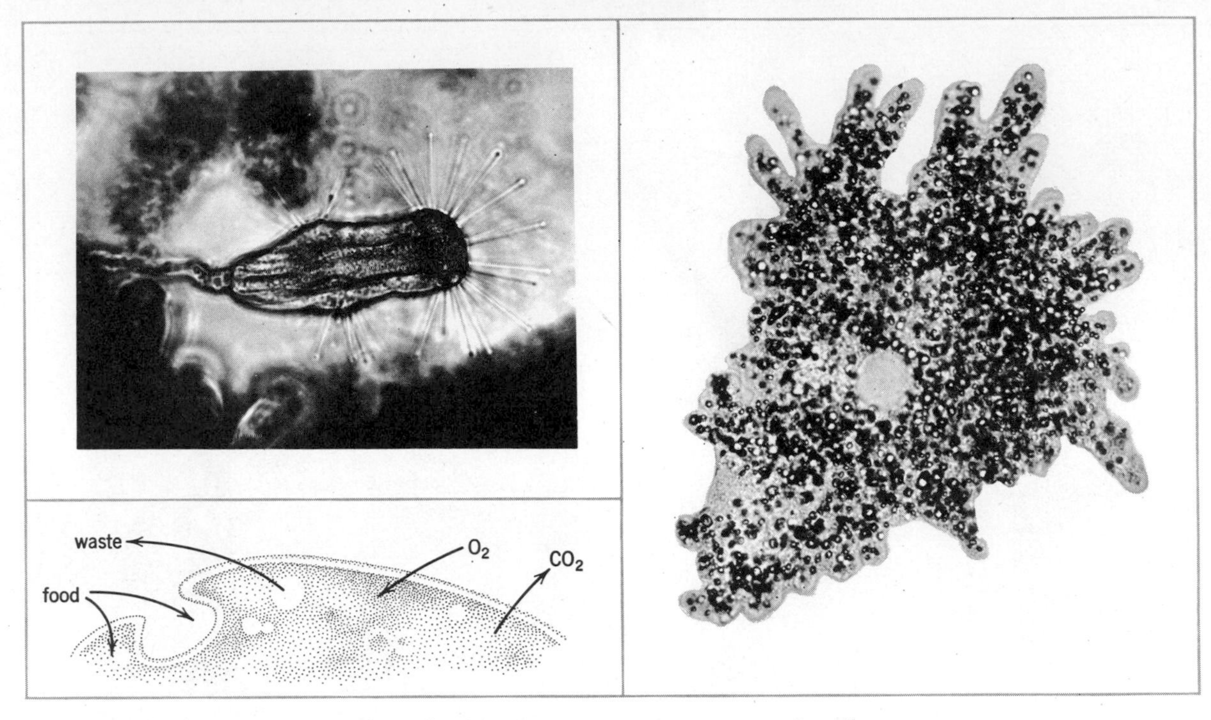

Figure 6.1. In protozoans (two are shown here) respiratory gas exchange occurs through the cell surface.

General Body Surface. Some organisms do not have any specialized respiratory structure. They simply exchange gases through the body surface. This is usually the case in small, less active organisms that are restricted to moist environments. All single-celled organisms (Figure 6.1) and some multicellular organisms such as the rotifers and *Hydra* are members of this group (Figure 6.2). These organisms must remain moist for respiration to occur because gases must dissolve in water to pass through their thin body membranes.

Larger animals like the earthworm and some amphibians, which live in moist environments, also exchange respiratory gases through the general body surface. In these animals, blood is transported to the body surface where it exchanges respiratory gases with the environment. The blood then carries the gases through the body.

Gills. Many animals that live in water or inhabit moist environments possess gills (Figure 6.3). These respiratory structures can be considered as outfoldings of the body surface; they increase the amount of surface available for respiration and allow more rapid exchange. Gill systems lose water rapidly in a dry environment; hence, they often contain adaptive structures that close them off from a temporarily unfavorable environment and prevent water loss for a period of time.

Lungs and Tracheae. Land-dwelling animals that are not restricted to moist environments generally have either lungs or tracheae. *Lungs* may be considered infoldings of the body surface (Figure 6.4). Like the outfolding of the gills, this infolding results in a larger area for gas exchange. However, the fact that the lungs are inside the animal protects their moist surfaces from a large water loss. In the lungs, exchange of gases is rapid.

Tracheae are a series of tubes extending throughout the body of insects, spiders, and certain other related arthropods (Figure 6.5). As air moves in and out of the tracheae, it dissolves in the fluid that fills the ends of the tubes. Water loss is minimal and, incidentally, there is no need for the blood to transport respiratory gases, because the tubes pass near every cell in the body. An animal that possesses tracheae cannot be very large, for the weight of such

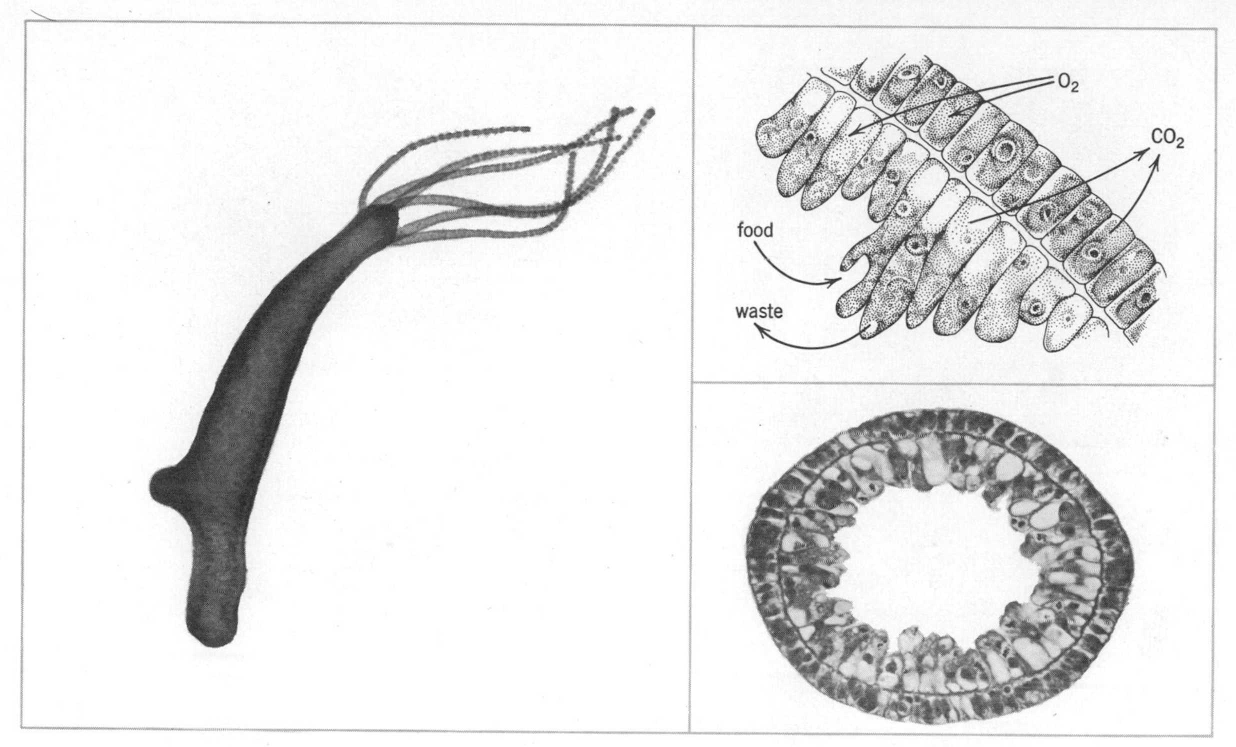

Figure 6.2. In Hydra, *a multicellular animal, respiration occurs through the body surface. Notice in the cross section (lower right) that* Hydra *has only two cell layers.*

reinforced tubes quickly becomes prohibitive with increasing size.

Intake of Materials: Digestion

Since respiratory gases are composed of small molecules, they pass readily through moist membranes. Before food materials can cross membranes, however, they must usually be broken down to simple sugars, amino acids, fatty acids, glycerol, or other small molecular forms.

Digestive processes, which are hydrolytic reactions, are found in all types of organisms. Hydrolysis is characteristic of digestion inside and outside cells and involves specific digestive enzymes.

Vacuoles. The digestion that occurs inside cells (intracellular digestion) breaks down molecules that have been synthesized within the cell and those brought into the cell (Figure 6.1). Digestive enzymes are secreted into food vacuoles in the cytoplasm, where hydrolysis to simple molecules occurs. This type of digestion is found in many kinds of animals. *Amoeba* and *Paramecium* (Figure 6.6) are two examples that carry on only intracellular digestion.

External Digestion. Although intracellular digestion might be considered the simplest form, many fungi (molds) have no digestive system and actually secrete enzymes onto the food outside their bodies and absorb materials that have already been digested to simple molecules (Figure 6.7). Of course, many internal parasites, like the tapeworm, absorb predigested food from the environment and also lack digestive systems.

Gastrovascular Cavity. Some animals, like *Hydra* and planaria (Figure 6.8), have incomplete digestive tracts. These tracts with only one opening are associated with either intracellular digestion or a combination of intracellular and extracellular digestion. Cells lining the digestive tract engulf undigested or partially digested particles of food. The tracts do not show specialization of areas for different functions. The digestive tract is called a gastrovascular cavity because it may also serve as a transport system.

Complete Digestive Tracts. Most animals above the *Hydra* and planaria level of complexity contain digestive systems with a mouth, digestive tube, and

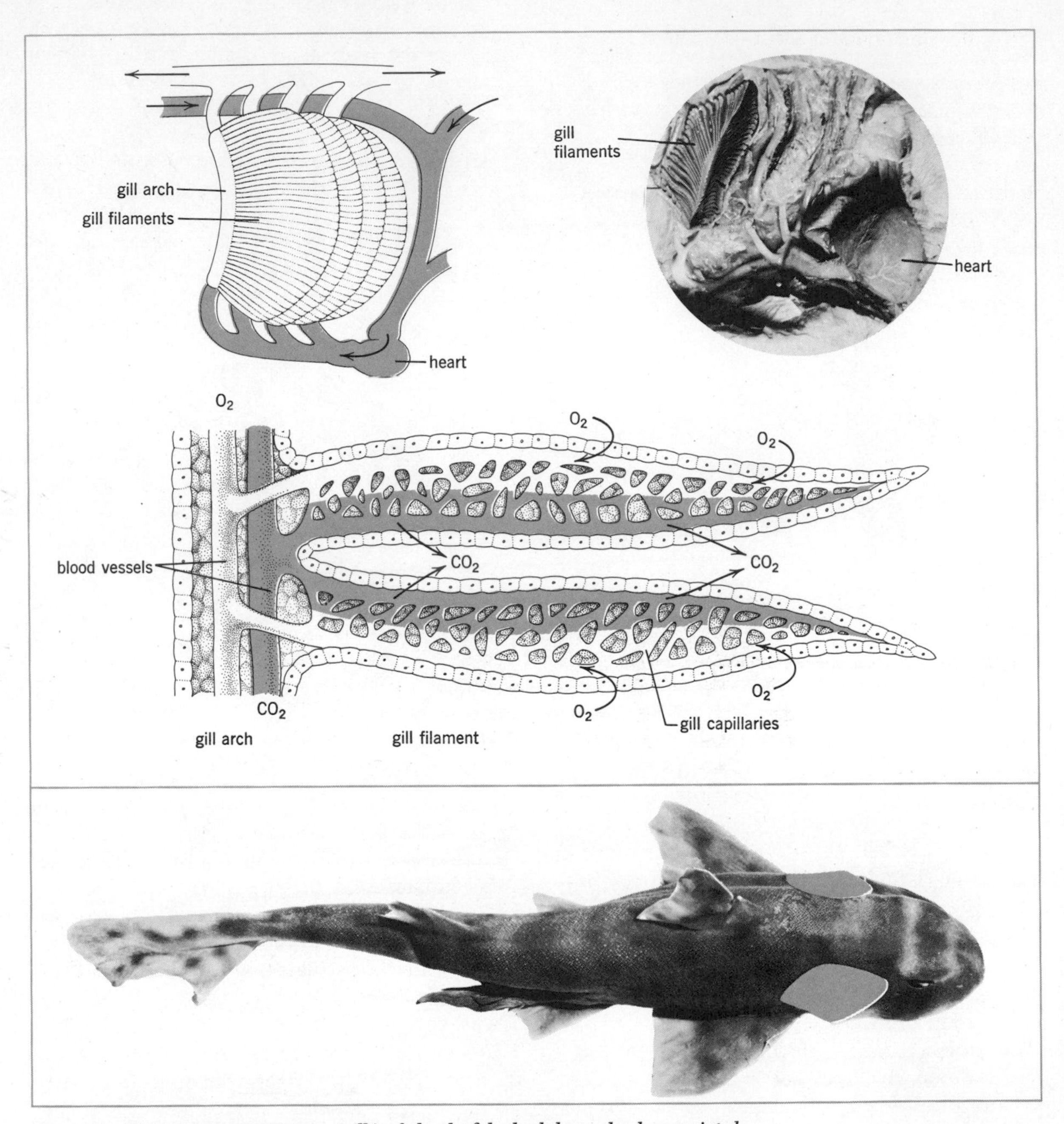

Figure 6.3. The respiratory structures (gills) of the dogfish shark have closely associated blood vessels. The middle drawing shows detail in two gill filaments.

anus (Figure 6.9). This is termed a complete digestive tract. It lets food material move in one direction and permits specialization of regions of the tract.

For example, the liver and pancreas are specialized parts of the tract that secrete important digestive juices into the gut. Thus, the initial stages of digestion, such as mechanical dissolution, are separated from the later stages of digestion and absorption. Finally, there may also be an area for treatment of the remaining contents that are largely waste materials. Removal of water and uptake of some ions are found in this area.

The specialization and consequent complexity of the digestive tract allow a more nearly constant intake of digested substances from the tract.

Transport Systems

After food has been reduced to particles that can be moved through the membranes of the gut into the organism itself, the materials must be transported to the areas where they are needed. The mechanisms for this transport also generally remove waste materials and carry products from the cells. Some organisms need special transport systems while

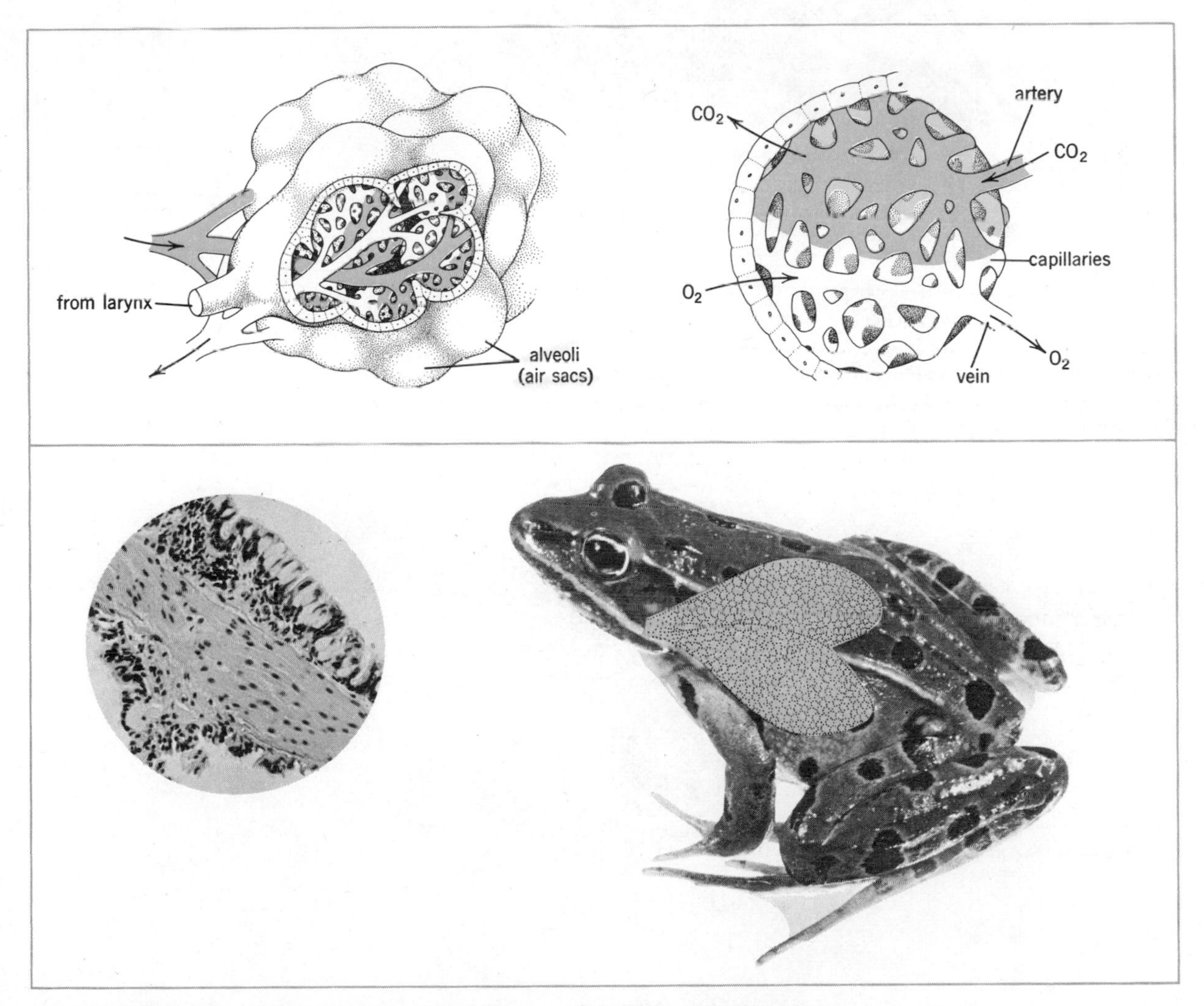

Figure 6.4. Frogs, like most land animals, possess lungs. A photomicrograph of frog lung tissue is shown at lower left. Usually lungs are divided into alveoli (upper left). The pattern of gas exchange in an alveolus is shown at upper right (Photograph of frog courtesy of Dr. John A. Moore).

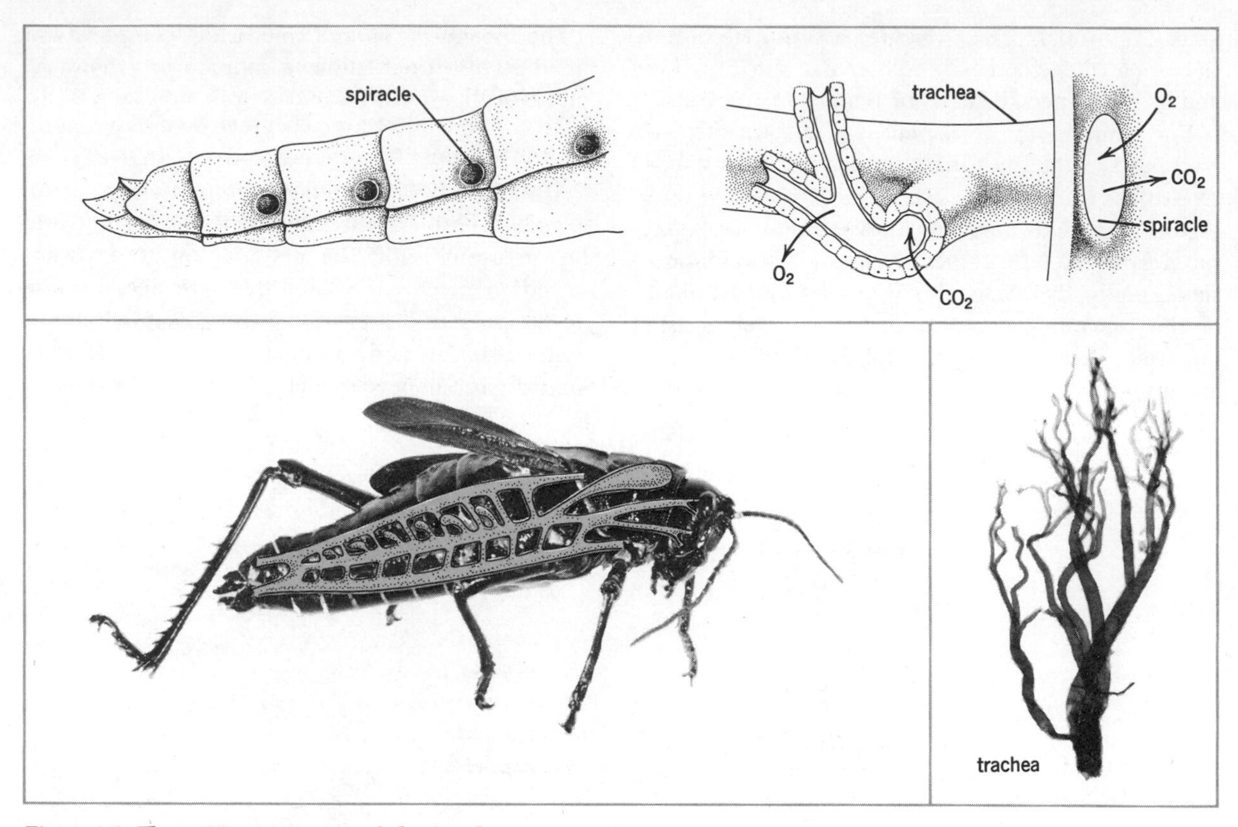

Figure 6.5. The respiratory system of the grasshopper and other insects consists of an interconnected series of tubules called tracheae. These open to the outside of the insect by small openings called spiracles (upper left). Gas exchange occurs in small branches of the tracheae (upper right).

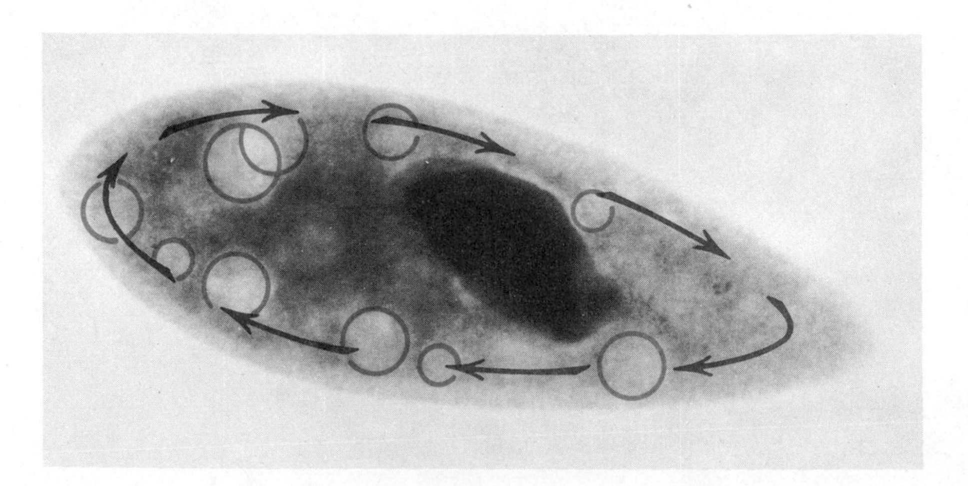

Figure 6.6. The movement of food vacuoles (yellow circles) in Paramecium *follows a definite path (CCM: General Biological, Inc., Chicago).*

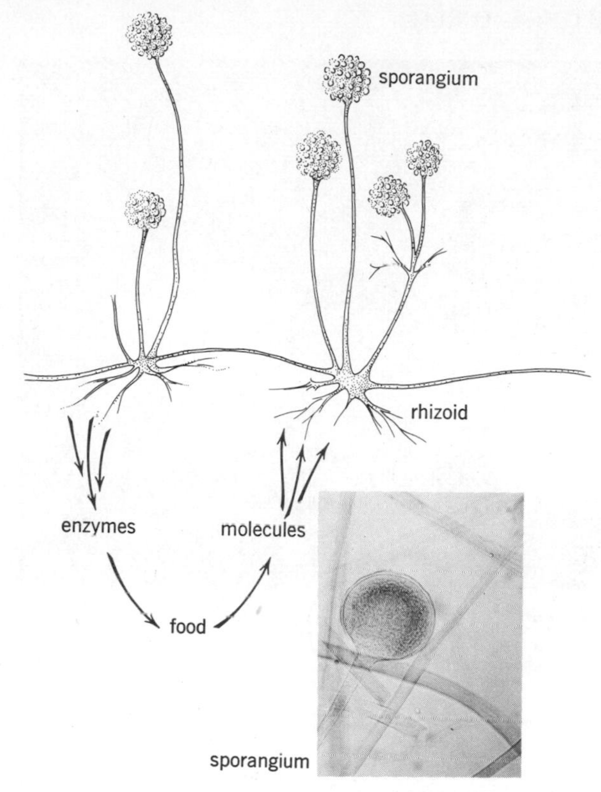

Figure 6.7. The bread mold exhibits extracellular digestion. The sporangia are reproductive structures.

others do not. A consideration of the size of the organism illustrates some of the problems involved.

When all of an organism's cells are close to its surface, simple diffusion may move materials rapidly enough to supply the organism's needs. In most cases no transport system is present. Single-celled organisms and thin, flat organisms are examples. When there are tissues that are more than a few cells removed from the surface, a speedier means of material transport is needed; without one, the tissues will die.

Gastrovascular Cavity. As mentioned previously, some animals like planaria and *Hydra* have a gastrovascular cavity that penetrates to within a short distance of every cell (Figures 6.2 and 6.8). In such organisms, diffusion effects the transfer of materials over the remaining distance to the cells. Within the cavity, movement of materials is limited to "sloshing" of the fluid and partially digested food. Complete mixing of the contents results, and the efficiency of transport is low. The animals having this system are small and the level of complexity and rate of activity are low.

Transport Systems in Plants. The transport system of higher plants is composed of two distinct sets

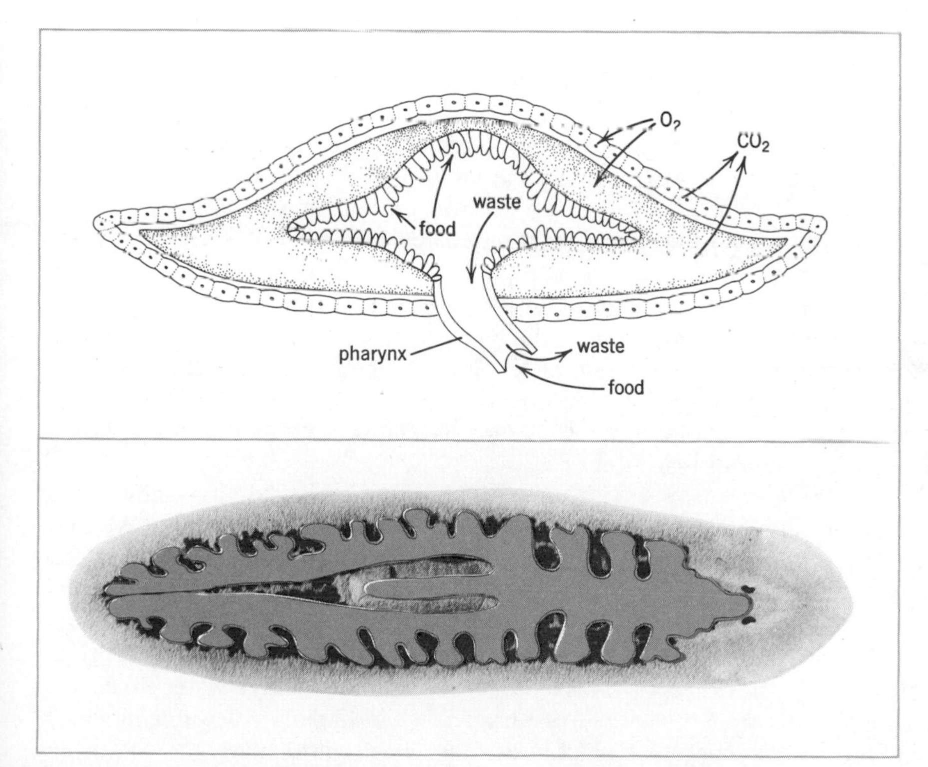

Figure 6.8. The extensive incomplete digestive tract of planaria is shown in blue. The intake of food, elimination of undigested waste, and exchange of respiratory gases are shown in the cross section at top.

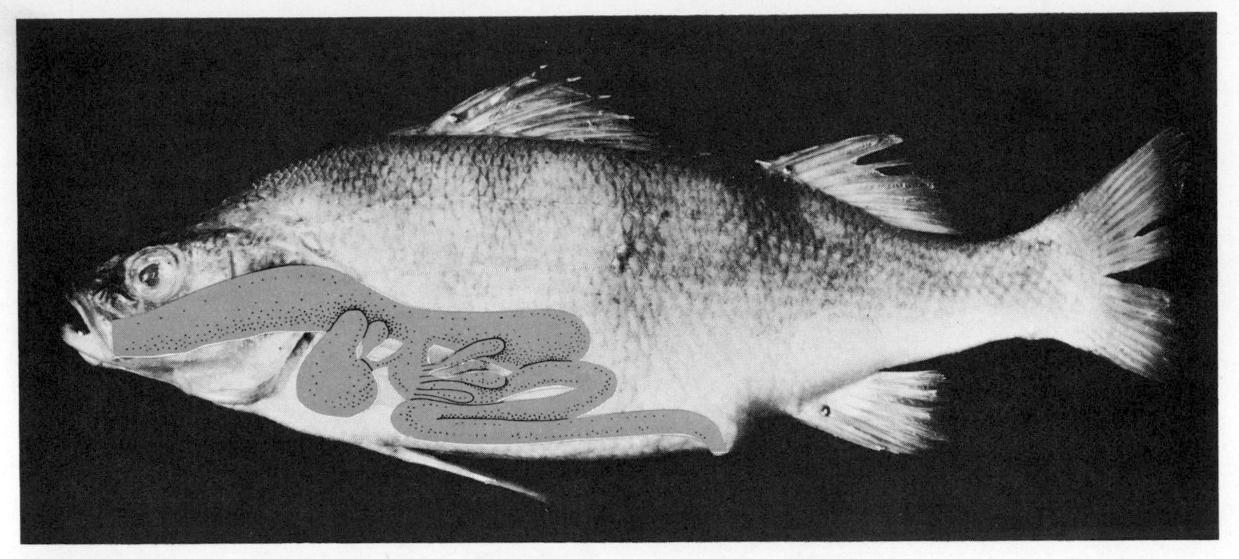

Figure 6.9. The digestive system of the fish is complete, having both a mouth and an anus. Specialized regions for digestion and absorption of foodstuffs are also present.

of tubular elements (Figures 6.10 and 6.11). At maturity, one of these sets, the *xylem*, is composed of the walls of dead cells and carries mostly water and minerals. The cells of the other set, the *phloem*, are living and carry organic materials.

The water-bearing xylem is associated with the wood of the plant. This dead-cell system presents a relatively open set of tubes through which the water may pass but which is incapable of expending any energy to help move the water. Two forces from other sources move the water: *root pressure* and *transpiration pull.*

Root hairs near the tips of roots extend into the soil. Water molecules are usually found in higher concentration in the soil than in the root hairs. Therefore, water molecules move by osmosis from the soil into the root hair. The resulting increase in water concentration in the root hair causes water molecules to move into the next cell, and so on in sequential fashion to the xylem at the center of the root.

Water is also carried from the soil into the root by active transport. Therefore, root pressure results partly from the concentration gradient from the soil to the cells in the center of the root and partly from the active transport of water by the living root cells.

Transpiration pull is brought about by the interplay of three factors: evaporation of water from the leaves (transpiration), osmosis, and the cohesion of water. Transpiration is constantly occurring. This lowers the concentration of water in the exposed cells of the leaves. In response to the lower concentration, water moves into these cells from the veins by osmosis. The whole column of water in the xylem of the veins is pulled upward because water molecules cohere, that is, tend to cling to one another.

Minerals from the soil are brought to the xylem of the root by active transport and then moved with the water. This active transport of minerals helps to enhance the osmosis of water in the xylem. The motions described are all in one direction, up the plant. The roots pick up the material from the soil and transport it to the rest of the plant, particularly to the leaves, where the materials are used or lost. Since the source is at one end and the site of utilization or loss is at the other end, no return pathway is necessary. The transport system in plants is not a circulatory system as is the system found in animals.

The method by which the living cells of the phloem transport organic materials is not well understood. Diffusion, osmosis, and cytoplasmic flow probably play some part. In any case, it is active transport that moves sugar molecules and other large molecules into phloem cells at the site of their origin. This motion decreases the relative water concentration in these cells, and additional water moves in by osmosis. Since a cell (particularly those with cell

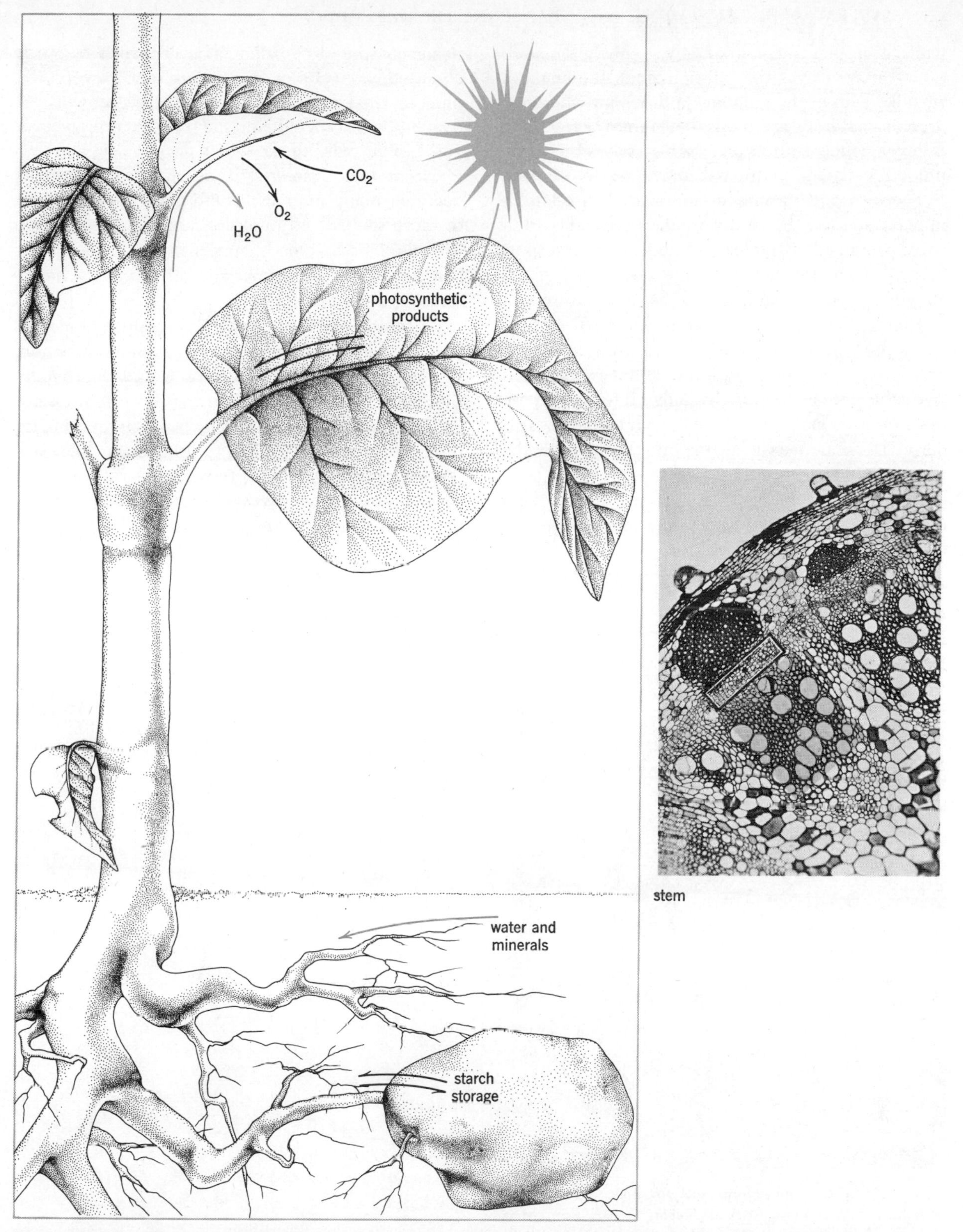

Figure 6.10. The transport system of flowering plants. The direction of transport of materials is indicated by the arrows. The relationship of xylem (blue) and phloem (black) is shown at right in the photomicrograph of a stem cross section (photograph: Ward's Natural Science Establishment, Inc.).

walls, as in plants) can hold only so much, the additional water drives the solution along the chain of tubelike cells of the phloem. At the point where the organic materials are used, their removal results in a lower concentration of dissolved particles in the phloem cells and water will move out by osmosis.

The transport in phloem, unlike that in xylem, is in two directions. Since the organic material is originally produced in the leaves, most of the transport is away from the leaves to the rest of the plant. However, other organic substances, such as growth regulators, are produced from carbohydrates in many parts of the plant, and move from the site of origin to other regions. In some instances, materials from the phloem eventually move into the xylem. For example, maple sap, from which maple sugar is derived, is obtained by cutting into the xylem of the maple tree.

In most trees, phloem tissue forms a rather thin inner layer of the bark, the remainder of the tree being composed of xylem. If a groove is cut so it completely encircles the trunk and severs the phloem, the tree gradually dies. If the groove is cut deeper, the leaves wilt and the tree dies rapidly. Can you explain why these events occur?

Circulatory Systems. True circulatory systems occur in many animals. These generally have one or more specialized pumping vessels called hearts. The fluid base (blood) moves from the heart (or hearts) through the tissues and returns to the heart, forming a complete circuit.

Circulatory systems may be either open or closed. In an *open system,* the blood is pumped a short distance from the heart in vessels and then enters the spaces between cells (Figure 6.12 top). It bathes the cells directly, exchanges materials with them by diffusion, and then returns to the heart. Open systems have a low blood pressure and rate of flow. In addition, the relative amounts of blood flowing to various structures are poorly controlled. This sys-

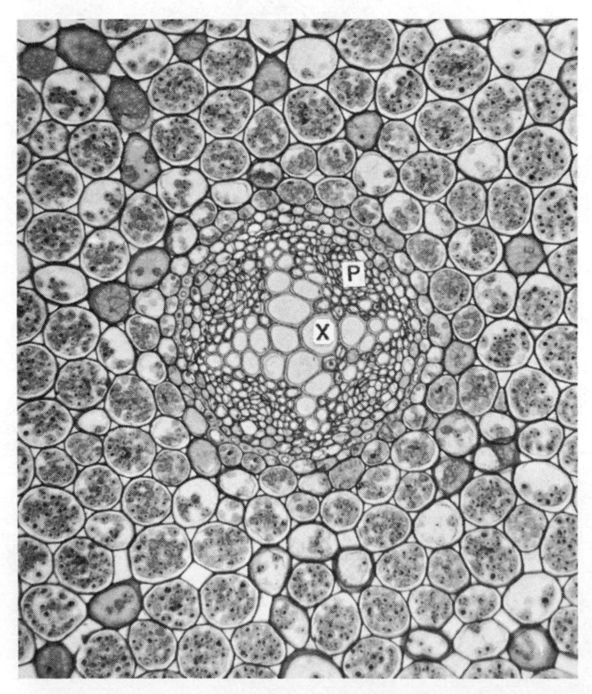

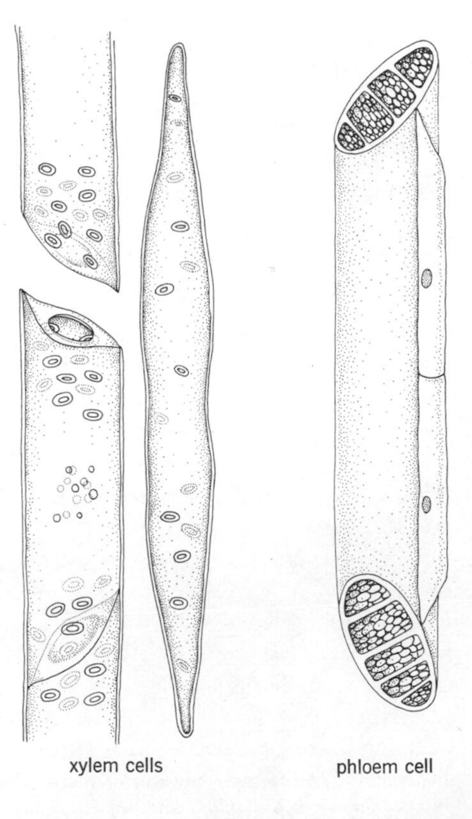

Figure 6.11. Xylem and phloem cells are shown schematically at right. At left is a cross section showing these cells in a root. (Xylem cell redrawn from Harry J. Fuller, The Plant World, *third ed., Holt, Rinehart and Winston, New York, 1955; photomicrograph of* Ranunculus *root, Ward's Natural Science Establishment, Inc.)*

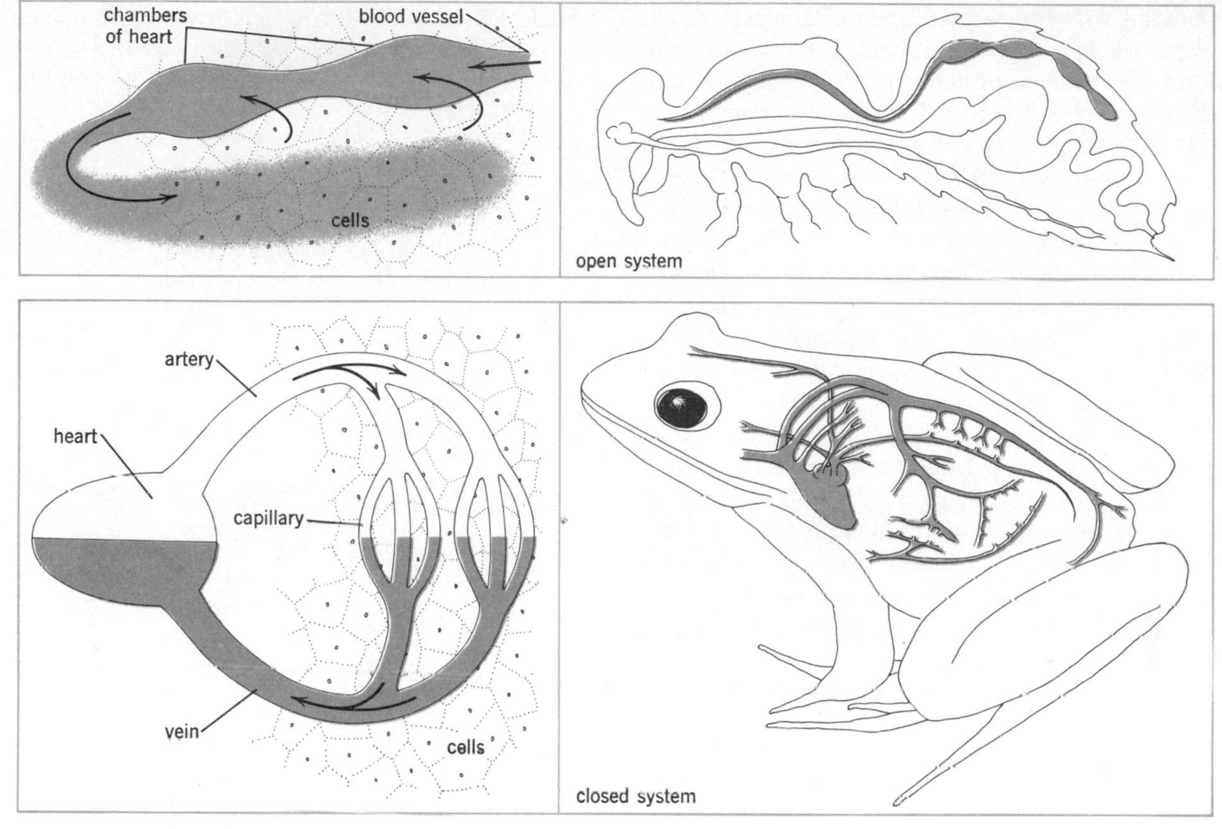

Figure 6.12. The major vessels of open (upper right) and closed (lower right) circulatory systems. The pattern of blood flow in each is shown at left. Notice that the blood in the closed system is always contained in vessels.

tem, however, works satisfactorily for insects and many other invertebrates.

In a *closed system*, the blood circulates in vessels (Figure 6.12 bottom). Consequently, the cells of the body are separated from the blood not only by the walls of the blood vessels but also by a space between the vessels and the cells. It is in this space around the cells, filled with a watery fluid called *lymph*, that exchange of materials between the cells and the blood must occur.

Blood pressure drives all constituents of the blood (except blood cells and most proteins) through the walls of small vessels called *capillaries* into the lymph. By a combination of diffusion and active transport, each cell obtains from the lymph such substances as glucose, amino acids, and oxygen. Cellular waste products such as carbon dioxide and ammonia diffuse into the lymph and then into the capillaries to become part of the venous blood. Not all of the lymph, however, is carried off by the capillaries. Some enters small tubes called lymph vessels found in the spaces between cells.

These small vessels coalesce to form a few large ones that ultimately empty the lymph into the blood near the heart. The lymph vessels constitute a vital part of the total circulatory system because fluid is continually entering the intercellular spaces from capillaries and must be removed. In addition, lymph nodes in the system trap bacteria and other foreign matter, and function in the body's manufacture of antibodies.

The lymphatic system also has another major transport function. In the walls of the small intestine, lymph vessels termed *lacteals* receive fats, especially the long-chain varieties, produced by the preliminary digestion that has occurred in the intestinal cavity. These fats and the lipid called cholesterol are carried to the bloodstream and utilized at various places in the body.

In the closed type of blood system, blood can be

pumped from the heart out into the body under considerable pressure. This pressure is important in forcing substances through capillary walls and into kidney tubules (Chapter X). Another advantage of the closed system is that blood vessels can be restricted or dilated in size so that a greater supply of materials is made available to those cells that need it.

The heart also shows many refinements and adaptations among different groups of animals. The most efficient and powerful is the four-chambered heart of birds and mammals. The advantage of this type is that freshly oxygenated blood from the lungs is kept entirely separate from the deoxygenated blood coming into the heart from the rest of the body. Also, this heart allows blood to be pumped to tissues at a higher pressure. Both of these adaptations are essential to maintaining the metabolism of extremely active creatures like birds and mammals.

The method by which materials are treated and transported depends on the following three characteristics of the organism: (1) its size and shape, (2) its activity rate, and (3) the habitat in which it lives. The structures an organism possesses are not distributed in random fashion but form a coordinated unit that allows the organism to succeed in its environment.

Principles

1. Transport systems are necessary when some of the organism's cells are too distant from the environment to obtain materials by diffusion.
2. The organism's activity, size, and habitat determine the nature of the digestive, respiratory, and transport systems.

Suggested Readings

Biddulph, Susann, and Orlin Biddulph, "The Circulatory System of Plants," *Scientific American,* Vol. 200 (February, 1959). Offprint No. 53, W. H. Freeman and Co., San Francisco.

Brooks, Steward M., *Basic Facts of Body Water and Ions.* Springer Publishing Co., New York, May, 1961.

Comroe, Julius H., Jr., "The Lung," *Scientific American,* Vol. 214 (February, 1966). Offprint No. 1034, W. H. Freeman and Co., San Francisco.

Kylstra, Johannes A., "Experiments in Water Breathing," *Scientific American,* Vol. 219 (August, 1968). Offprint No. 1123, W. H. Freeman and Co., San Francisco.

Mayerson, H. S., "The Lymphatic System," *Scientific American,* Vol. 208 (June, 1963). Offprint No. 158, W. H. Freeman and Co., San Francisco.

Ray, Peter Martin, *The Living Plant.* Prentice-Hall, Englewood Cliffs, N.J., 1963, pp. 44–59, 70–77.

Schmidt-Nielsen, Knut, *Animal Physiology.* Second edition. Prentice-Hall, Englewood Cliffs, N.J., 1964, pp. 1–36.

Wood, J. Edwin, "The Venous System," *Scientific American,* Vol. 218 (January, 1968). Offprint No. 1093, W. H. Freeman and Co., San Francisco.

Zimmermann, Martin H., "How Sap Moves in Trees," *Scientific American,* Vol. 208 (March, 1963). Offprint No. 154, W. H. Freeman and Co., San Francisco.

Zweifach, Benjamin W., "The Microcirculation of the Blood," *Scientific American,* Vol. 200 (January, 1959). Offprint No. 64, W. H. Freeman and Co., San Francisco.

Questions

1. Water dwellers live in an environment less rich in oxygen than do land inhabitants. What is the major problem an organism faces as an air breather in contrast to one that obtains its oxygen from water?
2. Classify the devices used by organisms for extracting oxygen from the environment.
3. Do plants exhibit any specializations for the exchange of respiratory gases? Explain.
4. If an organism ingested a mixture of glycerol, amino acids, and glucose, would digestion be necessary? Why?
5. A tapeworm has neither a digestive tract nor a respiratory system. How does this multicellular organism respire and obtain food materials?
6. Do plants have digestive organs or digestive tracts? Do they carry out the process of digestion? Where?
7. Classify the kinds of digestive systems found in organisms.
8. What advantages may a complete digestive tract have for animals that show high activity rates? Why do large animals show this type of system?
9. Do plants have a transport system? If so, describe it.
10. If xylem consists of dead cells, how can it function?
11. Why should the xylem and phloem not be termed a circulatory system?
12. What functions are served by circulatory systems?
13. What advantages does a closed-type vascular system have over the open type? Which type is characteristic of higher animals, e.g., vertebrates?

CHAPTER VII

Control within Cells

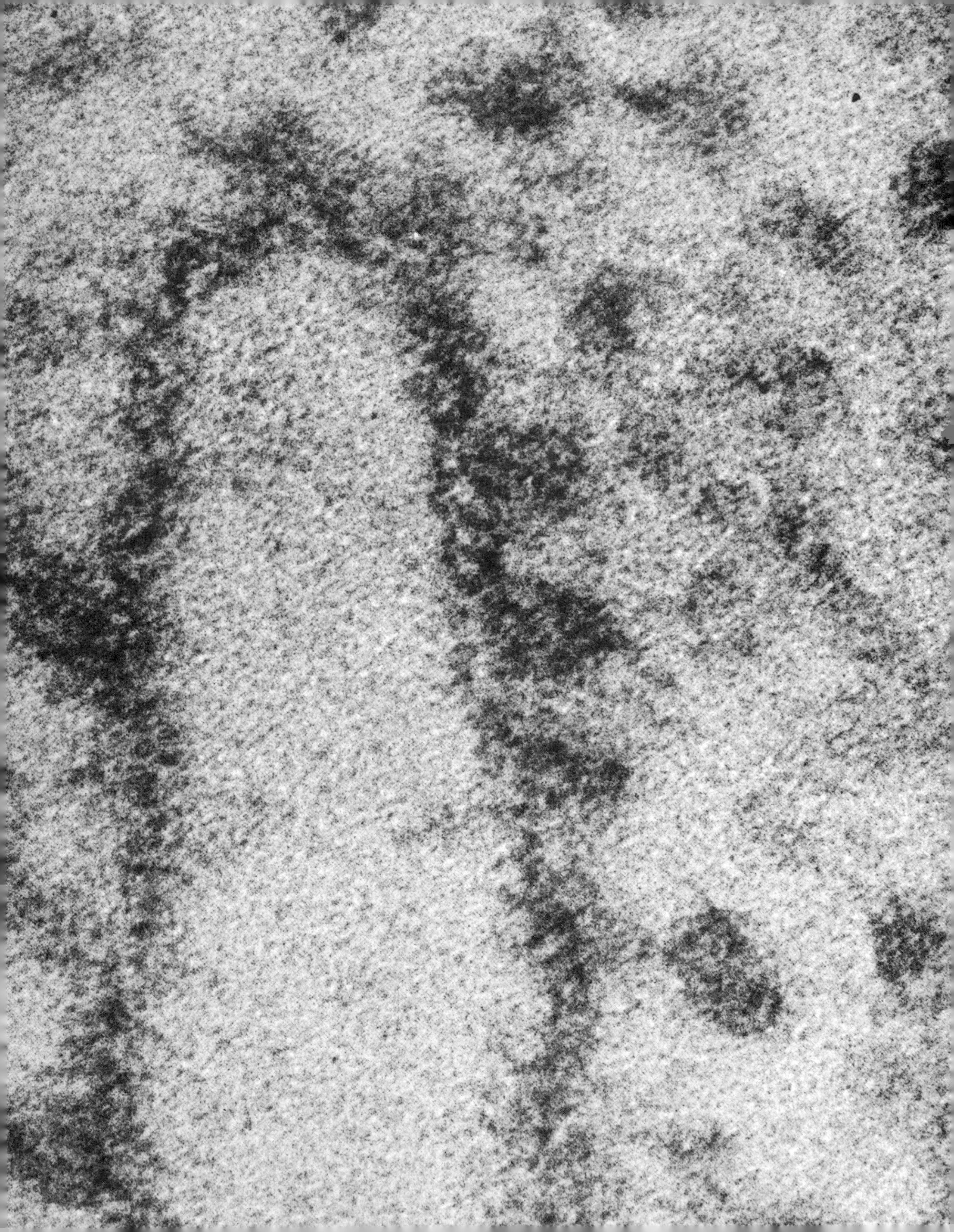

An electron photomicrograph of a small segment of the endoplasmic reticulum (dark loop) with several ribosomes (round bodies) (×150,000) (Dr. F. Sjostrand).

CHAPTER VII

Control within Cells

The cell is the seat of many complex and varied reactions. Those reactions that are part of the basic mechanisms of energy transfer, such as glycolysis and the Krebs cycle, occur with few variations in every living cell. Other reactions result from the particular needs and activities of the cell that are related to its specialization. Such reactions include the syntheses of hormones and special enzymes. What mechanisms or processes control all these reactions so that they are carried out in an orderly fashion? What mechanisms cause the available raw materials to follow one chemical pathway on one occasion and another on some other occasion? What mechanisms account for differences in the activities of individual cells even though the basic energy transfer phenomena are the same? For answers to these questions we must turn to some of the current concepts about control mechanisms within cells.

Law of Mass Action

One common type of control within cells is based on the amounts of interacting substances present at a particular moment. This type of control depends on the operation of the *law of mass action*, which states that the rate of a chemical reaction is proportional to the product of the concentrations of the reacting substances. This means that an increase in the concentration of the reactants increases the rate of the reaction.

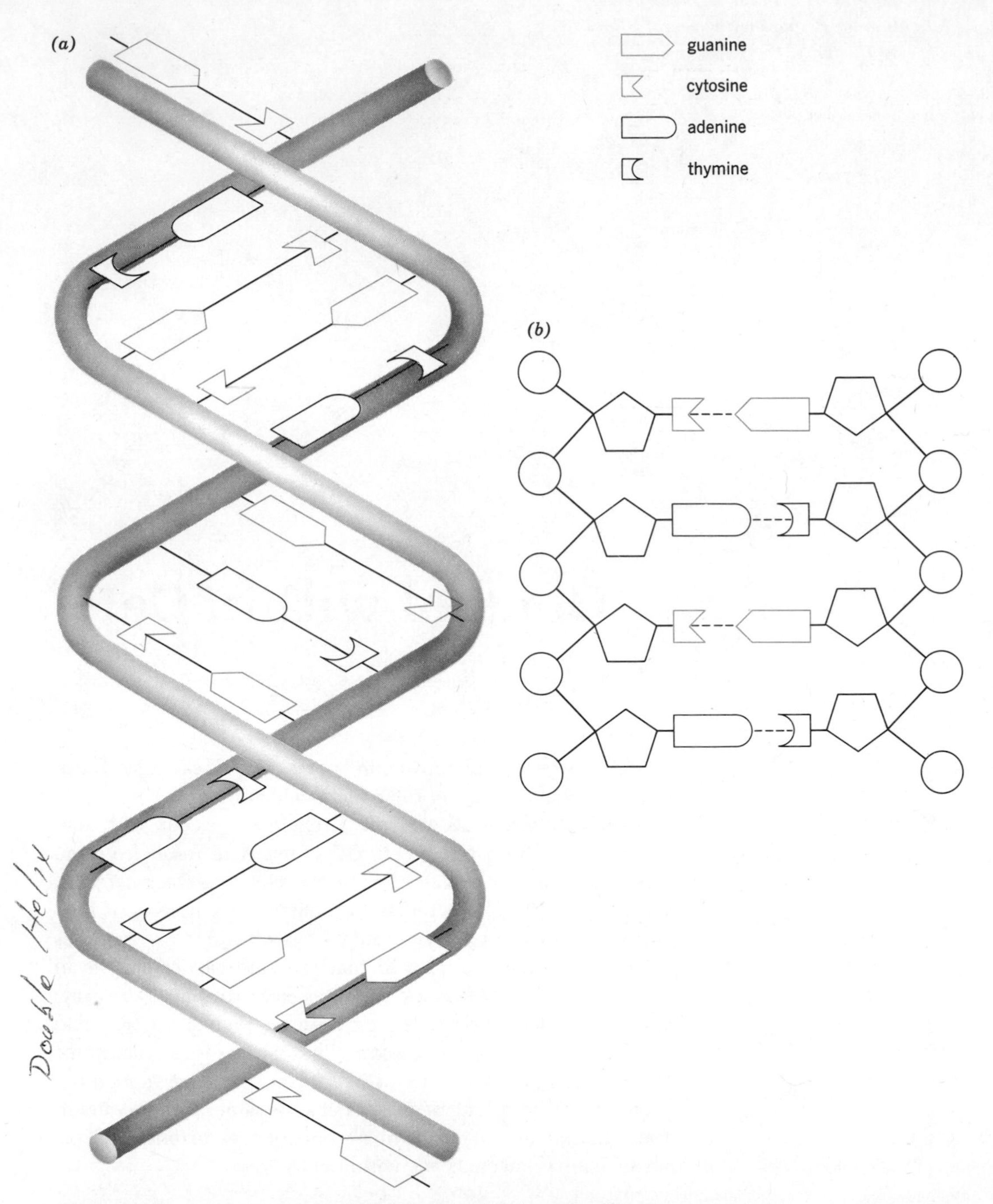

Figure 7.1. The structure of DNA (left). The detail of a small segment is shown at the right. Each pentagon represents a sugar molecule and each circle indicates a phosphate.

It also means that an increase in the concentration of the *end products* of the reaction will slow down the rate of the reaction. An example of this type of control is found in the reactions of cellular respiration (discussed in Chapter V) in which PGAL is converted into pyruvic acid. If pyruvic acid accumulates in the cell, the reaction slows down. As the reaction rate decreases, PGAL tends to accumulate. This again speeds up the reaction, unless, as often happens, some of the PGAL enters other kinds of reactions, such as synthesis into glucose or conversion into glycerol. These reactions are likewise subject to mass action control.

In a cell, this mass action control means that the accumulation of the end products of one reaction favors the occurrence of one or more other reactions. Conversely, the smaller the concentration of end products, the faster the reactions leading to it. This is a highly adaptive feature because the rate at which a substance will be produced is determined in part by the rate at which it is being used.

Reactions leading from PGAL to other products tend to move in one direction only. Why should this be so? Part of the answer depends on the observation that PGAL is continually being formed—from photosynthesis in plants and from the conversion of carbohydrates in animals. This provides a constant source or input of PGAL. In addition, there is a tremendous metabolic demand for PGAL in a variety of reactions. If we call these reactions that drain PGAL from the cell a "sink," then we can speak of the *source-sink phenomenon,* which results from the combination of these two trends. This phenomenon tends to favor reactions leading away from PGAL.

The other part of the answer involves the amount of energy present in molecules before and after a reaction. Every reaction favors the formation of substances containing a smaller amount of energy than the reactants *unless* some force supplies energy to drive the reaction in the opposite direction. In this situation, molecules may be transformed into substances of higher energy content. The transformation of PGAL into glucose and complex carbohydrates such as starch and glycogen is a familiar example.

IMPORTANT SECTION!

Genetic Control

Not all cells show the same energy transfer reactions. Remember the differences between anaerobic respiration in yeast and muscle cells. Why do these differences occur? A simple answer would be that not all cells have the same enzymes and therefore do not show the same reactions. But this answer only begs the question. Why do not all cells have the same enzymes? The answer to this involves a second major type of cellular control, the interaction between the genetic material and the rest of the cell.

DNA Structure. The genetic material of the cell constitutes the basic input of information that determines the activities of the cell. It is composed of very long DNA (deoxyribonucleic acid) molecules surrounded by a protein coat. This DNA-protein complex is located in the nucleus of the cell, as described in Chapter III. The information that controls the cell lies in the structure of the DNA, not in the protein. Some of the evidence for this statement will be given in Chapter XVII.

Every molecule of DNA is comprised of many nucleotides linked in long chains (Figure 7.1). Each nucleotide contains a phosphate group, a five-carbon sugar (deoxyribose), and an organic base (see Figure 7.2). Bonds form between the phosphate group of one nucleotide and the sugar of the next nucleotide. The phosphate of this second nucleotide is bonded to the sugar of the third, and so forth—thus forming a long chain of nucleotide units each bonded from phosphate to sugar.

The bases of the nucleotides extend out from the phosphate-sugar chain. Four kinds of these bases are found in DNA: *adenine, cytosine, guanine,* and *thymine.* Although the four bases are all fairly similar, their sequence in the chains forms a highly significant chemical code. A DNA molecule is composed of two of these chains of nucleotides spiraling around a common center to form a double helix. The bases of one chain are bonded by weak bonds (hydrogen bonds) to the bases of the other chain. These bonds form only between specific pairs of bases. Adenine, in one chain, is always bonded to thymine in the other chain, and cytosine is always bonded to guanine (Figure 7.3). Because of the specific pairing of bases, the sequence of bases in one chain determines the sequence of bases in the other. Therefore, we refer to the two chains as complements of each other. This theory for the structure of DNA was proposed by James Watson, an American biologist, and Francis Crick, an English scientist. For this work they were awarded a Nobel prize in 1962.

DNA Function. DNA controls cell activities by

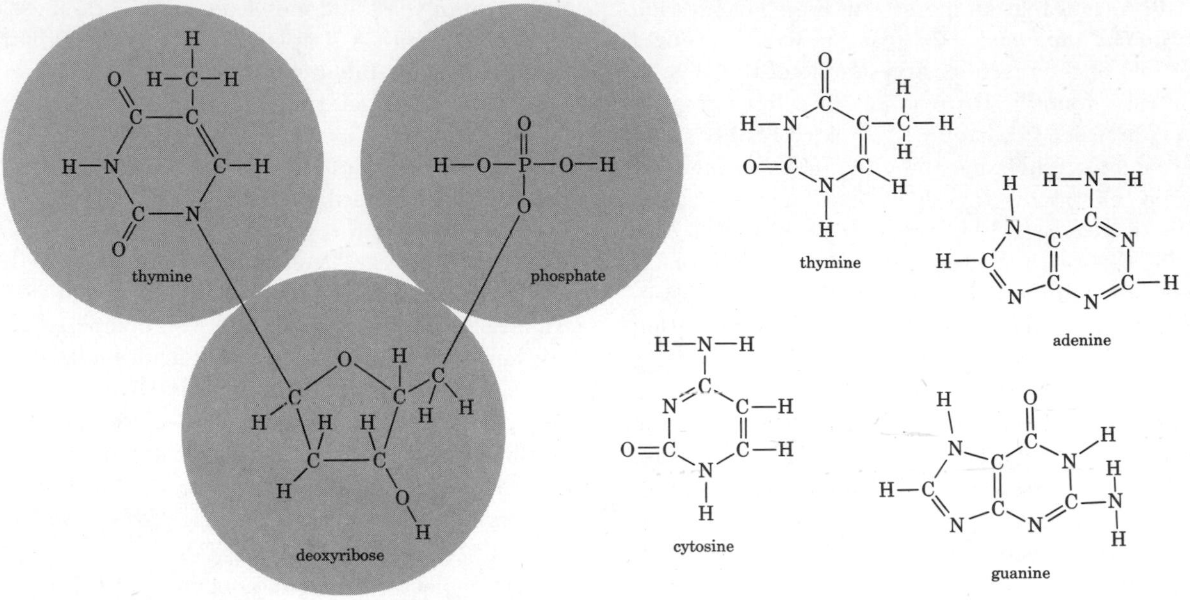

thymine

phosphate

deoxyribose

thymine

adenine

cytosine

guanine

nucleotide: deoxythymidine

bases

Figure 7.2. The structural details of a nucleotide. The four organic bases that occur in DNA are shown on the right.

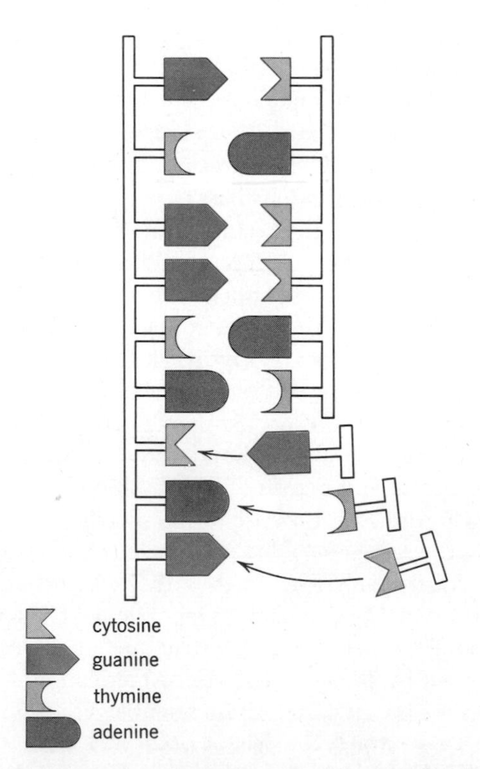

Figure 7.3. Formation of specific pairs of bases in DNA.

directing the synthesis of all proteins in the cell. This control is exercised indirectly, however, since most DNA is confined to the nucleus and most proteins are made in the cytoplasm. The chemical code (sequence of bases) in DNA is transcribed into another molecule called RNA (ribonucleic acid). The RNA then leaves the nucleus and takes its copied code into the rest of the cell. This RNA is appropriately called *messenger* RNA.

If DNA is to serve as a pattern for RNA, the two nucleotide chains of the DNA helix must separate from each other along part of the molecule. One or both chains can then serve as a pattern for making an RNA chain. The RNA chain is made by matching ribose-containing nucleotides (from a supply in the nucleus) with the appropriate bases on the DNA chain. In this event, cytosine pairs with guanine and thymine (in the DNA) with adenine. However, in RNA, uracil is used in place of thymine, so that the DNA adenine must pair with uracil in making RNA. To complete the RNA molecule, adjacent RNA nucleotides must be linked by phosphate-to-sugar bonds (Figure 7.4). This RNA synthesis requires ATP energy and is controlled by an enzyme, RNA polymerase, found in the nucleus.

After the RNA is completed, it breaks its transitory union with DNA and moves through the nuclear membrane into the cytoplasm by processes that are not yet understood.

Messenger RNA. Messenger RNA (mRNA) is a long, single-stranded molecule containing the nucleotides adenine, guanine, cytosine, and uracil. The sugar present in RNA (ribose) is only slightly different from the deoxyribose present in DNA.

Because mRNA was synthesized as a complementary copy of a segment of DNA (Figure 7.5), the mRNA contains in coded form the same information for making proteins. The function of mRNA is to attach to one or more ribosomes in the cytoplasm and become the pattern on which a protein can be put together from amino acids. Messenger RNA is unstable and extremely variable in size. Presumably the length of the molecule relates to the size of the protein it is to synthesize.

The specific nature of the instructions carried by mRNA has been clearly demonstrated by experiments with immature red blood cells, which are mainly concerned with producing hemoglobin. Using an ultracentrifuge, experimenters separated the messenger RNA and ribosomes from the other parts of the cells and placed them in an environment containing ATP, another variety of RNA (called transfer RNA), and amino acids. The mRNA continued to synthesize hemoglobin from amino acids—but only hemoglobin and no other proteins. This is evidence that a specific mRNA controls the production of each kind of protein.

The genetic code transcribed into mRNA from DNA is evidently a triplet code, that is, the bases on mRNA are read in groups of three. Each group specifies an amino acid. A simple calculation tells us that the four bases yield 64 possible triplet codes.

Transfer RNA. Transfer RNA, formed under the direction of a small portion of DNA, consists of a single strand of nucleotides that is much shorter than

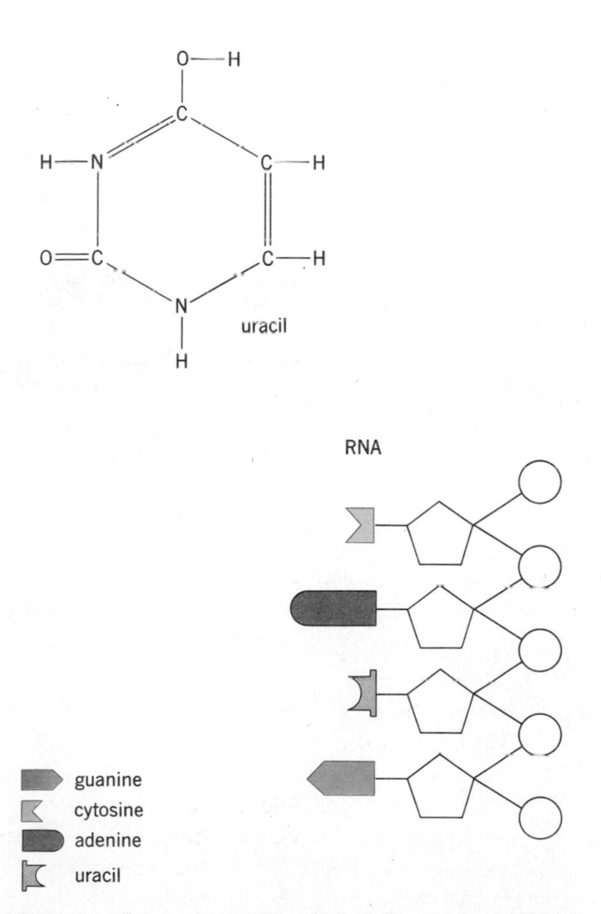

Figure 7.4. The structure of uracil and of a segment of RNA.

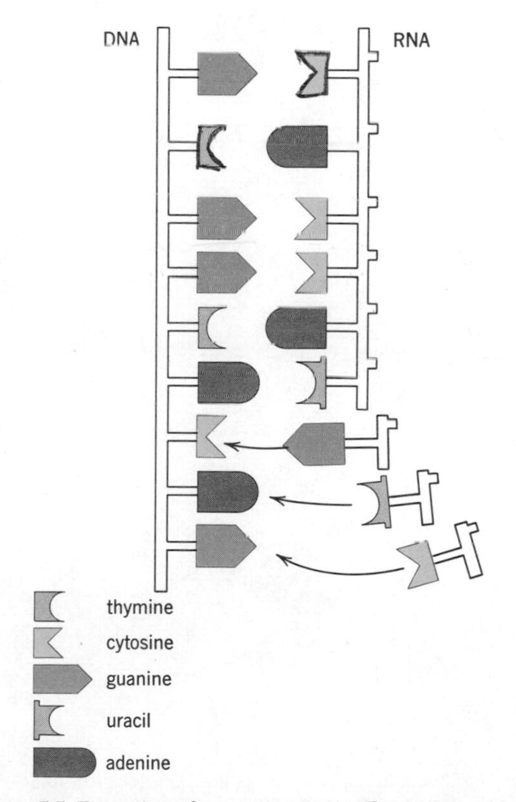

Figure 7.5. Formation of messenger RNA. The pattern of bases in messenger RNA is controlled by the sequence of bases in the DNA molecule.

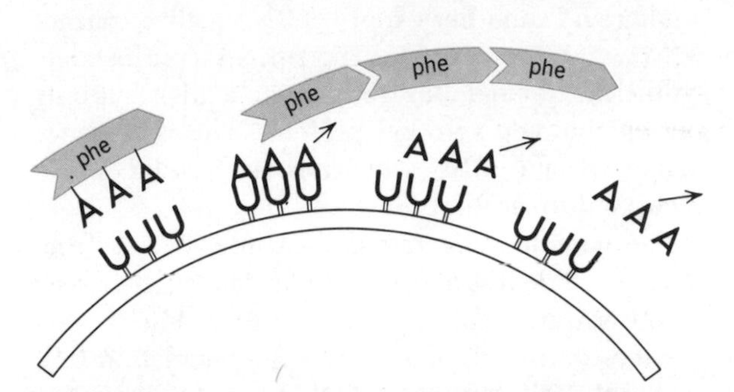

Figure 7.6. Results of an experiment using a messenger RNA containing only uracil (U). Only the amino acid phenylalanine (blue symbols) will be incorporated into protein. Each set of three As represents a transfer RNA.

mRNA. In addition, the strand is coiled about itself. Its function is to transport amino acids from the cytoplasm to the ribosomal site of an mRNA molecule; hence the term *transfer* RNA. Each tRNA is apparently coded for a particular amino acid by means of three nitrogen bases at one end of the molecule. For example, three consecutive adenines at one end, AAA, specifies the amino acid phenylalanine (Figure 7.6). A tRNA with AAA at one end will combine with and transport only this amino acid. In addition, the triplet on the tRNA will only match the correct triplet of complementary bases on the mRNA (in this instance, uracil, uracil, uracil). The tRNA, then, serves two functions: specifying an amino acid and identifying a location on the mRNA.

Since there are twenty amino acids and 64 triplet codes, some of the amino acids must be represented by more than one code. Perhaps this adds needed flexibility to the coding system. Two of the triplets do not appear to recognize any of the amino acids.

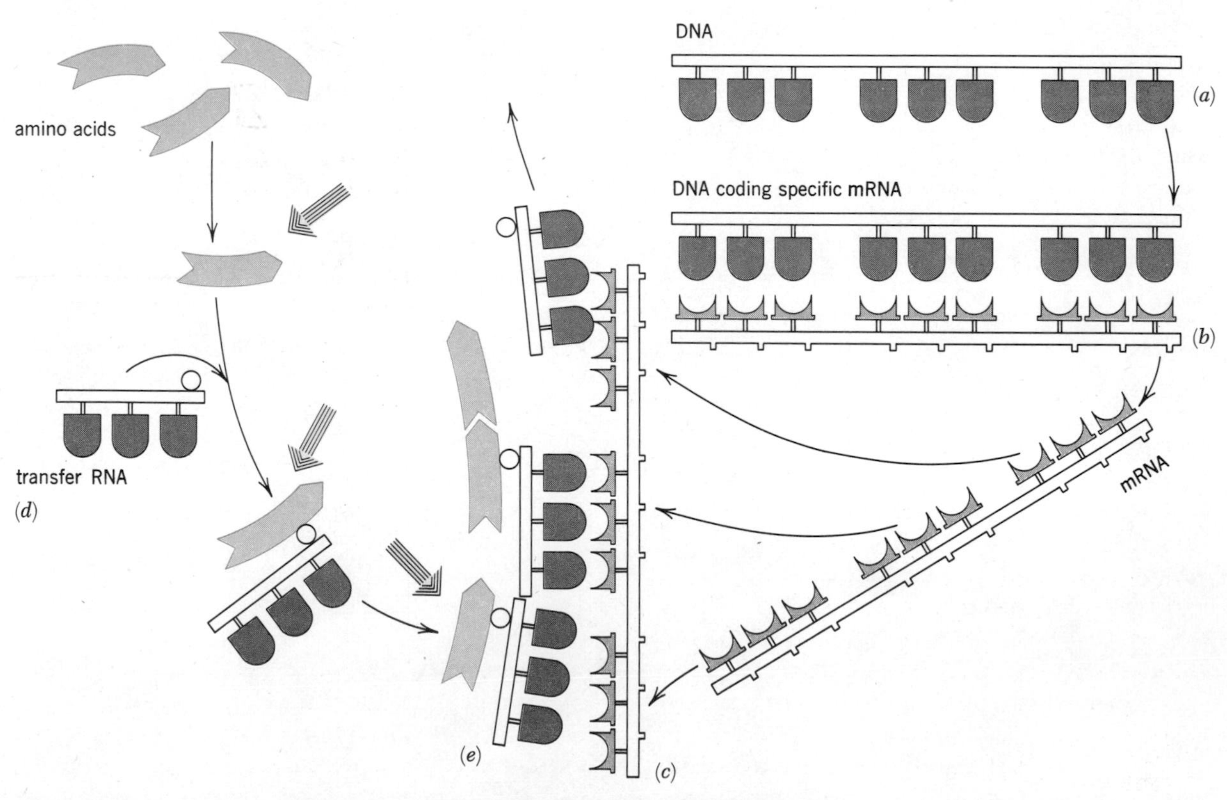

Figure 7.7. Summary of protein synthesis. (a) *The information for the formation of a protein molecule resides in DNA.* (b) *Transmission of information to mRNA.* (c) *mRNA moves into the cytoplasm.* (d) *Transfer RNA accepts a specific amino acid.* (e) *Transfer RNA-amino acid complex comes in contact with the messenger RNA and protein is formed.*

They may serve as "punctuation" codes for ending the protein synthesis.

Ribosomes—Sites of Protein Synthesis. Ribosomes have been known for some time to be the site of protein synthesis. Visible only with an electron microscope, they are nevertheless abundant in all cells. They are spherical and composed of about equal amounts of RNA and protein.

Ribosomes function in small groups as if they were machines on an assembly line. One after another, the members of a group attach to a single messenger RNA. Moving in only one direction along the mRNA, the ribosomes evidently play a part in attaching tRNAs to appropriate sites on the mRNA. Ribosomal action has been compared to that of the slider on a zipper. In some manner, ribosomes are able to "read" the code of the mRNA as they pass along it. When the protein is completed, the ribosomes leave the mRNA and also liberate the protein. It takes roughly one minute for ribosomes to assemble a hemoglobin molecule in this way. This model of ribosome function has not been fully confirmed, although experimental evidence supports the general idea.

Protein Synthesis Summary. The directions for making proteins reside in DNA in a chemical code using four nucleotides. In a process involving enzymes and the expenditure of energy, the code is copied as a template in mRNA, which transmits the code into the cytoplasm of the cell (Figure 7.7). There, mRNA associates with a group of ribosomes. Under the direction of enzymes, tRNAs pick up specific kinds of amino acids from a supply that is free in the cytoplasm. This tRNA-amino acid combination is a reactive substance because some of the energy used in combining the two components remains in the molecule. The tRNAs move the amino acids to the locality of an mRNA-ribosome complex. With the aid of ribosomes, the amino acids are linked into a protein. In other words, the code in the mRNA has been translated into a protein. The protein is released from both tRNA and mRNA. This frees both kinds of RNA to repeat their functions.

The deciphering of the genetic code has been a major area of research since about 1960. In order to identify the meaning of the 64 possible triplets in mRNA, investigators like Marshall Nirenberg and his associates at the National Institutes of Health devised many ingenious experiments.

Nirenberg's group began by synthesizing artificial mRNAs composed of only three nucleotides. In this system, the order of the nucleotides may be controlled so that only one kind of mRNA is produced in each case. Each type of artificial mRNA was then mixed with ribosomes by grinding up colon bacilli (*Escherichia coli*) and then separating the various structures of the cell with an ultracentrifuge (Figure 7.8).

In the presence of an energy source and the proper enzymes, transfer RNAs were then mixed with amino acids, one of which had been labeled with carbon 14. This mixture was then added to the mRNA-ribosome solution. After allowing a short time for the reactions to occur, the entire contents were poured through a cellulose nitrate filter that retained the ribosomes and anything attached to them, but allowed the other constituents to pass through.

After several rinses, the filter was examined for radioactivity. If the artificial mRNA had been coded for the amino acid that contained the C^{14} label, the filter retained a marked amount of radioactive material. If the mRNA was coded for some other, virtually no radioactivity was found there.

By tireless repetition of this basic technique, Nirenberg's group has identified the meaning of nearly all possible triplets. Figure 7.8 shows some of these.

Proteins and Cell Activity

It should be clear from the foregoing description that different DNA base sequences produce different kinds of proteins. This reveals part of the reason why all cells do not produce the same enzymes (proteins) or, consequently, show the same chemical activities. But we also need to examine the roles that proteins play in the cell if we are to answer this question completely.

Proteins function in three different ways in the cell: some as enzymes, some as structural proteins, and some as antibodies. We shall consider each class in turn.

Enzymes catalyze all cellular reactions, which makes them extremely important in cell activity. As mentioned in Chapter II, all enzymes are at least partly protein. The protein portion is often responsible for determining whether the enzyme will affect a certain chemical reaction. An example is the enzyme, lysozyme, which dissolves bacterial cells by digesting a complex sugar located in the cell wall.

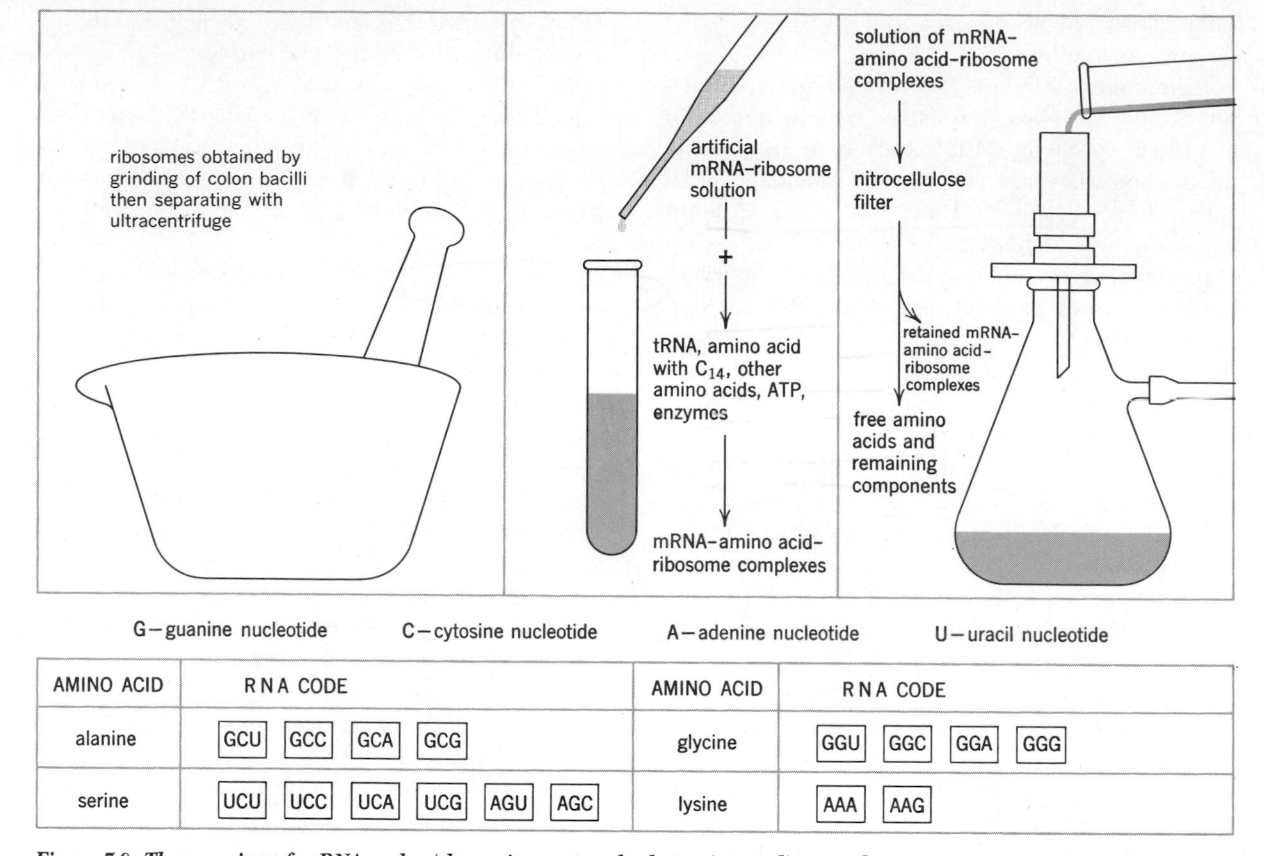

AMINO ACID	RNA CODE	AMINO ACID	RNA CODE
alanine	GCU GCC GCA GCG	glycine	GGU GGC GGA GGG
serine	UCU UCC UCA UCG AGU AGC	lysine	AAA AAG

Figure 7.8. The meaning of mRNA codes (shown in part at the bottom) was discovered by Nirenberg's group by the methods depicted at the top. The amino acid encoded by the artificial mRNA was retained by the filter.

This enzyme contains 129 amino acid subunits in a chain that folds upon itself in a three-dimensional form. The sugar fits into a pocket in the structure of the enzyme.

In the chain of events leading from DNA to protein, an alteration in any of the steps may result in a different enzyme being produced. A change in enzymes, of course, changes the activity of the cell.

Structural proteins are major constituents of skeletal and muscle tissue. Also, all membranous structures contain a structural protein component. How this type of protein affects cell activity is illustrated in the following example. Muscle cells contain two specific proteins called *actin* and *myosin*. If the cell is prevented from producing either one, it cannot contract. Similarly, modification of a protein that plays a role in any structure of a cell may modify

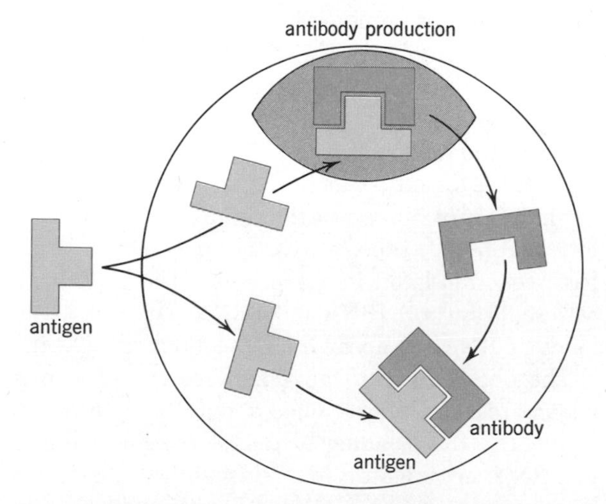

Figure 7.9. When an antigen enters the body it stimulates antibody production. The antibodies are then available to interact with subsequent antigen molecules (below).

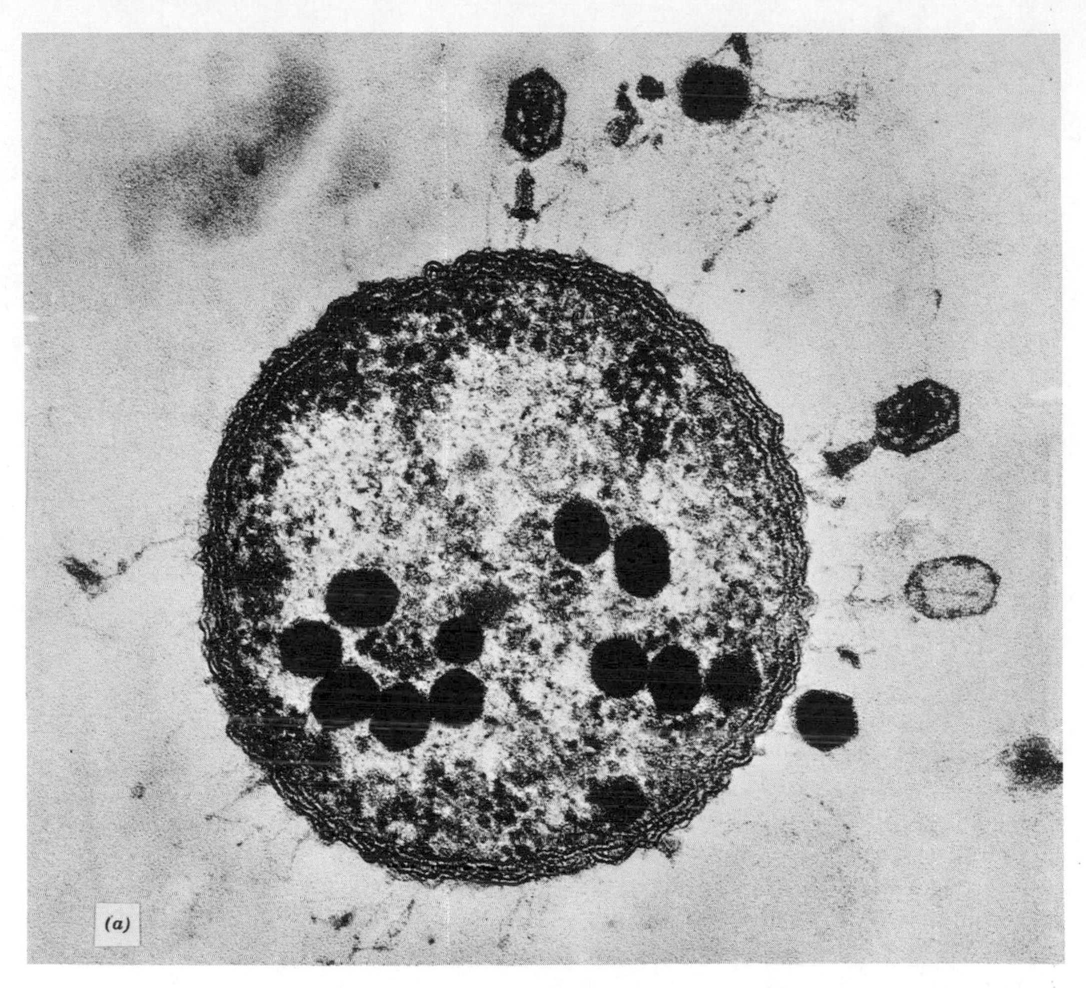

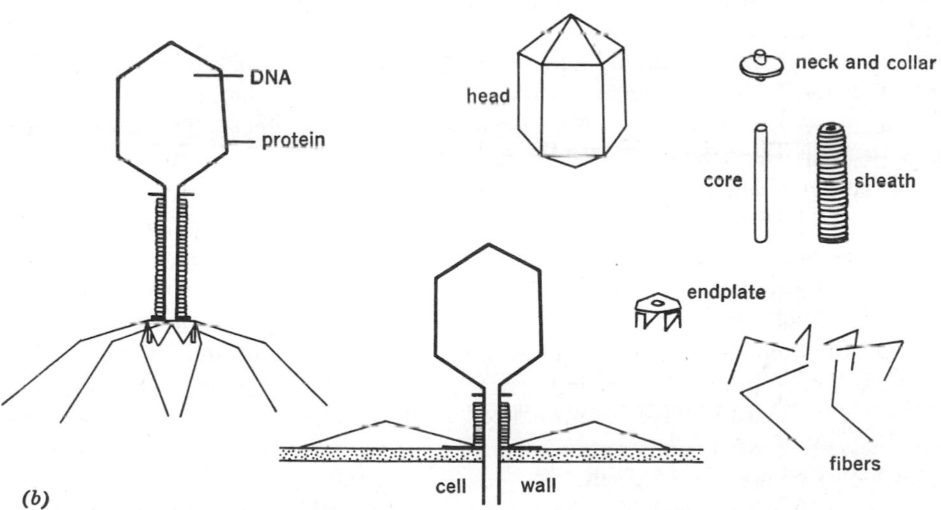

Figure 7.10. The photomicrograph shows a section of a bacterial cell from a culture infected with a virus. Clearly visible within the cell are condensed viral DNA cores (black hexagons). Attached to the cell surface are three viruses, one empty and two still partially filled. The virus at the top shows long tail fibers and the spikes of the tail plate in contact with the cell wall. Below is a diagram illustrating the parts of the virus and its attachment to the bacterial wall (center) (×150,000) (Dr. L. Simon).

the activity of that cell. Think of the importance a change in one of the proteins involved in membrane structure might have for many of a cell's components.

Antigens and Antibodies. When a new protein, a large carbohydrate, or a bacterium containing these molecules enters an organism, the organism's cells react by producing an antibody to combat the foreign material (called an antigen) (Figure 7.9). This antigen-antibody reaction plays a major role in defense against disease. If the proper antibodies are present, they may completely inactivate the foreign organism and prevent any disease from developing.

Antibodies are highly specific for the antigen that they deactivate. In some way, not yet clarified, a cell's protein-making machinery recognizes the foreign body, then synthesizes the particular protein that will inactivate it. A surprising aspect of antibodies is that all are identical by present-day chemical and physical tests. They can only be isolated or separated from one another by observing their action with specific antigens.

This antigen-antibody reaction may occasionally malfunction. In one type of human leukemia, for instance, a change in the DNA of developing white blood cells causes them to produce antibodies against red blood cells. This causes mass destruction of red blood cells and the consequent extreme anemia that is a symptom of many leukemias.

Viral Effects. The changes in cell activity brought about by viral diseases can be understood in the context of the DNA-RNA-protein synthesis scheme of control. A virus is a DNA or RNA molecule with a protein coat (Figure 7.10). When a virus infects a bacterial cell, only its DNA (or RNA) actually enters the cell (Figure 7.11). The additional nucleic acid acts like new genetic information, causing the cell to produce new viruses (Figure 7.12). The usual synthetic and energy-supply mechanisms are all diverted from maintenance and growth of the cell to production of viruses. After a large number of new viruses have been produced in the cell, the cell breaks open, releasing viral particles capable of invading other cells.

To exert this effect on the cell, the virus must cause the production of its own messenger RNA. This RNA must be able to carry out synthetic reactions at a higher rate than the host cell's own RNA. Experiments with the messenger RNA from one particular virus show that this RNA does direct the

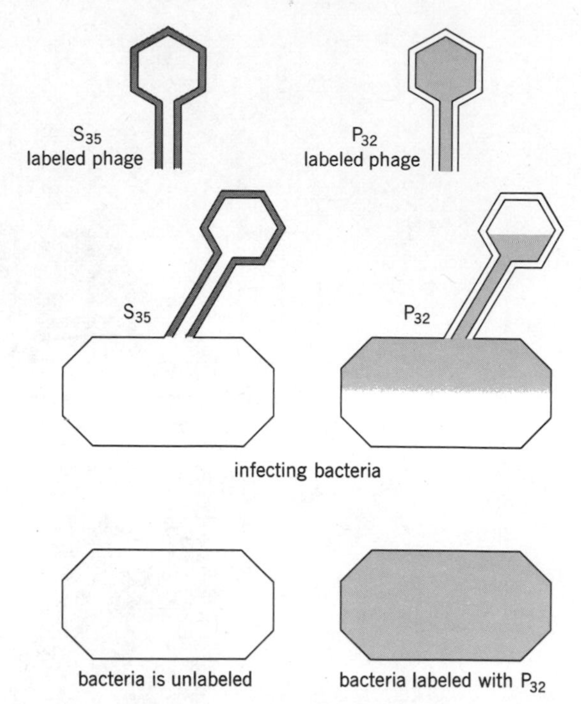

Figure 7.11. An experiment showing that only the nucleic acid enters the bacterial cell. The protein of a viral coat contains sulfur but no phosphorus whereas the nucleic acid contains phosphorus but no sulfur. When the virus is labeled with radioactive sulfur (S_{35}) and allowed to infect bacteria, the bacteria show no radioactivity. When the virus is labeled with radioactive phosphorus (P_{32}), the infected bacteria show radioactivity.

incorporation of amino acids into protein much more rapidly than the RNA from the bacteria that the virus infects. These experiments have been carried out using bacteria infected by viruses. However, evidence indicates that viral diseases in humans involve similar phenomena.

Cancer. In cancer, the cells grow and divide without apparent control. This brings them into sharp conflict with the rest of the organism's cells, which are under normal control. Although this pattern is very different from the pattern of viral infection we have just discussed, the cause of many sorts of cancers is known to be a virus. Even leukemia may fall into this category.

In viral cancers, a partial takeover by a new nucleic acid control system occurs. This is similar to the events following ordinary viral infections. It is interesting to consider the effects of this system on cell activity. Familiar effects of cancer are increased protein synthesis resulting in more rapid cell growth

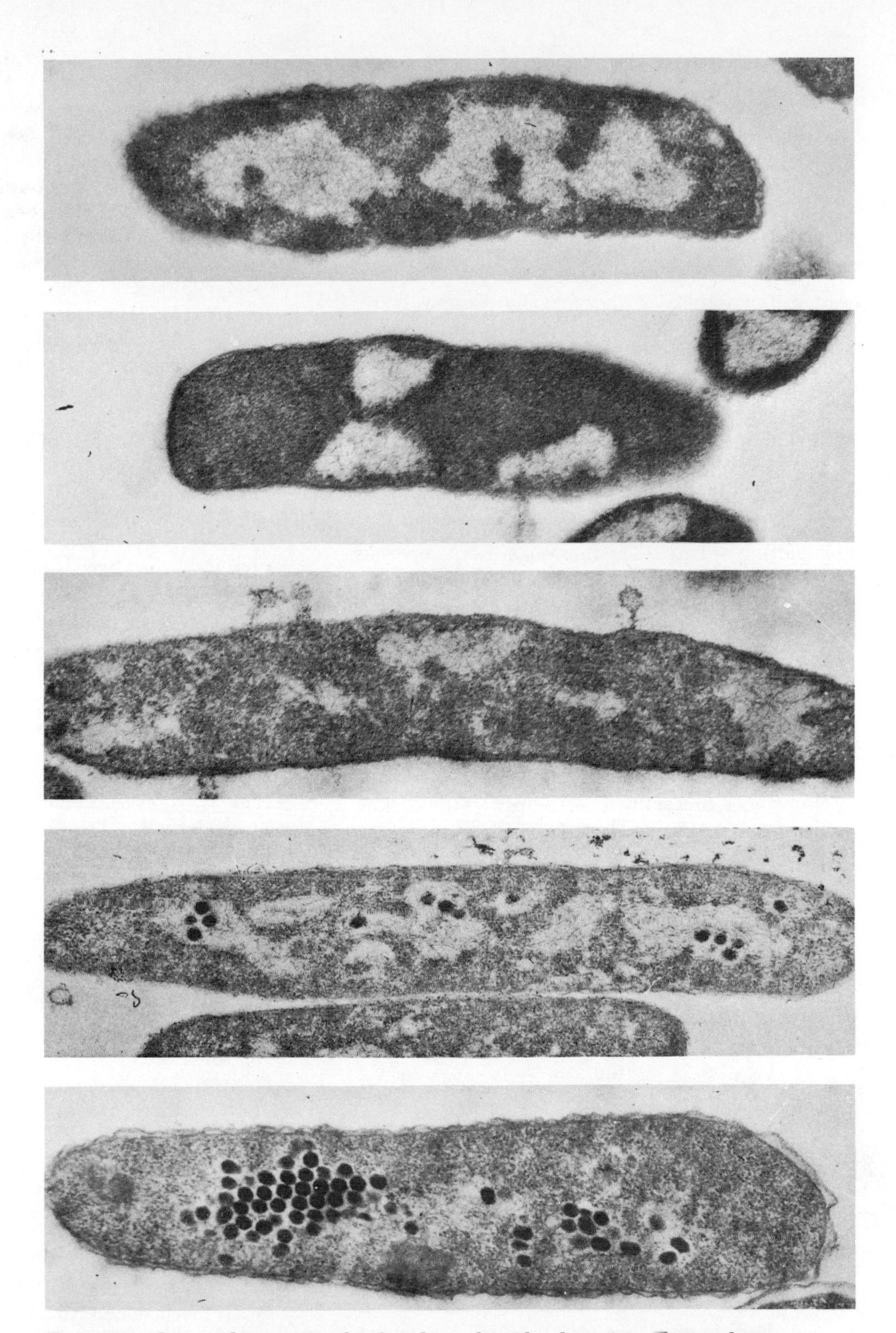

Figure 7.12. Electron photomicrographs of viral growth inside a bacterium. The top photograph shows a colon bacillus before infection. The second shows vacuoles forming along the cell wall four minutes following infection. The third photograph, taken ten minutes after infection, shows the protein coats of the phage on the bacterial surface. The cellular interior has been reorganized by the virus. In the fourth photograph (twelve minutes after infection) several new virus components have appeared. In the bottom photograph (thirty minutes after infection) many viruses are visible. These will soon be liberated by rupture of the cell (Dr. E. Kellenberger).

and division, a higher metabolic rate, and lack of coordination with surrounding cells. One of the most instructive examples of such change may be found in some forms of cancer cells. It was observed that nearly all respiration in these cells occurred anaerobically, regardless of the amount of oxygen available. Later work showed that there was a marked shift in the activity of the enzymes present. Here, a change in the control of protein production resulted in a modification in enzyme activity and in abnormal metabolism in the cell.

In summary, two very different types of control exist within cells. One type is genetic, and determines what the cell will do within fairly narrow limits. The information contained in triplet code in DNA is translated into messenger RNA by the specific pairing of bases. The messenger RNA serves as a pattern that is translated into a specific kind of protein by the interaction of a triplet of bases in each transfer RNA with those in the messenger RNA.

In the other type of control, changes in the environment or changed conditions within the cell modify the cell's activity within these genetically imposed limits. Mass-action phenomena present one example of the modifications by which the amount of a substance present in a cell helps determine how much of that substance is produced. In addition, the functions of other cells modify cell activity. Coordinated activity among many cells may result.

Principles

1. The relative concentrations of materials in a cell affect the rate and direction of metabolism.
2. DNA exercises the ultimate endogenous control over cell activities.
3. DNA's control is expressed by the synthesis of specific proteins. This control is mediated by messenger and transfer RNA and the ribosomes.

Suggested Readings

Beadle, George W., "Structure of the Genetic Material and the Concept of the Gene," in *This Is Life*, edited by W. H. Johnson and W. C. Steere. Holt, Rinehart and Winston, New York, 1962, pp. 185–211.

Crick, F. H. C., "The Structure of the Hereditary Material," *Scientific American*, Vol. 191 (October, 1954). Offprint No. 5, W. H. Freeman and Co., San Francisco.

Crick, F. H. C., "The Genetic Code," *Scientific American*, Vol. 207 (October, 1962). Offprint No. 123, W. H. Freeman and Co., San Francisco.

Crick, F. H. C., "The Genetic Code: III," *Scientific American*, Vol. 215 (October, 1966). Offprint No. 1052, W. H. Freeman and Co., San Francisco.

Dubecco, Renato, "The Induction of Cancer by Viruses," *Scientific American*, Vol. 216 (April, 1967). Offprint No. 1069, W. H. Freeman and Co., San Francisco.

Gorini, Luigi, "Antibiotics and the Genetic Code," *Scientific American*, Vol. 214 (April, 1966). Offprint No. 1041, W. H. Freeman and Co., San Francisco.

Holley, Robert W., "The Nucleotide Sequence of a Nucleic Acid," *Scientific American*, Vol. 214 (February, 1966). Offprint No. 1033, W. H. Freeman and Co., San Francisco.

Horsfall, Frank L., Jr., "Some Facts and Fancies about Cancer," *Perspectives in Biology and Medicine*, Vol. VIII, No. 2: 167–179. 1964.

Hurwitz, Jerard, and J. J. Furth, "Messenger RNA," *Scientific American*, Vol. 206 (February, 1962). Offprint No. 119, W. H. Freeman and Co., San Francisco.

Ingram, Vernon M., "How Do Genes Act?" *Scientific American*, Vol. 198 (January, 1958). Offprint No. 104, W. H. Freeman and Co., San Francisco.

Kornberg, Arthur, "The Synthesis of DNA," *Scientific American*, Vol. 219 (October, 1968). Offprint No. 1124, W. H. Freeman and Co., San Francisco.

Menninger, John R., "Transfer RNA," *Science Journal*, Vol. 1 (October, 1965).

Nirenberg, Marshall W., "The Genetic Code: I," *Scientific American*, Vol. 215 (October, 1966). Offprint No. 1052, W. H. Freeman and Co., San Francisco.

Nossal, G. J. V., "How Cells Make Antibodies," *Scientific American*, Vol. 211 (December, 1964). Offprint No. 199, W. H. Freeman and Co., San Francisco.

Rich, Alexander, "Polyribosomes," *Scientific American*, Vol. 209 (December, 1963). Offprint No. 171, W. H. Freeman and Co., San Francisco.

Rous, Peyton, "The Challenge to Man of the Neoplastic Cell," *Science*, Vol. 157, No. 3784: 24–28 (July 7, 1967).

Speirs, Robert S., "How Cells Attack Antigens," *Scientific American*, Vol. 210 (February, 1964). Offprint No. 176, W. H. Freeman and Co., San Francisco.

Spiegelman, S., "Hybrid Nucleic Acids," *Scientific American*, Vol. 210 (May, 1964). Offprint No. 183, W. H. Freeman and Co., San Francisco.

Wood, William B., and R. S. Edgar, "Building a Bacterial Virus," *Scientific American*, Vol. 217 (July, 1967). Offprint No. 1079, W. H. Freeman and Co., San Francisco.

Questions

1. Why is the analogy of source-sink an appropriate one for the law of mass action?
2. In what way is a DNA molecule analogous to a twisted ladder?
3. In what chemical sense is DNA a monotonous molecule? (Compare to protein.)
4. If you determined the amount of one of the bases, e.g., adenine, in a DNA sample, what would you know about the other bases?
5. The chapter specifies that there are sixty-four possible forms of transfer RNA. See if you can determine how this number was derived.
6. Trace the steps involved in the translation of a DNA code in the nucleus into an enzyme out in the cytoplasm.
7. Ribosomes are usually abundant in cells. Why is this necessary? Would you term this condition an adaptation?
8. Why did Nirenberg's group use mRNAs only three bases in length rather than ones about the size of natural ones?
9. What are three different kinds of proteins?
10. How does knowledge of the structure of DNA aid researchers in understanding the activity of viruses?
11. What is the consequence, in general terms, of altering a DNA or RNA "code" in a cell?
12. If DNA controls cellular activities, does the cellular environment play any role? Why?

CHAPTER VIII

Control by Chemical Agents

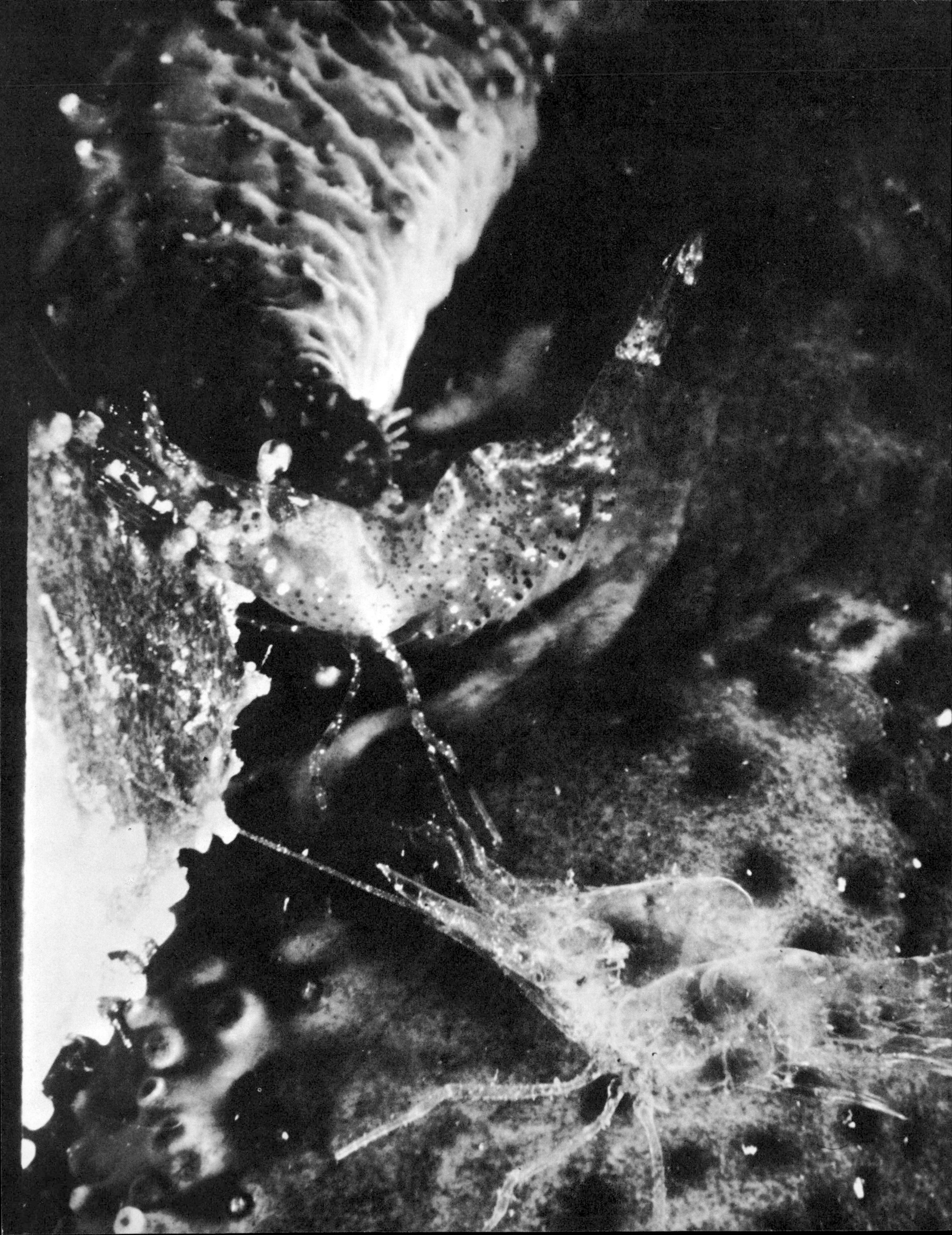

A recently molted shrimp (center). The old exoskeleton is in the lower portion of the photograph (Dr. K. R. H. Read).

CHAPTER

VIII

Control by Chemical Agents

In observing the ways a cell controls its many activities we learn a number of things about how its reactions are kept in harmony. Recall the types of control involving DNA, RNA, protein synthesis, and mass action phenomena discussed in the previous chapter. This type of regulation appears in all cells.

Another type of control appears when the masses of cells comprising a multicellular plant or animal must be coordinated in their actions. This coordination may be achieved in two ways: through chemical agents (hormones and related chemicals) and through nerve impulses.

Chemical agents that play a role in coordination show remarkable diversity in both their chemical nature and their mode of activity. Some are hormones, that is, chemicals produced by a cell or group of specialized cells and bringing about coordination elsewhere in the body. Many coordinators, however, do not fit the definition of a hormone. Some are clearly not hormones, although others are close to being so. For example, CO_2 is produced by all cells but has a regulatory function in controlling the breathing rate in mammals. Also, many organisms release hormonelike substances into the environment and influence the behavior of other organisms (see Chapter XI).

Before considering specific kinds of hormones and chemical agents, it may be helpful to review the major functions they serve:

1. Regulation of the internal environment of organisms. This includes many aspects of the treatment and transport of materials as described in Chapter VI, cellular respiration, regulation of internal fluids, and even the secretion of other hormones.
2. Regulation of growth, development, and specialization of tissues in both plants and animals.
3. Regulation of reproductive cycles.
4. Regulation of behavior.

Hormones in Animals

More is known about hormones in mammals than in other animals, mostly because there has been interest in the functioning of hormones in humans. In fact, the term *hormone* was proposed by E. H. Starling in 1905 to describe a chemical that stimulated the mammalian pancreas to secrete digestive juices. Enough information has accumulated over the years to make possible detailed descriptions of hormones in man, as shown in Table 8.1 (see also Figure 8.1). Other vertebrate groups, like birds and fishes, are known to have hormonal systems like man's, but much less is known about them. Generally, we can say that lower vertebrate groups like fishes have less well-defined endocrine organs (hormone-producing structures) than birds or mammals. However, the

Table 8.1. Major Endocrine Tissues and Hormones of Man

Gland or Tissue	Hormone	Major Function of Hormone
Thyroid	Thyroxine	Stimulates rate of oxidative metabolism and regulates general growth and development.
	Thyrocalcitonin	Lowers the level of calcium in the blood.
Parathyroid	Parathormone	Regulates the levels of calcium and phosphorus in the blood.
Pancreas (Islets of Langerhans)	Insulin	Decreases blood glucose level by promoting storage in the liver.
	Glucagon	Elevates blood glucose level.
Adrenal medulla	Epinephrine (adrenalin)	Various "emergency" effects on blood, muscle, temperature.
Adrenal cortex	Cortisone and related hormones	Control carbohydrate, protein, mineral, salt, and water metabolism.
Anterior pituitary	1. Thyrotropic	1. Stimulates thyroid gland functions.
	2. Adenocorticotropic	2. Stimulates development and secretion of adrenal cortex.
	3. Growth hormone	3. Stimulates body weight and rate of growth of skeleton.
	4. Gonadotropic (2 hormones)	4. Stimulate gonads.
	5. Prolactin	5. Stimulates lactation.
Posterior pituitary	1. Oxytocin	1. Causes contraction of some smooth muscle.
	2. Vasopressin	2. Inhibits excretion of water from the body by way of urine.
Ovary (follicle)	Estrogen	Influences development of sex organs and female characteristics.
Ovary (corpus luteum)	Progesterone	Influences menstrual cycle, prepares uterus for pregnancy; maintains pregnancy.
Uterus (placenta)	Estrogen and progesterone	Function in maintenance of pregnancy.
Testis	Androgens (testosterone)	Responsible for development and maintenance of sex organs and secondary male characteristics.
Digestive system	Several gastrointestinal	Integration of digestive processes.

Modified from A. Nason, *Modern Biology*, 1965, John Wiley and Sons, New York.

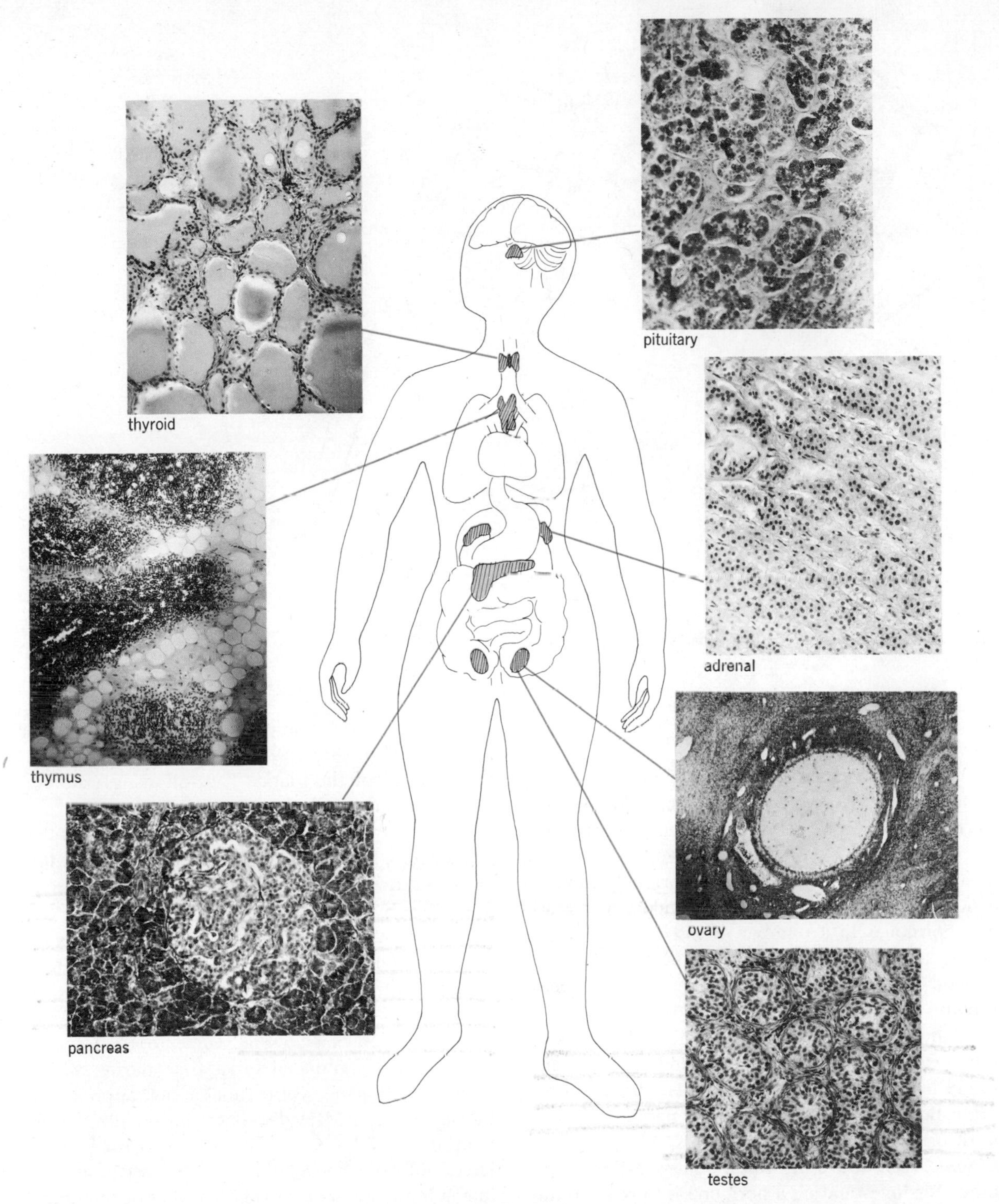

Figure 8.1. The location and microscopic structure of glands that secrete some important hormones in man.

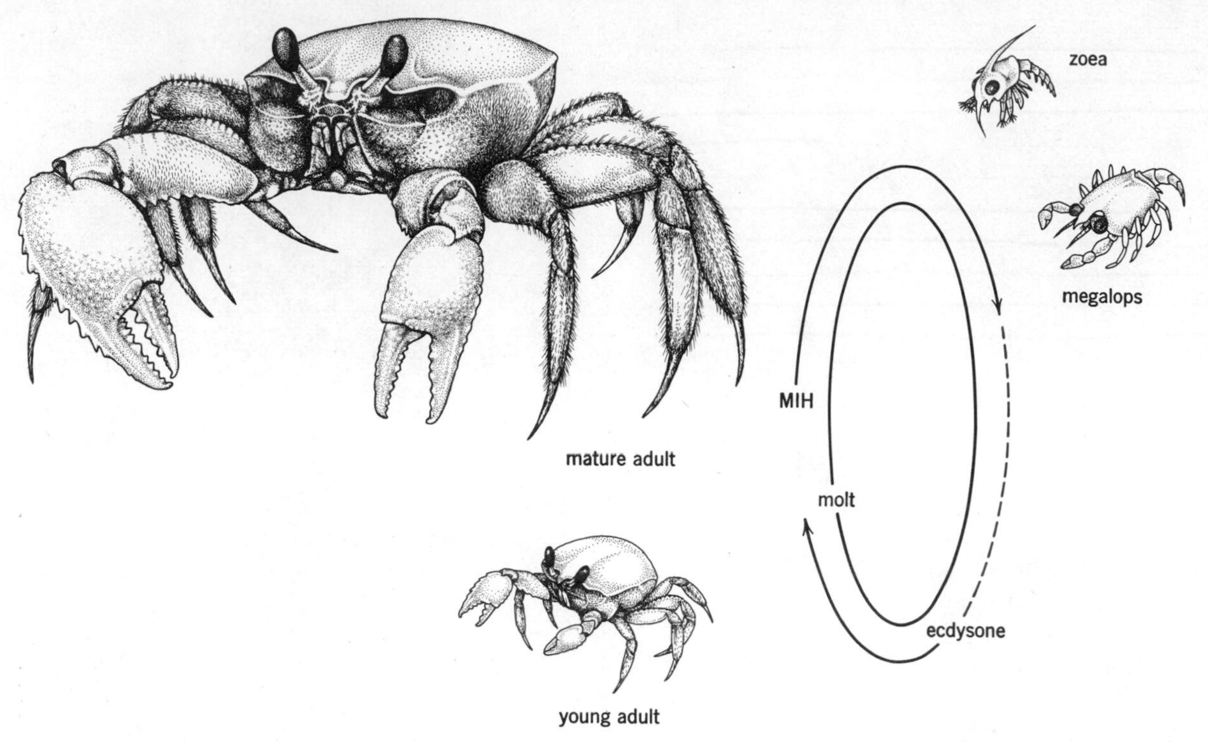

Figure 8.2. The molting cycle of a crab. Several molts are involved in the development from larval stages (right) to adult. The broken blue line indicates the buildup of ecdysone.

functional importance of hormones may be equally great in all groups.

Biologists have long suspected that many activities such as molting, coloration, and reproductive behavior in invertebrates are controlled by hormones. Only in recent years, however, have the presence of such hormones and their sites of origin been verified. Investigators have identified endocrine tissues and hormones in a number of insects and crustaceans, and have observed hormonal activity in annelid worms, molluscs, echinoderms, and cephalopods (octopi).

A notable feature of invertebrate endocrine systems is that most of the hormone-producing cells are found in the animal's nervous system. Some vertebrate hormones are produced by nerve cells, but most originate in nonneural glands.

Some Techniques of Study. The classical technique for investigating a hormone is to remove the gland that is suspected of being its source and observe what specific function in the organism ceases. This function should then be restored by administering an extract of the missing tissue to the organism. In addition, giving the extract to a normal animal should result in different, often opposite, effects than the removal of the gland. The following example illustrates these techniques.

In crabs, the hard outside covering, known as the exoskeleton, must be shed periodically to allow the crab to grow. This molting (called *ecdysis*) is controlled by two hormones (Figures 8.2 and 8.3). *Ecdysone*, produced by a gland called the Y-organ, causes a complex set of events that accompanies actual molting. *Molt-inhibiting hormone* (MIH), produced by a gland called the X-organ, prevents the production of ecdysone. Under normal conditions, MIH is produced for a time during which growth takes place. As the time of molt approaches, the secretion of MIH declines, thus ending the inhibition of ecdysone production. Ecdysone is released into the body fluid by the Y-organ, and its increasing amounts cause the molting of the crab's exoskeleton.

To confirm these functions, experiments have been performed in which the X-organ was removed. The crab then continuously produced ecdysone, molted

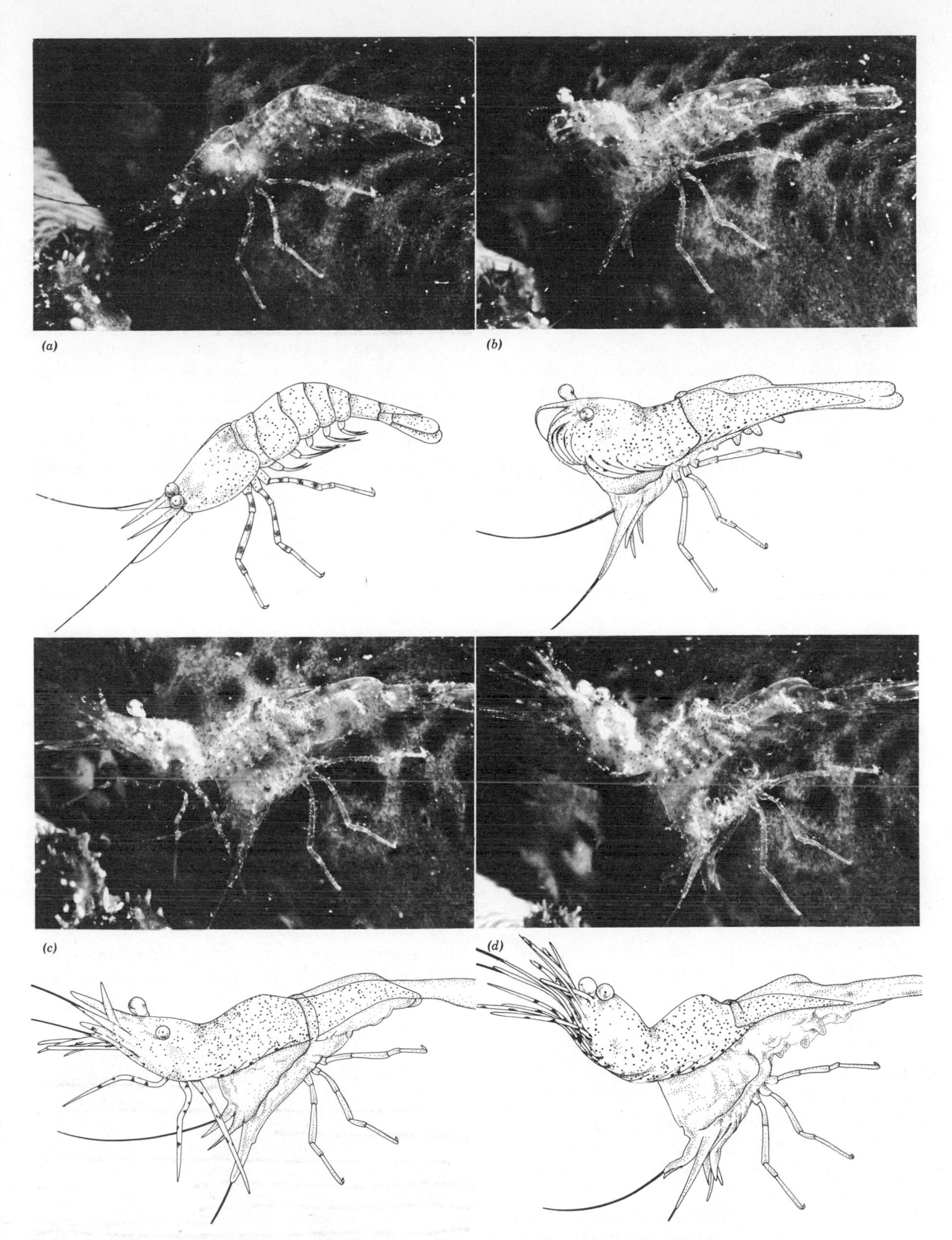

Figure 8.3. A shrimp molting. The sketches below the photos are provided to help in interpreting the event. The frontispiece for this chapter shows the molting process completed (Dr. K. R. H. Read).

in rapid succession, and finally died. If we grind up X-organs and inject their extract into a crab from which the X-organ has been removed, we find that molting ceases as long as the injections continue.

The opposite effect on molt may be obtained by removing the Y-organ. Growth continues until the animal fills the external skeleton. MIH production decreases but no molt occurs. Whenever we inject ecdysone into crabs, molt ensues. This shows that the Y-organ produces ecdysone, which stimulates molt, and that the MIH does not inhibit ecdysone's action but only its production.

The action of MIH in the control of crab molting introduces one further fact about chemical coordinating systems. Many chemical coordinators prevent some activity from occurring, that is, they are inhibitors. Coordination can thus be achieved through inhibition as well as through stimulation. Notice that the definition of a hormone does *not* state whether the chemical agent has an inhibitory or stimulatory function.

In these investigations, it is important to perform additional control experiments called sham operations. In such experiments the same areas in the animal are cut as during the removal of the gland but the gland is left in the animal. These operations show that it is not the damage done to the animal by the surgery, but the actual removal of the gland that brings about the observed effects.

Students sometimes wonder why hormones are not studied by analyzing body fluids, like blood, for their presence or absence. Usually this is not possible because the amount of a hormone in the fluids is too small to be isolated or even detected. However, the technique of bio-assay may be used in some instances. This consists of injecting the fluid that contains the hormone into a test animal and observing its specific action on the animal. For example, if urine from a pregnant human is injected into a female rabbit, the rabbit will ovulate. An even simpler test involves the use of toads, which will release eggs or sperm under these conditions. Such pregnancy tests, a type of bio-assay, work because the human uterus secretes hormones during the early stages of pregnancy. These hormones are capable of stimulating the maturation and release of eggs or sperm in other animals.

Vertebrate Hormones. The chemical structure of vertebrate hormones varies greatly. Three general groups are represented. The simplest hormones are composed of modified amino acids or small groups of amino acids. Examples of these are adrenalin, thyroxin, and the hormones released by the posterior pituitary gland. Somewhat larger than the amino acid hormones are the steroid hormones. These are relatives of the fats. Cortisone, estrogen, and testosterone are all members of this group. The third and largest group of hormones are proteins. They range in size from smaller molecules, such as insulin, to very large molecules like gonad-stimulating hormones and growth hormones. All of the hormones from the anterior pituitary gland are proteins.

In mammals, hormones coordinate a wide range of activities. Many of these coordinating roles are well known. An example is the part that insulin plays in carbohydrate metabolism. When the glucose concentration of the blood increases, more insulin is released by special cells in the pancreas. This release helps to move glucose from the blood into body cells, particularly the cells in the liver, and reduces the blood glucose concentration. Insulin is one of the few proteins whose structure we know in detail.

Some of the hormones produced by the adrenal cortex are important in the control of salt and water metabolism. Injections of cortisone, for example, reduce the swelling in arthritic joints. It is important to note that the result is only a relief of symptoms and not a cure for the disease.

Other types of activities coordinated by hormones involve many separate organs and mechanisms. One of these is reproduction which involves three different sites, the gonad, pituitary, and brain, all secreting hormones that are involved in this control.

Many of the things we associate with reproduction result from the hormones that are produced by gonads. For example, in a male vertebrate the testes produce androgens, male sex hormones, which are released into the blood stream. These hormones cause development of typical male structures, body form, and coloration as well as male behavior patterns for the species. That is, not only reproduction itself, but also the secondary sex characteristics are controlled by hormones.

Androgens also affect some cells in the part of the brain called the hypothalamus (Figure 8.4). These nerve cells produce a hormone (a neurosecretion) that has only one known function: to stimulate the cells in the anterior pituitary gland that produce gonadotropic (gonad-stimulating) hormones. These

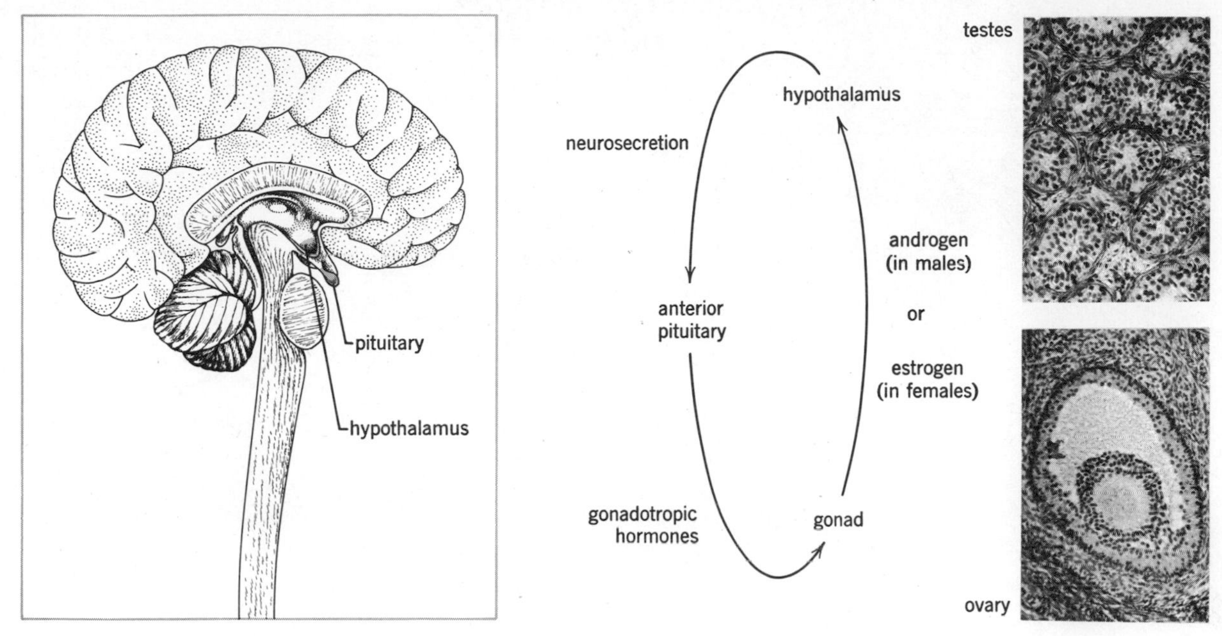

Figure 8.4. Relationships among glands and hormones involved in the control of reproduction in vertebrates. The diagram on the left indicates the position of both pituitary and hypothalamus. Photomicrographs show details of the tissues of the gonads.

hormones stimulate the testes to produce sperm and androgens. The androgens in turn inhibit the activity of the hypothalamic cells, thus completing the cycle of action.

Throughout the cycle, a nearly constant level of hormone production is maintained. If the amount of androgens rises above a certain level, the hypothalamic cells become inhibited and produce a smaller amount of hormone; there is less gonadotropic hormone from the pituitary and, consequently, less androgens are produced. When the level of androgens drops, the inhibition of the hypothalamic cells is reduced, and they stimulate the pituitary to release a little more gonadotropic hormone. Balance is thus achieved. Such a mechanism—by which the controlling element is itself controlled by the element it controls—is called a *feedback* mechanism. This type of regulation is typical of hormonal systems.

We know, however, that the level of hormones involved in reproduction does not always remain the same. Rather, there is usually a distinct reproductive season and then a season during which the reproductive structures are inactive. How is the hypothalamic-anterior pituitary-gonad system of reproductive control stimulated into activity at some times and inhibited at other times?

Some excellent research into this problem has been done with the initiation of breeding in birds. W. Rowan's experiments in the 1920s showed that the gonad development and breeding behavior of birds were stimulated by exposure to an increase in the number of hours of light per day. Since then, through the experiments of many biologists, it has been shown that the increased day length is detected by receptors in the retina of the eye (Figure 8.5). Through some unknown neural pathway, the nerve impulses from the retina stimulate the activity of the neurosecretory cells in the hypothalamus. The hormones produced then stimulate production of gonadotropic hormones. The factor that initiates the whole gamut of reproductive activities is the increase of day length.

Thus we see that two mechanisms are responsible for control of the hormones that coordinate reproductive activity. The first is the feedback mechanism, which helps maintain the hormones from the gonads, hypothalamus, and anterior pituitary at a constant level. Initiation, modification, and termination of these hormonal processes are controlled

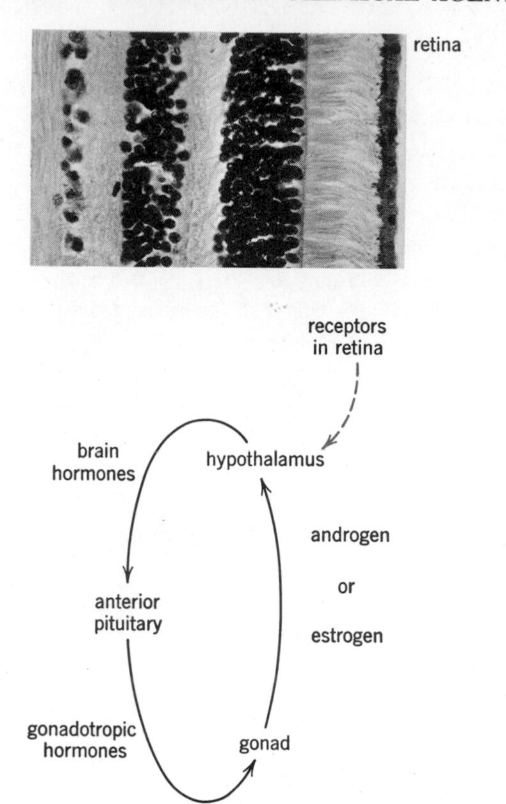

Figure 8.5. In many avian species the reproductive cycle is controlled by light that affects the hypothalamus via the retina.

by the second mechanism, the nervous system, through the secretion of hormones by certain nerve cells. Control exerted by the nervous system synchronizes the activity of chemical coordinating systems with environmental changes.

Insect Hormones. Among invertebrates, insect hormones are the best known. These include ecdysone and juvenile hormone.

Ecdysone functions in the vital growth and molting process in insects in much the same way as it does in the crab. Chemical analysis indicates that it is a steroid, and thus related to an important group of vertebrate hormones.

The mode of action of ecdysone may have far greater significance than any other aspect of the hormone. When ecdysone is applied to giant chromosomes, which are located in special cells in many fly larvae, puffy areas appear along some of the chromosomes a short time later (Figure 8.6). In addition, the size of the puffs is proportional to the dose of ecdysone. Further studies have shown that RNA synthesis and protein synthesis were greatly accelerated in the puffed regions. Such experiments suggest an attractive hypothesis: that ecdysone acts directly upon chromosomes to activate a gene or set of genes which in turn directs the synthesis of enzymes involved in molting. If this is the actual mechanism and also occurs with other hormones, then we have an answer to one of the major problems in contemporary biology: What controls the action of genes? A note of caution: there are alternative explanations for chromosome puffs. A variety of chemicals, including narcotics and some ions, cause a pattern of puffs identical to those produced by ecdysone. These chemicals do not cause molting.

Juvenile hormone is produced by a pair of small bodies that are found near part of the brain of an insect. Like ecdysone, it is involved in the control of growth. Its specific function is that of keeping the insect in its larval form for a certain number of molts. For example, the silkworm larva (caterpillar) molts four or five times before becoming a pupa.

Juvenile hormone keeps the creature in its caterpillar form until the last molt (Figures 8.6 and 8.7).

Chemically, juvenile hormone is related to the seemingly ubiquitous steroids. An interesting sidelight on this structure is that substances that mimic juvenile hormone action are widespread in nature. Thus, juvenile hormone activity has been found by bio-assay in tissue extracts from many invertebrates, vertebrates, microorganisms, and plants.

These hormone mimics may act as natural insecticides. Dr. Carroll Williams, an authority on insect hormones, observed during a trip that one of the main Amazon River tributaries was abnormally free of insect life. Study of the water showed the presence of a substance that inhibited insect development, possibly a juvenile hormone mimic from vegetation along the flooded river valley. This suggests the possibility of developing a commercial product for combating insects that would not be deadly to other forms of life, as are the present organic insecticides like DDT and parathion.

Hormones in Plants

The growth of plants and the specialization of their structures are controlled by chemicals termed plant hormones. These differ markedly from the typical animal hormones in several ways. They are produced by cells, not glands, and they travel from cell to cell rather than through the plant's vascular tissues. In addition, the effects of these substances on the plant are more generalized than the localized, highly specific activity of most animal hormones.

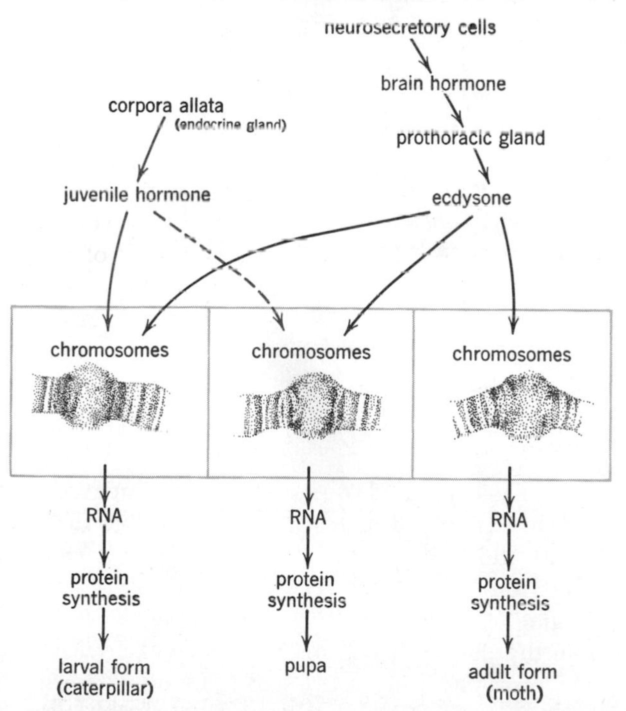

Figure 8.6. Juvenile hormone and ecdysone affect chromosomal puffing. The amount of juvenile hormone present determines the form of the individual following the molt.

The hormones that affect growth and differentiation in plants illustrate these differences dramatically. The three different classes that are usually recognized—auxins, gibberellins, and cytokinins—are easily distinguished on the basis of their chemical characteristics. The functions of these hormones show overlap between the classes in some cases, cooperation in others, and antagonism in still others. This confusing situation makes a functional differentiation difficult.

Auxins. More is known about auxins than other plant hormones. These agents originate in, and regulate the growth of, the stem tips of plants.

When the stem tip is removed, the plant stops growing. Replacement of the tip, or an agar block on which the tip has been allowed to stand, will cause growth to resume (Figure 8.8). An inverted stem tip has no stimulating effect, because the hormone only moves away from the original stem tip regardless of which way the stem is oriented.

When we look at the result of removal of the stem tip, many possible explanations come to mind. The cessation of growth could be caused by:

1. Damage to the cells by cutting.
2. Loss of the contact between the cells in the tip and in the rest of the stem.
3. Removal of the source of a chemical that stimulates growth.

An experiment must be designed to test each possibility. Since simple replacement of the cut tip results in the resumption of growth at a near normal rate, we can rule out the first explanation. As to the second possibility, we are able to achieve nearly normal growth from the stem again by inserting an agar block between the stem tip and the stump. Placing the stem tip on the agar block allows a chemical to diffuse into the block, and then into the stem to produce growth. Resumption of growth shows that contact between the stem and stem tip is unnecessary. Therefore, we are left with the third possibility: a chemical agent diffuses from the tip down the stem and stimulates growth.

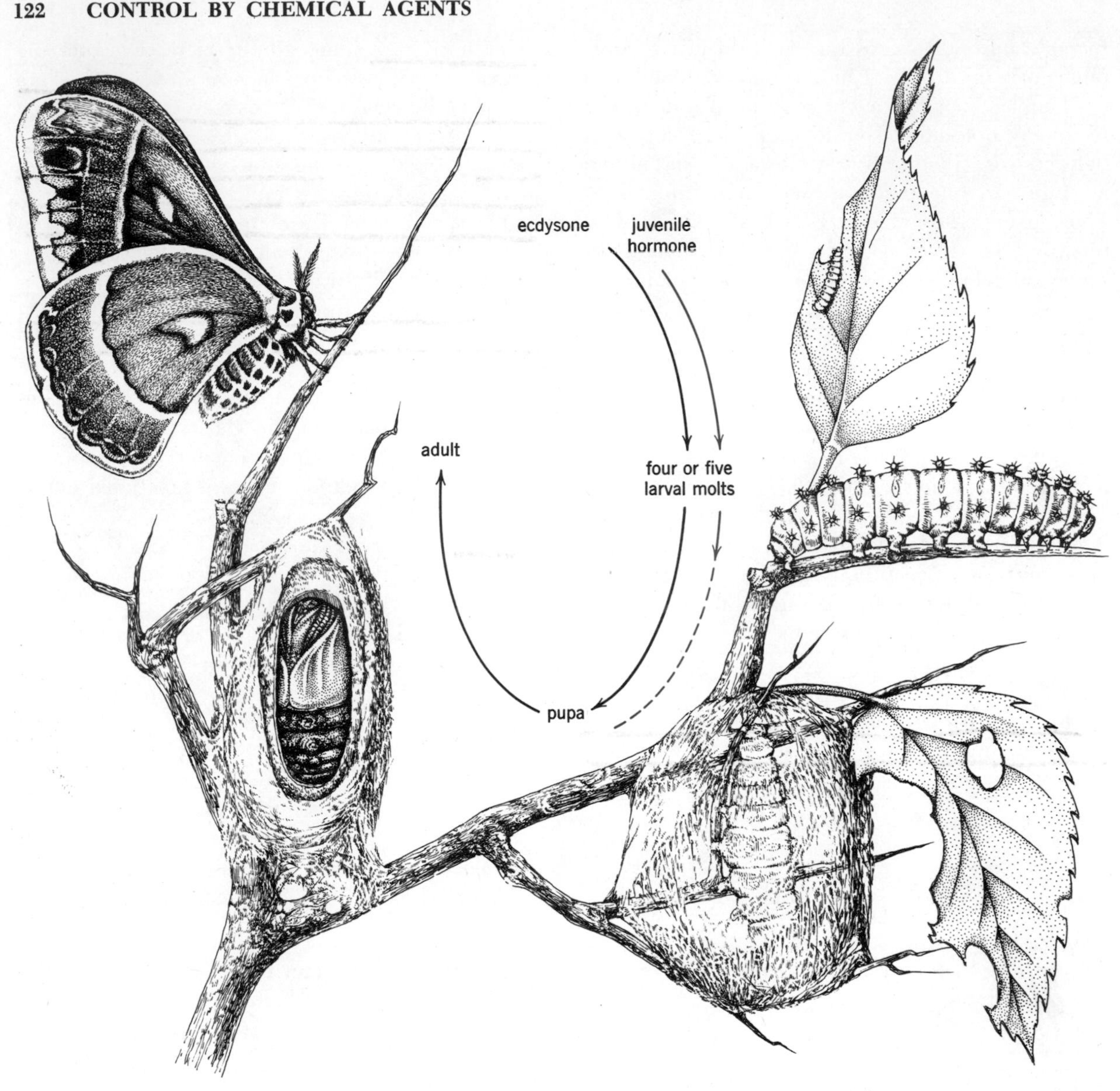

Figure 8.7. Hormonal control of growth and development in the life cycle of a moth.

The conclusive experiment involves placing agar blocks on two sets of stems. One group of blocks previously had stem tips resting on them, while the other group did not. (Compare Figure 8.8c, c′, and 8.8d, d′.) Since growth resumes only in stems of the first set, we can eliminate the possibility that the pressure of the agar block against the stem is the stimulating agent. Clearly, some chemical agent must be involved.

This concept, developed by Fritz W. Went in the 1920s, has been used in numerous experiments devised to explore the effects of auxin. In one of Went's investigations, agar blocks containing auxin were placed on one side of decapitated seedlings. The seedlings soon showed a growth curvature *away* from the side on which the block was placed. The amount of curvature was found to be directly proportional to the concentration of growth-promoting substance in the agar. This is now used as a bio-assay for auxin. With this technique an investigator can test a plant extract or a synthetic auxin for its growth hormone activity.

Such precise tests are not yet available for other plant hormones. These are studied instead by spraying them on plants or supplying them in a nutrient solution to the roots. Recently, tissue and cell culture techniques have been developed that enable an investigator to expose single cells or small masses of cells to various kinds of hormones.

Analysis of stem tip extracts by chromatographic methods (see Chapter II) have demonstrated the presence in plants of many different auxins. Many synthetic auxins, that is, chemicals with growth-stimulating abilities, also are known and some have commercial uses. Of these, one commonly known as 2,4-D is a widely used herbicide. In the proper concentration 2,4-D kills broadleafed plants rather than stimulating their growth.

Auxins normally function by controlling elongation of cells and cell proliferation, and by interacting with other plant hormones in coordinating general plant growth. However, like many other hormones, auxins have different effects on various parts of the organism.

A concentration of auxin that is great enough to cause maximal growth in the stem is so great that it inhibits the elongation of the root. When we consider the natural growth movements (tropisms) of plants, we find that these variations are adaptive.

Plants respond to stimuli such as light and gravity by growing toward or away from them. For example, the stem of a plant placed on its side soon grows upward away from the pull of gravity whereas the roots grow downward. At this time, more auxin is found on the underside of both the stem and the root. Stems grow toward a source of light and roots bend away. Auxin is found more concentrated on the shaded side of the organs. In all such tropistic growth patterns, one side of the organ grows faster than the other. Growth is increased in one side of the stem but decreased in the same side of the root by the higher concentration of auxin.

Natural and synthetic auxins have several horticultural uses today (Figure 8.9). One is used to promote root formation on pieces of stems used for propagating certain plants. Camellia and lemon cuttings form roots much more rapidly if dipped in auxin before planting. Apple and citrus trees will hold their fruit longer if sprayed with auxin. Seedless tomatoes, cucumbers, watermelons, and other fruits have been obtained by spraying the flowers with auxin. An entire field of pineapples can be induced to flower and form fruit simultaneously by the use of auxins. As a result, harvesting procedures are greatly simplified.

Gibberellins. This group of hormones was originally found in a fungus that infects rice plants and causes abnormally tall, spindly growth. When an extract containing the hormone is applied to other plants, unusual growth events often occur (Figure 8.9). Cabbage plants may grow to six feet in height and bean plants may change from a bushy type of growth into a climbing vine. It also causes many dwarf varieties of plants to grow into tall types. These abnormal growth patterns result from increased stem elongation.

It is now thought that gibberellins are normal constituents of all green plants, but in amounts far less than used in the experimental extracts described above. Evidence indicates that these hormones promote seed germination, stimulate the synthesis of auxin, induce root growth in seedlings, and interact with other hormones in regulating plant growth generally.

Cytokinins. Cytokinins are hormones that greatly accelerate cell division and differentiation

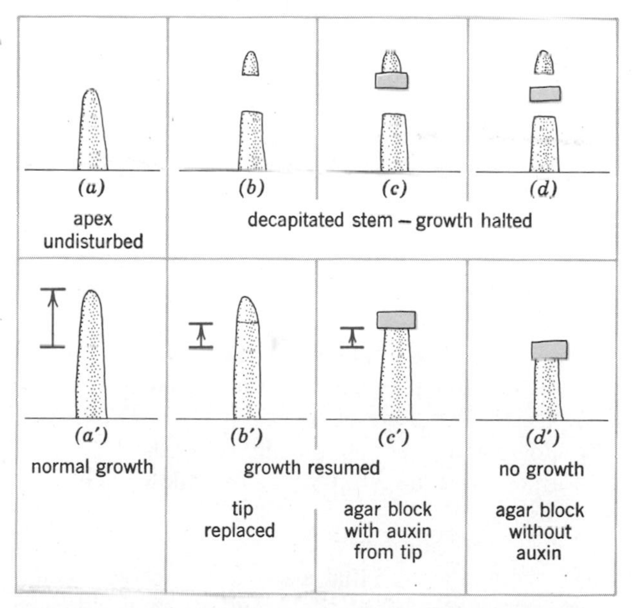

Figure 8.8. Experiments that demonstrate the presence of auxins in the stem tips. (*After* **Arthur W. Galston, The Life of the Green Plant**, *second ed., Prentice-Hall, Englewood Cliffs, N.J., 1964.*

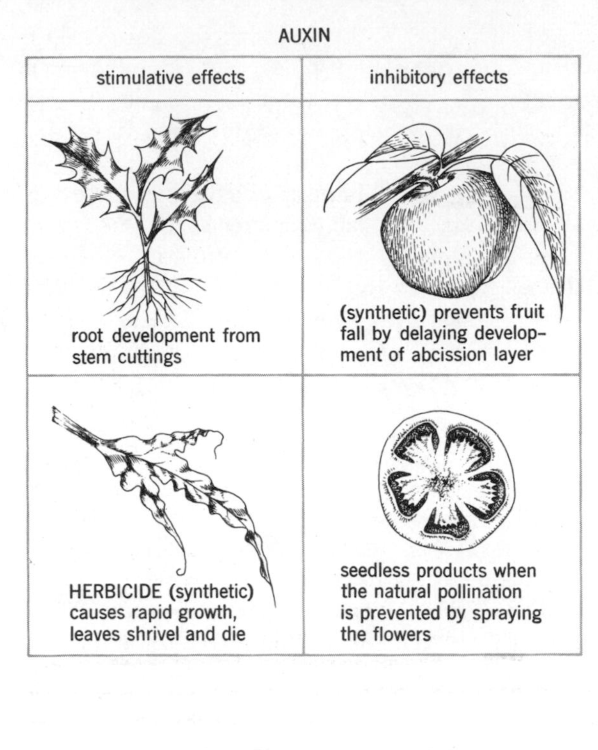

Figure 8.9. Some effects of auxins and gibberellins.

in plant tissues. Such hormones were first sought in coconut milk, because this milk is necessary for growing plant cells in laboratory glassware. Eventually, these potent growth hormones were found to be derivatives of purines and pyrimidines (Figure 8.10). They are found in seeds, coconut milk, and plant sap but only in extremely small quantities. The knowledge that cytokinins induce mitosis and differentiation and are related to the materials of which DNA and RNA are composed has generated much biological interest. A possibility exists that these hormones are the results of the breakdown of nucleic acids.

In concluding our discussion of plant hormones, we should mention that many plants secrete chemical agents into their surroundings that influence other organisms. These agents range from insect attractants produced by flowers and fruits, to substances that function like the insect juvenile hormones described previously. Some plants like the desert brittlebrush even prevent the growth of other plants nearby, thereby lessening competition for needed resources. The phytotoxic material from brittlebrush leaves enters the soil by rain or when leaves fall. Unlike some phytotoxic substances, this chemical does not prohibit the growth of other individuals of its own species.

DNA and Chemical Control

Evidence is beginning to indicate that in some cases the site of hormonal action is the gene. For example, it may be that ecdysone causes chromosome puffs by acting directly on the DNA. It is thought that hormones in general promote the synthesis of specific messenger RNA molecules, which in turn direct the synthesis of specific enzymes. By

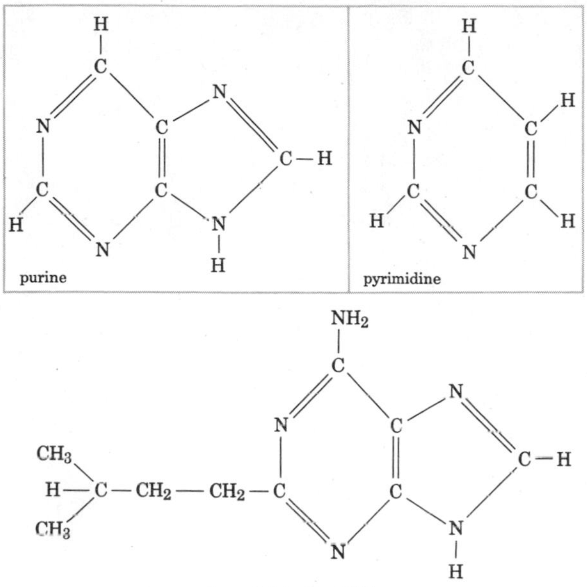

purine

pyrimidine

isopentyladenine (IPA)

Figure 8.10. Isopentyladenine is a cytokinin that is related to purine. Note the similarities in the molecular structure of these two molecules.

this mechanism a hormone could direct certain physiological activities in a cell. It has also been suggested that hormones may act at different sites along the DNA-RNA-enzyme synthesis chain of events.

An interesting demonstration of a mechanism that fits the above hypothesis is found in female mammals. The female sex hormones (estrogens and progesterones) produced by the ovary cause growth of the mammary glands. The messenger RNA content of the cells in these glands increases with an increase in sex hormones. The possible relationships of larger amounts of messenger RNA to more rapid protein synthesis and to greater growth is obvious. This hypothesis, that hormones in general act by increasing messenger RNA production, is an attractive one.

Principles

1. Chemical coordinating mechanisms control activities at all levels from the cell to the population.
2. These chemical agents are either carried by fluids or move through the environment.
3. The amount of a chemical coordinating agent is often controlled by some aspect of the process that it controls, that is, by a feedback mechanism.

Suggested Readings

Etkin, William, "How a Tadpole Becomes a Frog," *Scientific American,* Vol. 214 (May, 1966). Offprint No. 1042, W. H. Freeman and Co., San Francisco.

Galston, Arthur H., *The Life of the Green Plant.* Second edition. Prentice-Hall, Englewood Cliffs, N.J., 1964, pp. 54–80.

Jacobson, Martin and Morton Beroza, "Insect Attractants," *Scientific American,* Vol. 211 (August, 1964).

Naylor, Aubrey W., "The Control of Flowering," *Scientific American,* Vol. 186 (May, 1952). Offprint No. 113, W. H. Freeman and Co., San Francisco.

Overbeek, Johannes van, "The Control of Plant Growth," *Scientific American,* Vol. 219 (July, 1968). Offprint No. 1111, W. H. Freeman and Co., San Francisco.

Salisbury, Frank B., "Plant Growth Substances," *Scientific American,* Vol. 196 (April, 1957). Offprint No. 110, W. H. Freeman and Co., San Francisco.

Went, F. W., "Plant Growth and Plant Hormones," in *This Is Life,* edited by W. H. Johnson and W. C. Steere. Holt, Rinehart and Winston, New York, 1962, pp. 213–253.

Questions

1. At what level (cell, tissue, organ, etc.) do chemical coordinating substances operate? Support your answer with examples.
2. Why is it necessary for crabs to molt at intervals?
3. Are ecdysone and MIH antagonistic hormones? What causes each to be produced or to cease its action? Can you name other hormonal control mechanisms that function in this manner? Can you name any that do not?
4. Explain the role of feedback mechanisms in the control of reproduction.
5. Hormone production can be strongly influenced by the organism's environment. How do birds illustrate this?
6. What substances could be called natural insecticides? How do they act?
7. What evidence indicates that plants also have hormonelike control mechanisms?
8. How do these differ from the typical animal hormonal mechanisms?
9. What is a bio-assay? Give one example from plants and one from animals.
10. What is an example of a hormonelike chemical that travels through the environment rather than through body fluids?
11. What advantage does a desert plant gain by releasing toxic chemicals into the soil?

CHAPTER IX

Control by Nervous Systems

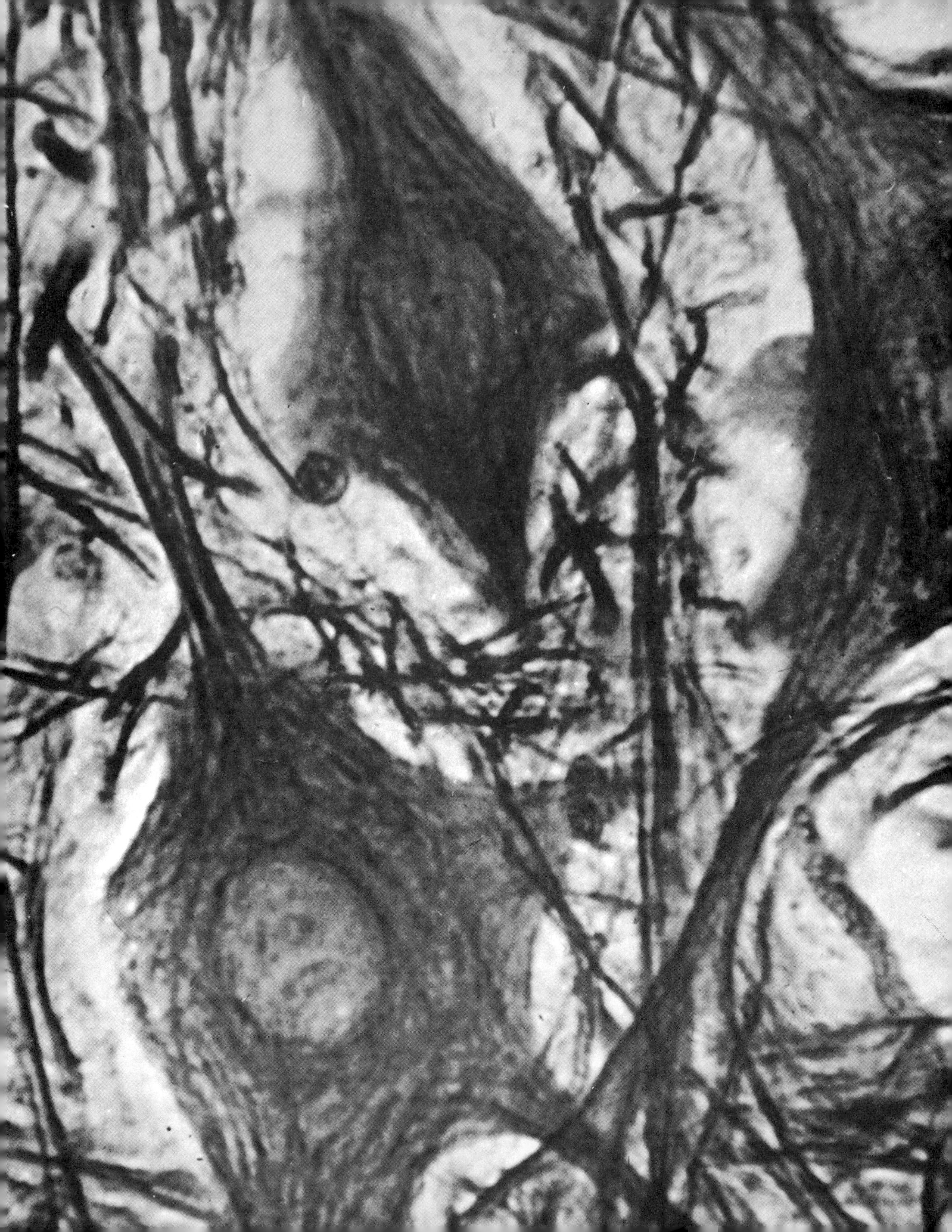

Cell bodies of three nerve cells.

CHAPTER IX

Control by Nervous Systems

Of the two types of coordinating systems, the nervous system is the more familiar to the general public. The operation of the nervous system and its relationship to the chemical coordinating systems are complex, but an understanding of both is essential for an insight into control phenomena in animals.

The specialized cells (neurons) that conduct nerve impulses in the nervous system are all somewhat similar in structure. Each neuron consists of a nerve cell body containing a nucleus, and of one or more cytoplasmic processes called fibers. These long fibers reflect the specialization of nerve cells for transmission of nerve impulses. The nerves found in higher animals are bundles of these fibers and their sheathing cells. The nerve cell bodies are found in or near the central nervous system.

The cells of the nervous system are not all alike, and they do not all have the same function (Figure 9.1). Certain ones have the task of picking up various changes in the environment (stimuli). These cells are called *receptors*. They change the energy of the stimulus into energy that is used to initiate nerve impulses. The first nerve cell receiving the impulse directly from the receptor is called the *sensory neuron*. Receptors are in contact with only one end of the sensory neuron, thus helping to insure the one-way transmission of the impulse through the nervous system. From the sensory neuron, the impulse may pass through a number of *association neurons*, which is the

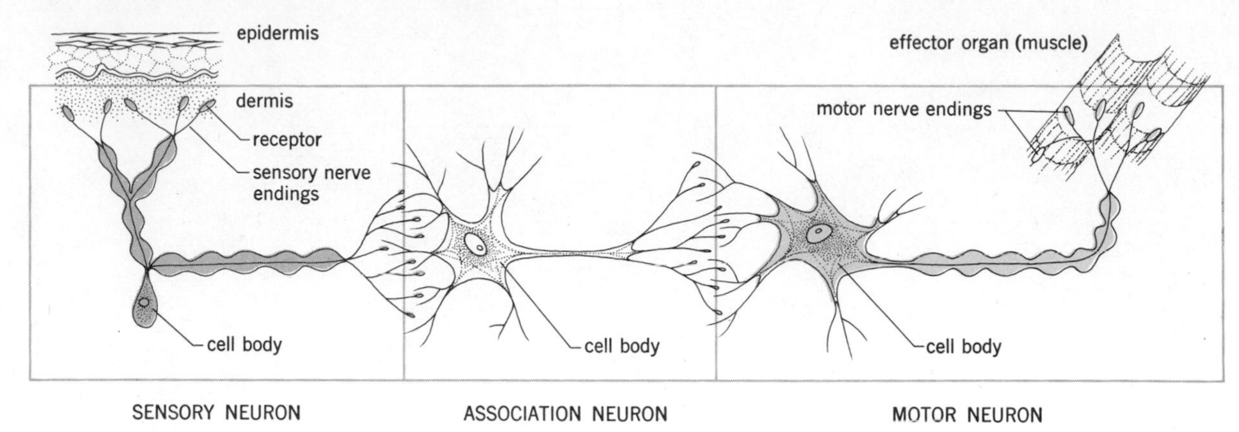

Figure 9.1. The structural components of a nervous system.

name applied to all nerve cells between the sensory neuron and the ultimate neuron. The final nerve cell, which carries out the appropriate action in transmitting the impulse to the effector organ—usually a muscle—is called the *motor neuron.*

The Nerve Impulse

When a nerve fiber is at rest, that is, when it is not carrying an impulse, the concentrations of the various kinds of ions are different inside the cell membrane than outside (Figure 9.2). Sodium ions (Na^+) are found in high concentration outside the cell and in low concentration inside the cell. Conversely, potassium (K^+) shows a high concentration inside the cell and a low concentration outside.

The outside of the fiber (surrounded by sodium ions) is positively charged in relation to the inside. The inside of the fiber is negatively charged, even though it contains potassium ions, because of the presence of chloride ions (Cl^-) and negatively charged organic molecules. This difference in electrical potential on the two sides of the nerve fiber membrane is termed the *membrane potential* or *resting potential.* This is the normal condition of the nerve fiber when it is not transmitting an impulse. The sodium and potassium ions tend to diffuse across the membrane. However, the cell maintains the resting potential by actively extruding sodium ions and tending to accumulate potassium ions.

As the train of events that we call the nerve impulse starts at any one place on the fiber, the membrane's permeability to sodium changes. Instead of letting very little sodium through, suddenly the membrane allows the sodium to pass in freely, and it rushes into the cell. Since each sodium ion is positively charged, this movement results in a positive charge on the inside of the cell as compared with the outside. This reversal of electrical charge is called *depolarization.*

Potassium ions, with their positive charge, now begin to flow out faster than the sodium flows in. When these positive charges are again moved to the outside of the cell, they initiate the process of *repolarization,* (the return to the resting state) and the inside of the cell again becomes negative. Repolarization is hastened as the membrane again becomes less permeable to sodium. The active transport of sodium out of the cell and the reestablishment of the potassium concentration inside it restores the cell to a resting state.

This describes only the events at one point on the nerve fiber. How does the impulse move along the fiber? As the sodium rushes into the fiber, the outside of the fiber briefly becomes negatively charged at that point. Adjacent to this point, a part of the fiber remains positively charged. This difference in potential between the areas is the *action potential.* Since opposite charges attract one another, the positive charges move toward the negative charges on both sides of the membrane. The result is a partial neutralizing of the charge or a small decrease in the electrical potential (a depolarization) of the membrane adjoining the site of the impulse. This depolarization causes the membrane to become more permeable. Sodium rushes into the cell and the impulse continues. Bit by bit, the impulse moves down

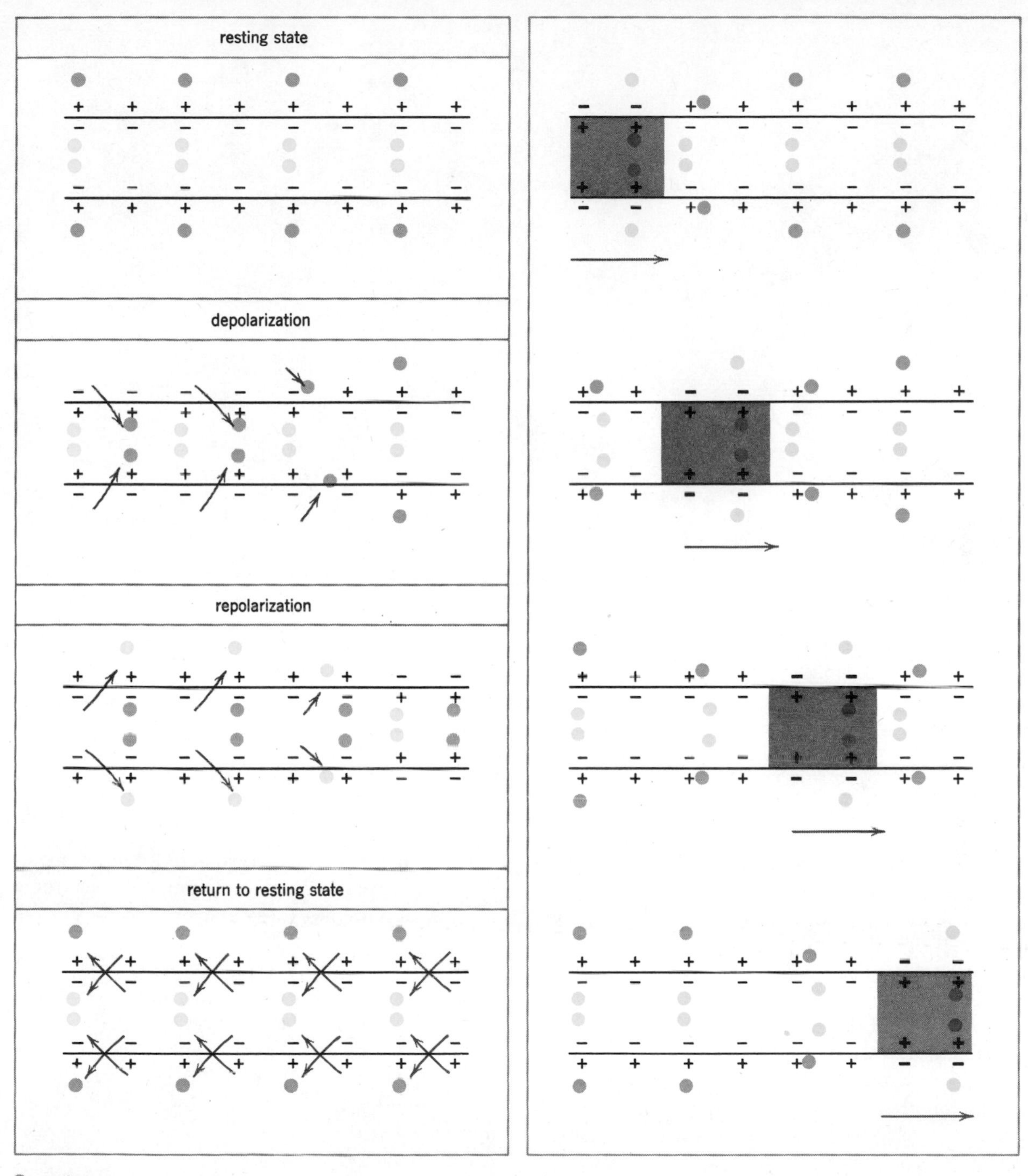

Figure 9.2. The events during conduction of a nervous impulse. The illustration on the left shows the movement of ions at a single point on the nerve fiber. Transmission of the impulse (gray) along the fiber is indicated at the right.

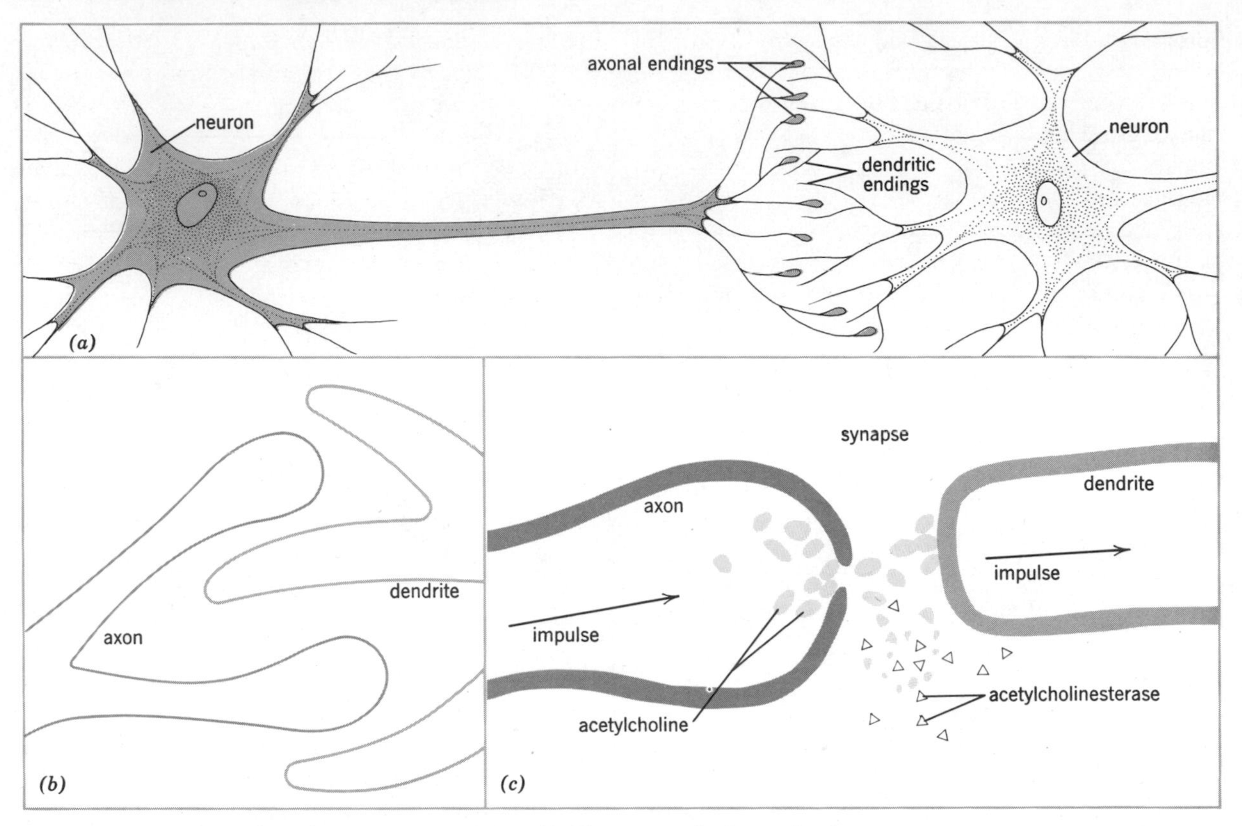

Figure 9.3. A representation of synaptic transmission. (a) *The structural relationship between axon and dendrite. Part* (b) *shows a higher magnification of the synaptic area. During the transmission of a nerve impulse* (c) *the axon produces acetylcholine, which then stimulates the dendrite to carry the impulse. Acetylcholinesterase inactivates the acetylcholine.*

the fiber—much more slowly than an electrical current and by a different means.

When an electrical current travels along a conductor, the conductor serves only as a passive carrier of the electrons that move in the same direction as the current because this movement of electrons *is* the current. But during transmission of a nerve impulse, ions are the only particles that move. Moreover, they move at right angles to the movement of the impulse, that is, they move in and out of the fiber. The nerve impulse is thus a self-propagating wave of depolarization followed by repolarization moving down the nerve fiber. Although a nerve impulse is not an electrical current flowing along a nerve fiber, the impulse is accompanied by measurable changes in electrical potential in the membrane. Much of the research on nerve action involves measuring these changes under controlled conditions. Tiny electrodes, placed on both sides of a membrane, detect electrical activity in the membrane as ions pass through it. Appropriate electronic equipment amplifies, measures, and records this activity.

On any one nerve fiber, the impulse never varies in strength. If the change in environmental conditions (the stimulus) is barely great enough (threshold strength) to cause the nerve fiber to carry an impulse, the impulse will be of the same strength as one incited by a stronger stimulus. *The all-or-none law says that if the nerve fiber carries any impulse, it will carry a full-strength impulse.*

Synaptic Transmission

As an impulse travels through the nervous system, it soon reaches the end of a fiber. Between the end of one fiber and the beginning of the next is a small gap, the *synapse* (Figure 9.3). The synapses are important parts of the nervous system because their properties determine many of the system's properties. Let us look at three of their characteristics: (1) the method of transmitting the impulse across the synapse; (2) the usual one-way transmission pattern; and (3) the modification of the transmission of nerve impulses by other nerve impulses.

The nerve impulse is transmitted across the gap by a different means than the one that moves it along the fiber. When the impulse reaches the end of the fiber, it causes the tip to secrete a chemical. This agent diffuses across the synapse and depolarizes the membrane of the next nerve fiber, causing it to become permeable to sodium and initiate an impulse. Notice that rather than the electrical properties of the impulse being responsible for synaptic transmission, a chemical is the transmitting agent.

Most of the synapses in our bodies have *acetylcholine* as the transmitter agent; the remaining synapses secrete other agents. As nerve impulse after nerve impulse causes the secretion of acetylcholine into the synapse, we might expect this chemical to become highly concentrated at the synapse and diffuse in all directions. Actually it does diffuse in all directions, but never gets very far or becomes very concentrated. An enzyme, *acetylcholinesterase*, which breaks down the transmitting agent, is present at all times in the vicinity of the synapses. It inactivates acetylcholine by splitting the molecule, thereby preventing continued depolarization of the next fiber or the inordinate spreading of the nerve impulse to other nerve fibers.

Under some conditions, this enzyme is kept from functioning. If a person receives a dose of an organic phosphate insecticide, such as Malathion or parathion, the activity of acetylcholinesterase is blocked. What would be the effects of a dose of Malathion on synaptic transmission? What behavioral effects would result?

At a synapse, only one of the two nerve cell endings is capable of secreting the chemical transmitter agent. If a nerve impulse was initiated somewhere in the middle of a nerve cell, it would proceed to both ends of the cell. But as only one end can produce acetylcholine, the impulse is able to cross the synapse to the next cell only at this point. At the other end of the cell, no chemical agent will be released and the impulse will be unable to proceed farther. This simple fact assures the one-way transmission of impulses through the system.

At some synapses, the amount of chemical substance produced when an impulse reaches it may not be sufficient to stimulate an impulse to the next neuron. However, if a series of such impulses reaches the synapse in rapid succession, enough acetylcholine may be released to initiate an impulse in the next cell. This phenomenon is called *summation*, because the effects of several nerve impulses or several fibers are added together to propagate the impulse in the next cell.

Differentiation of Stimuli

We have established that all nerve impulses are depolarizations along a membrane and that they are conducted at full intensity or not at all. If impulses are all alike, how then does the nervous system distinguish different types of stimuli like light and sound, or different intensities of the same stimulus? How is a gentle tap separated from a hard blow? There are several answers to these questions based on the transmission of nerve impulses, the nature of receptor organs, and the interpretation of nerve impulses by the brain.

The intensity or strength of a stimulus is differentiated in several ways. A strong stimulus, like a hard blow, initiates impulses in many nerve fibers, while a lighter stimulus involves only a few. Also, a strong stimulus usually produces a greater frequency of impulses along nerve fibers than a weak one does. Finally, summation provides a way in which subthreshold stimuli can be detected. All of these methods provide a potential coding device by which the nervous system can differentiate stimuli.

The retina of the eye illustrates some of these principles. Under normal circumstances all retinal cells respond only to light. Some respond, however, to lower light intensities (have lower thresholds) than others. You may have noticed that in a very dark room you see less by looking directly at an object than by looking out of the corner of your eye. The receptors at the edges of the retina have lower thresholds than those near the center of the eye. As the strength of the stimulus increases, more and more receptors respond, and more and more nerve fibers carry impulses to the brain.

Under abnormal circumstances a receptor may respond to a different kind of stimulus, but this will be interpreted by the brain as the kind of stimulus usually picked up by the receptor. If you have ever been hit on the eye, you may remember having seen a flash of light. The retina was stimulated mechani-

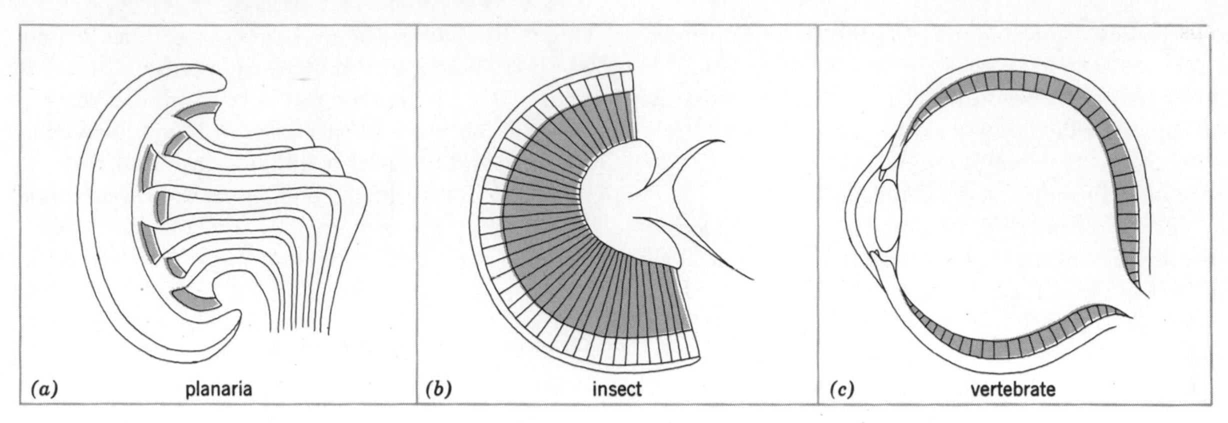

Figure 9.4. Structure of three types of light receptors. Light-sensitive areas are shown in blue.

cally, but the nerve impulses were interpreted as light. Any stimulus to these receptors is always interpreted as light.

The coordination of an animal's activities depends not only on the movements of impulses along nerve fibers, but also on the brain's interpretation of the information they contain and on their orderly transmission to the proper effectors. This is the function of an intact nervous system rather than of isolated parts.

Receptors. The task of a receptor is to detect changes in the environment (stimuli) and convert these into nerve impulses. All nervous systems, whether simple or complex, must contain some units that function as receptors. An obvious fundamental principle is that an organism can only detect those stimuli for which it has receptors. A fish cannot hear sound waves when in the air, and humans cannot see infrared light because receptors for these stimuli are not present in the organisms. Application of this concept also throws doubt on the ability of a human to perceive the brain waves of another individual, or for migrating birds to navigate by sensing the electromagnetic fields around the earth.

A survey of the types of receptors found throughout the animal world shows which kinds of environmental information animals most often use. These appear to be light waves, temperature, presence of chemical substances, sound, and gravity. Such a survey also shows a marked diversity in the types of receptors used for detecting one environmental variable. For example, even though a jellyfish has a gravity detector that enables it to maintain its equilibrium, this detector does not approach the anatomical complexity of the inner ear of a vertebrate. Nevertheless, it provides the same basic information to the animal.

Nearly all animals have light receptors, ranging from a tiny dab of light-sensitive pigment in unicellular forms to the complex eye of higher animals. In multicellular animals, light detection is based on specialized cells that convert light into nerve impulses. Many adaptive arrangements of these cells have occurred in the course of evolution. Planaria has a simple cuplike cluster of these cells on each side of its head (Figure 9.4*a*). The clusters provide Planaria with information on the presence of light and probably something of its intensity and the direction of the source.

Insects have evolved a compact mass of cells with accessory structures that function as lenses (Figure 9.4*b*). The large number of compartments that receive light from only a very small area allows the insect to detect movement efficiently.

Other lines of evolution led to the vertebrate type of eye (Figure 9.4*c*), with its various mechanisms for focusing light waves on the light sensitive cells (retina). Additional refinements such as color vision also occurred.

Nervous Systems

Trends. The degree of control exerted over body activities depends on the complexity of the nervous system. Many organisms lack a highly specialized nervous system. In some, such as *Hydra* and *Velella* (a small colonial jellyfish), we find only a net of nerve cells (Figure 9.5). These are so unspecialized that impulses may travel in any direction. Because of this simple nervous system, *Hydra* is not able to show complex behavior. Only a few responses, such as avoiding a stimulus by withdrawal or by bending away, are possible.

The type of nervous system found in earthworms and insects is typical of many invertebrates. Each body segment contains one or two clusters of nerve cells called ganglia. Sensory and motor fibers enter and leave each ganglion, innervating the structures that belong to that particular segment. Ganglia in adjacent segments communicate by nerve fibers. The overall structure, then, is a chain of interconnected ganglia extending the length of the animal's body. In many invertebrates, ganglia in the head region fuse to form a single ganglion called a brain. This type of brain is not the equivalent of a vertebrate brain since it does little more than coordinate the functions of the fused segments it represents. Even though the invertebrate nervous system is drastically different in structure from the vertebrate type, both systems share many functional features with respect to response to stimuli, nerve impulses, and general coordinating effects on the body. After all, many invertebrates engage in exceedingly complex motor activities like swimming or flying. It is of interest to note that even the highly specialized mammalian nervous system retains some vestige of a nerve net in the wall of the intestine and has an elaborate set of ganglia as part of its autonomic nervous system.

As we look at the more intricate nervous systems, we find several general patterns of changes in their structures as they gain complexity (Figure 9.6). These are:

1. Specialization of cells into receptors for particular stimuli, sensory neurons, etc., with transmission in only one direction along the nerve fibers.
2. The development of many association neurons.
3. The concentration of nerves into special struc-

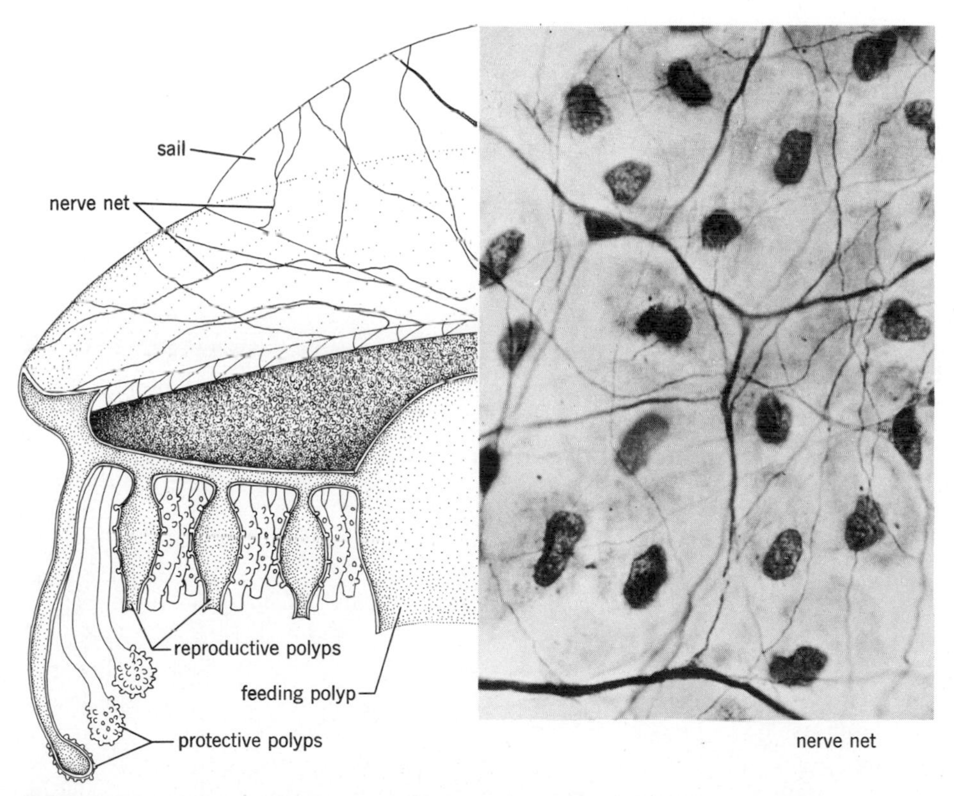

Figure 9.5. Nerve net of Velella. *Left, distribution of the nerve net. Right, a photomicrograph of part of the nerve net (Dr. G. O. Mackie).*

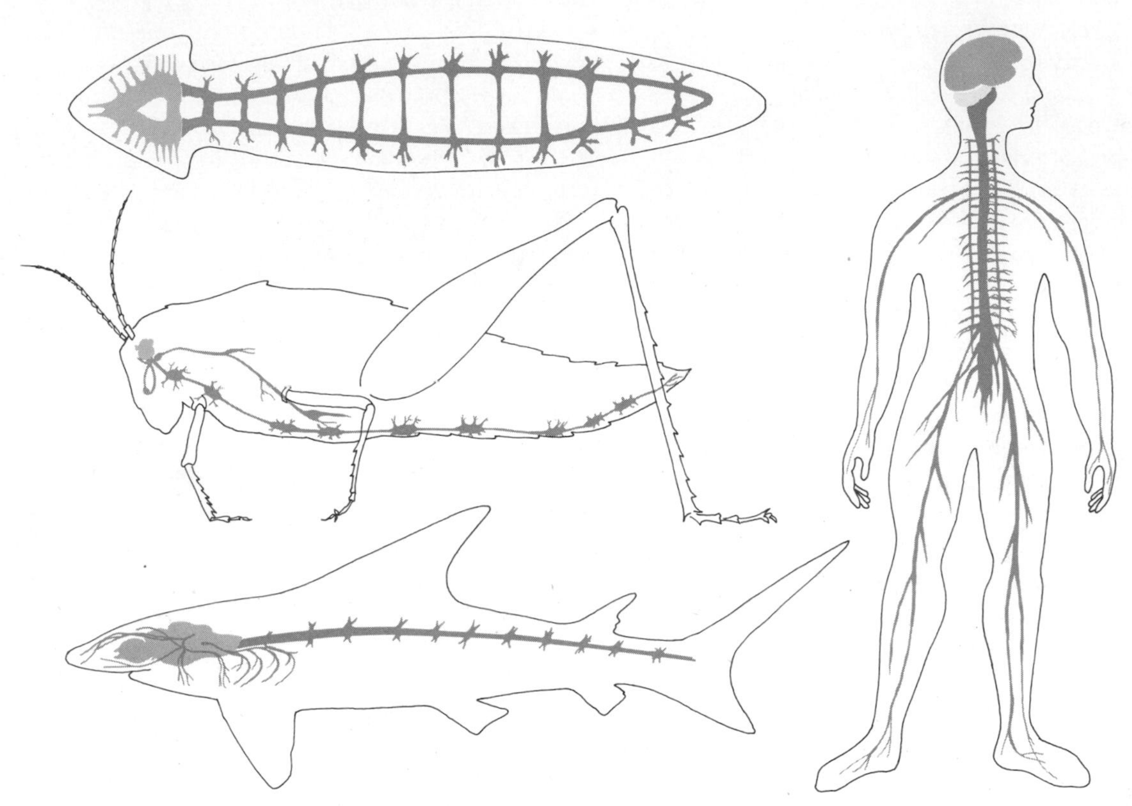

Figure 9.6. The larger elements of the nervous systems of several animals: Planaria, grasshopper, shark, and man.

tures such as nerve cords rather than a loose net.

4. The concentration of association neurons near the head end of the animal. This concentration results in the formation of a brain, and is called cephalization. As more complex receptors are developed, these are also usually found near the head end.

Why are these changes important? The advantages brought about by the first two types of changes are evident if we compare a system containing association neurons with a simple system having only one nerve cell between receptor and effector. In the simple system, each time a neuron carries an impulse it must carry it to all effectors reached by this neuron, and the response is always the same. The presence of more neurons between receptor and effector introduces the possibility of many different pathways. The examples used represent extremes. However, the principle also operates at the intermediate stages: the more complex the nervous system, the more complex the behavior may become.

The third and fourth trends reflect the changes that accompany the development of movement in animals. One part of the animal is constantly coming into contact with new areas before the rest of the animal. It is important to have the sensory structures located here so that adverse conditions can be detected by the animal before it has reached them or before much of its body has made contact with them. It is advantageous for the nerve centers associated with these sense organs to be located near them. Consequently, the nerve elements become concentrated at the leading end, and the structures that we call head and brain result.

Mammalian Nervous Systems. As a result of these types of changes, highly specialized nervous systems have developed from simple ones. The nervous system possessed by mammals is the most fa-

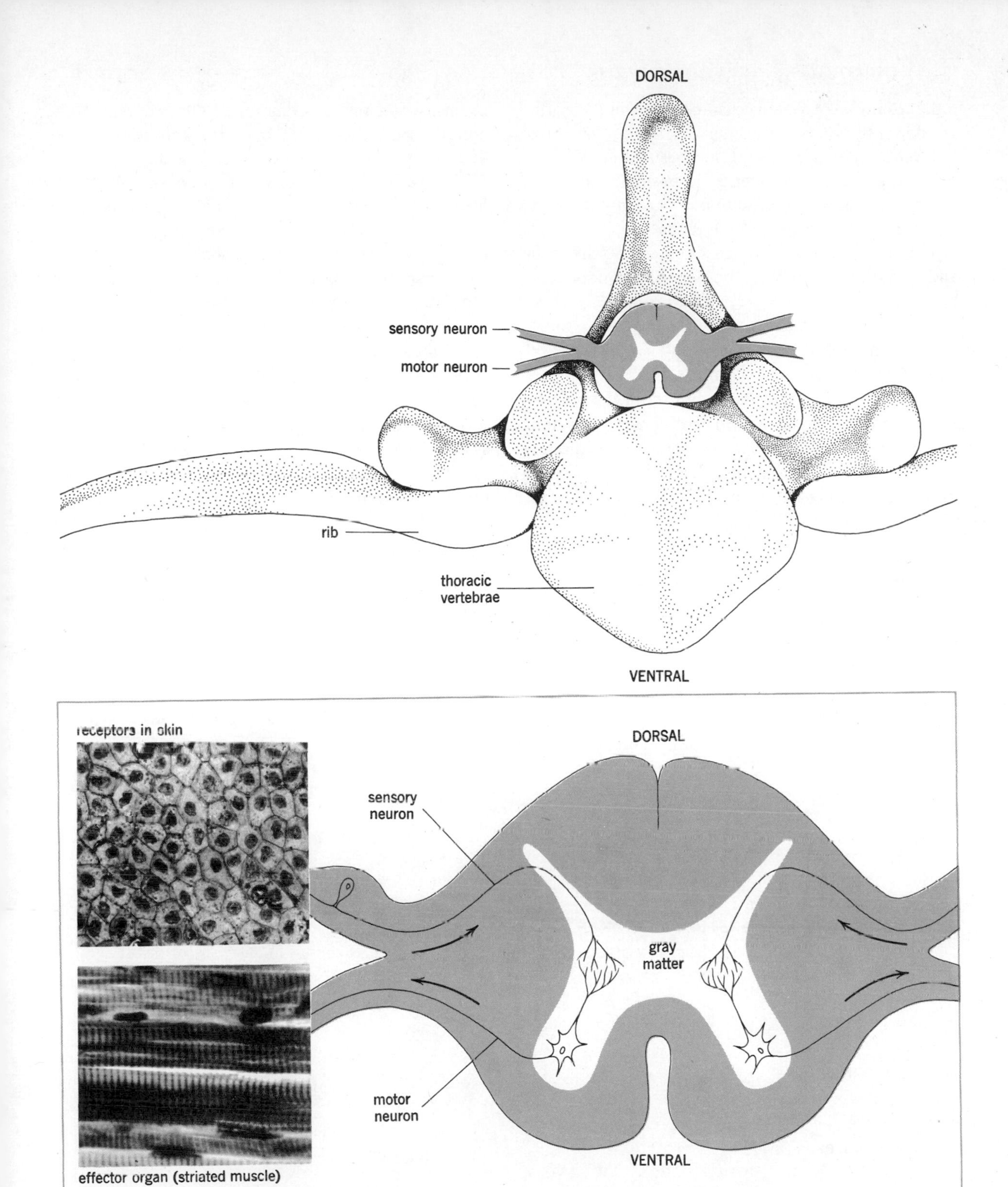

Figure 9.7. The spinal cord lies within vertebral elements (top). The lower drawing is an enlargement of the spinal cord showing some of the structural details involved in the spinal reflex. Each pathway consists of receptor (upper left), neurons, and effectors (lower left).

miliar example of a complex nervous system. Broadly speaking, the mammalian nervous system has three different parts: the central nervous system, the peripheral nervous system, and the autonomic nervous system. Although these three parts are interdependent, each of them has different functions.

The central nervous system, composed of the brain and spinal cord, contains the majority of nerve cell bodies in the nervous system. Sensory neurons carry information from the receptors into the central nervous system. Here the various inputs are interpreted and integrated, and a coordinated set of impulses is sent along the motor nerves, a part of the peripheral system, running from the central nervous system to the effectors. For simple responses, this integration often occurs in the spinal cord (Figure 9.7). An example of this is the common knee jerk reflex that occurs when the patellar tendon is tapped. More complex activities result from the activities of one or more parts of the brain (Figure 9.8).

The peripheral nervous system consists of nerve fibers whose cell bodies lie near the spinal cord or within the central nervous system. Since this system has no synapses with association fibers within it, there can be no interpretation or integration. It functions solely to carry impulses to and from the central nervous system.

Although the autonomic nervous system is often considered a part of the peripheral nervous system (it is outside the central nervous system; hence the term "peripheral"), we shall treat it as a distinct system. This system actually comprises two antagonistic systems, the sympathetic and parasympathetic nervous systems (Figure 9.9), which provide the

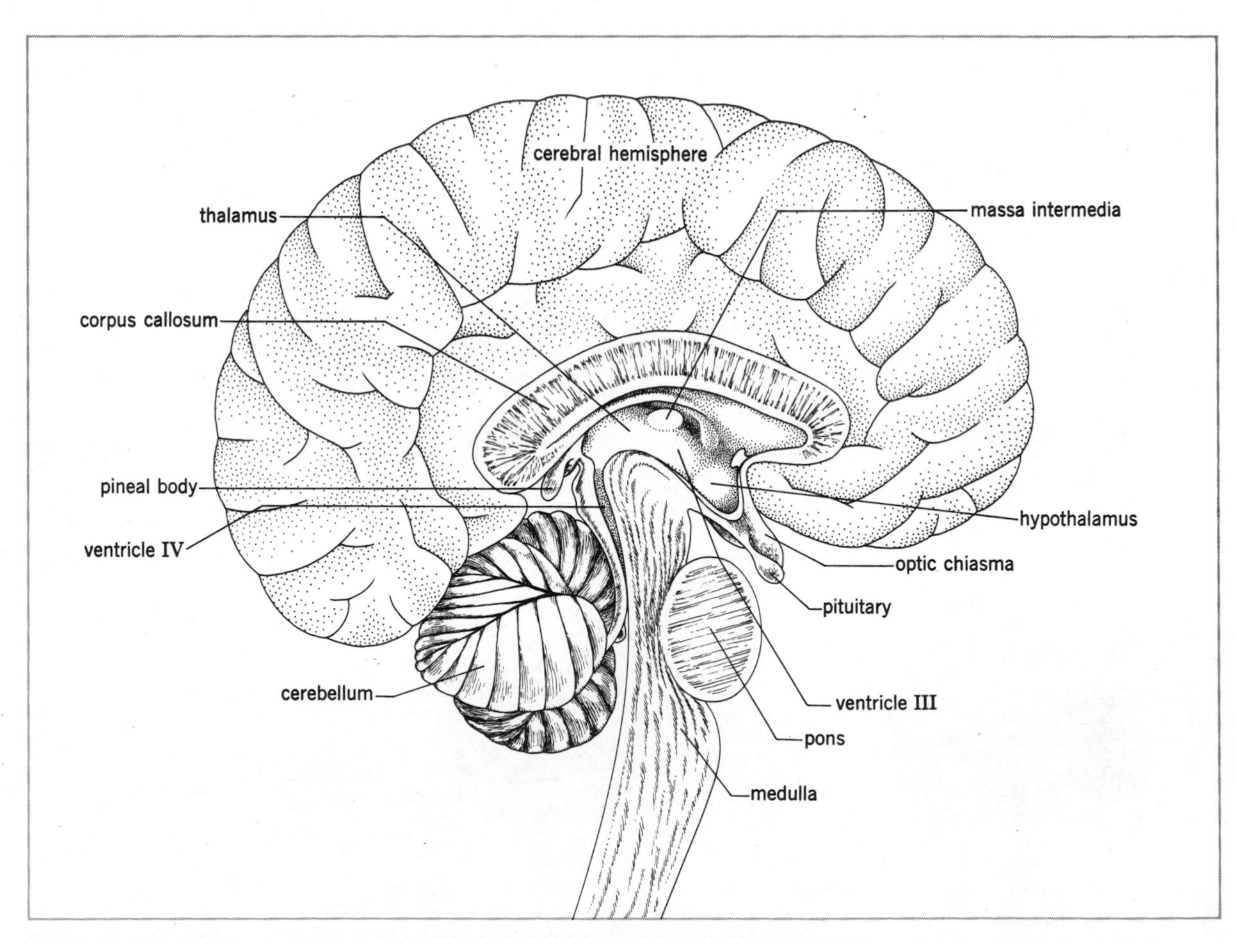

Figure 9.8. The structural complexity of the mammalian brain parallels its functional characteristics.

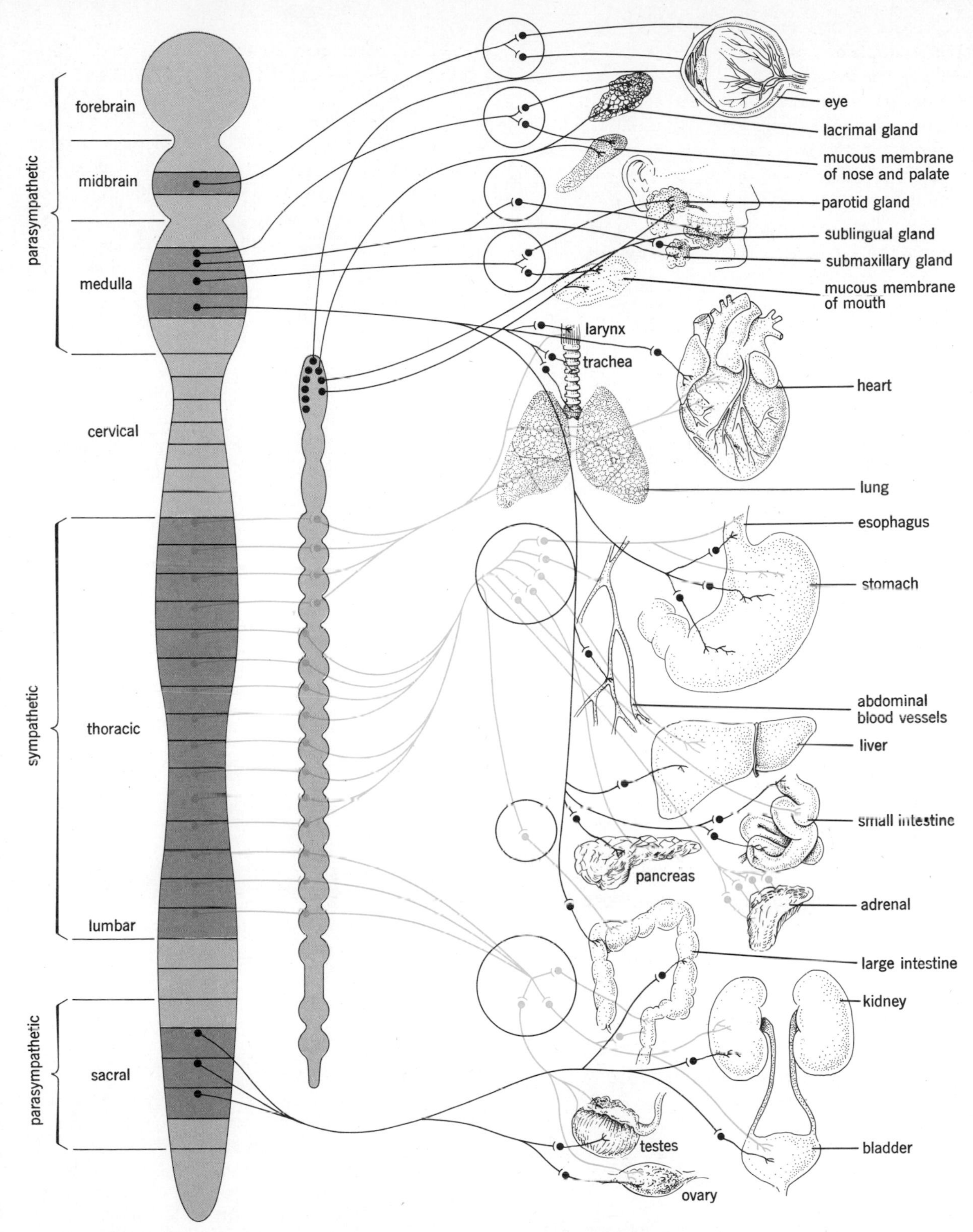

Figure 9.9. The autonomic nervous system and the structures it innervates. (Modified after *A. Nason,* Textbook of Modern Biology, *1965, John Wiley and Sons, New York.*)

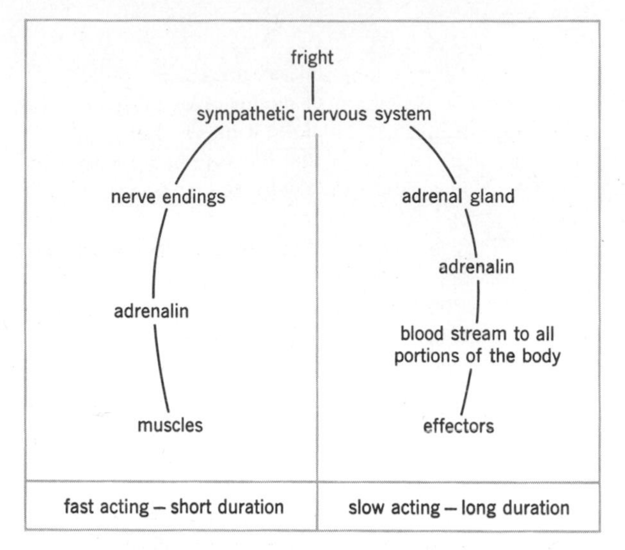

Figure 9.10. The dual pathways leading to adrenalin release during fright.

nerve supplies for the visceral organs such as the digestive tract, lungs, heart, and bladder, and a few other structures like the iris of the eye and some blood vessels. In controlling the activities of these organs, the autonomic nervous system is semi-independent of the central nervous system, since many of its activities are integrated within its bounds and, hence, are involuntary.

The action of the sympathetic nervous system, which releases adrenalin as a transmitter agent at the nerve ends, has been characterized as preparing the animal for "fight or flight." That is, the functions that are needed for immediate and extensive muscular activity, such as heart and respiration rate, are increased; those involved in longer-term energy needs, such as digestion, are decreased. The parasympathetic nervous system affects these organs in exactly the opposite way by releasing acetylcholine at its nerve endings.

If we judiciously choose an action controlled by these systems, we can achieve the same results either by blocking the activity of one system or by stimulating the activity of the other. For some years, a drug called atropine was used in eye examinations. It caused dilation of the pupil by blocking the action of acetylcholine on the nerve cell membranes. However, the long-lasting effects of this drug made its use somewhat undesirable. Now the same dilation of the pupil is accomplished by mimicking the action of the sympathetic nervous system with a synthetic adrenalin, called neosynephrine, which does not have such long-lasting effects.

The stimulation of the entire sympathetic nervous system, as in fright or anger, not only causes the secretion of adrenalin at the nerve endings but also releases large amounts of adrenalin into the blood stream from the interior of the adrenal gland. This adrenalin, a hormone, moves to all portions of the body, reinforcing the activity of the adrenalin released at the nerve endings. The apparent overlap points out one area of relationship between the neural and hormonal systems. The two systems may bring about the same effects but differ in the time aspects of coordination. The nervous system acts quickly because the transmitter agents are released so close to the site of action, but its action is of short duration; hormones take some time to reach the site of action through the bloodstream, but continue to be effective for a longer period of time (Figure 9.10).

Another area of overlap between the coordinating systems has been discovered recently. For some time it has been known that nerves sometimes stimulate the release of a hormone (adrenalin from the adrenal gland), but it is now known that some nerves actually secrete hormones. The hormones produced by the hypothalamus, which help coordinate reproductive activity, are good examples of neurosecretions that control the production or release of other hormones by more typical glands (Chapter VIII). Some neurohormones act directly on effector organs to coordinate body functions.

This overlap and interaction of the nervous and endocrine systems help to achieve the proper coordination of body functions and the maintenance of a satisfactory level of material necessary for continued existence.

Principles

1. Neural coordinating mechanisms are found in most multicellular animals and are characterized structurally by specialized cells: neurons and receptors.
2. A nerve impulse is a depolarization and repolarization of the neuron that occurs sequentially along the cell.
3. Chemicals produced by the ends of nerve cells facilitate the passage of the impulse between nerve fibers.
4. Neural and hormonal coordinating systems overlap functionally, but the neural coordination acts more rapidly and has a shorter duration.

Suggested Readings

Agranoff, Bernard W., "Memory and Protein Synthesis," *Scientific American*, Vol. 216 (June, 1967). Offprint No. 1077, W. H. Freeman and Co., San Francisco.

Baker, Peter F., "The Nerve Axon," *Scientific American*, Vol. 214 (March, 1966). Offprint No. 1038, W. H. Freeman and Co., San Francisco.

Eccles, Sir John, "The Synapse," *Scientific American*, Vol. 212 (January, 1965). Offprint No. 1001, W. H. Freeman and Co., San Francisco.

Galambos, Robert, *Nerves and Muscles*. Doubleday and Co., Garden City, N.Y., 1962.

Keynes, Richard D., "The Nerve Impulse and the Squid," *Scientific American*, Vol. 199 (December, 1958). Offprint No. 58, W. H. Freeman and Co., San Francisco.

Krnjevic, K., "Chemical Transmission in the Central Nervous System," *Endeavor*, Vol. XXV, No. 94 (1966).

McGeer, Patrick L., "Mind, Drugs, and Behavior," *American Scientist*, Vol. 50, No. 2 (1962).

Questions

1. Can a neuron function as an effector? Why?
2. Describe the chain of events that constitutes the movement of an impulse along a nerve fiber.
3. Some texts maintain that no one really knows what a nerve impulse is. Explain why this statement is or is not accurate.
4. What aspect of nerve impulse transmission requires the expenditure of ATP?
5. Why do we not consider a nerve impulse just a type of slow electricity flowing along a fiber?
6. What aspect of neural function illustrates neural and chemical control systems operating together?
7. Does acetylcholine fit the definition of a hormone given in Chapter VIII? Why?
8. What is the basic function of a receptor?
9. In general, why do complex receptors occur only in relatively complex organisms?
10. Can you think of any organism that has receptors that humans lack?
11. Why is it convenient or necessary, from a structural standpoint, for animals to have an autonomic nervous system as well as a central nervous system?

CHAPTER X

The Interaction of Control Systems: Homeostasis

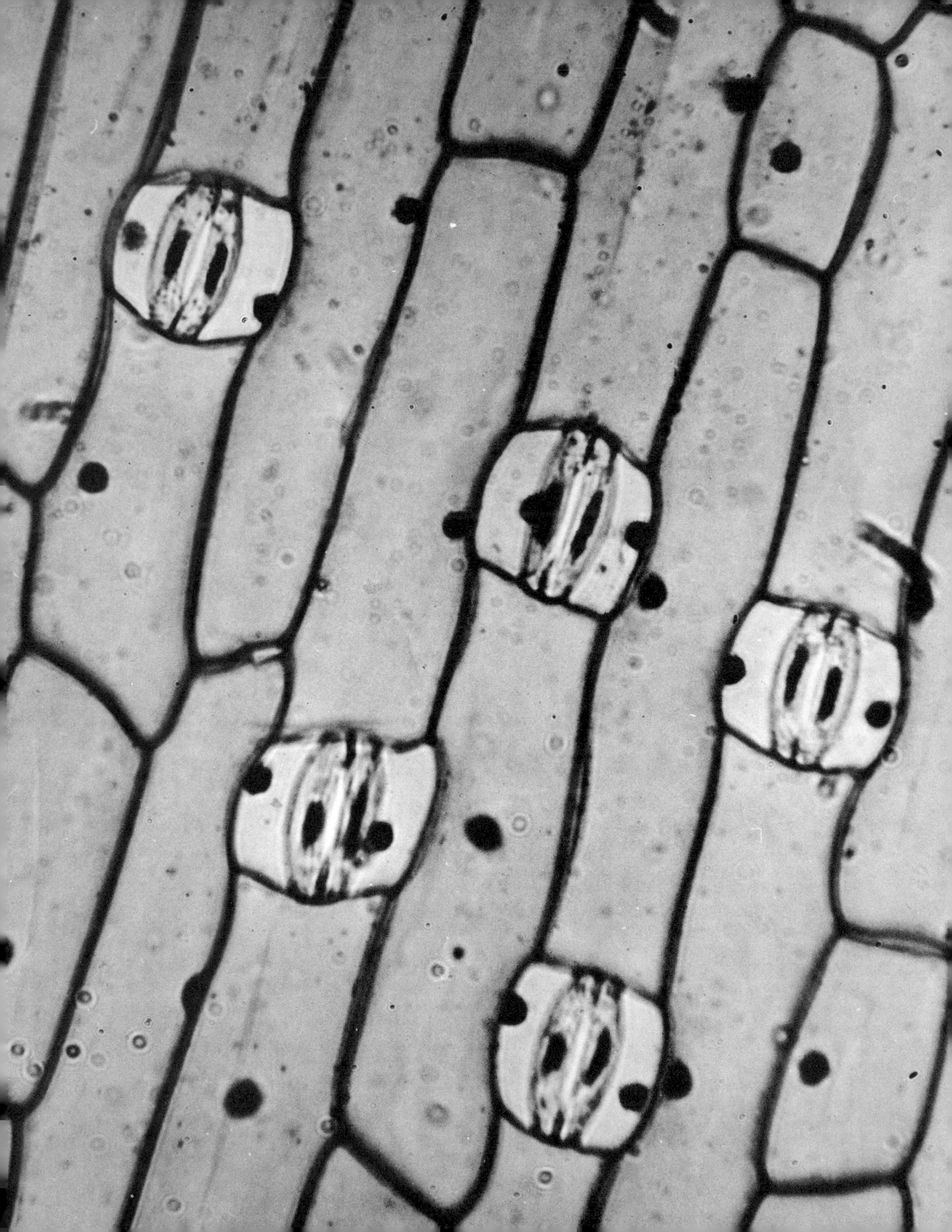

Stomata are openings that perforate the epidermis and lead to the interior of the leaf. Guard cells (small spindle-shaped cells) are located on either side of each stoma.

CHAPTER X

The Interaction of Control Systems: Homeostasis

In the nineteenth century Claude Bernard, a French physiologist, observed that the internal environment of organisms was maintained within narrow limits. For example, he noted that the chemical composition of blood fluctuated only slightly regardless of dietary intake or other activities of the organism. Bernard termed this maintenance process *homeostasis.* The concept of homeostasis, or *steady-state* control, is now recognized as one of the fundamental principles of biology. A high percentage of the energy expended by any organism is utilized in the maintenance of the steady state. If the energy supply is cut off, disorder ensues.

Principles

Steady state does not refer to a static or unchanging condition. Rather, it signifies a dynamic equilibrium. When we speak of a normal body temperature, blood sugar level, or heartbeat rate, we allude to values that fluctuate slightly around a norm. These fluctuations occur because a functionally desirable level is maintained through equalization of the input and output of the materials. The regulation of these fluctuations involves various steady-state controls (homeostatic mechanisms), and these occur on all levels, from the concentration of molecules within the cell (remember the effects of mass action) to

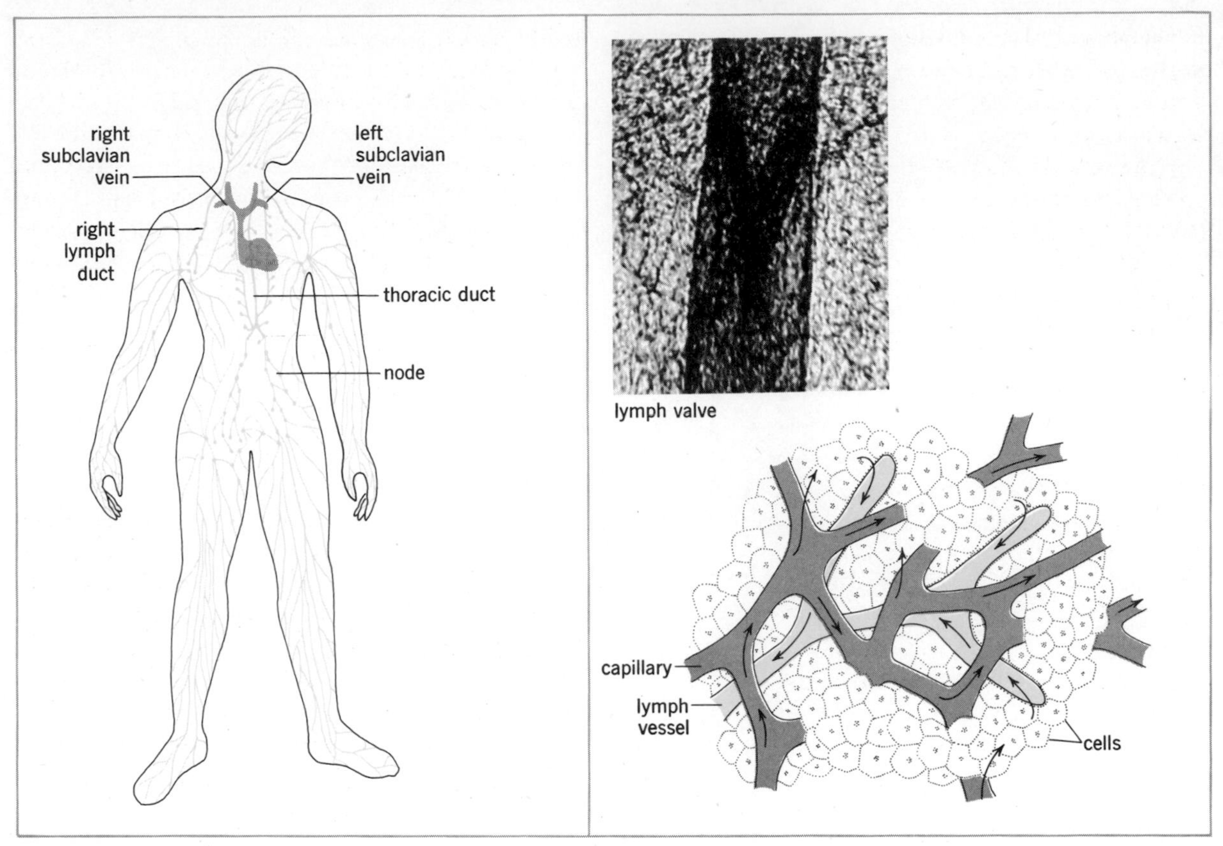

Figure 10.1. Man's lymphatic system and its relationship to cells. Valves shown in the photograph prevent backflow in the lymph vessels. The yellow vessels are lymph capillaries; the blue are blood capillaries.

the numbers of each species in a community.

The study of homeostatic mechanisms impresses two principles upon us. First, the maintenance of a steady state generally necessitates the interactions of the chemical and neural coordinating systems. Second, a steady state is necessary for the maintenance of life.

The remainder of the chapter consists of examples of homeostasis. As you consider each example, apply the principles for homeostatic mechanisms and notice the modifications that the basic makeup of the organism imposes in each case.

Examples

Lymph Volume. Among the general features of the lymphatic system, described in Chapter VI, is the fact that humans and other mammals normally maintain a nearly constant volume of lymph fluid between the cells. A steady input adds material to the fluid as blood pressure drives materials out of the capillaries. There are two routes for the return of materials pushed in by blood pressure: capillaries and lymph vessels (Figure 10.1). If either route for fluid return is impeded in any way, or if the fluid coming from the capillaries contains too much protein, the resulting accumulation of fluid in the intercellular spaces causes edema (swelling) of the tissues.

In the tropical disease called elephantiasis, small roundworms block the lymph vessels and prevent the return of fluid. The affected limbs swell to a grotesque extent, as shown in Figure 10.2. In elephantiasis, the normal homeostatic mechanism for removing intercellular fluid at the same rate at which new fluid enters the spaces from the capillaries has been upset. This illustrates the concept that diseases in general may be regarded as alterations of homeostasis.

Stomata. Although the relationships among the

different control systems in homeostasis can be easily illustrated with animal examples, comparable examples in plants are difficult to find. The absence in plants of complex coordinating systems, such as nervous systems, necessitates simpler mechanisms.

The actions of the guard cells around the openings (stomata) in the leaf surface show some characteristics of a homeostatic mechanism. In most plants, stomata are open during daylight hours and closed at night. This action is caused by guard cells becoming turgid when exposed to light and returning to a less turgid condition in the absence of light. This is a highly adaptive homeostatic function: it permits the interior of the leaf to exchange gases with the atmosphere when photosynthesis takes place.

Many hypotheses have been proposed to account for stomatal activity. A current one is that when light falls on the leaf, the concentration of carbon dioxide molecules near the stomata decreases as carbon dioxide is utilized in photosynthesis. The decrease in carbon dioxide concentration causes water to move into the guard cells by osmosis (Figure 10.3). When the cells fill with water, that is, become turgid, their shape and structural relationships force the stomata to open. In the absence of light, these events are reversed and the stomata close.

Another current hypothesis states that photosynthetic products synthesized in the guard cells themselves bring about the increased concentration of particles that leads to turgidity. As is so often the case in areas of contemporary research, one can interpret experimental evidence to support either of these hypotheses.

Stomatal opening has one drawback for the plant; it allows water to evaporate rapidly from the interior of the leaf. If this loss is greater than the amount supplied by the roots, it may activate a water conservation mechanism in the leaf. As water decreases in the leaf, the cells become relatively limp and the leaf wilts. Wilting helps prevent further water loss by reducing the leaf surface exposed to the air. Prolonged wilting, however, is detrimental to plants.

Thus, a balance is found between two opposing needs of the plant, the maintenance of carbon dioxide concentrations for photosynthesis and the maintenance of a suitable concentration of water. Notice that no definite system is involved in this homeostatic mechanism. Although it relies on relatively simple physical phenomena, life continues through the maintenance, within narrow limits, of the internal environment in the leaf.

Blood Gases. The control of carbon dioxide and oxygen in the blood stream is another example of homeostasis. This control involves some sense receptors outside the central nervous system, and other receptors in the brain itself that affect coordinating centers in two different parts of the brain. These centers in turn stimulate action in two different effectors to accomplish one goal, maintenance of the proper amount of oxygen in the blood. We also see a principle present in most homeostatic mechanisms: the fluctuations in the concentration of a material around its proper level maintain that level.

The basic metabolic relationship between oxygen and carbon dioxide makes it possible for mammals to control indirectly the oxygen content of the blood stream. Remember that if the cell's utilization of oxygen increases, the cell will generally produce a greater amount of carbon dioxide. Conversely, in the lungs the loss of carbon dioxide from the bloodstream accompanies increase in oxygen intake.

Two factors, the rate of respiration and the rate

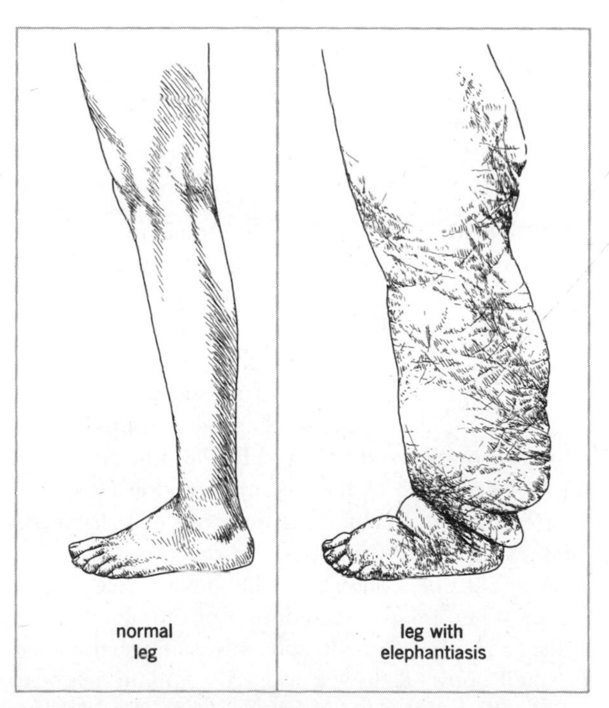

Figure 10.2. Elephantiasis involves tremendous swelling of connective tissue, as can be seen at right.

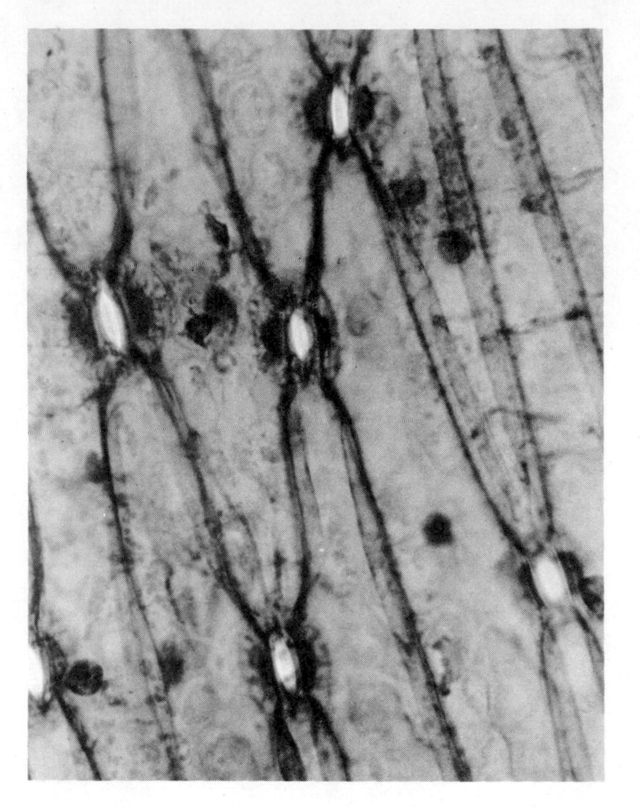

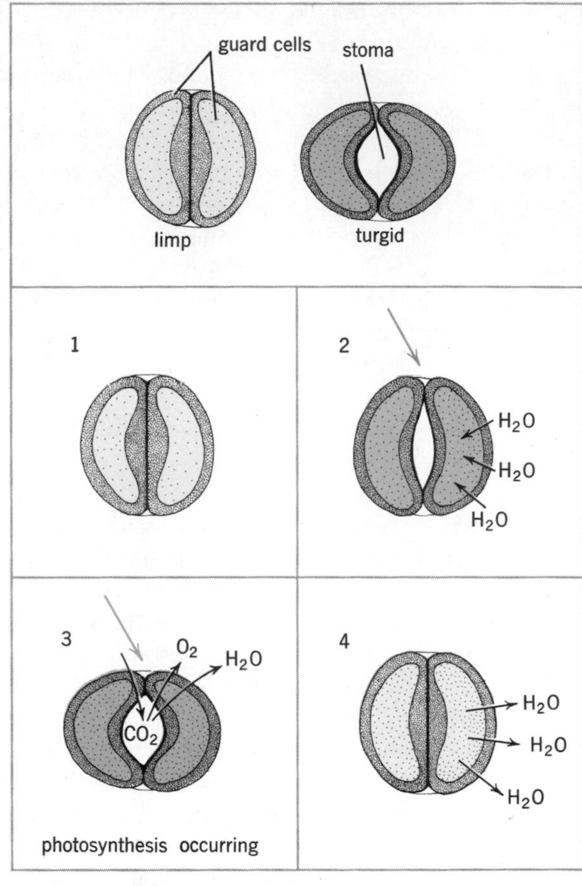

Figure 10.3. The structure and function of stomata, shown open in the photomicrograph. At right, the upper diagram depicts closed and open stomata. Diagrams 1 to 4 illustrate a normal cycle of stomatal activity.

of heartbeat, influence the carbon dioxide and oxygen content of the blood. The more frequently the air in the lungs is expelled, the more oxygen is available to the blood. In addition, the fresh supply of air will contain less carbon dioxide, resulting in a more rapid diffusion of carbon dioxide from the bloodstream. A more rapid heartbeat propels more blood through the lungs, and this also increases the rate of exchange of the respiratory gases.

Most of the mechanisms involved in the control of these gases are initiated by the changes in the carbon dioxide content of the blood. (See Figure 10.4.) When the oxygen content of the bloodstream falls, the carbon dioxide concentration usually rises. A small body on the carotid artery (main artery to the head) senses the increased carbon dioxide concentration and sends impulses to the brain which result in other impulses being sent to the heart to step up its rate of beat. Furthermore, the increased carbon dioxide concentration has a direct effect on a nerve center in the brain stem, which causes the respiratory rate to increase through nerve impulses to the diaphragm and other muscles of the rib cage. Thus, an increase in carbon dioxide concentration produces both a faster heart rate and a higher respiratory rate.

The operation of this homeostatic system can be observed when a person hyperventilates (forces himself to breathe rapidly). As the carbon dioxide level decreases in the blood, heartbeat and blood pressure drop. Skin divers sometimes hyperventilate in order to stay under water longer, but in doing so they take the risk of blacking out and possibly drowning.

Osmoregulation. Many homeostatic mecha-

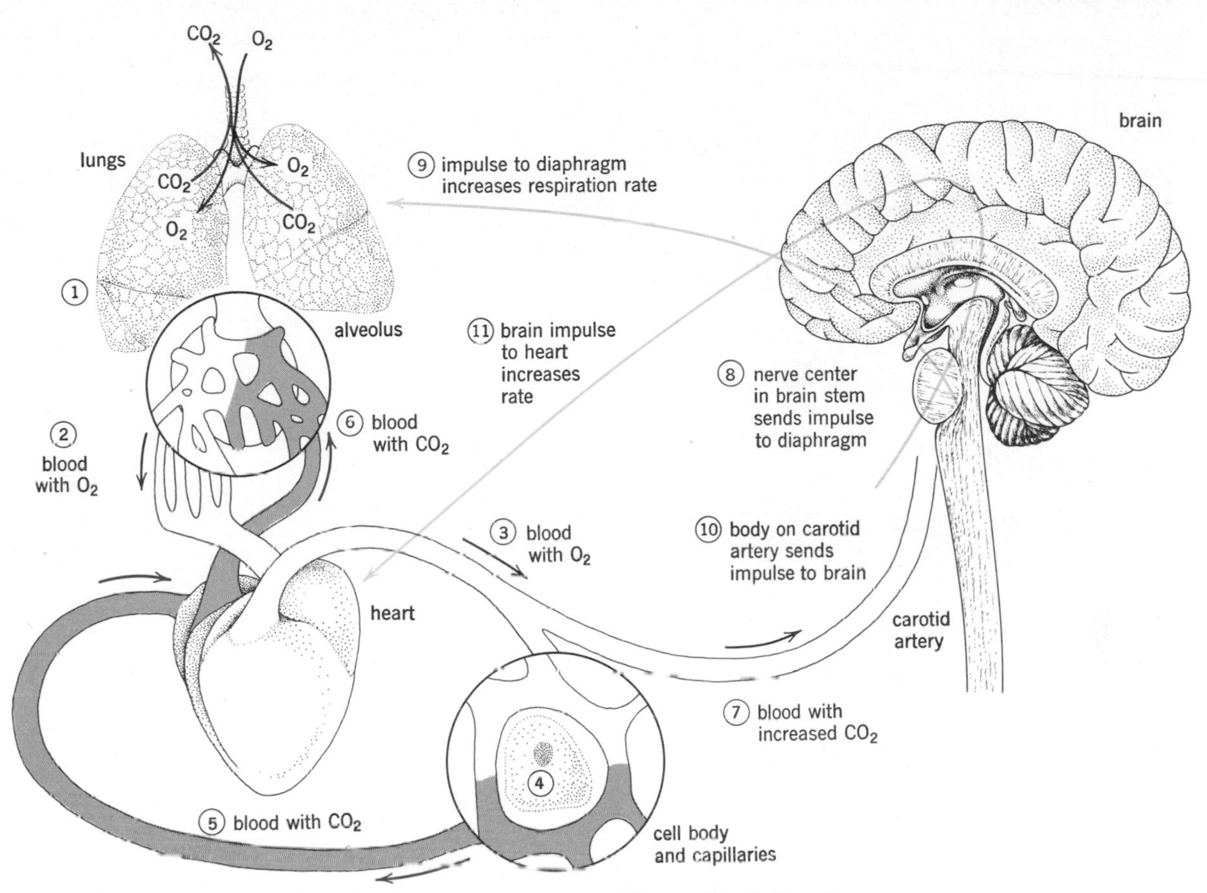

Figure 10.4. The control of carbon dioxide and oxygen content of human blood. Parts 1 to 6 show the movement of the gases within the circulatory system. Parts 7 to 11 depict the sequence of events following an increase in carbon dioxide concentration in the blood.

nisms involve still more complex interactions of coordinating systems. To understand the complexities and the variations among different organisms in accomplishing the same end, we must examine the same homeostatic mechanism in several organisms.

Many organisms are capable of controlling the concentration of water and salts in the cells, that is, they are capable of osmoregulation. Control of these materials is essential if the organism is to live in freshwater or dry land environments. Although the level of complexity varies in different organisms, the same general processes occur in the osmoregulatory organs (kidneys) of most larger animals. First, the organ receives a massive input of materials from the body fluid including both useful and waste products. This is followed by a massive return of most of these materials to the body fluid. Some of the materials that still remain in the organ may be useful to the organism at one time and present in excess amounts at another. The final event is the selective adjustment of the concentration of these materials consistent with the conditions in the organism. The fluid left in the organ after this adjustment is eliminated from the body.

The kidney found in crayfish (Figure 10.5) clearly shows these basic processes. Its low blood pressure is sufficient to drive fluid from the body cavity into each kidney, starting with the coelomic sac. This fluid contains both useful and waste products. In the first part of the kidney tube, called the labyrinth, most of the materials are reabsorbed into the blood stream by active transport and diffusion. This, of course, requires the expenditure of energy by the cells of the kidney. In the next portion of the tube

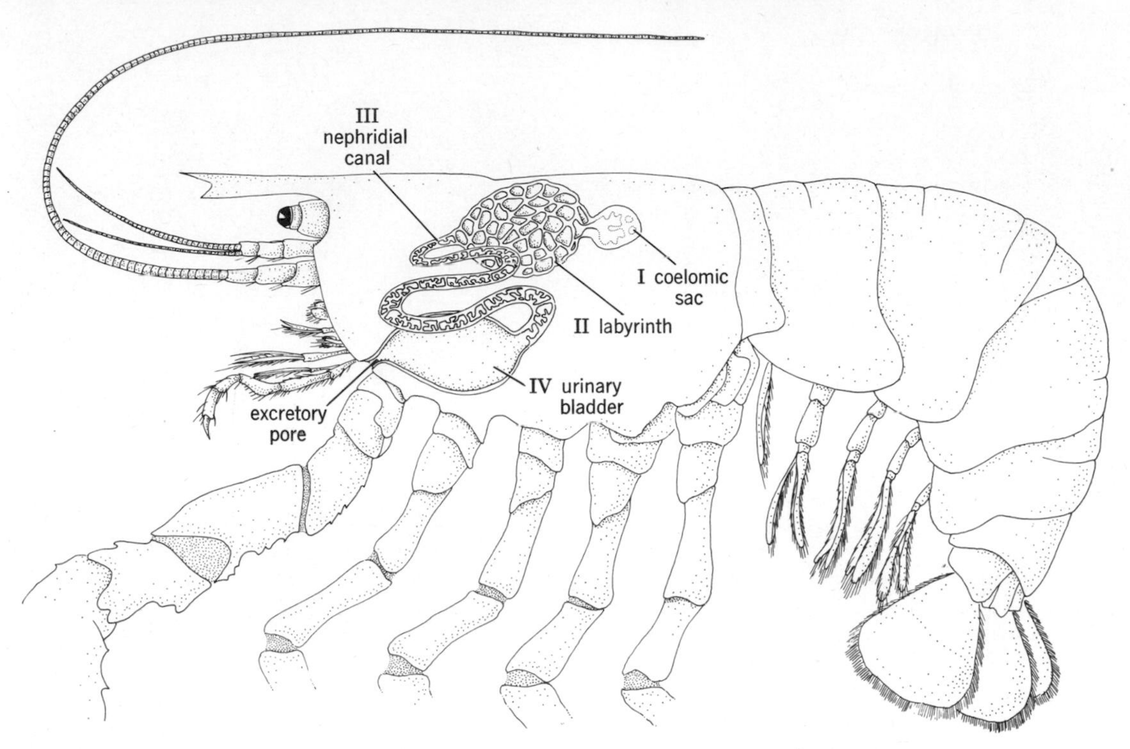

Figure 10.5. Crayfish kidney. I to IV indicate areas for intake, massive return, selective adjustment, and storage, respectively. (Modified from C. L. Prosser and F. A. Brown, Jr. (after A. Krogh and H. Peters), **Comparative Animal Physiology,** *second ed., 1961, W. B. Saunders Co., Philadelphia).*

(the nephridial canal) some materials that are conserved by the crayfish are reabsorbed from the remaining fluid. Other materials are transported from the blood to the fluid inside the tube. Now the fluid can be called urine and is simply stored in the bladder and then released to the outside of the organism.

Although the basic unit of the vertebrate kidney, the nephron, differs from the crayfish kidney in external appearance (Figures 10.6 and 10.7), the general processes are parallel in both. We find an area of massive input of fluid (*glomerulus* and *Bowman's capsule*), an area of relatively nonselective reabsorption (*proximal convoluted tubule*), an area for selective reabsorption (*Henle's loop* and *distal convoluted tubule*) and an area (*collecting duct*) for the final water uptake and for removal of the remaining materials, the urine, from the nephron. The only marked functional difference between the nephron and the crayfish kidney is the presence of the area of final water uptake in the former.

The fluid filters into the tubule through the glomerulus and Bowman's capsule under the force of blood pressure. Most of the materials in the blood are composed of molecules small enough to pass through the capillary wall. Only large molecules, like proteins, and blood cells fail to pass into the kidney tubule. Unless some disease or injury modifies or breaks the capillary walls, neither proteins nor blood cells ever appear in the urine. The materials that enter the tubule are waste products such as urea, a nitrogenous waste product that is less toxic than ammonia, and useful substances such as water, glucose, salts, and amino acids.

About two thirds of the volume of blood and intercellular fluid passes into the kidney tubules each hour. Because the amount of fluid passing into the tubule is so large, it is essential that most of this fluid be reabsorbed. The reabsorption occurs in the proximal convoluted tubule. Here, under normal conditions, all the glucose and amino acids are reabsorbed

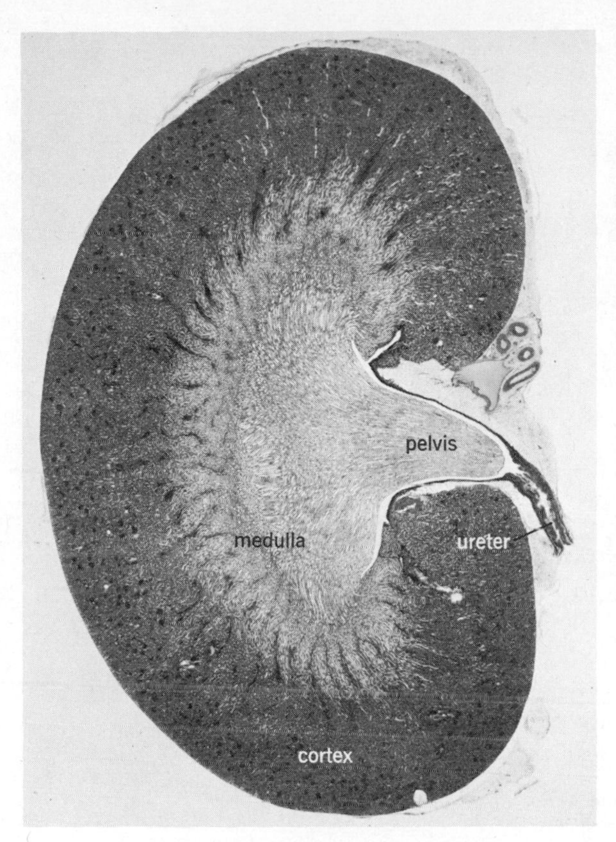

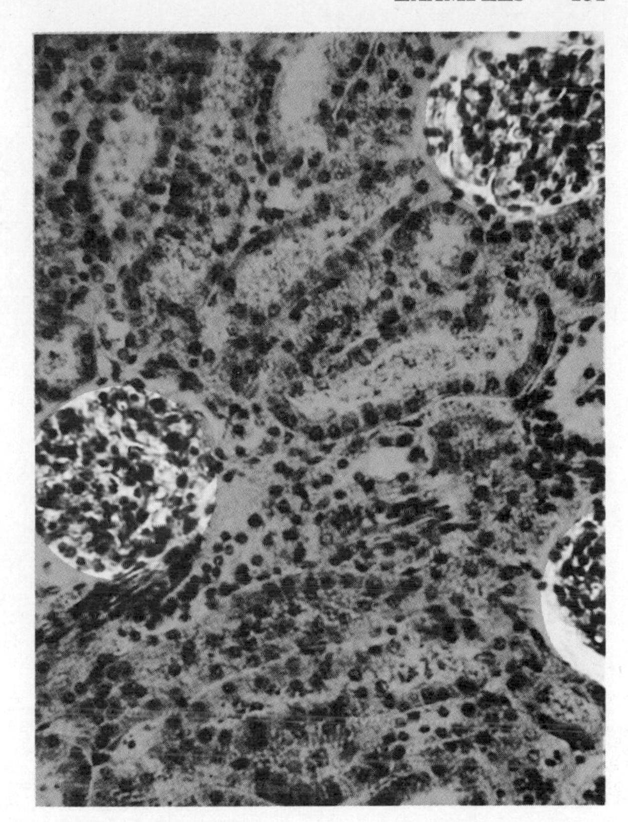

Figure 10.6. The general structure of the rat kidney (left) (CCM: General Biological, Inc.). Tubule cells (blue) and glomeruli are shown at the right.

by active transport into the capillaries surrounding the tubule. In addition, active transport removes about 99 percent of the salts. Water and urea diffuse from the kidney tubule as the other materials are transported out.

In addition, a few other minor modifications occur in this convoluted tubule. Some molecules too large to filter through the capillary wall are carried by active transport from the blood into the tubule. Most of these molecules, like penicillin, do not normally appear in the animal. Because the tubule is usually selective, urine is a highly useful excretory product for indicating certain kinds of homeostatic failures in humans. The presence of glucose may indicate diabetes, for example, and proteins in the urine may indicate certain kidney diseases.

The reabsorption of water in the proximal convoluted tubule occurs without regard for the shortage or excess of water in the animal. In contrast, sections of the tubule from Henle's loop through the collecting duct carry out processes that result in the selective reabsorption of water. The exact amount reabsorbed depends on the concentration in the blood.

If the blood has a low amount of water, cells in the hypothalamus sense this shortage (Figure 10.8). They cause the release of a hormone from the posterior pituitary gland into the bloodstream. This hormone increases the permeability of the distal convoluted tubule to water. Water leaves the tubule more freely and consequently a greater amount is reabsorbed here.

Once the amount of water in the blood reaches a normal level, or higher, the same cells in the hypothalamus no longer cause the release of as much of the hormone from the posterior pituitary gland, and the permeability of the distal convoluted tubule to water drops to its normal level, or below. Less water is reabsorbed, thus causing a tendency to maintain a constant amount of water in the body.

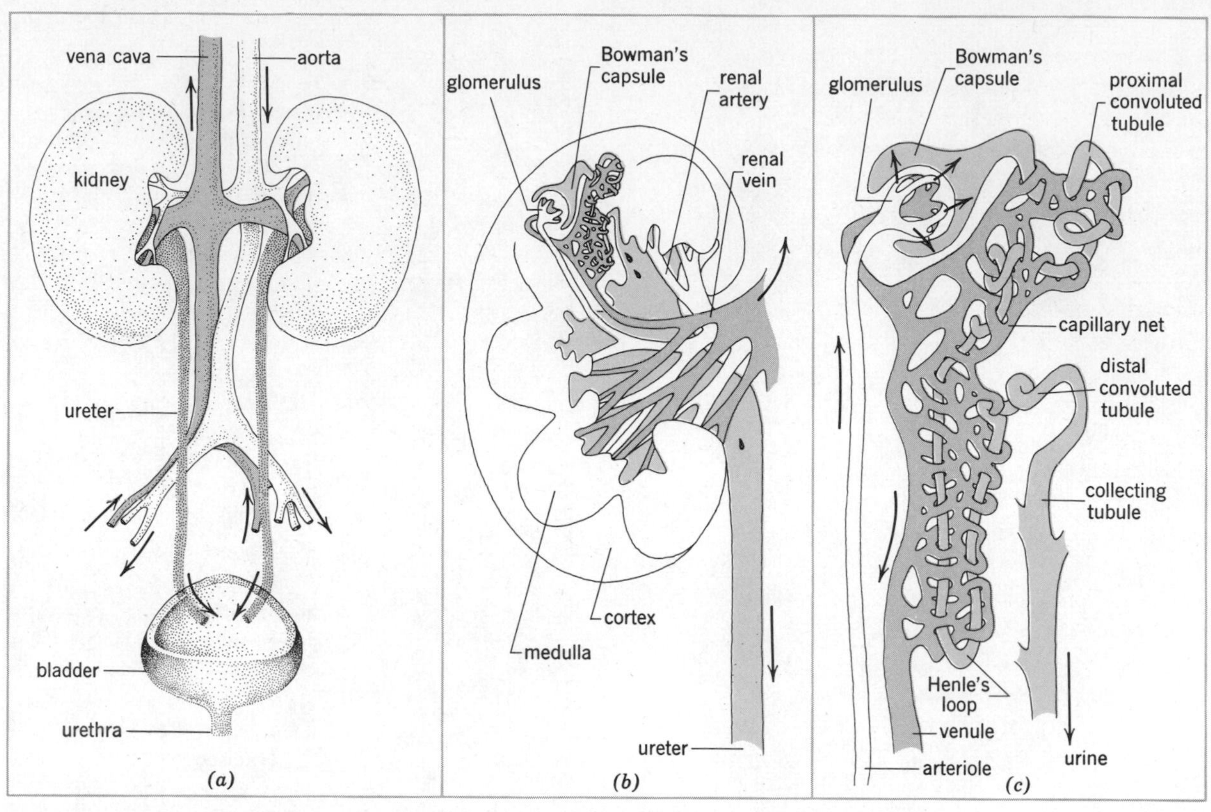

Figure 10.7. (a) ***The mammalian excretory system.*** (b) ***A longitudinal section of the kidney showing the position of a nephron.*** (c) ***Detailed structure of the nephron.***

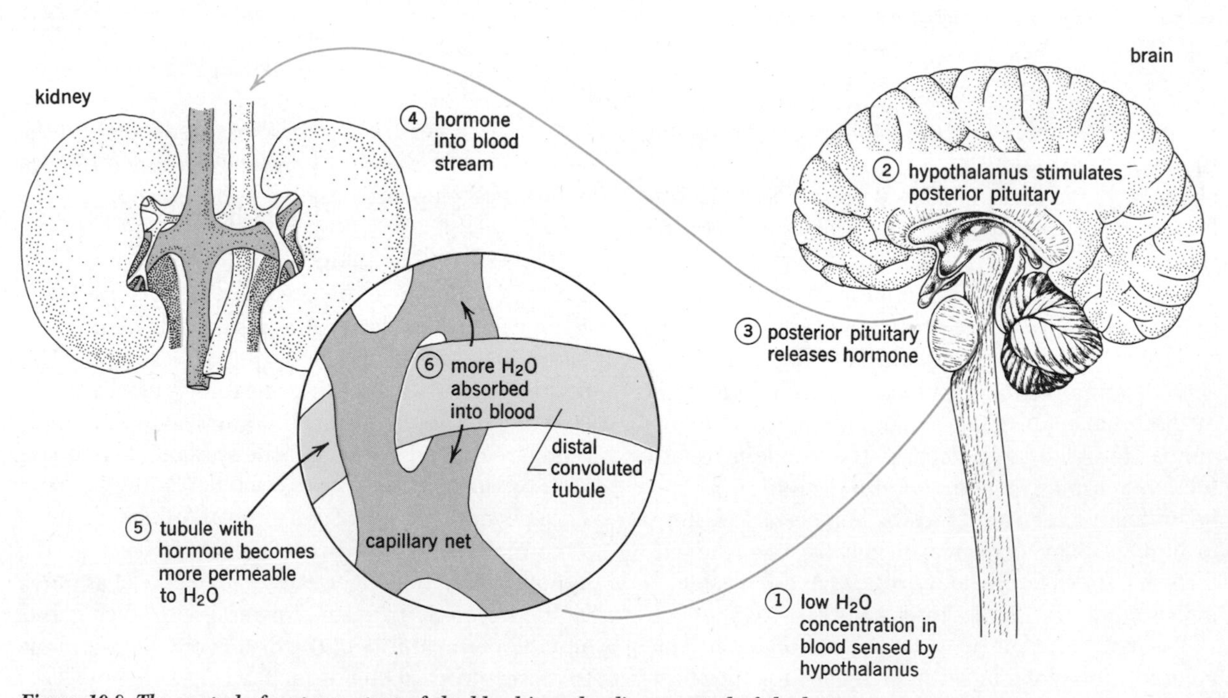

Figure 10.8. The control of water content of the blood is under direct control of the hypothalamus.

The reabsorption of salts in the distal convoluted tubule is also subject to hormonal control. Some of the steroid hormones from the adrenal gland tend to increase both reabsorption of sodium and excretion of potassium in the tubule. When potassium increases in the body, more of the hormones are released from the adrenal gland. A return to normal salt concentrations results in a reduction in the amount of adrenal hormones released. Although the control of the amount of water involves neural and hormonal elements, the mechanisms for reabsorbing salt are apparently partially controlled by purely hormonal coordination. The kidney is therefore a good example of the interplay of several parts of different coordinating systems to bring about homeostasis or steady state of the body fluids.

The examples presented here all emphasize the single universal characteristic of homeostatic mechanisms: their tendency to preserve life. After these examples, one may be tempted to call life itself a steady state. Most biologists would agree.

Principles

1. Organisms maintain their internal environment within narrow limits by regulating intake and loss of necessary materials.
2. Most of the regulatory mechanisms involved in homeostasis comprise neural and hormonal processes.

Suggested Readings

Adolph, E. F., "The Heart's Pacemaker," *Scientific American*, Vol. 216 (March, 1967). Offprint No. 1067, W. H. Freeman and Co., San Francisco.

Benzinger, T. H., "The Human Thermostat," *Scientific American*, Vol. 204 (January, 1961). Offprint No. 129, W. H. Freeman and Co., San Francisco.

Chapman, Carleton B., and Jere H. Mitchell, "The Physiology of Exercise," *Scientific American*, Vol. 212 (May, 1965). Offprint No. 1011, W. H. Freeman and Co., San Francisco.

Langley, L. L., *Homeostasis*. Reinhold Publishing Co., New York, 1965.

Merrill, John P., "The Artificial Kidney," *Scientific American*, Vol. 205 (July, 1961).

Schmidt-Nielson, Knut, *Animal Physiology*. Second edition. Prentice-Hall, Englewood Cliffs, N.J., 1964, pp. 47–67.

Scholander, P. F., "The Master Switch of Life," *Scientific American*, Vol. 208 (December, 1963). Offprint No. 172, W. H. Freeman and Co., San Francisco.

Smith, Homer W., "The Kidney," *Scientific American*, Vol. 188 (January, 1953). Offprint No. 37, W. H. Freeman and Co., San Francisco.

Questions

1. What is health from a biological viewpoint? Is it a homeostatic state?
2. Why is it difficult to find examples of steady-state systems in plants?
3. Do you think that the plant growth substances described in Chapter VIII function as a homeostatic system? Explain.
4. In what sense are stomata homeostatic devices?
5. Explain how the amount of carbon dioxide in the blood regulates the rate of heartbeat and breathing.
6. How does this mechanism affect the length of time that a person can hold his breath?
7. Explain why the kidney is often used as an example of a homeostatic organ?
8. Which of the following seems to play the most important function in steady-state systems: nervous system, hormonal system, or a combination of the two?

CHAPTER XI

Communication: The Basis for Interorganismic Coordination

A honeybee returning to the nest (© Walt Disney Productions).

CHAPTER
XI

Communication: The Basis for Interorganismic Coordination

The maintenance of life requires more than the maintenance of a constant environment within an organism. Every organism necessarily interacts to some degree with other organisms. Survival often depends on whether this interaction is properly coordinated. One means of attaining this is for information to be communicated from one organism to another. Examples of vital information that needs to be passed along are physiological readiness for mating, presence of food, and danger of predators. If this communication does not occur, the life of the individual and of the species will be endangered.

Communication may be accomplished by four different types of signals: *acoustic, visual, chemical,* and *tactile.* Although these are nearly self-explanatory terms, examples help to indicate the kinds of signals in each category. We must keep in mind that the complexity of communication possible between organisms is dependent on the complexity of their nervous systems.

In the examples that follow, some specific color or movement, sound, smell, or touch serves as the vehicle for transferring information from one organism to another. Three general conditions must be met. First, if these signals are not to be confused with others, there must be *effectors* available that can produce exactly the same signal each time. Second, the organism that receives the signal must have *receptors* capable of distinguishing it from other similar signals. Third, a system must be present that can interpret and coordinate the information received.

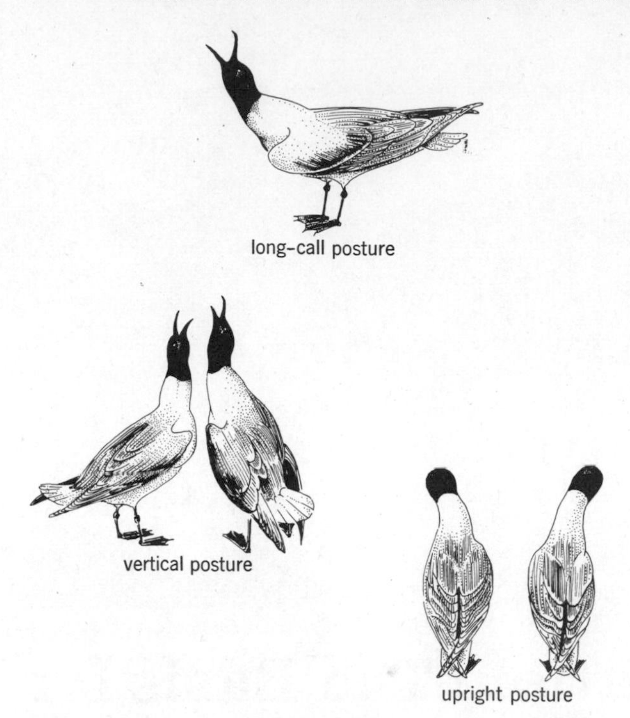

Figure 11.1. Combinations of auditory and visual signals in the Little Gull. The long call and vertical postures are signals that enable the gull to recognise members of its species. The upright posture is a ritualized stance used before mating by this species of gull. (*Adapted from an illustration by Louis Darling for* Time-Life Books, *1963, Time Inc.*)

As a rule, more complex behavior results from the greater number of distinctive behavior patterns available to an animal with a more advanced nervous system. Since these animals can make different responses to slightly different stimuli, they generally employ more intricate sets of signals for communication.

Types of Signals

Until recently, most of the studies on animal communication concentrated on the kinds of signals used by man himself, namely auditory and visual. However, with the discovery of pheromones (see p. 159) as well as a heightened general interest in animal behavior, the importance of other modes of communication has become evident. For example, most, if not all, animals transmit information via specific chemical signals. This mode of communication is probably used more extensively than any other single kind. Man with his poorly developed olfactory sense was not aware of this chemical language until very recently. Perhaps his perfumes and deodorants are a crude form of this communication medium!

The sense of touch, as a means of passing along information, is vital to some animals but again is used only slightly by man.

Acoustic Signals. Acoustic signals are used by a variety of organisms, including some insect groups and many of the vertebrates. A majority of the invertebrate groups, however, do not produce sound and must depend on other means of communicating. We usually think of sound being produced vocally, but insects vibrate various parts of their bodies to make specific signals, and male woodpeckers of some species drum on hollow objects during their courtship of females. In bats and porpoises, ultrasonic signals are used as echo-location devices as well as communication.

Bird songs are familiar acoustic signals and consequently have been investigated extensively. By singing, a male bird may inform other birds as to his species, the fact that he is a male, and that he is in breeding condition on a territory (any defended area). All of this information is important to other birds. It helps females to find males of the same species who are ready to mate. It helps reduce unnecessary fighting between males by notifying other males of the presence of a resident bird in the territory. Singing is also important in bringing the female of some species into breeding condition. These activities help to maintain the species by furthering reproduction and reducing losses from competition. From even this brief description it is evident that a wealth of information is transmitted from bird to bird by vocal means.

In nearly all organisms the ability to produce specific sounds or other kinds of signals and to interpret them correctly is primarily hereditary. In a sense, the production of the signal or the reaction to it is automatic, almost involuntary. For example, a tape recording of the distress calls of a starling will alarm and set in flight a flock of starlings even though the signal obviously arises from a machine and not from an injured bird.

Visual Signals. Birds also communicate by color or pattern of plumage, by specific body movements, and by assuming certain body forms. For example, the male flicker has a black "whisker" mark on his face which is lacking in the female. Otherwise, the plumage appears to be identical in the two sexes. Normally, males court females and try to drive other males from their territories. If a male is caught by an experimenter and the black whisker marks painted out, he will be treated by the rest of the population as a female. Other males will allow him on their territories and will attempt to court him. Females will ignore him. Likewise, if a whisker is painted on a female, she is treated as a male by other flickers. Here, one small difference in coloration is obviously very important to the species.

Colors and color patterns function in the communication systems of nearly all animals, frequently as part of their courtship behavior. Thus the male green anole (a lizard frequently mistaken for a chameleon) extends a red-orange flap of skin from its throat when encountering another anole. This simple signal identifies the species, sex, and breeding condition of the animal. It attracts females and establishes the territory of the male.

Body form and movement are also frequently used for transmitting information as illustrated in Figure 11.1. Everyone understands the signal conveyed by a dog with bared teeth and erected back hair!

Chemical Signals. Chemical signaling is evidently widespread among all levels of the animal kingdom and is even used by a few plants. Much of this signaling involves the use of *pheromones,* which are chemicals released into the environment that influence the behavior of another organism of the same species.

Pheromones have been studied in detail in insects. Here they are often sex attractants produced by the female, are effective in fantastically minute amounts, and attract only males of the same species. In the case of the gypsy moth, females produce a chemical called gyplure, which the wind carries great distances. Males perceive the odor and follow the "sex trail" to the female. This enables the sexes to find each other at night and reduces the exposure of the moths to predators. This pheromone has been synthesized commercially and used to trap the moth, whose caterpillars are extremely destructive to trees in New England. Undoubtedly, similar lures will be widely used eventually for such pests as cockroaches.

In ants, the laying down of a chemical to mark the pathway to a new food source is another example of the communication of exact information in animals. This chemical is secreted by a special gland and applied to the ground by means of a stinger. Other ants following the first ant may reinforce the trail. This enables ants to move to and from the food along a well-marked pathway, rather than to wander about searching for food. Since the trail disappears as the chemical evaporates, only relatively new trails will be followed.

This type of communication among ants points out a unique feature of the use of pheromones: the signal persists and continues to convey information for a short time after its producer has departed. In other words, an individual can communicate with another in the future.

Our final example of chemical communication involves a genus of water molds called *Achlya.* Here chemicals travel between male and female *Achlya* fungi to synchronize their reproduction (Figure 11.2). Each of the four chemical substances involved in the sequential interactions between two fungi causes a specific behavior. In this way the chemicals released by *Achlya* function like pheromones in animals.

Tactile Signals. Touch receptors form part of the basic anatomy of all animals, functioning in such diverse activities as avoiding obstacles, fighting, and copulating. It is not unexpected to find this system also involved in the transfer of information. Many animals use tactile signals in their courtship behavior. In turtles, the male often strokes or scratches the female during courtship. The male three-spined

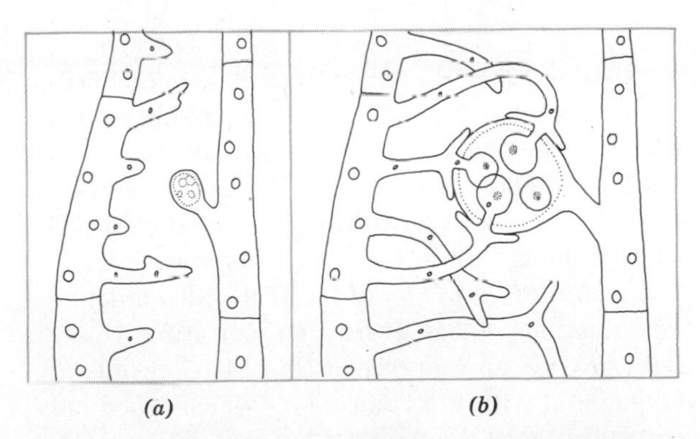

Figure 11.2. Chemical communication and reproduction in Achlya. *In* (a) *the male strand (at left) releases a chemical into the water that stimulates the female strand (right) to form an egg. A secretion from the female causes the male to grow finger-like extensions toward the egg. In* (b) *the reciprocal chemical signaling between male and female strands has brought about envelopment of the mature egg by gamete-producing extensions of the male strand. Fertilization follows.*

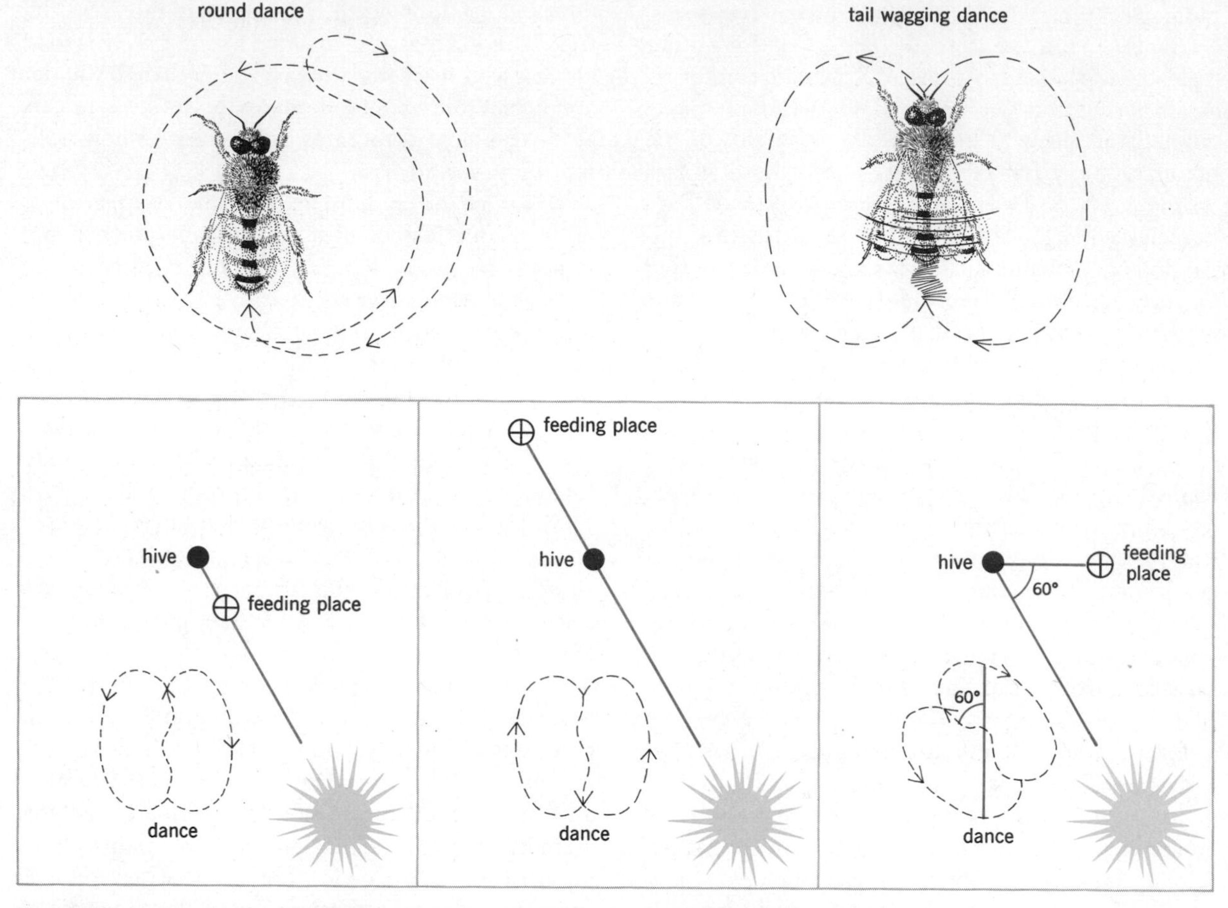

Figure 11.3. Communication by dances in honeybees. The lower diagrams show how direction is communicated. (*Based on experiments and observations by Prof. Karl von Frisch.*)

stickleback fish nudges the tail of the female to trigger her release of eggs. In the courtship of field crickets, the male touches the female repeatedly with his antennae. Certain fishes generate electricity that is often used as protection but may also function as a signaling device under certain conditions.

Combinations of Signals. Nearly all animals use combinations of signals in communicating. Honeybees provide an example of this. Many people are acquainted with the "dance" by which a bee indicates to other bees the distance and direction to a source of food or a suitable nest site (Figure 11.3). If the food is near the hive, the bee that discovers the source performs a round dance in the hive. This gives no information on the direction to the food but simply says that it may be found within a short distance of the hive. When the food is located farther from the hive, the bee performs a wagging dance which follows a modified figure-eight pattern. Two variations in the dance occur—one giving the direction to the source, the other the distance.

The direction is always given in terms of the angle from the sun. If the central portion of the figure eight is vertical on the comb in the hive, the food lies in the direction of the sun. The angle from the vertical axis indicates the angle from the sun at which the food is located. The central portion of the dance is the only part during which the bee waggles, that is, moves its abdomen from side to side. The number of times this waggling occurs per unit of time indicates the distance to the source. The closer the source is, the more rapid the waggling. The precision of the information conveyed about the location of the food is demonstrated by man's ability to find the food source on the basis of observations of the frequency and direction of the dance.

As other bees fly out, gather food, and return to the hive, they also dance. As the food source becomes depleted, the dancing for that source diminishes and the attention of bees is drawn to some other food source. Taking into account the fact that bees are attracted to the dancing that has the greatest activity, what kind of food source would draw the most bees if dances for two sources were being carried out at one time?

Studies on different varieties of honeybees have shown some interesting variations of the dances. The most unusual is the variation in the coding of the distance to the food in different varieties of bees. All varieties use the same basic code but differ in the distance at which they change from the round dance to the waggle dance (10–275 feet) and also in the distances coded by various rates of waggling. Four waggle dances in 15 seconds may range in meaning from 1000 feet in one variety to well over 2000 feet in another. Hives with a mixture of varieties of bees show interesting results in the misinterpretation of the dances.

The emphasis has been on the visual elements of the dance. When we remember that the dance is usually performed in the hive, which may be nearly dark, the importance of tactile and auditory elements of communication is apparent. Recent experiments show that the frequency of a sound produced during the dance alters with a change in distance to the food source. In addition, the presence of the food on the dancing bees communicates the type of food available.

Most of our knowledge about the language of the bees has been derived from experiments performed by an Austrian biologist, Karl von Frisch, and his associates. Recently, however, two American biologists, experimenting with the food-finding abilities of honeybees, concluded that some aspects of von Frisch's dance-language hypothesis are too complex. According to their interpretation, bees do not use information gained from the dance in determining distance to a food site. Instead, they use other information after leaving the hive, such as the odor of their hive mates, in searching out a particular food source.

This scholarly controversy, including von Frisch's defense of his hypothesis, can be read in the February 17, 1967 and November 24, 1967 issues of *Science*. It provides an example of the kind of disagreement that often leads to additional advances in research.

Deception in Communication

In the examples of communication discussed so far, information was passed from one individual to another of the same species. Communication is also found between individuals of different species. Many species of insects, such as the monarch butterfly, which are impalatable or capable of hurting the predator, advertise their presence by distinctive coloration. Once a predator discovers the bad taste or the sting, it will avoid other individuals that look similar. Occasionally, other species look very much like the noxious species but do not have the sting or the bad taste. This similarity in appearance is called *mimicry* (Figure 11.4). Individuals of these mimic species are also avoided by predators and are

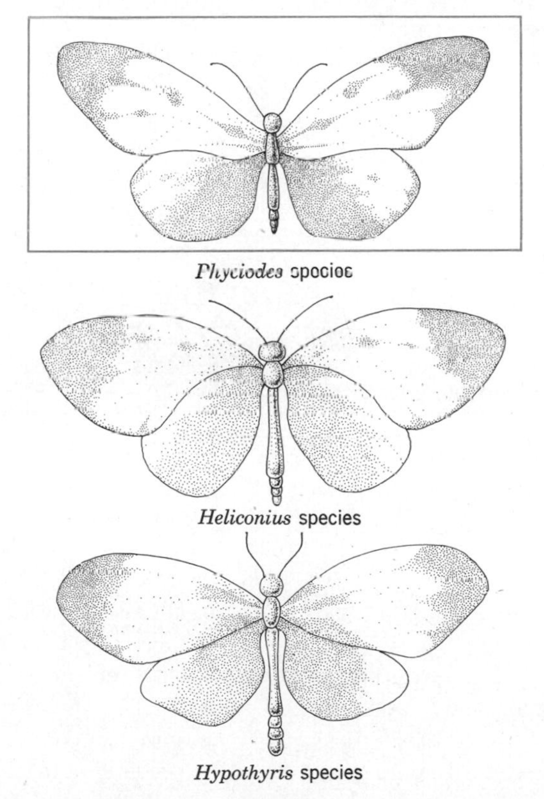

Figure 11.4. Mimicry. The lower two species of butterflies are inedible. The third species is not eaten by predators even though it is edible. (Adapted from an illustration by Margaret L. Estey for Time-Life Books, *1964, Time Inc.)*

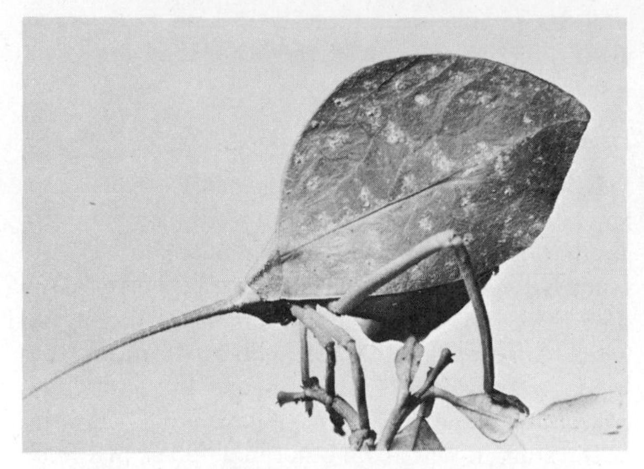

Figure 11.5. This leaflike insect is extremely difficult to locate when in its natural surroundings. Its actual body length is about $1\frac{1}{2}$ inches. (Dr. E. S. Ross).

Figure 11.6. Arctic hare. This mammal closely matches its snowy background (© Walt Disney Productions).

protected in this way. Notice that the information communicated to the predator in this case is false.

Coloration may serve to communicate false information in another way. Many organisms are protectively colored, that is, they match their normal background. Here, the predator seems to see only the background but no prey. (See Figures 11.5 and 11.6.)

Scent is also a means of communication between species. Many prey are found by predators through their odor. The skunk (Figure 11.7) presents the opposite picture. We know the effect that the smell of skunk has on our own behavior. Many other animals react in similar fashion.

Figure 11.7. A skunk communicates its presence by scent and distinctive coloration (Dr. L. Ingles).

Most bird calls or songs are very specific, relaying information only to other members of that species. However, some types of warning notes affect individuals of many species. If a hawk appears, a single warning note may silence the activity of nearly all the birds in the area. The same nonspecific warning note may be given by many species and recognized as a warning note by many small birds. Rattlesnakes pose an interesting problem in this respect, since they usually vibrate their rattles while pursuing prey. Whom are they warning if, in fact, it is a warning signal? After all, it seems self-defeating for the snake to alert the animal it intends to eat.

Sound may also serve as a carrier of false information. The hognosed snake, or "puffing adder," when threatened will spread its neck and hiss (Figure 11.8). It also makes false striking motions. This fierce act seems to be relatively successful in misleading other animals. Even people think the puffing adder is dangerous. Those who have kept hognosed snakes in captivity have discovered that they are really docile.

Figure 11.8. The hognosed snake communicates threatening information by its behavior. In reality, this snake is docile when handled (CCM: General Biological, Inc., Chicago).

In this chapter we have presented only one aspect of a recently developed field in biology, that of animal behavior. Pioneers in this field were mostly Europeans such as Konrad Lorenz, Karl von Frisch, and N. Tinbergen, but interest is now worldwide. New discoveries and concepts in this field hold great promise in helping man better understand his own highly complex behavior.

Principles

1. Acoustic, visual, chemical, and tactile signals are the means by which organisms communicate information to one another to coordinate their activity.
2. The intricacy of the signals used is linked to the complexity of the nervous system.

Suggested Readings

Bonner, John Tyler, "How Slime Molds Communicate," *Scientific American*, Vol. 209 (August, 1963). Offprint No. 164, W. H. Freeman and Co., San Francisco.

Esch, Harold, "The Evolution of Bee Language," *Scientific American*, Vol. 216 (April, 1967). Offprint No. 1071, W. H. Freeman and Co., San Francisco.

Frings, Hubert and Mable, "The Language of Crows," *Scientific American*, Vol. 201 (November, 1959).

Frings, Hubert and Mable, *Animal Communication*. Blaisdell Publishing Co., Boston, 1964.

Krogh, August, "The Language of the Bees," *Scientific American*, Vol. 179 (August, 1948). Offprint No. 21, W. H. Freeman and Co., San Francisco.

Lorenz, Konrad, *King Solomon's Ring*. Thomas Y. Crowell Co., New York, 1952.

Marler, Peter, "Animal Communication Signals," *Science*, Vol. 157, No. 3790 (August 18, 1967).

Mykytowycz, Roman, "Territorial Marking by Rabbits," *Scientific American*, Vol. 218 (May, 1968). Offprint No. 1108, W. H. Freeman and Co., San Francisco.

Sebeok, Thomas A., "Animal Communication," *Science*, Vol. 147, No. 3661 (February 26, 1965).

Tinbergen, N., "The Curious Behavior of the Stickleback," *Scientific American*, Vol. 187 (December, 1952). Offprint No. 414, W. H. Freeman and Co., San Francisco.

von Frisch, Karl, "Dialects in the Language of the Bees," *Scientific American*, Vol. 207 (August, 1962). Offprint No. 130, W. H. Freeman and Co., San Francisco.

von Frisch, Karl, Adrian M. Wenner, and Dennis L. Johnson, "Honeybees: Do They Use Direction and Distance Information Provided by Their Dancers?" *Science,* Vol. 158, No. 3804 (November 24, 1967).

Wenner, Adrian M., "Sound Communication in Honeybees," *Scientific American,* Vol. 210 (April, 1964). Offprint No. 181, W. H. Freeman and Co., San Francisco.

Wenner, Adrian M., "Honeybees: Do They Use The Distance Information Contained in Their Dance Maneuvers?" *Science,* Vol. 155, No. 3764 (February 17, 1967).

Wilson, Edward O., "Pheromones," *Scientific American,* Vol. 208 (May, 1963). Offprint No. 157, W. H. Freeman and Co., San Francisco.

Questions

1. Name the types of signals used by animals for communication. Which of these are used by humans? Are any used by plants?
2. Do you see a relationship between the kinds of receptors an animal has and the types of communication it utilizes?
3. Can you provide some examples of visual communication not described in the chapter?
4. How do you suppose that male and female mockingbirds, which are identical in appearance, communicate their sex differences to each other?
5. What functions are served by communication among organisms that belong to different species?

CHAPTER XII

Asexual Reproduction

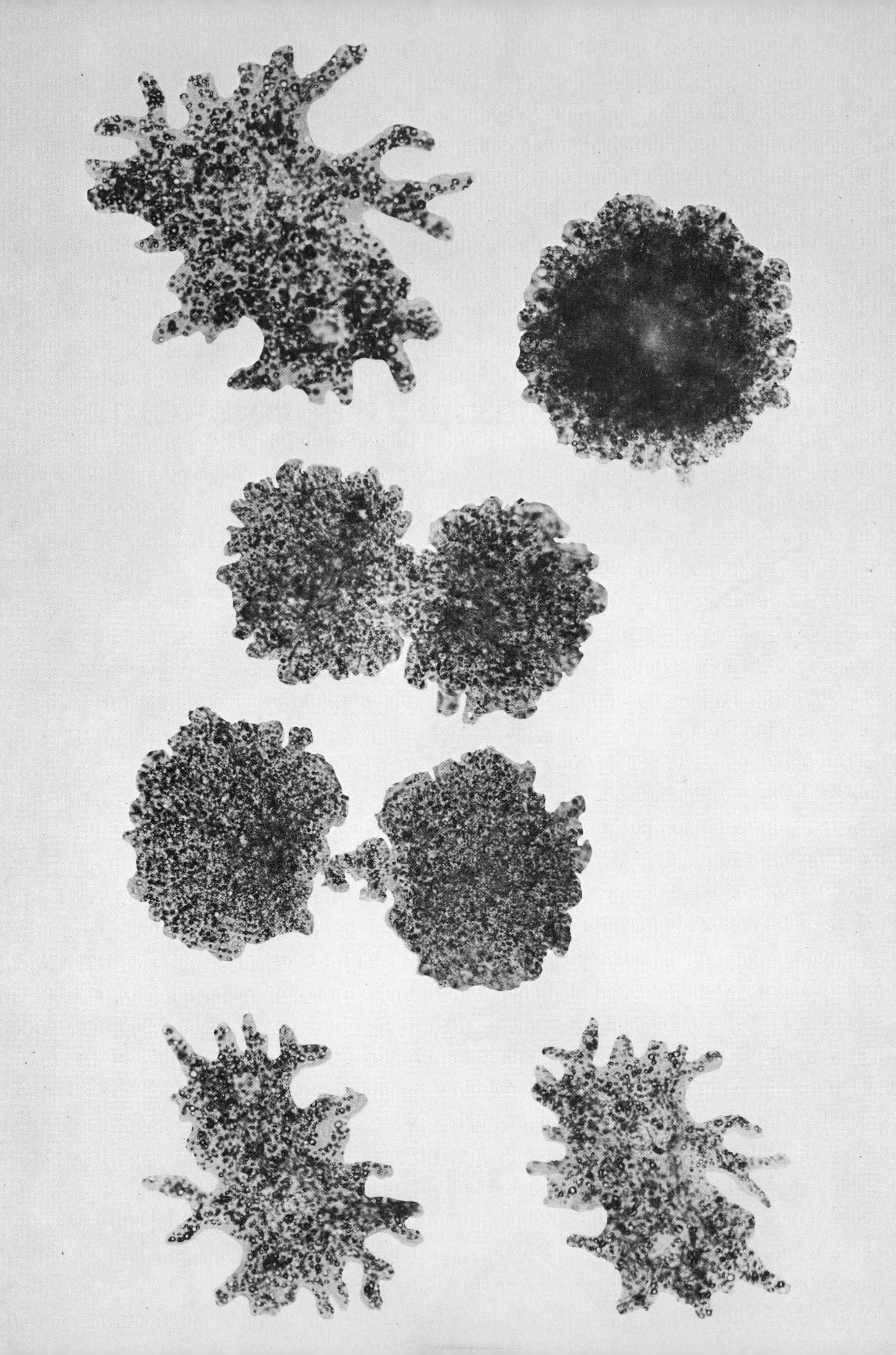

Amoeba proteus ***undergoing binary fission. The sequence goes from upper left to upper right and then down (×110).***

CHAPTER XII

Asexual Reproduction

Up to this point we have covered topics pertaining to the maintenance of the individual, such as energy utilization, coordination, and communication. Important as these are, they require another major factor of life, namely reproduction, in order to bridge the gap between generations.

In simple terms, reproduction is the biological process that provides new individuals. In the context of Chapter X, reproduction may also be considered a means of species steady state, that is, a means of maintaining population levels and genetic continuity. Whenever natural populations are studied over a period of time, their numbers are often observed to fluctuate around a mean value, a phenomenon that usually suggests a homeostatic mechanism at work.

When one individual, one "parent," produces offspring without the participation of special sex cells, the process is known as *asexual reproduction.* In contrast to this, sexual reproduction, taken up in the next chapter, involves the fusion of two specialized nuclei to produce a fertilized egg.

Reproduction is first and fundamentally a cellular process because it originates at this level. Thus, to understand the mechanics by which new individuals are produced we must examine the way in which new cells are produced.

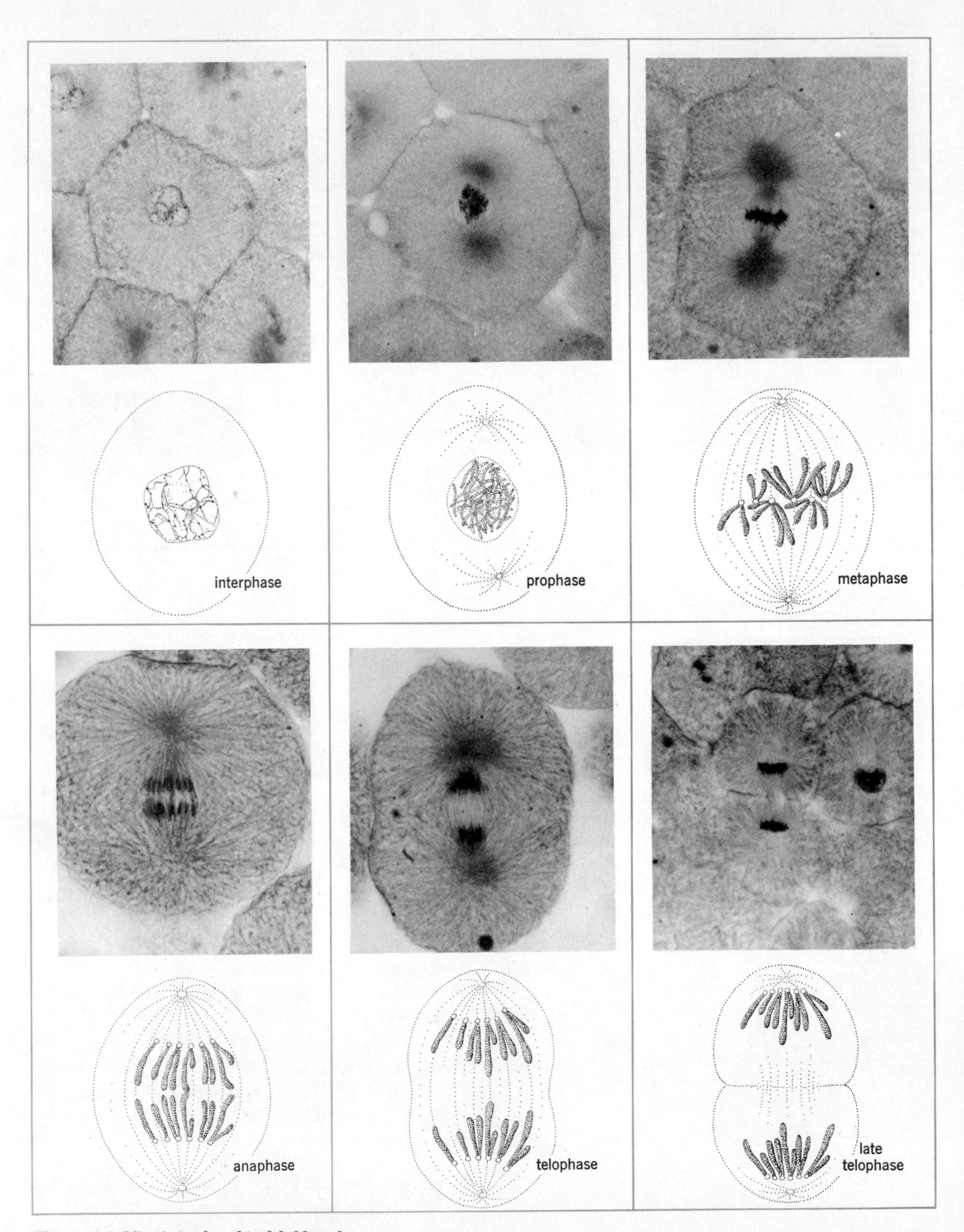

Figure 12.1. Mitosis in the whitefish blastula.

Cellular Reproduction: Mitosis

Reproduction, as a cellular activity, comprises a series of orderly events in the nucleus termed *mitosis.* As a consequence of these events, the nucleus divides into two nuclei containing identical kinds and amounts of hereditary substance. Strictly speaking, mitosis refers only to events in the nucleus; the division of the cytoplasm is called *cytokinesis.* Figure 12.1 illustrates cell division as a series of stages called prophase, metaphase, anaphase, and telophase. Each of these stages is described in the following paragraphs.

Prophase. One of the first observable indications of the beginning of mitosis occurs when threadlike structures, called chromosomes, become visible. These "threads" gradually coil up to become shortened, distinct bodies. This has the effect of changing a tangled mat of threads into short, maneuverable rods. Careful observation reveals that each chromosome at this time consists of a pair of threads known as *chromatids* or sister chromosomes; the significance of this does not become evident until later.

As the chromosomes take form, other events occur. The nuclear membrane and the nucleoli gradually disappear and, simultaneously, a new structure, the *spindle,* appears. It consists of slender strands extending nearly the length of the cell and expanded in its middle region like two cones placed base to base. The spindle is composed of long protein fibers that derive from the cytoplasm. In animal cells, the centrioles appear to control formation of the spindle, with one centriole located at each pole. Cells of most plants do not have centrioles, and the control device for spindle organization in these is not known. After the spindle forms, the chromosomes begin to orient themselves to it. Each chromosome attaches to a spindle fiber at a special region in the chromosome called a *centromere.*

The many events of prophase expend ATP energy. If the cell's respiratory machinery is halted early in prophase, mitosis ceases. Later, in prophase, this is not true—the cell appears to somehow accumulate sufficient energy to complete its division.

Metaphase. After attaching to spindle fibers, the chromosomes gather toward the equator (middle) of the spindle. This state, with the chromosomes lined up across the middle of the spindle, is termed *metaphase.* This is a particularly useful stage for the cytologist, since chromosomes are in their most distinct form and can be seen and counted more easily than in other stages (see Figure 12.1). About 1935 it was found that the drug colchicine interfered with spindle formation and thus stopped mitosis at metaphase. The drug became a useful tool that is still widely applied when cytologists wish to study chromosome shapes and numbers in an organism.

Anaphase. Following metaphase, the centromere of each chromosome divides. This allows the paired chromatids that form each chromosome to be pulled apart lengthwise. One chromatid of each chromosome moves toward one pole of the spindle, and the other chromatid toward the other pole. It has been suggested that the spindle contracts and pulls the chromosomal threads apart, and it has also been suggested that some force repels the threads. The contractile idea seems more reasonable in view of the protein structure of the spindle fibers.

Since all chromosomes do this at the same time, identical amounts and *kinds* of chromosomal material (including DNA) move to opposite ends of the spindle. Anaphase ends when the chromosomal material reaches the ends of the spindle.

Telophase. In *telophase,* the last stage of mitosis, the chromosomes begin to return to the threadlike stage. They uncoil and eventually become as indistinct as they were at the beginning of prophase. The spindle gradually disintegrates, the nucleoli reappear, and a nuclear membrane reforms about each of the two masses of nuclear material, which are then spoken of as *daughter nuclei.* Cell division is not completed, however, since we have yet to deal with the division of the cytoplasmic portion of the cell (cytokinesis). In animal cells, during telophase, the cytoplasm begins to constrict through the equator of the spindle until two daughter cells are formed. In plant cells, a *cell plate* starts to form in the middle of the spindle and then appears to work out to the outside of the cell until two daughter cells form. In both plants and animals, the result is the same: two daughter cells are formed, genetically identical to each other and to the mother cell from which they arose. These daughter cells undergo a period of growth and assume their role with other cells in their environment. Once a cell differentiates, it usually never divides again.

The duration of mitosis varies greatly depending on the tissue in which it occurs and the condition under which it is measured. For example, in root tips where growth is rapid, the mitotic cycle occurs within a few hours. Mammalian cells in tissue culture may require twelve hours. Bacteria may complete cell division in twenty minutes. Prophase and telophase are the longest stages, with anaphase taking the least amount of time.

Interphase. The stage during which a cell is not visibly reproducing is called *interphase.* Cells normally remain in interphase longer than in mitosis. This is a significant stage in relation to mitosis because it is during this period that the DNA is duplicated. Evidence for this was obtained from cells that were exposed to radioactively tagged substances, such as thymidine, used in the synthesis of new DNA. Only during interphase did a significant amount of the tagged material show up in the cell's DNA. Although we describe mitosis in terms of what chromosomes do, since they can be observed under the microscope, the primary function of mitosis is the quantitative separation of the duplicated DNA molecules into two equal masses.

Mitosis is a universal biological phenomenon and one of the most fundamental life processes. Not only does it produce new individuals but in addition it replaces dead cells, heals wounds, and regenerates lost or injured parts of the body in some cases. For many decades, biologists have attempted, without success, to find the system that regulates mitosis. Recent research indicates that cells produce mitotic inhibiting chemicals called *chalones.* Normally, the cells of a tissue produce a quantity of chalone that reduces the mitotic rate. If the tissue is injured or part of the cells are removed, the concentration of chalone drops and the rate of mitosis increases until the normal number of cells is restored. Sometimes cell division appears to get out of control, resulting in abnormal growths like cancers and tumors. At least in some of these cases, it appears that the cells produce markedly less chalone than do normal cells. This results in a high rate of cell division.

Advantages and Disadvantages of Asexual Reproduction

Reproduction of the nonsexual type is common among lower plants and animals and is advantageous in various ways. It is a relatively simple process involving only mitosis, in contrast to the complexities of egg and sperm formation found in sexual reproduction. Only one individual or parent is involved, and hence there is no necessity for complex mating procedures. Large numbers of offspring can be produced simultaneously. For example, one sporangium of a bread mold can produce thousands of spores, each of which can grow into a new mold. Organisms often use asexual reproduction as a means of dispersal, as is the case with spores. These are frequently microscopic bodies that are light enough to be blown about in the air.

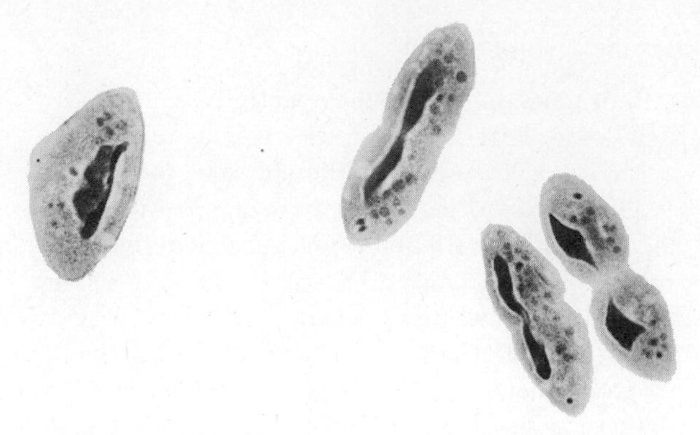

Figure 12.2. Fission in Paramecium *(Ward's Natural Science Establishment, Inc.).*

Offspring produced by asexual means are nearly exact copies of the parent and consequently exhibit almost no variability. This is an advantage when the parent is well adapted to a stable environment. On the other hand, a species that employs only asexual reproduction has a limited range of adaptability and cannot adjust rapidly to changing environmental situations. This is the major disadvantage of the process.

Major Categories of Asexual Reproduction

In considering the variations of this kind of reproduction, we have chosen to subdivide it into categories of fission, budding, sporulation, and vegetative reproduction. It could be argued that *all* cell divisions, including the growth of embryos or tissues, constitute reproduction. But we shall restrict our discussion to the four categories named.

Fission. In fission, the nucleus divides mitotically and then the entire organism divides to form two separate organisms. This is a common means of reproduction among many of the unicellular organisms such as protozoans (Figure 12.2), bacteria, yeasts, and numerous algae. Few multicellular organisms reproduce by fission because of their complexity. However, some forms like flatworms show a modified form of fission in which their bodies

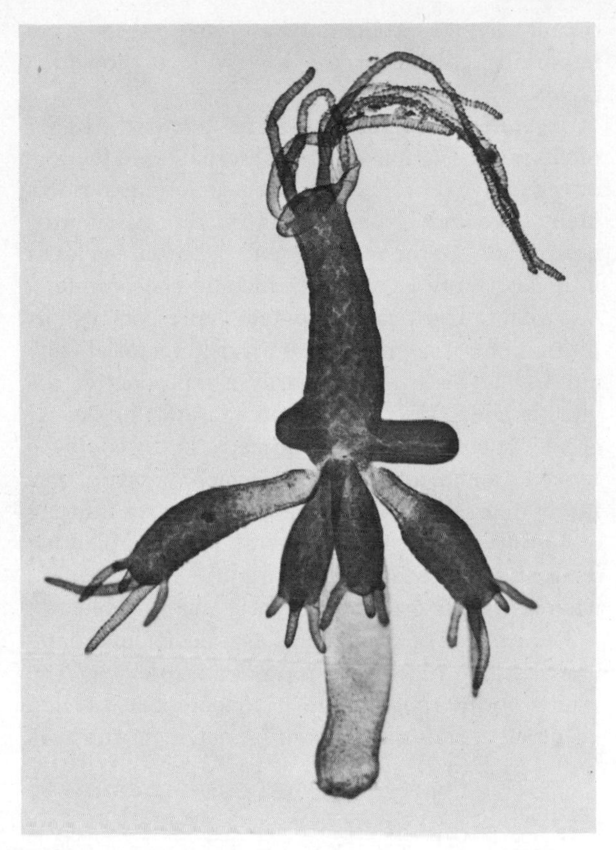

Figure 12.3. Budding Hydra *(CCM: General Biological, Inc., Chicago).*

Figure 12.4. Budding in the jellyfish Aurelia. *The branchlike structures are the tentacles of the various young jellyfish.*

constrict into two or more pieces, each of which grows into another worm.

Budding. In budding, a portion of the parent's body grows by repeated cell division to form an appendage or region that will eventually become another organism. There are many variations of this type of reproduction. Yeast cells, for example, form new members in this way. In the small, freshwater animal called *Hydra,* a portion of the body of an adult grows out and develops a mouth and tentacles (Figure 12.3). This offspring now resembles the adult and eventually breaks free to pursue its own fate. The small liverwort plant *Marchantia* grows small cellular masses (buds) in cups on its surface. These buds float away in films of dew or rainwater to start new liverwort plants in other locations. Some of the jellyfish utilize a specialization of the budding process. In one stage of the life cycle, a stalklike body, attached to the substrate, buds tiny disklike jellyfish (Figure 12.4). Budding also occurs in a number of other forms, including some of the protozoans and sponges.

An additional aspect of asexual reproduction is what might be termed reproduction by *regeneration.* Here a portion of an organism, removed accidentally, is capable of growing into adult form. The arms of a starfish, a piece of an earthworm, and the plants already mentioned exemplify this phenomenon.

Sporulation. Sporulation, the formation of spores, is common in the plant world. The basic idea is the formation of cells that can develop into new plants under favorable conditions. Spores are often surrounded by a tough, protective coat that enables them to withstand adverse conditions (Figure 12.5). Fungi are singularly proficient at this since many of them, like the mushroom, can release tens of thousands of tiny airborne spores. This ancient and suc-

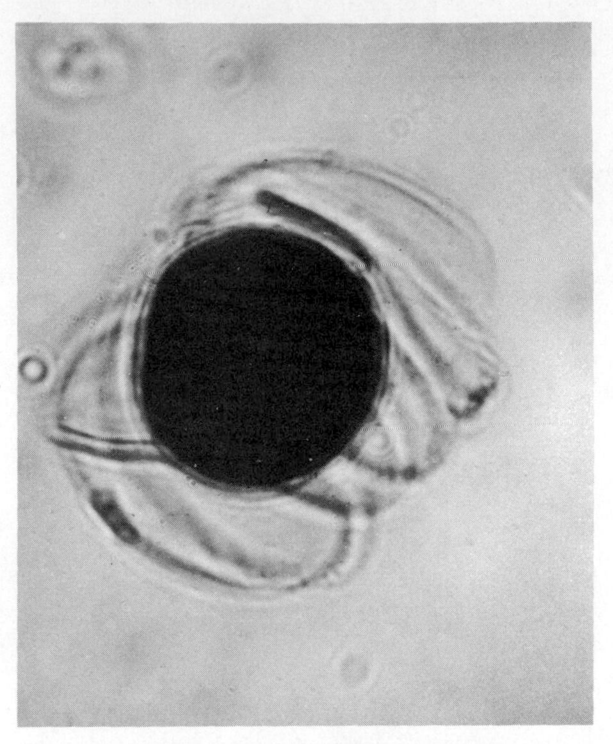

Figure 12.5. A spore showing the protective covering.

cessful way of reproduction has even persisted as an important portion of the life cycle of flowering plants.

Vegetative Reproduction in Plants. Higher plants possess various means of asexual reproduction, or vegetative reproduction as it is sometimes termed. Many have underground stems (rhizomes) that grow horizontally in the soil and send up aerial leaves at intervals. Any piece of the rhizome can become a new plant. The common potato represents a short portion of a rhizome filled with food material (Figure 12.6). This is not primarily a reproductive device, but man discovered long ago that the "eyes" of the potato, which are actually buds, could be removed and planted to produce more potato plants. In potatoes, at least, this is a better way of insuring uniformity in the next generation than planting seeds. Each eye produces a replica of its parent, whereas plants grown from seeds are variable.

The runners of grass plants are special horizontal branches that function as reproductive devices. The runner forms roots where it touches the soil and eventually sends out its own branches. In this way

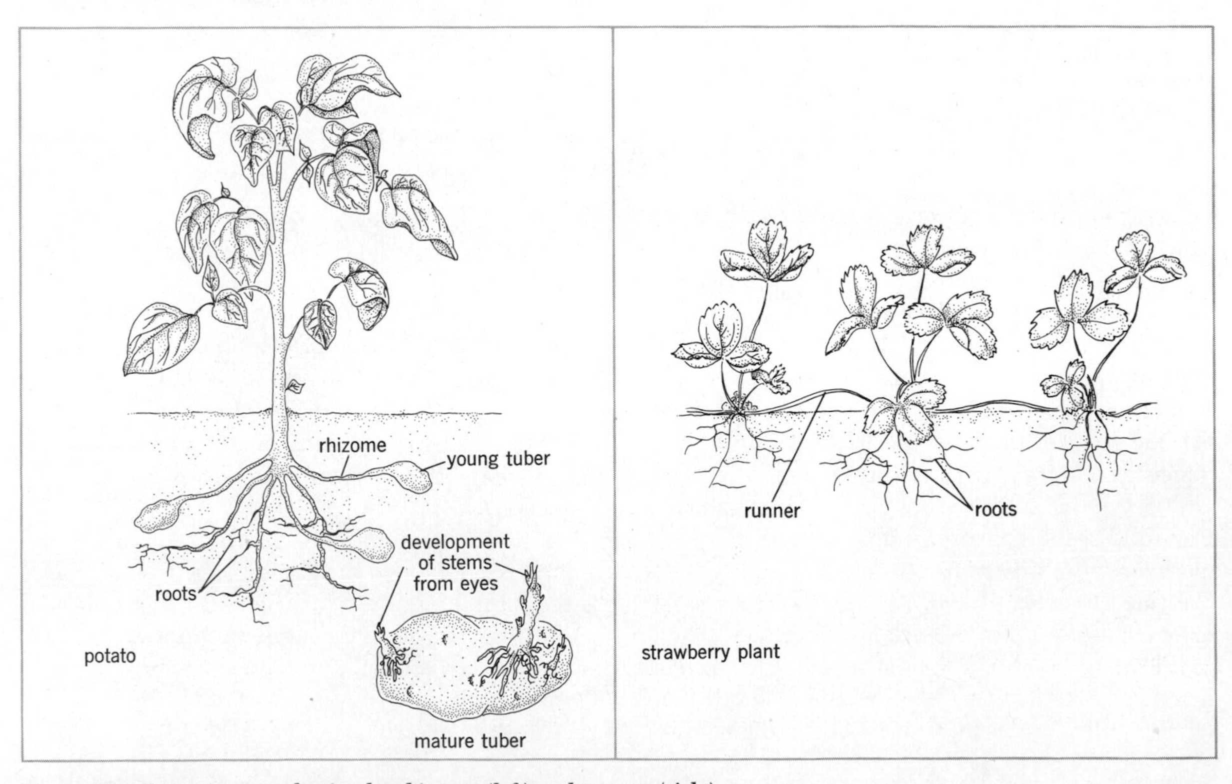

Figure 12.6. Vegetative reproduction by rhizomes (left) and runners (right).

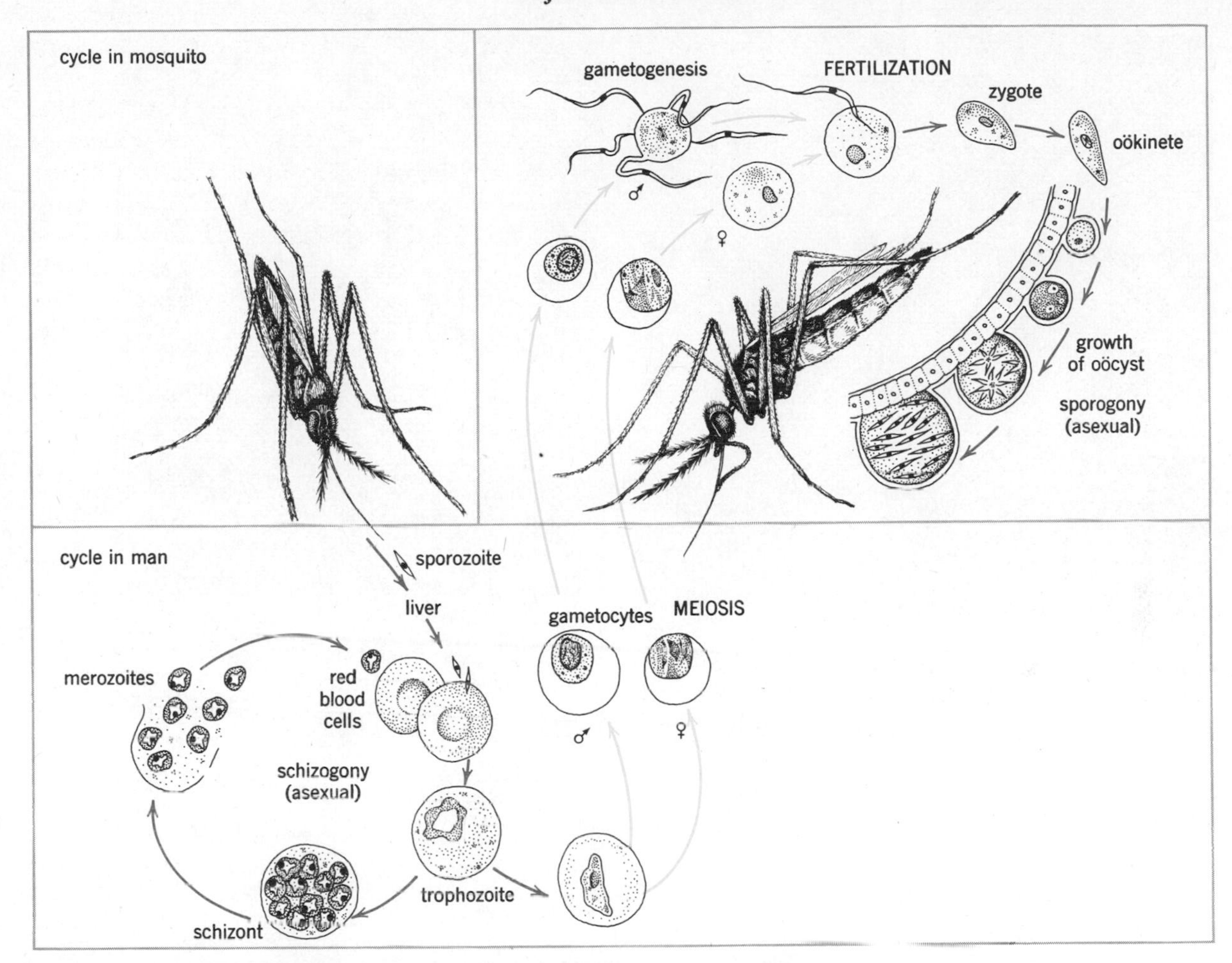

Figure 12.7. Life cycle of a malarial parasite. The blue arrows represent the asexual phase and the yellow arrows indicate the sexual phase.

the grass spreads. Nurserymen take advantage of this in planting lawn grasses like Bermuda and St. Augustine. They plant short sprigs of the grass which soon root and reproduce to form a lawn. Pieces of the stems (cuttings) of many plants and even their leaves are capable of growing into new plants. This is a widely used technique for reproducing ornamental plants that are difficult to grow from seeds.

A horticultural application of the versatility of plant reproduction is to bud or graft a portion of one plant onto a different, but related, plant. In this way, a plant that bears desirable fruit can be grafted to a rootstock that is resistant to soil diseases. Most citrus and other domestic fruits are grown this way. It is even possible to have a citrus tree that simultaneously bears lemons, oranges, and grapefruit!

Finally, it is noteworthy that a considerable number of organisms, including nearly all plants, reproduce both asexually and sexually during their life cycles. Many internal parasites such as malarial organisms (Figure 12.7), tapeworms, and flukes (Figure 12.8) reproduce asexually during certain parts of their life cycles. This increases their numbers significantly, improves their chances of dispersal to other hosts, and frequently insures their own survival.

Thus we find that asexual reproduction is widespread among members of the plant and animal kingdoms, even persisting among forms that also reproduce sexually. This is another way of stating that the adaptive advantages of asexual reproduction may outweigh the disadvantage of reduced varia-

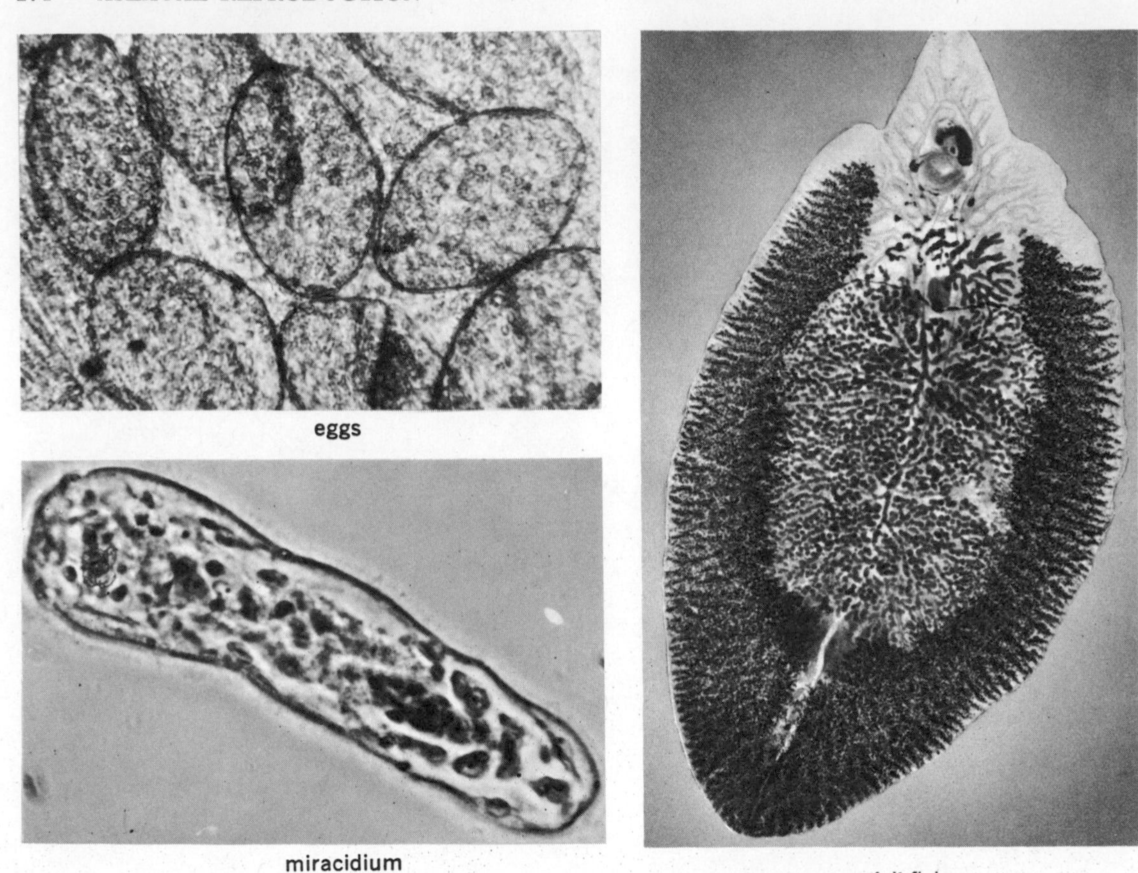

Figure 12.8. The illustration on the opposite page depicts the life cycle of the sheep liver fluke. The photograph at the right is of the adult fluke found in the bile ducts of the sheep. Two photographs on the left show the eggs and a free-swimming miracidium.

bility. In Chapter XIII, we shall consider the significance of sexual reproduction and see how some organisms utilize both means of reproduction during their life cycles.

Principles

1. Asexual reproduction derives its characteristics from mitosis.
2. Mitosis produces identical daughter cells because each receives identical hereditary material.
3. The similarity of asexually produced offspring is advantageous in stable environments, but disadvantageous in changing ones.

Suggested Readings

Bullough, W. S., "The Chalones," *Science Journal,* Vol. 5, No. 4 (April, 1969).

Duryee, William R., "Modern Aspects of Cell Division," *Frontiers of Modern Biology.* Houghton Mifflin Co., Boston, 1962.

Mazia, Daniel, "Cell Division," *Scientific American,* Vol. 189 (August, 1953). Offprint No. 27, W. H. Freeman and Co., San Francisco.

Mazia, Daniel, "How Cells Divide," *Scientific American,* Vol. 205 (September, 1961). Offprint No. 93, W. H. Freeman and Co., San Francisco.

Singer, Marcus, "The Regeneration of Body Parts," *Scientific American,* Vol. 199 (October, 1958). Offprint No. 105, W. H. Freeman and Co., San Francisco.

Swanson, Carl P., *The Cell.* Second edition. Prentice-Hall, Englewood Cliffs, N.J., 1964, pp. 62–77.

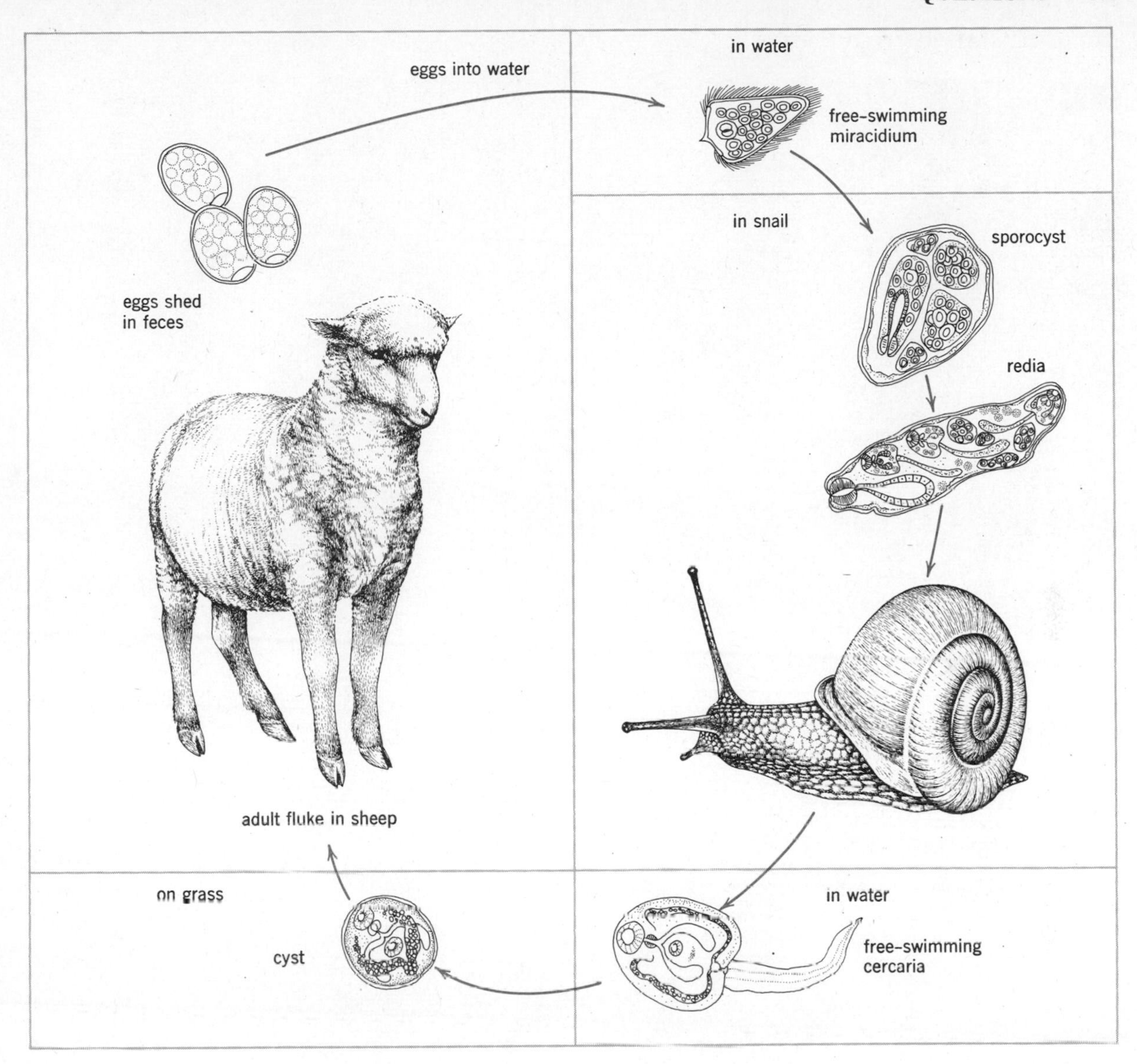

Questions

1. What is the basic difference between asexual and sexual reproduction?

2. Reproduction is fundamentally a cellular event. Can you defend this statement?

3. The term mitosis is derived from the Greek *mitos,* a thread. How does this fact relate to the process of mitosis?

4. Describe in your own words the events that occur during cell division.

5. What evidences indicate that the DNA is duplicated during interphase of mitosis?

6. What is the most important end result of mitosis?

7. What advantages does asexual reproduction provide for an organism?

8. What is the major disadvantage of asexual reproduction? Is this always a disadvantage?

9. List the major types of asexual reproduction and provide an example of each one from your local area.

CHAPTER XIII

Sexual Reproduction

Queen bee tended by workers (© Walt Disney Productions).

CHAPTER XIII

Sexual Reproduction

Sexual reproduction is the uniting of two specialized cells or nuclei known as *gametes* to form a fertilized egg or *zygote*. Although sexual reproduction is a complex biological activity, it occurs among nearly all groups of living organisms. We might conclude from this that there is a marked adaptive advantage to reproducing in this manner.

Advantages and Disadvantages of Sexual Reproduction

Probably the greatest biological advantage of sexual reproduction is the increased variability that results from uniting the hereditary material of two organisms. This creates a truly *new* organism, similar but never identical to either parent. Such variability is of tremendous evolutionary importance because it provides species with means of adapting over a long period of time to new environmental challenges. Even on a short-term basis this variability is advantageous since a population may encounter changes in its surroundings within relatively few generations. In this sense it gives a population a necessary plasticity.

There are three ways in which sexual reproduction seems to be a disadvantage: two parents are usually required, various courtship procedures are involved (in many animals), and the gametes must be brought together in nonmotile organisms such

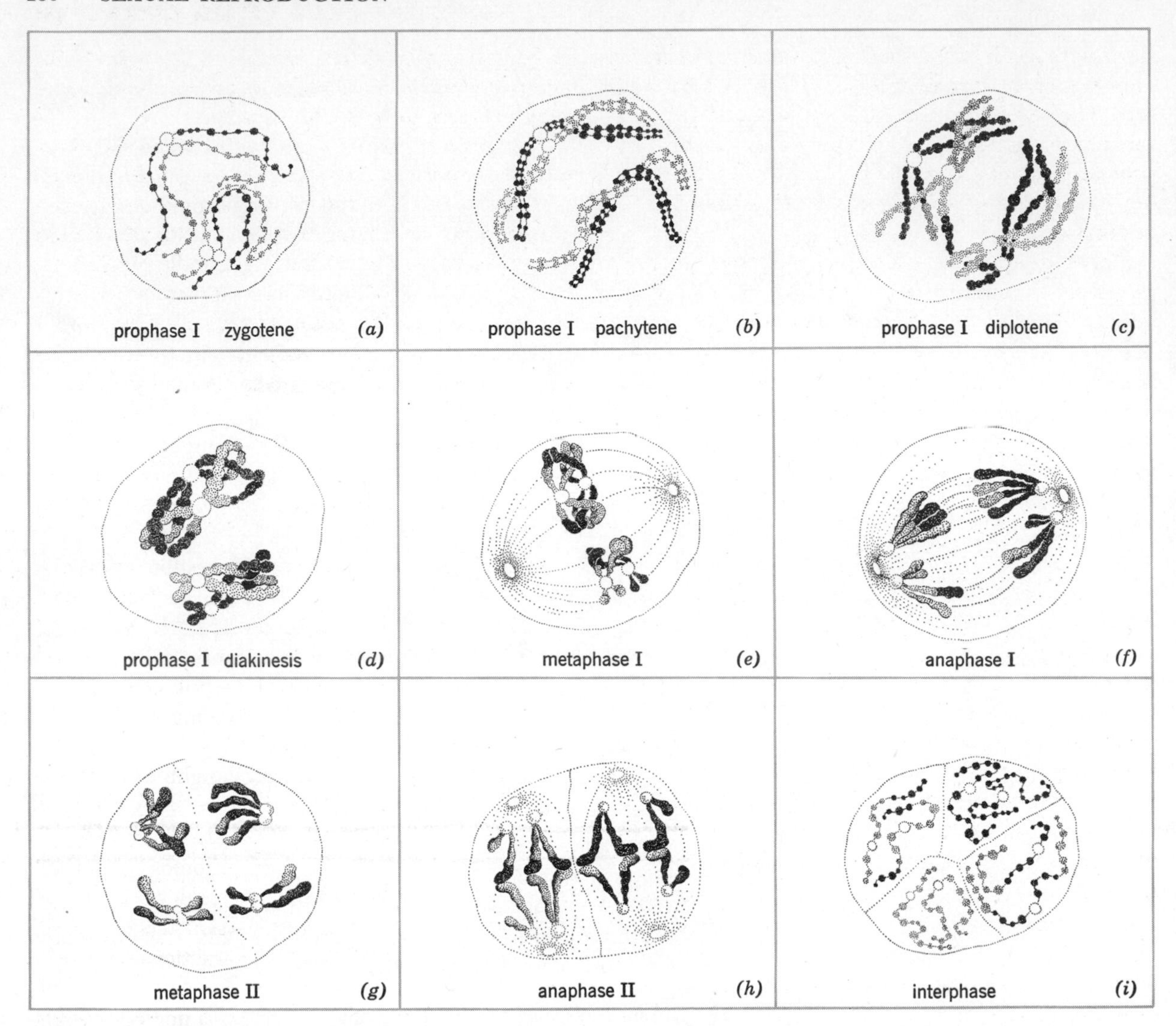

Figure 13.1. Stages of meiosis.

as plants. Despite these complexities, sexual reproduction prevails at all levels of life. It is so important, in fact, that a major part of an organism's life cycle is usually devoted to bringing about successful reproduction.

Meiosis

The Necessity for Chromosome Reduction. If sexual reproduction is the fusion of two specialized sex cells, what prevents the nuclear material from doubling in amount in each new generation? The answer lies in a special type of nuclear division called *meiosis.* This division precedes the formation of the gametes so that each gamete contains only half as much hereditary material as the parent cell from which it derives. Thus, when two gametes unite, the resulting zygote has the full or normal complement of hereditary substance. For example, human body cells contain 46 chromosomes (23 pairs): human eggs and sperm contain 23 chromosomes each (one member of each pair) as a consequence of meiosis. We refer to this reduced number as the *haploid* number and to the full complement of chromosomes as the *diploid* number.

Since meiosis is so important in sexual reproduction, we need to examine it in greater detail. Meiosis starts with a cell containing the full number of chromosomes characteristic of the organism (two haploid sets). The cell undergoes two successive divisions but the chromosomes are duplicated only once. This process results in four daughter cells, each containing one half the number of chromosomes in the original cell.

Stages of Meiosis. The first stage of meiosis, *prophase I,* initially resembles the beginning of mitosis. The diffuse chromosomal mass in the nucleus becomes a discrete network of threadlike bodies (Figure 13.1*a*). The nuclear membrane gradually disappears and the spindle forms. As the chromosomes take shape, each one lines up with its homologue, that is, with another one like itself, and the two become closely entwined (Figures 13.1*a* and 13.1*b*). This event is unique to meiosis; it does not occur in mitosis. An analogy to this pairing process can be made by pressing the palms of one's hands together. If you consider your fingers as chromosomes, then you have pairs of "homologous chromosomes." Each chromosome consists of two strands of hereditary material (two *chromatids*); thus a chromosome pair consists of four chromatids, as indicated in Figures 13.1*c* and 13.1*d*.

The chromosomes in each pair then move slightly apart as though the pairing process were ending. As this happens, some of the chromatids adhere to each other at one or more sites rather than separate (Figures 13.1*c* and 13.1*d*). These sites represent places where the chromatids have exchanged parts with each other. This activity, known as *crossing over,* has considerable hereditary importance, as we shall see in Chapter XV.

As prophase I terminates, the chromosomes in their paired arrangement begin to orient themselves on the spindle. Each pair has two centromeres, one for each homologous chromosome.

Metaphase I. In this stage, the chromosome pairs are arranged around the equatorial plane of the spindle (called the *metaphase plate*) and are attached to spindle fibers by their centromeres. The centromeres of each pair, however, lie opposite each other on either side of the imaginary plane as indicated in Figure 13.1*e*.

Anaphase I. Homologous chromosome pairs separate from each other, one member of each pair moving to each end of the cell (Figure 13.1*f*). The exact placement of the centromeres, one on either side of the equatorial plate, makes this segregation of homologues possible. Also, each centromere remains undivided as it pulls an entire chromosome with it to one pole of the spindle.

Telophase I. This phase begins when the chromosomes reach opposite ends of the spindle. At each pole there is a haploid set of chromosomes, each chromosome containing two chromatids attached to a single centromere. Often the spindle disappears, chromosomes uncoil, and a nuclear membrane forms around each set. In some organisms, telophase I is greatly shortened or omitted. In the latter instance, prophase II takes place immediately after anaphase I.

Prophase II. In each of the daughter cells produced in the first division, a spindle forms, and chromosomes coil up and begin to move toward the middle of the spindle.

Metaphase II. The chromosomes line up on the metaphase plate, attached by their centromeres to spindle fibers. Each chromosome clearly consists of two chromatids at this time (Figure 13.1*g*).

Anaphase II. In anaphase II, the centromeres divide and move apart, each carrying a chromatid to the poles.

Telophase II. Four haploid daughter nuclei are organized into four haploid daughter cells, the final outcome of meiosis (Figure 13.1*i*). Each haploid cell has one of each of the kinds of chromosomes originally found in the parent cell although only half the total number. This means that each daughter cell has a complete set of genetic instructions for growing into another organism.

Like mitosis, meiosis is virtually a universal process in the world of living things. The student of biology must understand the mechanisms and significance of both processes to develop an adequate appreciation of other fundamental biological events such as genetics. Table 13.1 briefly summarizes the differences between meiosis and mitosis.

Differentiation of Sex Cells

The haploid cells produced in meiosis are not mature sex cells. In animals further specialization, known as gametogenesis, is required to form sperm or egg cells.

In the differentiation of sperm cells in the testes, the cytoplasm is modified into a motile tail and a collar enclosing many mitochondria. The nucleus becomes the head of the sperm (Figure 13.2).

In the case of egg formation in the ovary, meiosis

Table 13.1. How Meiosis Differs from Mitosis

Meiosis	Mitosis
Occurs in specialized reproductive tissue only	Occurs in all growing tissues
Chromosomes pair in early prophase I	Chromosomes do not pair in prophase
Chromosomes exchange parts of chromatids while paired	Chromosomes do not exchange parts
Centromeres lie on either side of metaphase plate in division I.	Centromeres lie on the metaphase plate
Centromeres do not divide during anaphase I	Centromeres divide at anaphase
Haploid daughter cells formed at end of first division	Diploid daughter cells, formed at end of division, conclude mitosis
Another division produces a total of four haploid daughter cells to conclude meiosis	

and specialization into an egg take place concurrently. During this set of events, only one daughter cell survives to become the functional egg. The other three daughter nuclei are cast off as polar bodies but contribute their cytoplasmic material to the egg. This provides a large amount of cytoplasm and yolk for the development of an embryo (Figure 13.2).

In plants, mature sex cells are derived from the meiotic products by widely divergent means. Several examples will be treated under the discussion of life cycles.

Fertilization. The union of an egg and a sperm is called fertilization, although strictly speaking the term refers to the uniting of the egg and sperm nuclei. Normally, only one sperm nucleus enters an egg, and usually an egg of the same species. Biologists have long suspected that there is some kind of specific chemical recognition between egg and sperm but they have been unable to isolate such a substance. The sperm not only contributes a set of chromosomes to the egg but also stimulates the egg to begin developing. This is evidently a mechanical stimulus since the same result is often obtained if an egg is pricked with a needle. Additional details about the development of the fertilized egg are taken up in Chapter XVI.

Reproduction in the Protista

Paramecium illustrates the basic tenet of sexual reproduction, the fusion of nuclei. Under certain physiological conditions, two paramecia unite side by side (Figure 13.3). A series of nuclear activities occurs within each animal; eventually, each exchanges a portion of its nuclear material with the other. Then they separate and undergo two fissions. This exchange of nuclear, and presumably hereditary, material satisfies the general definition of sexual reproduction even though no gametes are formed and the different sexes are not recognizable.

Chlamydomonas is a motile, single-celled alga

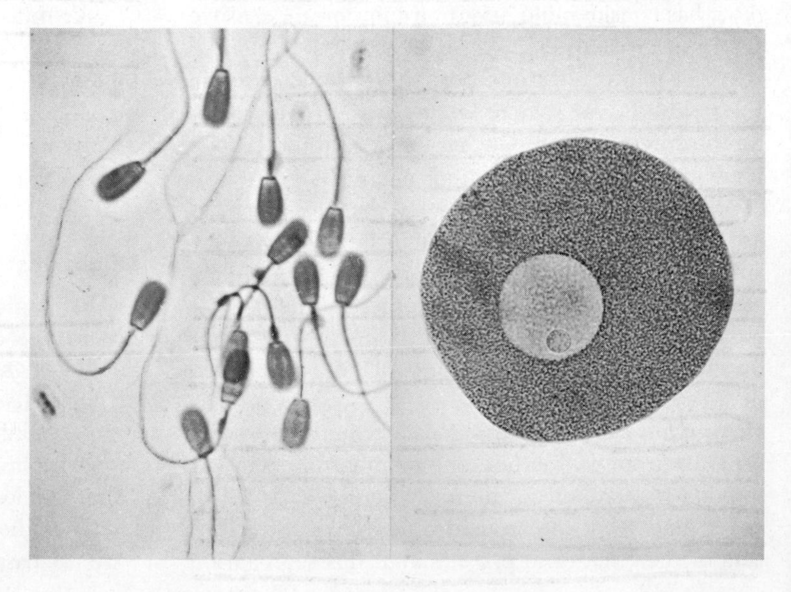

Figure 13.2. Mature sperm (left) and egg (right).

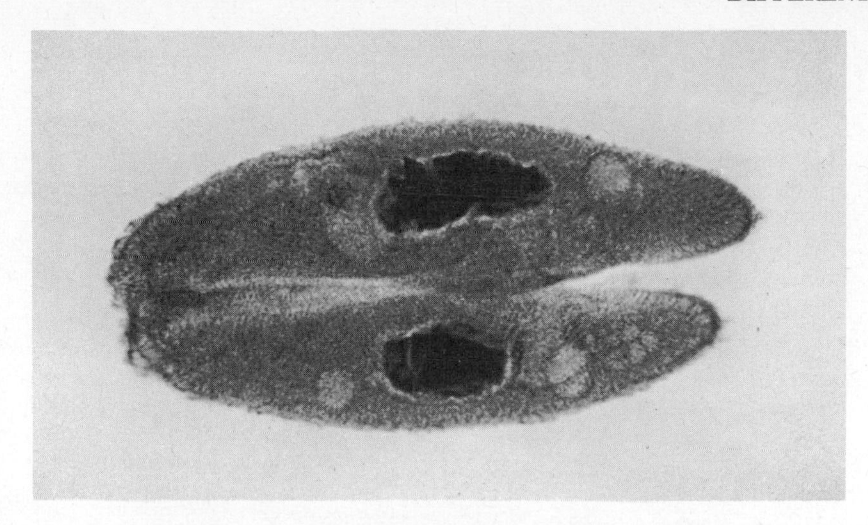

Figure 13.3. Two paramecia in the process of exchanging nuclear material (CCM: General Biological, Inc., Chicago).

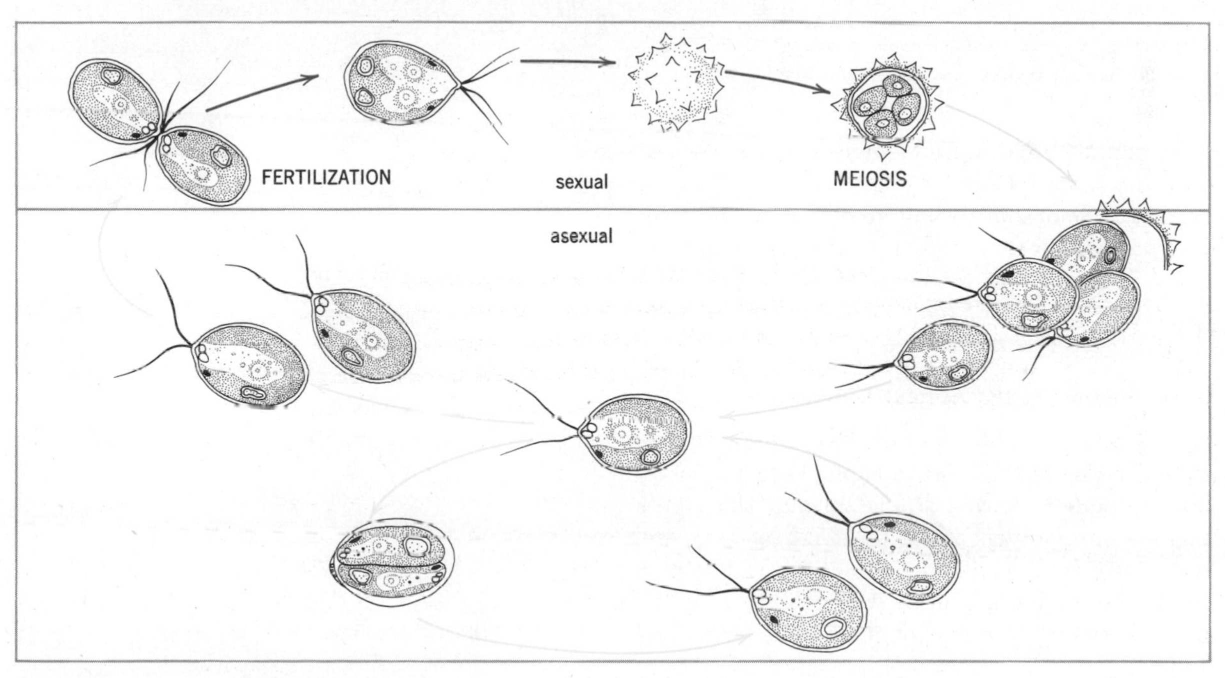

Figure 13.4. Life cycle of the alga Chlamydomonas.

that is a common freshwater inhabitant. It reproduces asexually by a form of fission (Figure 13.4, bottom). In addition, two cells may fuse to form a zygote, sometimes termed a *zygospore* since it is surrounded by a protective coat and can resist unfavorable conditions like drying (Figure 13.4, top). A zygospore undergoes meiosis to produce four haploid cells. These cells enlarge slightly to become the adult form of *Chlamydomonas*.

Two points are notable here. First, the organism spends most of its life cycle in the haploid state; second, the two cells that perform the function of gametes are identical at least in size and appearance. The presence of like gametes (*isogametes*) is common among algae and fungi, but not among members of the plant kingdom. As with animals, specialization of haploid cells into eggs or sperm (*heterogametes*) is the rule. Some biologists have speculated that the

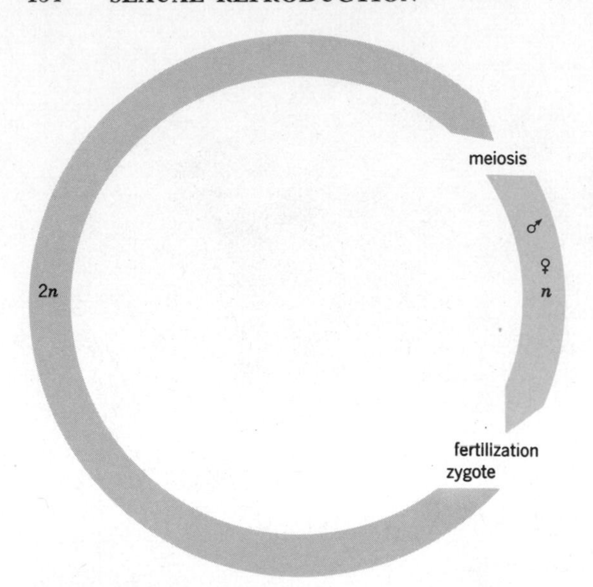

Figure 13.5. A generalized life cycle of animals. Yellow represents the haploid portion and blue is the diploid.

evolution of sexual reproduction could have begun with the fusion of similar cells (isogametes) during the life cycle of some primitive organism. Once this was successfully established, further specialization of the gametes into motile cells (sperm) and cells containing a store of food materials (eggs) would make the process even more efficient.

Reproduction in the Animal Kingdom

Most animals conform to the basic life cycle plan shown in Figure 13.5. Here a haploid egg and sperm nucleus unite to form a diploid zygote. The zygote matures into another gamete-forming generation and so the life cycle continues. Variations in this plan are almost endless as can be seen by examining the reproductive cycles of several types of plants and animals.

Hydra, the small, many-celled freshwater animal mentioned in Chapter XII, exemplifies several aspects of reproduction. Even in this relatively primitive form we find the gametes specialized into two types. Some species are *hermaphroditic* in that ovaries and testes form on the same animal (Figure 13.6). Sperm are liberated into the water around the animal and swim about until they encounter a mature egg. This method of liberating gametes into the surroundings is common among water dwellers. Jellyfish, for example, do the same thing although they are not hermaphroditic. In the life cycle of

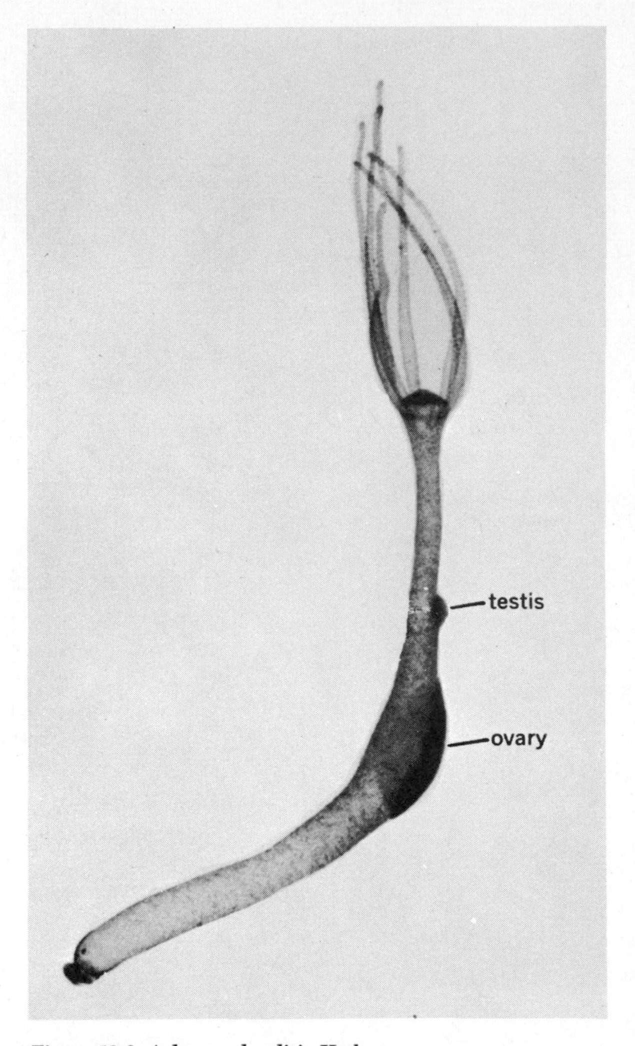

Figure 13.6. A hermaphroditic Hydra.

jellyfish, both the asexual and sexual phases are diploid but involve two different forms, the medusa and the strobila (Figure 13.7).

Earthworms contain both ovaries and testes but must mate in order to bring the gametes together. During the mating process each worm emits sperm which flow down special grooves to be stored by the other worm in a special sperm receptacle. After the worms separate, each produces eggs that are fertilized with the sperm obtained earlier from the other worm. Crossfertilization, and hence a mixing of hereditary material from two individuals, is thus achieved.

Hermaphroditism is common but not universal

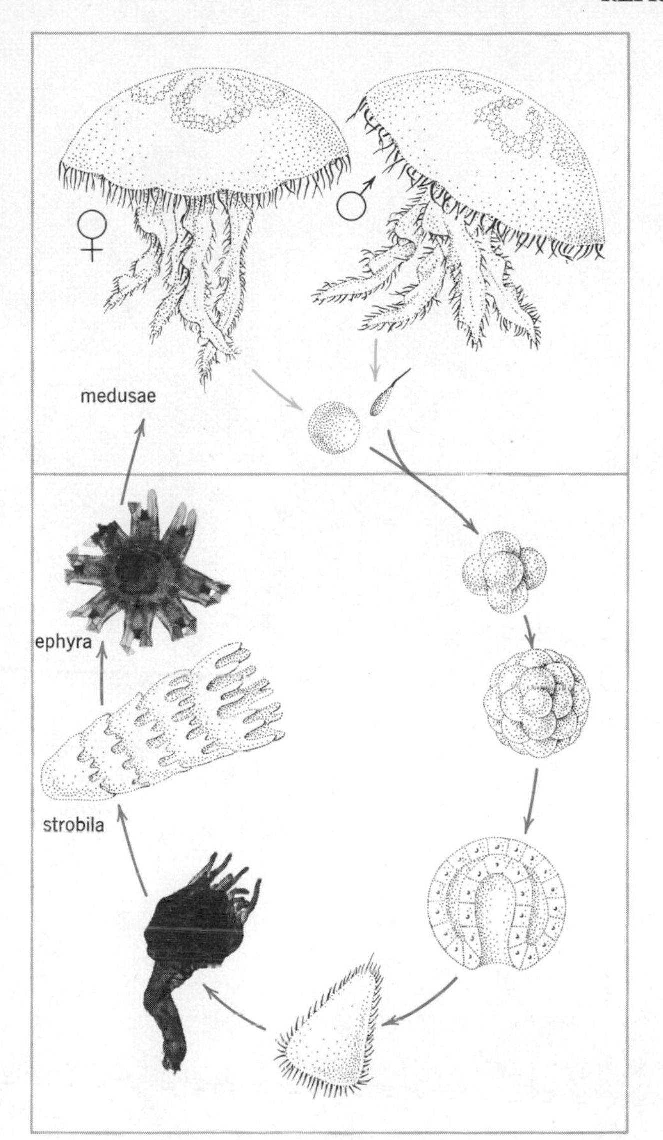

Figure 13.7. Stages in the life cycle of the jellyfish, Aurelia. *The ephyra is a juvenile form that matures into the adult medusa.*

among invertebrates. Some groups like the arthropods seldom have hermaphroditic species, whereas others like the mollusks have many. Generally, it seems to be found among organisms that have limited means of locomotion or that may seldom encounter one another. In all of these groups it is not unusual for the female, or each member of a hermaphroditic pair, to store sperm cells for long periods of time and use them on successive batches of eggs. Most hermaphroditic forms also have adaptations that encourage crossfertilization, although a few animals (for example, tapeworms) and a large number of plants are known to fertilize themselves.

Another adaptation frequently found in invertebrates is the ability of unfertilized eggs to develop into adults. This is called *parthenogenesis* and constitutes part of the normal reproductive cycle of many insects such as aphids and honeybees. In some species of aphids, the females lay eggs throughout the spring and summer which develop parthe-

Figure 13.8. In honeybees, the queen (a) *produces eggs which, if fertilized, develop into female workers* (b). *If the eggs are not fertilized, the resulting bees are male drones* (c).

nogenetically into more females. In the fall, some of the eggs hatch into males which mature and then mate with the females. They in turn begin laying diploid eggs. These eggs survive through the winter to start up new generations of aphids the following spring. In other species of aphids, the males have never been found and may not even exist.

In honeybees, the queen mates only once and stores the sperm from this mating for use during the remainder of her lifetime. As she lays eggs in the cells of the honeycomb, some are fertilized and others are not. The fertilized eggs develop into females, mostly workers, and the haploid eggs become males (drones). (See Figure 13.8.) Here we have a rather extreme case of parthenogenesis as well as sex determination by chromosome number.

Parthenogenesis is nonsexual in that only one cell is involved. Nevertheless, it involves a special structure, the egg, which normally takes part in sexual reproduction. Hence, some authors term it an asexual event while others treat it as a variation of sexual reproduction.

The reproductive patterns in groups of vertebrates also show considerable diversity. The bulk of fishes and amphibians have external fertilization with little or no parental care of the eggs and young. On rainy spring evenings, every roadside ditch seems to be full of singing frogs. The males are making this noise, in a sense advertising their presence to the females. As the females join the singing males the mating activity takes place. A male clasps a female from behind, and this behavior in turn stimulates the female to lay eggs. The male sheds sperm cells over the eggs as they emerge. The fertilized eggs are left in large masses in the water to develop with no further attention from the parents.

Most vertebrates, except mammals, are hatched from eggs and even here we encounter variations from the gelatinous coated eggs typical of amphibians to the shelled eggs of birds. In mammals, in contrast, the sperm cells are always deposited into the reproductive tract of the female, and there is a long period of parental care of the young. In addition, behavior related to reproduction, such as courtship between mates, becomes quite elaborate in birds and mammals.

Reproduction in the Plant Kingdom

The plant kingdom presents some other variations of the sexual reproduction theme. Figure 13.9 illustrates a generalized life cycle scheme for plants. As in the animal life cycle, a haploid egg and sperm unite to form a zygote which matures. Here the plant scheme takes a radical departure from the animal life cycle. This diploid plant forms haploid spores, rather than gametes, by meiosis. For this reason, it is called the *sporophyte generation*. The spores eventually become a gamete-forming generation, the *gametophyte generation*. Since the spores are already haploid, the gametes they produce by mitosis are haploid. This cycle bridges two generations, one haploid and one diploid, a plan basic to all plants, fungi, and algae. Meiosis occurs in the sporophyte generation but *not*, as in animals, in the gamete-producing organism. This is a situation peculiar to plants; it is often present as an adaptation for dispersal.

Moss. If we examine the life cycle of an advanced plant, a moss, we find still further adaptations of the reproductive process (Figure 13.10). The microscopic, haploid spores of the moss plant are transported by wind or water until they are deposited in a suitable environment, such as moist soil. A spore then germinates into a small branching filament which later becomes a moss plant. The

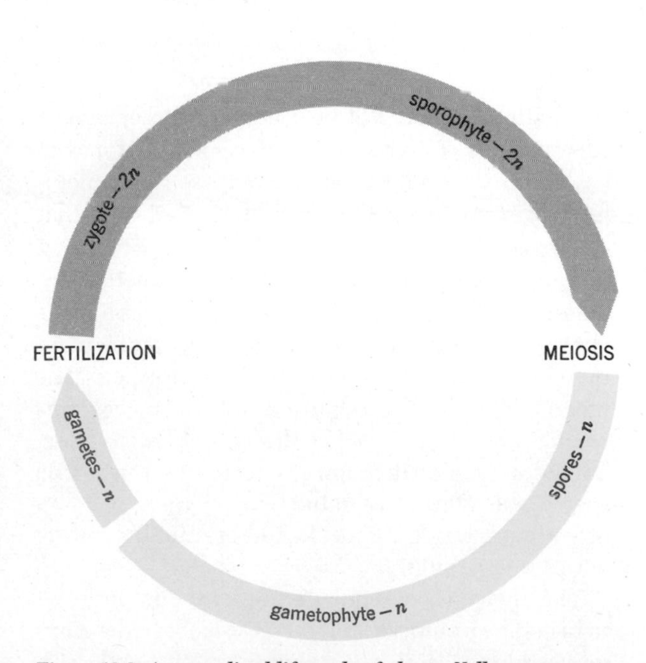

Figure 13.9. A generalized life cycle of plants. Yellow represents the haploid stages and blue the diploid stages in the cycle.

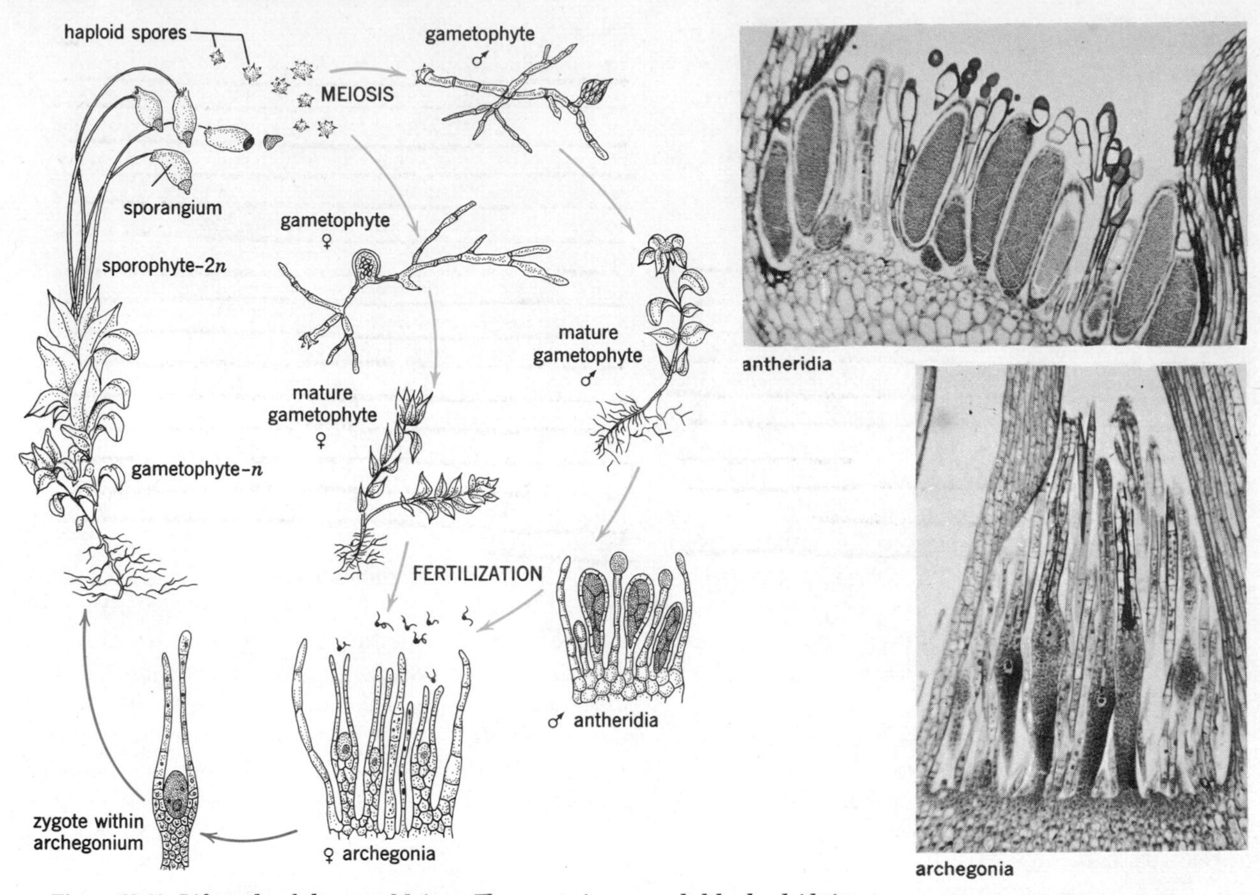

Figure 13.10. Life cycle of the moss Mnium. *The zygote is surrounded by haploid tissue.*

plants thus formed will bear either male or female reproductive cells, although some species of mosses bear both. These moss plants compose the gametophyte generation of the moss life cycle. Note that each plant grows from a haploid spore; hence it is haploid and its sexual products (gametes) are automatically haploid. The male reproductive tissue consists of a stalklike body near the tip of the plant in which numerous sperm cells are formed. The female reproductive organ consists of an egg surrounded by a thick jacket drawn out into a long, hollow neck. Fertilization is effected by the motile sperm cells swimming to the female organs in a film of water, entering the neck, and eventually uniting with the egg cells.

The *zygote* is consequently diploid. First it grows to become an embryo, and then the embryo develops into a long stalk with a capsule at the end. This structure grows out of the end of the female gametophyte. This new diploid structure is termed a sporophyte because a mass of cells inside the capsule forms haploid spores by meiosis. When mature, the capsule opens so that the spores are dispersed.

In mosses the two generations, sporophyte and gametophyte, are about equally prominent in the life cycle. Among seed plants, the sporophyte is the more dominant part of the cycle, while the gametophyte is reduced to a brief, often microscopic, part. This reduction was evidently a major trend in the evolution of the plant kingdom.

Flowering Plants. The cycle of a flowering plant nicely illustrates this last comment (Figure 13.11). Our common herbs, shrubs, and trees consist almost entirely of the spore-producing generation. The *stamens* (Figure 13.11, lower right) in the flower contain spore-forming bodies called *anthers.* Meiosis inside the anthers gives rise to haploid spores. Each spore undergoes a certain amount of differentiation

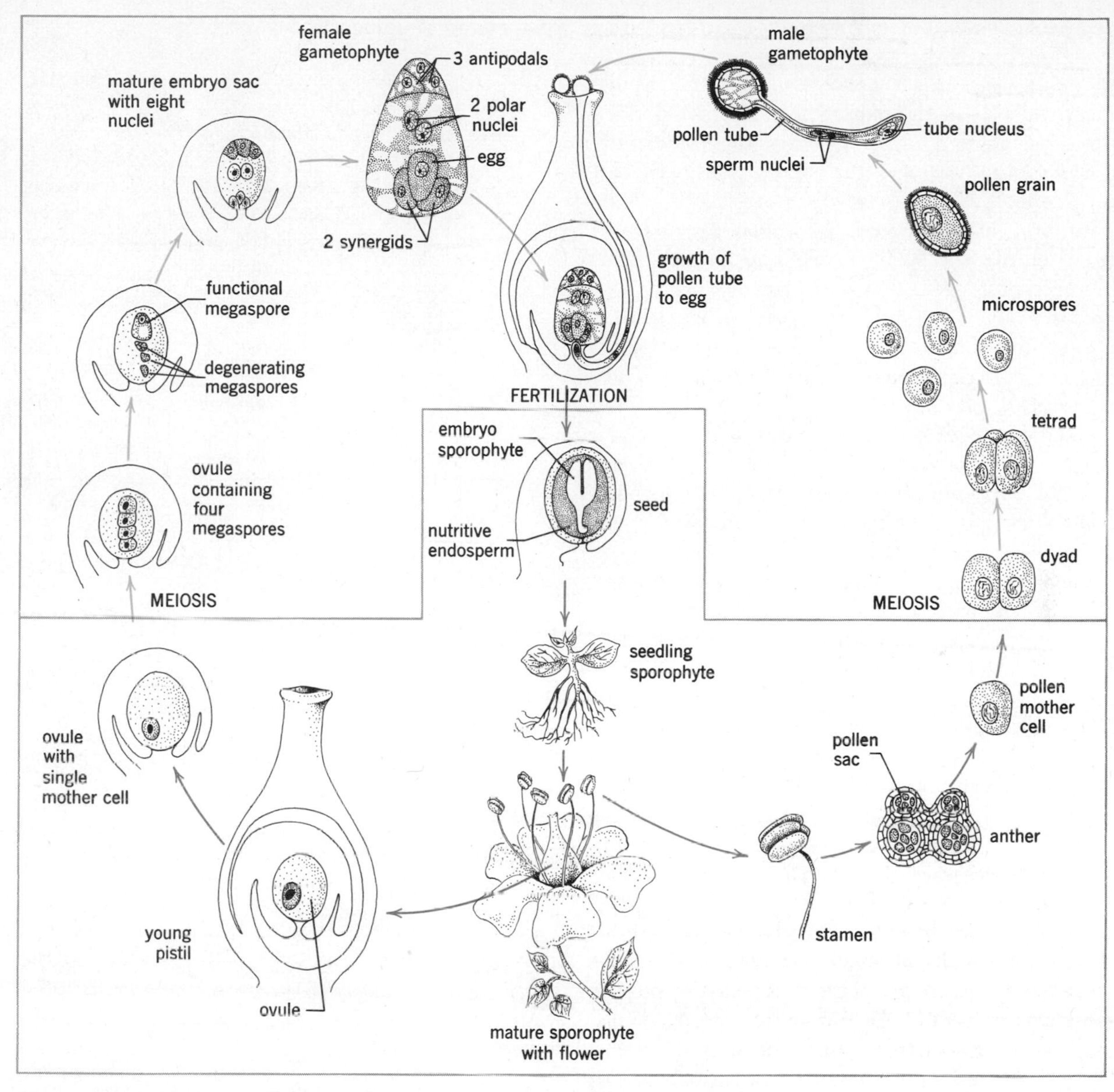

Figure 13.11. Life cycle of a flowering plant. In the female gametophyte, the antipodals and synergids have no known function.

(the nucleus divides once or twice) to become a male gametophyte or *pollen grain*.

The other type of spore-forming organ, also in the flower, is located in the base (ovary) of a structure called the *pistil* (Figure 13.11, lower left). Here cells again enter meiotic divisions to form haploid spores. Some of these, after a series of additional divisions and specialization, form female gametophytes called embryo sacs, each containing an egg.

If a pollen grain is transported to the stigma of the pistil, it grows a pollen tube down through the pistil to the embryo sac in the ovary. The contents of the pollen grain, usually two sperm nuclei, pass down the pollen tube into the sac where one sperm nucleus fertilizes an egg nucleus. The second sperm nucleus unites with another nucleus or other nuclei in the sac to form, after several mitoses, a tissue known as endosperm which nourishes the developing zygote and embryo. A plant embryo, surrounded by layers of food material and a tough outer coat, is termed a *seed*. This structure can exist and survive independently of the parent plant for considerable periods of time. It has all the capabilities of forming a new sporophyte when the environment permits it to germinate.

In most instances, seeds are not simply liberated from the plant as such, but are encased in various modified parts of the flower or plant as a *fruit*. These enclosing layers are either juicy or dry when the fruit is ripe. Dry fruits occur in a wide variety of plants including the milkweed pod with its feathery seeds, legumes like peas and beans, cocklebur with its spiny appendages, pecans, and the grains such as rice and wheat. These have various adaptations to facilitate dispersal by wind (milkweed), attaching to animal coats (cocklebur), passing through the digestive tract of animals and floating on streams (many nuts). Fleshy fruits are also widespread and include many familiar ones used by man for food: tomato, orange, watermelon, squash, peach, apple, and blackberry. Fleshy fruits in nature are used for food by many animals which in turn disperse the seeds in their droppings. In short, the fruit functions as an inducement to dispersal for the seeds within it and only rarely as nourishment or protection for them.

The examples of plant and animal life cycles presented here should give some idea of the many variations that exist. It is worth emphasizing that the basic plan of egg + sperm = zygote is nearly universal. Even those forms with isogametes foreshadow this idea. The many variations of this plan are adaptations to make the reproductive process more workable within the limitations of a particular kind of organism in a particular kind of environment. That is, a considerable number of the behavioral, anatomical, and physiological adaptations found in all organisms are built around their reproductive processes.

Principles

1. Sexual reproduction derives its characteristics from meiosis and fertilization.
2. Meiosis produces cells that vary in genetic makeup through a reduction in chromosome number.
3. The variation found among sexually produced offspring is advantageous in changing environments but less advantageous in stable ones.

Suggested Readings

Allen, R. D., "The Moment of Fertilization," *Scientific American*, Vol. 201 (July, 1959).

Edwards, R. G., "Mammalian Eggs in the Laboratory," *Scientific American*, Vol. 215 (August, 1966). Offprint No. 1047, W. H. Freeman and Co., San Francisco.

Jones, Jack Colvard, "The Sexual Life of a Mosquito," *Scientific American*, Vol. 218 (April, 1968). Offprint No. 1106, W. H. Freeman and Co., San Francisco.

Loomis, W. F., "The Sex Gas of Hydra," *Scientific American*, Vol. 200 (April, 1959).

Metz, Charles B., "Fertilization," *Frontiers of Modern Biology*. Houghton Mifflin Co., Boston, 1962.

Pincus, C., "Fertilization in Mammals," *Scientific American*, Vol. 184 (March, 1951).

Proctor, Vernon W., "Long-Distance Dispersal of Seeds in the Digestive Tract of Birds," *Science*, Vol. 160, No. 3825 (April 19, 1968).

Smith, C. Lavett, "Hermaphroditism in Bahama Groupers," *Natural History*, Vol. LXXVIII (June–July, 1964).

Swanson, Carl P., *The Cell*. Second edition. Prentice-Hall, Englewood Cliffs, N.J., 1964, pp. 78–93.

Questions

1. What major advantage does sexual reproduction provide for a population?
2. In what ways is it a drawback?
3. Why is it important that reduction in chromosome number accompany sexual reproduction?
4. Describe in your own words the events that constitute meiosis.
5. In what respect does a sperm cell illustrate the concept from Chapter III that specialization in form takes place primarily in the cytoplasm? How does this apply to an egg cell?
6. An organism originating by parthenogenesis must be identical to which parent? Why?
7. In what major aspect do the reproductive cycles of plants differ from those of animals? Is this an adaptation? For what?
8. Compare the reproductive cycle of a moss to that of a flowering plant, noting the basic similarities and differences.
9. The size, color, and scent of flowers are adaptations for what?
10. What important adaptive advantage do seed-forming plants have over those that form only spores?
11. Consider several types of fruit familiar to you and make an "educated guess" as to their means of dispersal in nature.
12. Is man an important dispersal agent? Why?

CHAPTER XIV

Mendel and the Beginning of Genetics

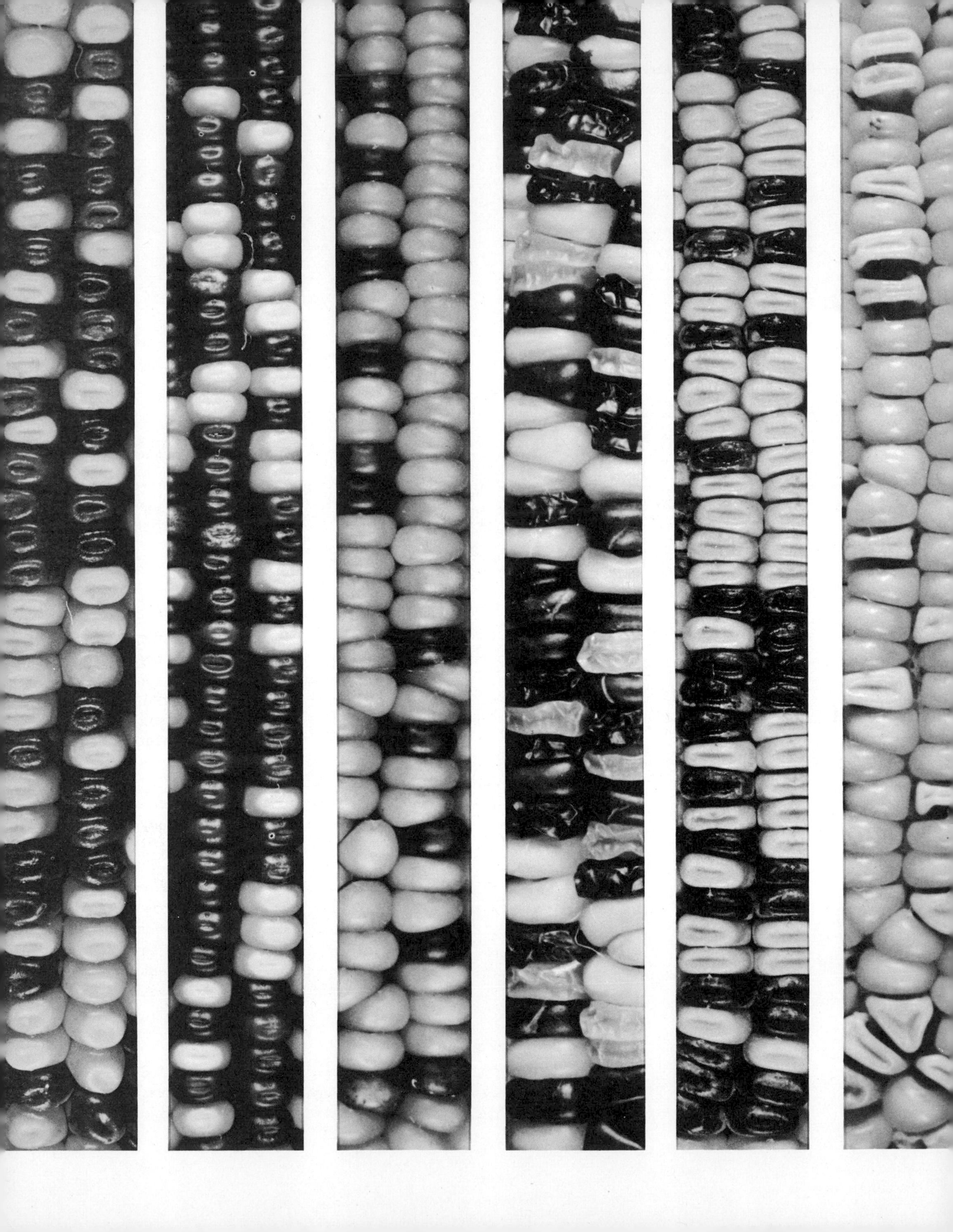

These segments of corn ears demonstrate Mendel's principles of heredity, and include monohybrid and dihybrid phenotypic ratios.

CHAPTER XIV

Mendel and the Beginning of Genetics

Sexual reproduction, as we have seen, is characterized by the uniting of two haploid nuclei which bring together hereditary material (DNA) from two parent organisms. The result, a zygote, then develops under the control of this DNA material. It will not be identical to either parent but instead will be a new, and in a sense unique, organism. Consider for a moment the differences between children of the same parents and you gain some notion of the concept of individual differences.

On the other hand, the zygote invariably develops into something quite *similar* to its parents. We can predict with certainty that dogs will produce puppies and human beings will beget human beings.

It is evident that this phenomenon is of universal importance in the living world. To understand it, we must explore how hereditary material passes from parents to offspring, how this material expresses itself in new combinations, and whether principles can be formulated that apply to so complex an event. This exploration is the study of heredity, the branch of biology known as *genetics*.

Historical Background

Most of our scientific knowledge of heredity has accumulated since 1900. As a subject of general interest and wide application, it is much older. After all, people do not have to understand

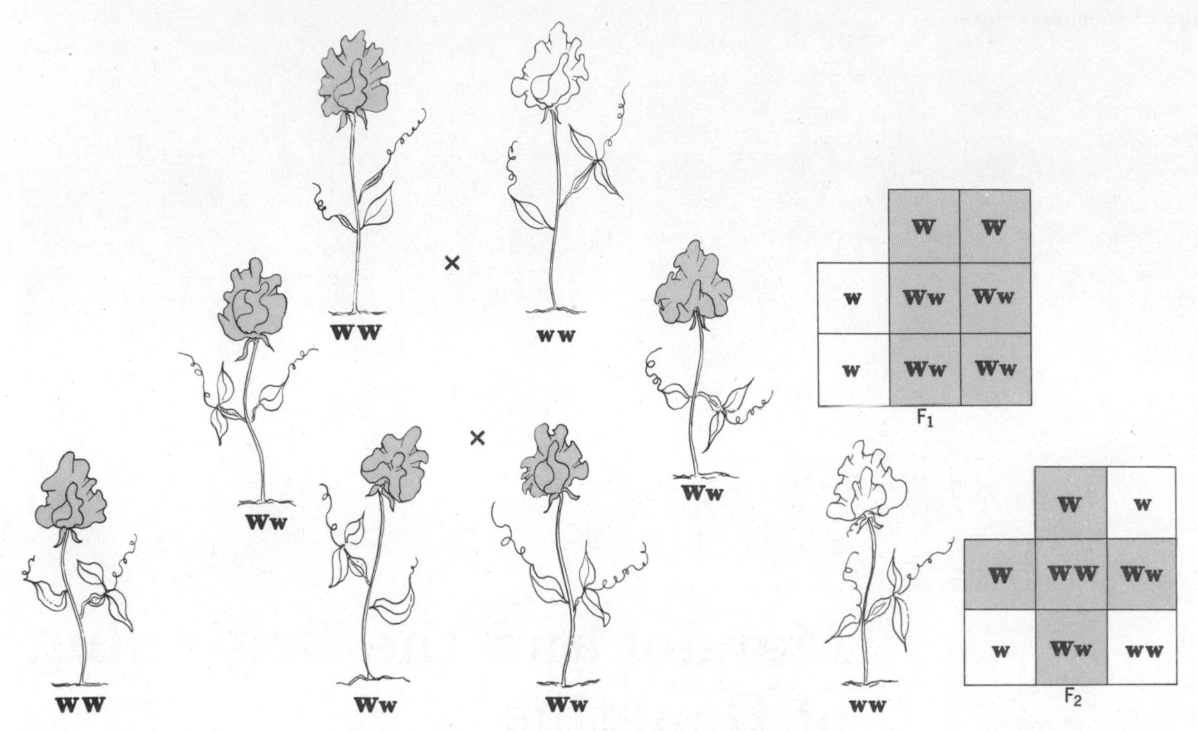

Figure 14.1. A graphic representation of Mendel's cross of red- and white-flowered peas.

the principles of heredity to breed farm animals and attempt to improve crops. It is in these areas of practical usage that some sort of crude applied genetics probably goes back to antiquity. Certainly many domesticated plants and animals were present thousands of years ago, which indicates they had an even earlier origin.

During this long history of practical experience, many misconceptions became associated with heredity, some of which persist. Probably the most widespread erroneous concept has to do with the influence of environment in changing or shaping new generations of organisms. Superficially it may seem reasonable to consider this influence a cause-and-effect relationship. Animals that try to hang from tree limbs by their tails should eventually give rise to young with prehensile tails; plants forced to grow in dry habitats should give rise to offspring with reduced leaves; what a mother does or feels should influence her unborn child, and so on. None of this is true, however, as we shall see shortly, because generally the environment has no effect on the structure of the hereditary material. Another common belief misconstrues blood as a hereditary factor, as reflected by terms such as "bloodline," "blue blood," "blood will tell," and so on. Again, this is a misconception arising from ignorance of how hereditary mechanisms really operate.

Gregor Mendel. Some of the basic concepts of heredity grew out of experiments performed by Gregor Mendel (1822–1884) in the middle 1800s. Unfortunately, the value of his work was not realized until 1900, so Mendel only received acclaim after his death.

Mendel spent most of his lifetime as a monk in an Austrian monastery. He was also educated in natural science and for many years was a schoolmaster. During this time he cultivated garden plants, primarily peas, in a series of experiments designed to study inheritance in plants. Owing to Mendel's good judgment and some degree of luck, garden peas turned out to be an excellent choice for this type of study. A number of distinct varieties exist—different flower colors, dwarf and tall plants, etc., and they can be cross-pollinated by hand fairly easily, but are self-pollinating otherwise. Moreover, they bear numerous progeny (seeds). In addition, Mendel displayed the wisdom of studying one trait at a time, repeating his experiments, and maintaining careful records of his findings.

For a better understanding of Mendel's work, it is helpful to study the technique of his experiments with the pea plants. If he wanted to cross a red-flowered plant with a white-flowered one, he first collected seeds from each of the two varieties. He assumed that these seeds always grew true to their respective flower colors, since pea flowers are shaped in such a way that the pollen cannot escape from them. Pea flowers then are self-pollinating, which means that, under normal circumstances, there is no cross-pollination between varieties. This is a handy method for assuring that one is working with purebred lines of descent.

Mendel then planted these seeds and allowed the plants to grow. Before the flowers matured, he removed the anthers from the flowers of one of the varieties and covered them with small sacks. When the female part of these flowers matured, he pollinated them by hand with pollen from another variety. Later he collected the seeds (the peas) which resulted from this artificial crossing. But, as you probably realize, he still had to plant these seeds and observe *their* flowers before he knew the results of this one cross.

On the surface, this appeared to be a simple experiment, but take a second look at what was involved.

1. Careful technique in hand-pollinating the flowers at just the right time of anther and pistil development, as well as prevention of accidental pollination from the wrong variety.
2. Close attention to and care for the plants between generations, that is, from seed to seed.
3. Careful record keeping as to the number of plants involved, types of crosses, and number and kind of offspring produced.

Now let us follow one of his experiments and then see if we can reach the same conclusions as Mendel did. When he wished to study the inheritance of flower color, he crossed red-flowered peas with white-flowered ones: the seeds he obtained grew into plants which bore red flowers only. He then let this new generation of peas with the red flowers fertilize itself, collected the seeds, planted them, and observed the results. In one experiment, the results turned out to be 705 red-flowered pea plants and 224 white-flowered ones (Figure 14.1). This is illustrated diagrammatically as follows. The parental generation (P_1) of plants is bred:

P_1: red-flowered × white-flowered

This cross yields seeds that grow into the *first filial* generation (F_1):

F_1: plants bearing red flowers only

When plants of the F_1 generation are crossed,

red-flowered × red-flowered

the yield is seeds that grow into the *second filial* generation (F_2):

F_2: 705 red- and 224 white-flowered plants (approximately a 3:1 ratio)

Mendel performed numerous experiments similar to this one covering various characteristics such as tall and dwarf plants, smooth and wrinkled seeds (Figure 14.2), yellow and green seed color, and so on. Over a period of eight years of experimentation, he used twenty varieties of garden peas. Eventually he studied his extensive data, noted the consistent occurrence of certain ratios of offspring, and realized that an orderly process had to be operating. Mendel finally arrived at some significant conclusions.

1. Hereditary traits are controlled by discrete units that pass unchanged from generation to generation. For example, the trait "white-flower" seems to disappear in the F_1 generation, but reappears in the F_2 progeny. Note also that there are no intermediate colors, only red or white.
2. Each trait is produced by *two* hereditary factors. This is a necessary assumption to account for the way in which a trait such as flower color appears in successive generations in a predictable ratio.
3. When two *contrasting* hereditary factors are present in an organism, such as red-flower color and white-flower color, only one will be expressed. One will be *dominant* and the other *recessive*. In the case of the peas, red-flower color is dominant to white-flower color since only the hereditary trait of red shows in the F_1 generation.
4. Each parent contributes only *one* of the two hereditary factors to each gamete. When the egg or sperm cells are formed, there is a separation or *segregation* of hereditary factors. For example, an egg may contain either a factor for red color or one for white color, but never both. (This is called *Mendel's law of segregation.*) Moreover, as a con-

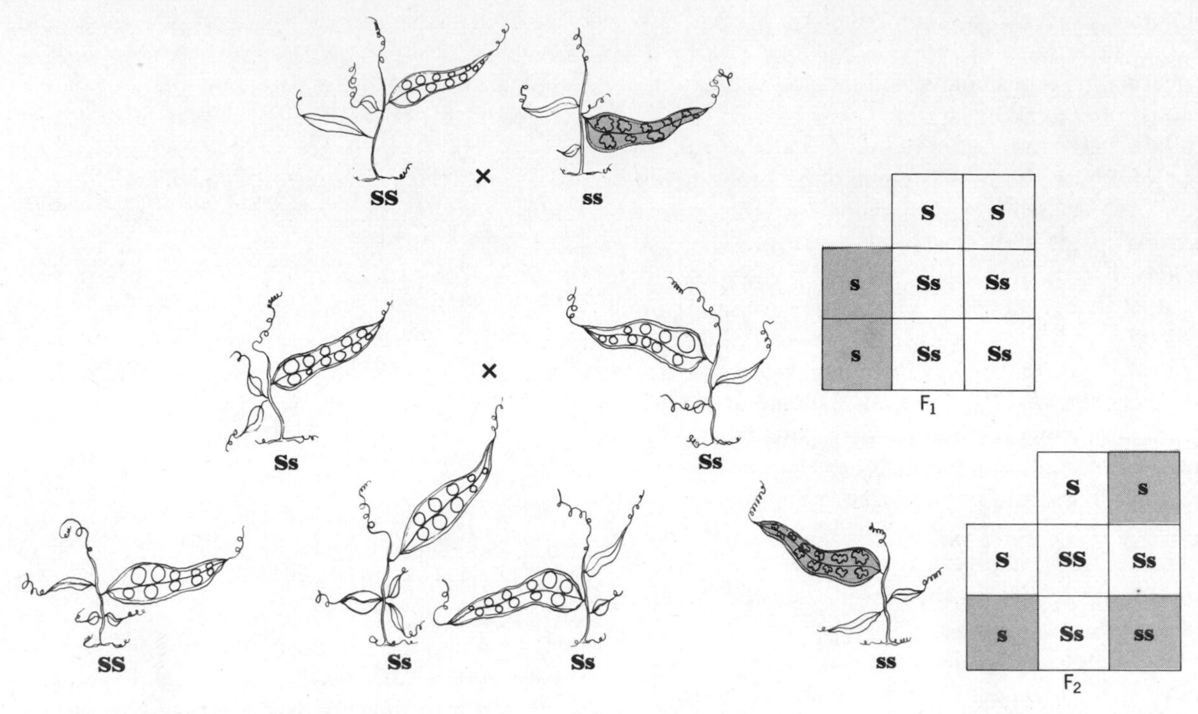

Figure 14.2. Mendel's experiment involving smooth- and wrinkled-seeded peas.

sequence of segregation, equal numbers of gametes of each kind are formed.

5. When gametes unite at fertilization, the two hereditary factors are brought together and again exist in pairs. Fertilization is a random union in the sense that equal numbers of the different kinds of gametes are produced, and it is a matter of chance how they will pair. This being true, it should be possible, on a probability basis, to predict the ratio of various characteristics in the offspring. In the flower-color experiment, for example, the ratio of approximately three red-flowered plants to each white-flowered plant conforms to the expected or predicted ratio of three to one.

Based on these conclusions, the cross between white- and red-flowered pea plants can be diagrammed with symbols. Mendel suggested using capital letters for dominant factors and lowercase letters for their recessive counterparts. For example, if **W** represents the hereditary factor for red and **w** the hereditary factor for white, then a plant could be **WW** (red), **Ww** (red), or **ww** (white).

P_1 generation: **WW** × **ww**
(red) (white)

(gametes) Ⓦ and Ⓦ × ⓦ and ⓦ

F_1 generation: **Ww**
(red offspring only)

Crossing F_1's: **Ww** × **Ww**

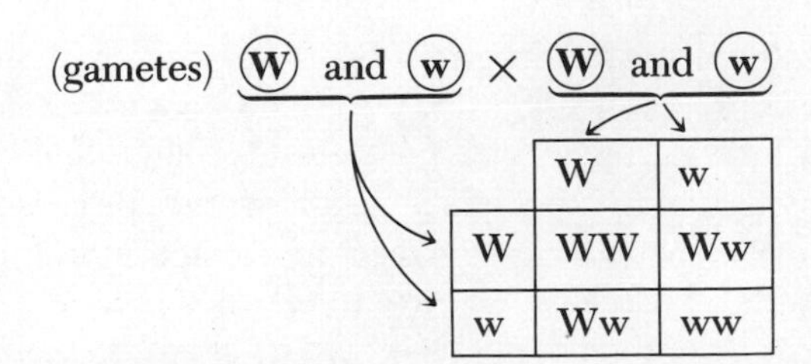

(This set of boxes is termed a Punnett Square and is useful in finding all possible gametic combinations.) Summarizing from the Punnett Square we obtain

F_2 generation: 1 **WW** + 2 **Ww** + 1 **ww**
(red) (red) (white)
3 red 1 white
Hence a 3:1 ratio

By using symbols we have illustrated how the five points just listed operate. The same rules of procedure apply in crosses dealing with either plants or animals. For example, crossing black and white guinea pigs through the second generation produces a ratio of about three black offspring to each white one.

We need to introduce a few more terms that will be useful in the topics that follow. Hereditary factors are commonly termed *genes*. Hence **WW** represents a pair of genes for red-flower color. Also, when the two genes for a trait are alike, as **WW** or **ww**, they are said to be *homozygous*. **WW** is the homozygous dominant state, and **ww** is the homozygous recessive condition. When contrasting genes occur, as in **Ww**, the condition is *heterozygous*. The two letters describe the composition of a gene pair, a *genotype*. **AA, Aa,** and **aa** are genotypes. The observed characteristic produced by a set of genes is called a *phenotype*. Red-flower color is the phenotypic characteristic observable from either genotype **WW** or genotype **Ww**. By way of contrast, there is only one genotype, ww, for white-flower color. Finally, contrasting genes such as the ones for red-flower color and white-flower color are termed *alleles*. To constitute an allelic pair, two genes must occupy the same sites on a homologous chromosome pair.

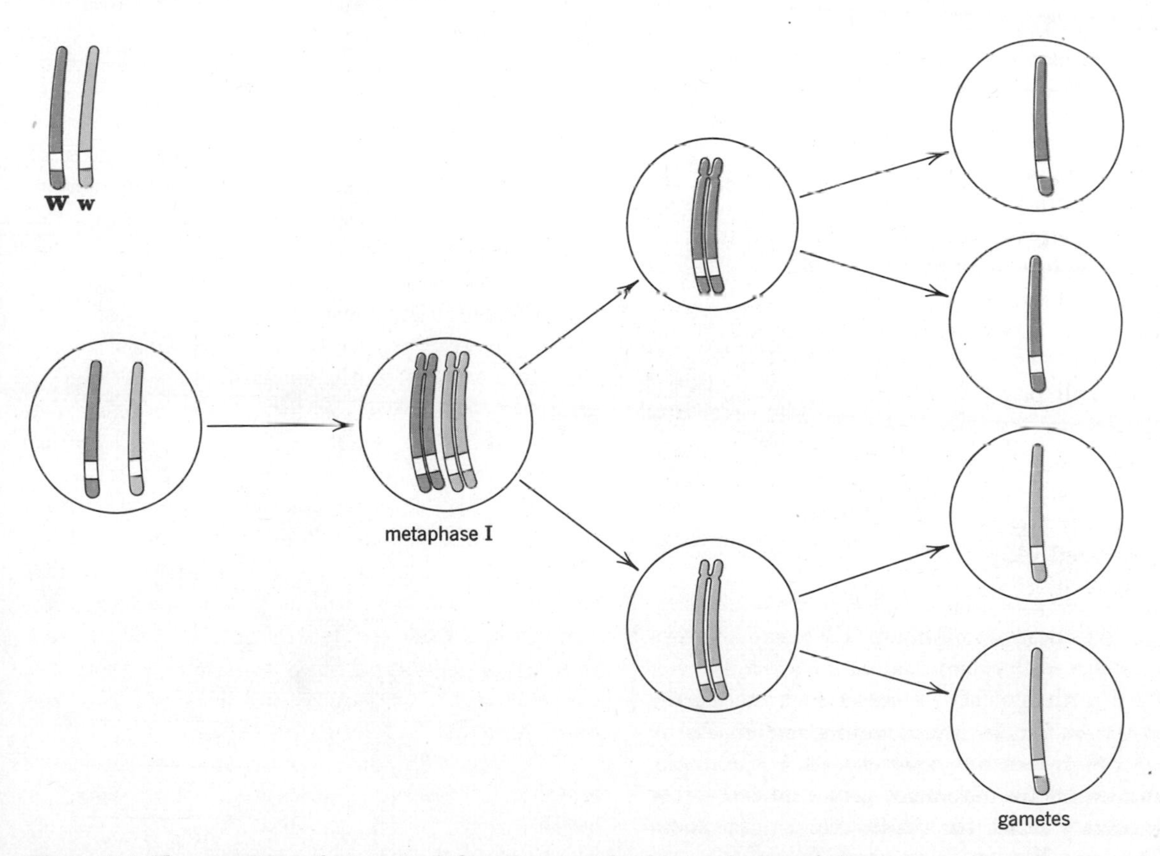

Figure 14.3. The segregation of one pair of chromosomes to daughter nuclei by meiosis.

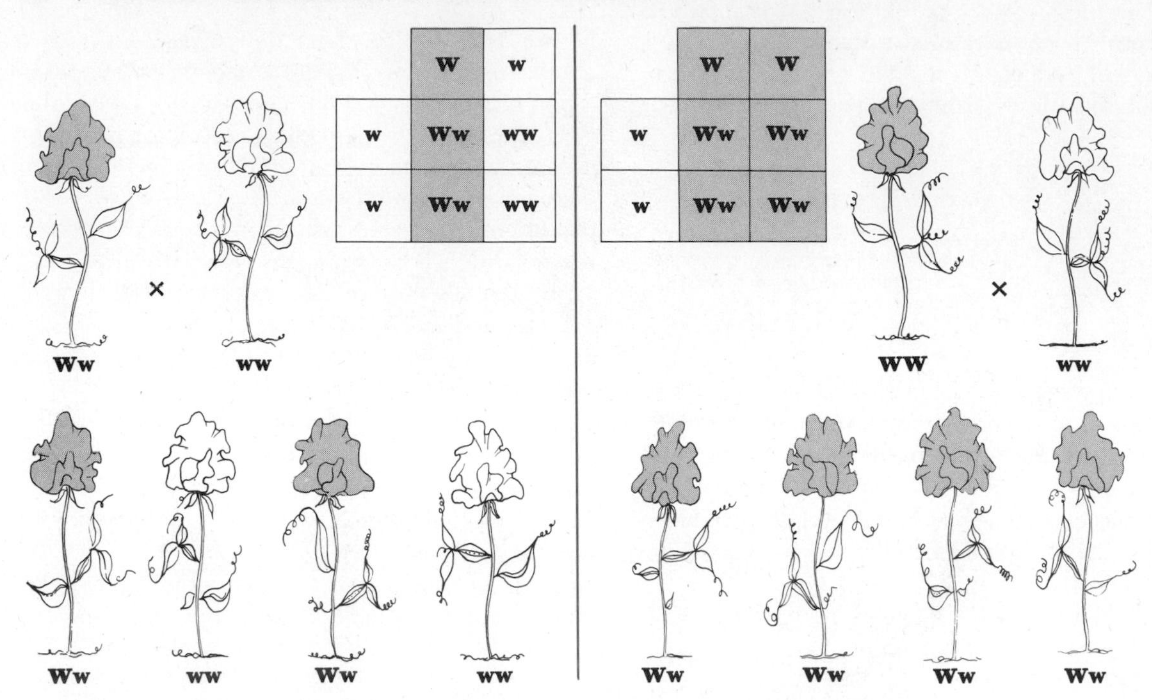

Figure 14.4. A test cross involving heterozygous (left) and homozygous (right) red-flowered peas.

Heredity and Meiosis

At this point it is necessary to relate Mendel's findings to meiosis, since meiosis is the actual mechanism on which hereditary transmission depends. Mendel, of course, was not aware of the events of meiosis.

Recall that chromosomes occur in homologues in cells; since the genes are located on the chromosomes, the genes also occur in duplicate. Thus the symbol **Ww** means that there is a gene for red color on one chromosome and a gene for white color on its homologue. Only one of these genes, however, will express itself in the organism; this is termed the dominant gene.

In prophase I of meiosis, homologous chromosomes (and their genes) pair up. Each pair of chromosomes lines up on the metaphase plate independently of all other pairs. At anaphase I, these pairs separate or segregate into daughter nuclei (Figure 14.3). Because the manner in which one pair segregates does not determine the way in which any other pair segregates, this process is called *independent assortment.* At the end of meiosis, two of the four resulting gametes contain a **W** gene and two carry a **w** gene. In other words, equal numbers of gametes carrying the **W** gene and the **w** gene are produced: in the diagram on p. 196 a single **W** or w represents either an egg or a sperm. Furthermore, the combining of male and female gametes (fertilization) is a random process so that all possible combinations that might occur must be found.

We can predict the most probable outcome of various genetic crosses because hereditary factors segregate independently, equal numbers of different kinds of gametes form, and there is an equal probability of different gametic unions (eggs and sperm). The ratio that is most likely to occur can easily be determined by counting the different combinations in the Punnett Square. Thus, in the example on p. 196, we can see from the Punnett Square that one fourth of the flowers will be white, one fourth homozygous red, and two fourths heterozygous for red. In terms of phenotype, out of every four F_2 progeny, one will bear white flowers and three will bear red ones—thus the 3:1 ratio.

This raises the question of how to determine whether an observed phenotype is homozygous or heterozygous. When one has a red-flowered pea plant, it is **WW** or **Ww**? To determine this, an additional cross has to be made between the red-flowered pea and a *homozygous recessive* (white-flowered) one. If the phenotype in question is heterozygous,

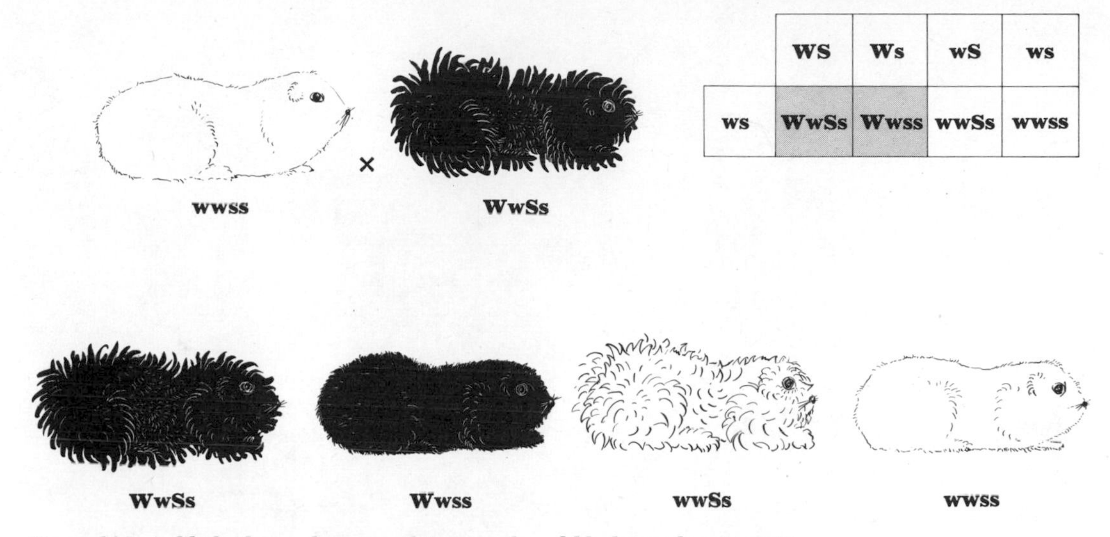

	WS	Ws	wS	ws
ws	WwSs	Wwss	wwSs	wwss

Figure 14.5. A dihybrid cross between white, smooth and black, rough guinea pigs.

two types of progeny should be produced in approximately equal numbers, as may be seen from the diagram.

	Ww	×	**ww**
Gametes:	**W** and **w** ×		**w**
Offspring:	**Ww** (red)	+	**ww** (white)

If, on the other hand, the red-flowered pea is homozygous (**WW**), then *all* of the progeny would have to be red flowered (Figure 14.4). This type of mating or cross is variously termed a *test-cross* or *progeny test,* and is applicable to both plants and animals. It has obvious practical uses for individuals engaged in trying to produce purebred varieties of livestock or farm crops.

Some Other Hereditary Patterns

Dihybrid Cross. Thus far we have followed the transmission and expression of one pair of genes, a *monohybrid* cross. What happens if we follow *two* hereditary traits at the same time, that is, a *dihybrid* cross? Actually the principles of the cross remain the same, but now we must work with *two* pairs of genes on separate chromosomes. Mendel performed such experiments and followed the simultaneous transmission of such combinations as flower color (red or white) and height of plants (dwarf or tall).

In order to demonstrate the mechanics of this type of cross and to show that it also applies to animals, let us consider guinea pigs. In these animals, black fur color (**W**) is dominant to white fur (**w**) and rough-appearing fur (**S**) is dominant to smooth fur (**s**) (Figure 14.5). The genes for these traits are located on two pairs of chromosomes. Thus various combinations of these genes may be indicated in the following way:

WWSS: a black guinea pig with rough coat
wwss: white, smooth coat
WwSs: black, rough coat
Wwss: black, smooth coat
etc.

The way in which these genes segregate and then recombine is illustrated in the following diagram, which shows mating between a male white-colored, smooth-coated guinea pig and a female that is heterozygous for both characteristics:

P_1:

♂ (male) ♀ (female)
wwss × **WwSs**

Gametes:

(ws) × **(WS)**, **(Ws)**, **(wS)**, **(ws)**

(Note that a gamete can contain only *one* of each kind of gene.)

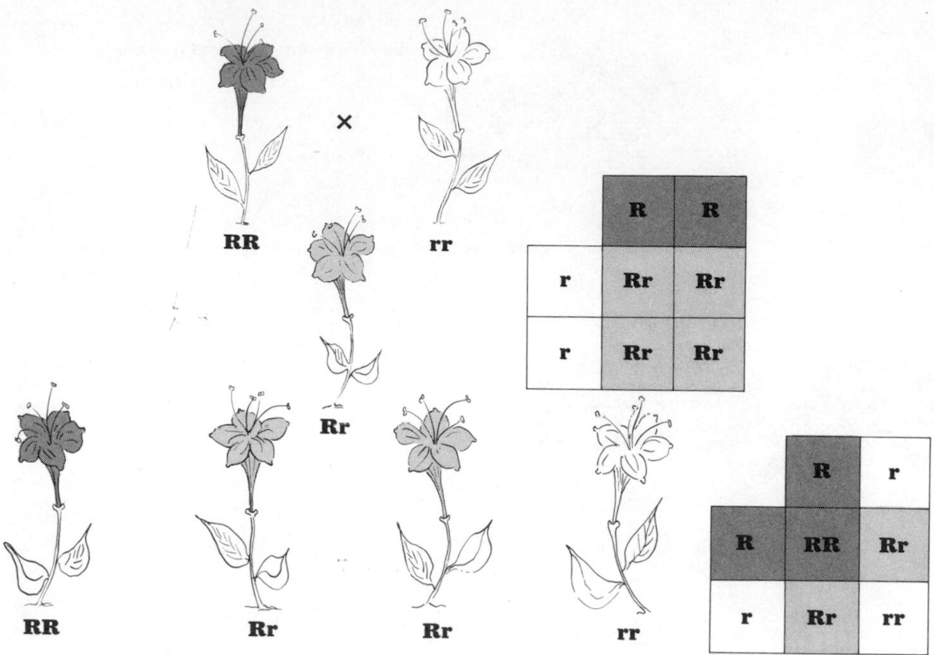

Figure 14.6. Inheritance of flower color in four o'clocks, an example of incomplete dominance.

F_1:

Mother's Gametes

Father's gametes		WS	Ws	wS	ws
	ws	WwSs	Wwss	wwSs	wwss

To summarize, the offspring may be one of the following:

- **WwSs:** black, rough
- **Wwss:** black, smooth
- **wwSs:** white, rough
- **wwss:** white, smooth

According to the Punnett Square, the proportion of offspring is 1 : 1 : 1 : 1; that is, the probability of any one of these combinations in a guinea pig offspring is one chance in four.

The principles operating here are the same ones used in monohybrid crosses. We are simply working with an additional pair of genes. These principles also apply to *trihybrid* inheritance (three pairs of genes) or as many gene pairs as we wish to consider. Beyond three pairs of genes, however, the mechanics become too cumbersome for work on the gametic combinations to be practical.

Incomplete Dominance. We have assumed that genes are always either dominant or recessive, but in reality other genetic interactions operate in relation to a pair of genes. It is not uncommon to find examples of a lack of dominance between two genes as in the common garden flower called four o'clocks. Here, a cross between red four o'clocks and white ones yields F_1's (heterozygous) which bear *pink* flowers (Figure 14.6).

A similar situation is met in tailless cats. If these animals are mated with long-tailed cats, the kittens (F_1) all have short tails. Short tails thus represent the consequence of the reaction between the gene for no tail and the gene for long tail. Can you work out the offspring from a cross between two short-tailed cats?

Analysis of a Human Pedigree. A considerable amount of knowledge has accumulated concerning hereditary traits in human beings. These traits are governed by the same hereditary concepts that we have applied to guinea pigs and garden peas. Of course, a basic problem arises here because experimental matings and test crosses cannot be performed under the control of the investigator. Consequently, human geneticists must utilize indirect sources of evidence such as hospital records, studies of identical

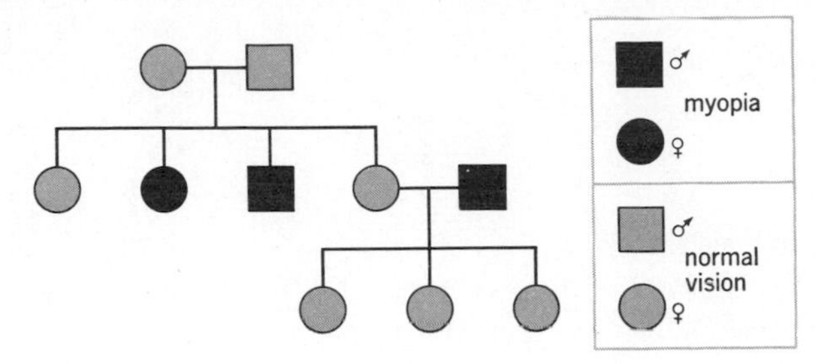

Figure 14.7. The occurrence of myopia in a family.

and fraternal twins, and even family histories. These sources often make it possible to follow a characteristic through several generations on a *pedigree chart*, as indicated in Figure 14.7. By studying a number of pedigree charts for myopia, an investigator can form an hypothesis about the type of inheritance involved. For example, it should be obvious that myopia does not result from a dominant gene and is not restricted to either sex. See if you can formulate an hypothesis to fit the facts observable on the pedigree.

Many familiar human characteristics show simple inheritance patterns. Freckles result from a dominant gene, red hair is recessive to nonred hair, blond hair is recessive to darker hair, wavy hair results from incomplete dominance between the gene for straight hair and the gene for curly hair, and left-handedness appears to be recessive to right-handedness.

On the serious side, a number of hereditary abnormalities are known in humans. One of these, phenylketonuria (PKU disease), occurs when the body is unable to metabolize the amino acid phenylalanine, leading to serious disorders including mental retardation. The disease evidently results from a pair of recessive genes. If the disease is detected early in infancy, its damaging effects can be prevented by eliminating phenylalanine from the baby's diet.

One type of dwarfism is controlled by a dominant gene that affects the growth of cartilage in the limbs. The head and body are normal sized but the limbs are greatly shortened. These dwarfs are normal in all other respects. Can you explain why two dwarf parents often produce normally proportioned children?

The concepts presented in this chapter should permit a general grasp of the basic hereditary principles discovered by Mendel. In Chapter XV we build on these Mendelian principles and examine some more recent developments in genetics.

Principles

1. Genetics is the study of the transmission and expression of traits in successive generations of organisms.
2. Gregor Mendel established the existence of discrete hereditary units and some of their simpler interactions, including dominance, segregation, and recombination.

Suggested Readings

Mendel, Gregor, "Letter to Karl Nageli," in *Great Experiments in Biology*, edited by M. L. Gabriel and S. Fogel. Prentice-Hall, Englewood Cliffs, N. J., 1955, pp. 228–233.

Mendel, Gregor, "Experiments in Plant-Hybridization," in *Classic Papers in Genetics*, edited by James A. Peters. Prentice-Hall, Englewood Cliffs, N.J., 1959, pp. 1–20.

Montagu, Ashley, *Human Heredity*. Mentor Books. The New American Library of World Literature, Inc., New York, 1960.

Sturtevant, A. H., "Social Implications of the Genetics of Man," in *Classic Papers in Genetics*, edited by James A. Peters. Prentice-Hall, Englewood Cliffs, N.J., 1959, pp. 259–263.

Questions

1. What is the hereditary significance of fertilization?
2. Can you think of some erroneous hereditary concepts not listed in the chapter?
3. Did the development of the basic hereditary principles (Mendel's laws) depend on the improvement in tools and technology as was true of studies of the cell? Defend your answer.
4. For what reasons would Mendel probably have been unsuccessful if he had used animals, such as rats or dogs, instead of garden peas for his studies on heredity?
5. What is the difference between independent assortment and segregation?
6. Make a list of the aspects of meiosis that are essential to the understanding of Mendel's principles.
7. In what way are the principles of heredity related to the idea of probability?
8. More is known about the inheritance patterns of abnormal human traits than normal ones. Why?

Post-Mendelian Genetics

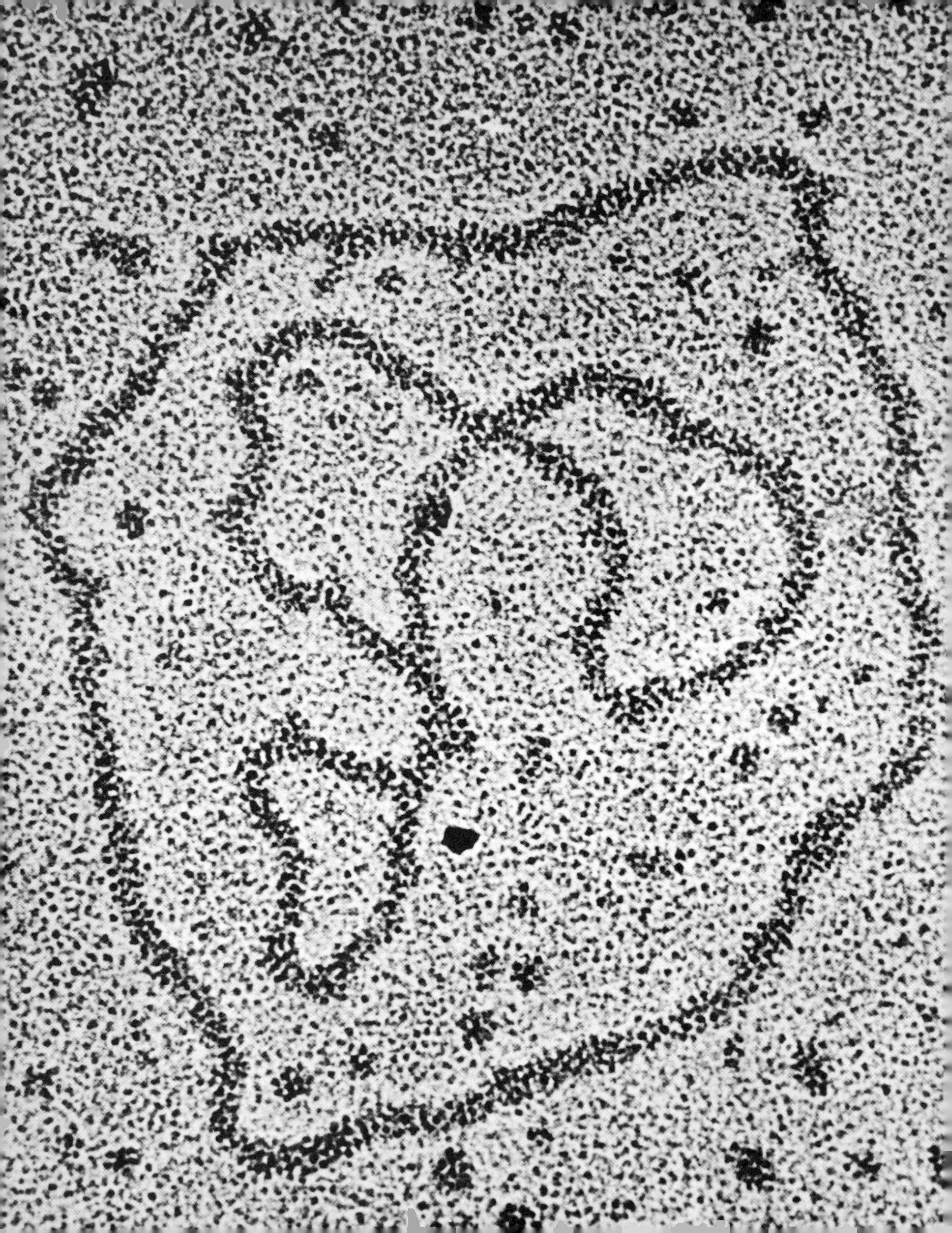

Bacterial chromosomes (×*218,000*)
(*Dr. A. Kornberg*).

CHAPTER
XV

Post-Mendelian Genetics

Mendel published his results in 1866 under the title "Experiments in Plant Hybridization," but no one realized their significance until 1900. In that year three geneticists, Hugo de Vries in Holland, Carl Correns in Germany, and Erich Tschermak of Austria independently discovered Mendel's publication and linked his explanations with their own findings. From then on genetics advanced rapidly; hence the title of this chapter, *post-*Mendelian genetics.

Chromosome Theory

Shortly after the rediscovery of Mendel's publication, Walter S. Sutton, a graduate student at Columbia University, began to study the relation of meiosis to the inheritance of Mendel's heredity factors. He noted that chromosomes differed qualitatively, that homologous chromosomes paired at the beginning of meiosis, and that they assorted randomly on the metaphase spindle. He concluded, "Thus the phenomena of germ-cell division and of heredity are seen to have the same essential features, *viz.*, purity of units (chromosomes, characters) and the independent transmission of the same. . . ." Stated in simple terms, Sutton was proposing that chromosomes were the physical carriers of hereditary units, and that these units were transmitted with the chromosomes. He also reasoned that since hereditary factors were more numerous than chromosomes, each

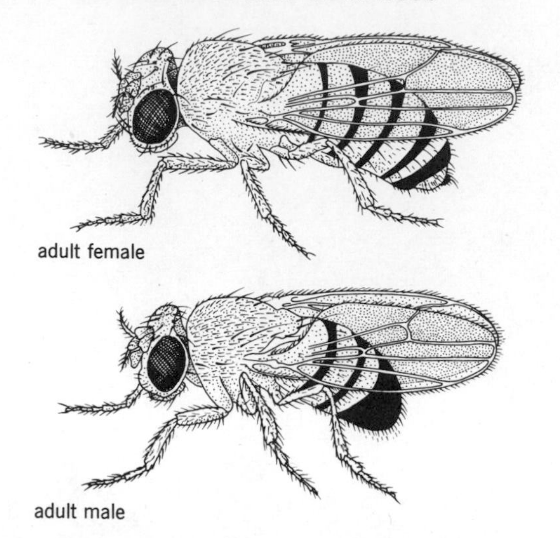

Figure 15.1. Male and female of the fruit fly Drosophila.

chromosome had to contain many such units. These basic concepts are known as Sutton's chromosome theory of inheritance. Although we accept these ideas as commonplace today, they were major contributions to genetics when Sutton published them in 1903.

Shortly after this (1910), Thomas Hunt Morgan began using the small, common fruit fly, *Drosophila*, in breeding experiments in his laboratory at Columbia University. Many rapid advances in genetics soon followed; in fact, many of the modern principles of heredity were worked out using this tiny fly. *Drosophila* can be maintained fairly easily in laboratory cultures, it reproduces prolifically, and it exhibits a variety of distinguishable phenotypic features (Figure 15.1). In addition, cells in its salivary glands contain giant chromosomes (Figure 15.2). These chromosomes are convenient for such experimental uses as correlating genes with specific loci (sites) on chromosomes or for studying the general structure of chromosomes.

While discussing genes and chromosomes we can clarify a matter that frequently puzzles students—what is the relationship between genes and chromosomes? A chromosome usually consists of DNA, RNA, and several types of proteins. This is termed a nucleoprotein complex. The DNA portion of this complex is the actual genetic material and will be discussed further in Chapter XVII. It is customary in biology to speak of genes being "in" or "on" chromosomes with the understanding that genes are integral parts of these nuclear structures.

Sex Determination

One of Morgan's many contributions concerned the discovery of a pair of chromosomes in *Drosophila* cells that appeared to determine sex; these are termed *sex chromosomes*. The two in the cell of a female are alike and are known as **X**-chromosomes. The two in male cells are not alike; one is an **X**-chromosome, as in the female, but its partner is different in shape and is termed a **Y**-chromosome. The four chromosomes segregate during meiosis and recombine at fertilization as follows:

	Female		Male
	XX	crossed with	**XY**
Gametes:	(**X**)	crossed with	(**X**) and (**Y**)
F_1:		**XX** and **YY**	

One half of the offspring receive the **XX** combination and one half receive the **XY** combination. This accounts for the expected 50–50 ratio of males to females in crosses.

It is convenient to speak of the nonsex chromosomes, as distinct from the sex chromosomes, as *autosomes*. A human being, for example, has 22 pairs of autosomes and one pair of sex chromosomes for a total of 23 pairs or 46 chromosomes.

Human beings and other mammals have the same type of sex chromosome mechanism, but it is not universal. In birds, the sex chromosomes are alike (**ZZ**) in the male but they differ in the female (**ZW**). In bees, sex is determined by chromosome number in that females are diploid and males haploid.

In recent years several important discoveries have been made about human sex chromosomes. A cell from a female may be identified by a tiny speck of nuclear material termed sex chromatin or Barr body (Figure 15.3*b*). It is thought to represent one of the female's **X**-chromosomes. Also, certain of the female's white blood cells have a small body called a "drumstick" attached to the nucleus (Figure 15.3*b*). Cells from a normal male, that is, an **XY** male, never show Barr bodies or drumsticks (Figure 15.3*a*). These *sex indicators* are observed in nondividing cells; hence they provide a convenient means of determining whether a sample of human tissue is derived from a male or a female. These indicators are also extremely useful in confirming the diagnosis of the unusual cases where a person has

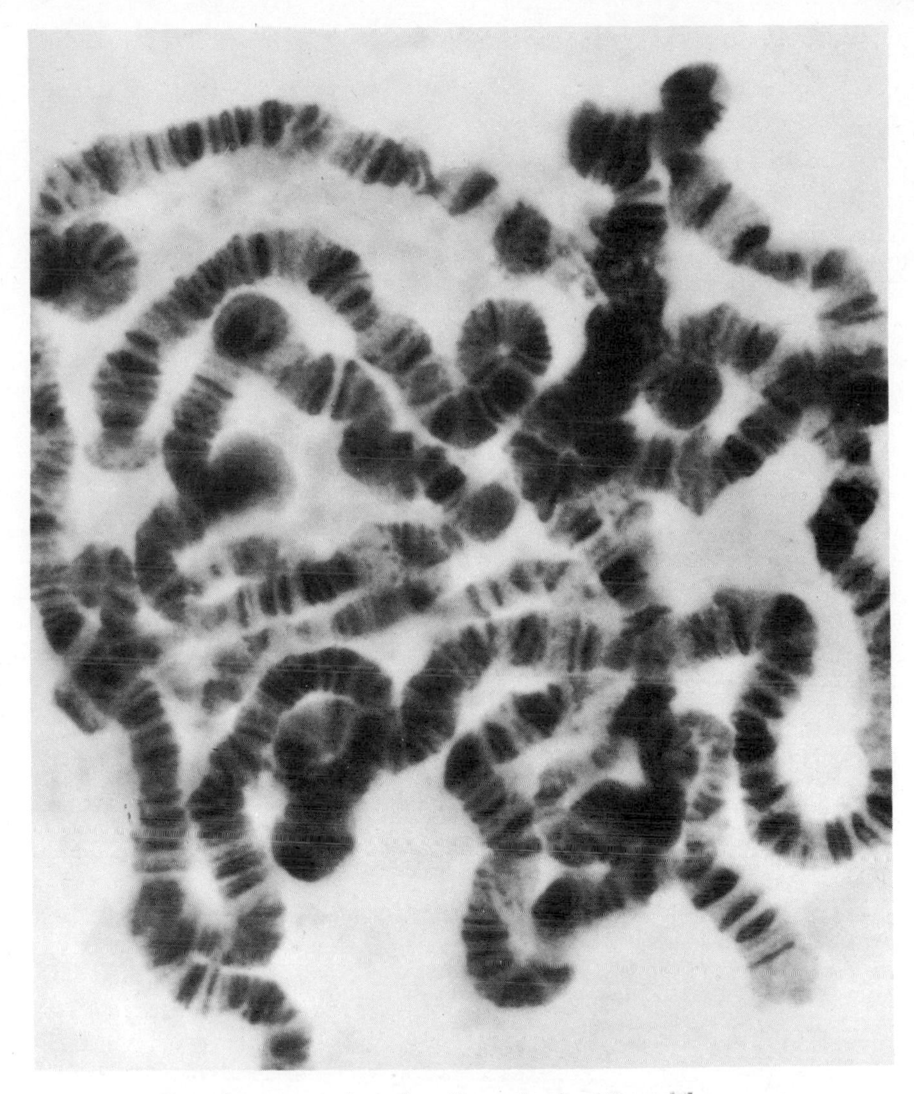

Figure 15.2. Giant chromosomes from the salivary glands of Drosophila.

an abnormal number of sex chromosomes. A male, for example, may have an extra **X**-chromosome, that is, be of an **XXY** genotype (a total of 47 chromosomes). These individuals typically have male genitalia, sparse body hair, and some female-like breast development. Cells from these males show Barr bodies and drumsticks (Figure 15.3*b*), indicators of the extra **X**-chromosome. Less common are the females whose cells show *no* Barr bodies or drumsticks. Such individuals have only one **X**-chromosome and no **Y**-chromosome for a total of 45 chromosomes (Figure 15.3*a*). Typical features are female external genitalia, underdeveloped breasts, short stature, and rudimentary ovaries. In both instances described here the individuals often have severe mental handicaps. Other known sex chromosome anomalies are **XXX** genotypes and even **XXXX** females and **XXXXY** males (Figures 15.3 *c-d*). As the number of **X**-chromosomes increases, so does the number of Barr bodies and the number of drumsticks. All of these unusual cases raise the question of what determines a male or a female. In humans, at least, a **Y**-chromosome evidently specifies maleness regardless of the number of **X**-chromosomes accompanying it. In the absence of a **Y**-chromosome, a human is a female.

How do such abnormalities arise? Part of the answer lies in the failure of sex chromosome homo-

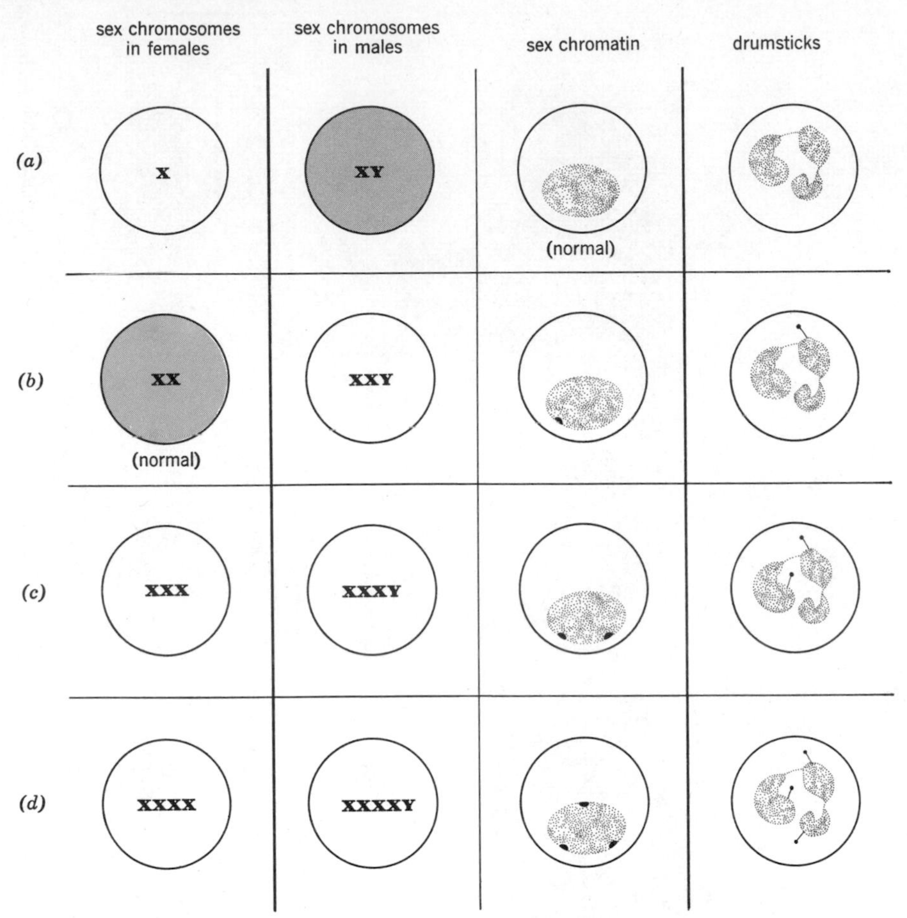

Figure 15.3. Sex differences in male and female blood cells as indicated by sex chromatin bodies and drumsticks (small pin-shaped forms). (Modified from "Sex Differences in Cells," by Ursula Mittwoch. Copyright © 1963 by Scientific American, Inc. All rights reserved.)

logues to segregate in prophase I of meiosis. Thus, if an **X**- and a **Y**-chromosome end up in a sperm cell, at fertilization the zygote would contain an **XXY** combination. Other abnormalities such as **XXX** can occur by similar aberrant chromosomal behavior in meiosis.

Sex-Linked Genes. Early in his work Morgan discovered that some traits in *Drosophila* were found more frequently in males than in females. White eye color is an example. This gene for eye color is located on the **X**-chromosome, that is, it is *sex linked.* The **Y**-chromosome carries few genes and has none for eye color. Thus we can consider it devoid of genes in our discussion. The gene for red eye (**W**) is dominant over the gene for white eye (**w**). Since they are located on the **X**-chromosome, we can use the symbols $\mathbf{X^W}$ and $\mathbf{X^w}$ for them. A cross between a white-eyed female fly and a red-eyed male is diagrammed as follows:

	Female		Male
	$\mathbf{X^wX^w}$	crossed with	$\mathbf{X^WY}$
Gametes:	$\mathbf{X^w}$	crossed with	$\mathbf{X^W}$ and $\mathbf{Y}$
F_1:	$\mathbf{X^WX^w}$ and $\mathbf{X^wY}$		

Note that the male F_1's are now white eyed and the females are red eyed (Figure 15.4). The **X**-chromosome in males has no homologue so that every gene on it, recessive or dominant, is expressed.

Many sex-linked traits have been investigated in humans. Familiar ones include hemophilia, red-green color blindness, and certain types of muscular

Figure 15.4. A cross involving the sex-linked trait of white eye color in Drosophila.

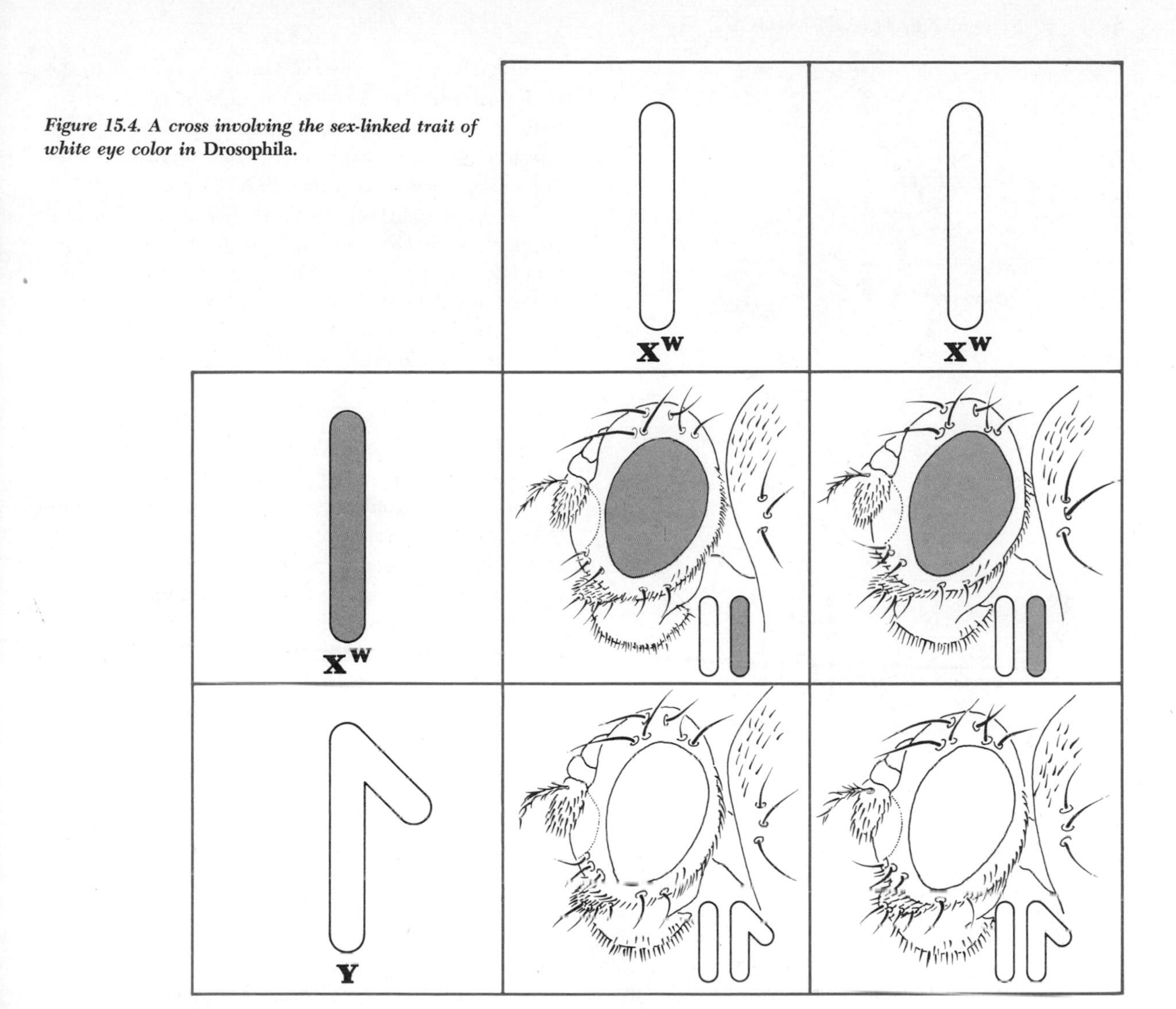

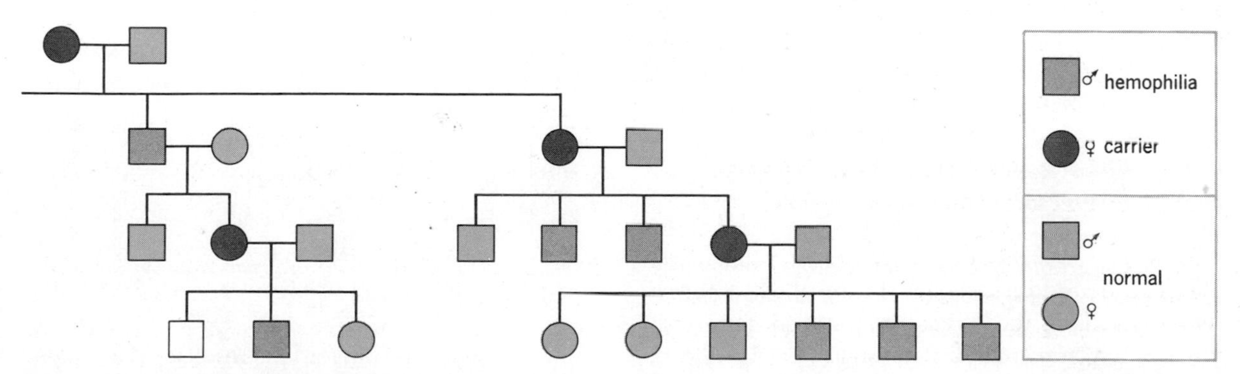

Figure 15.5. A pedigree of hemophilia in humans.

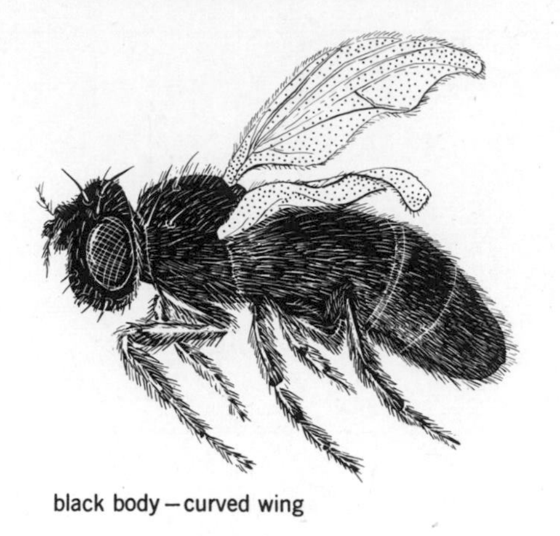

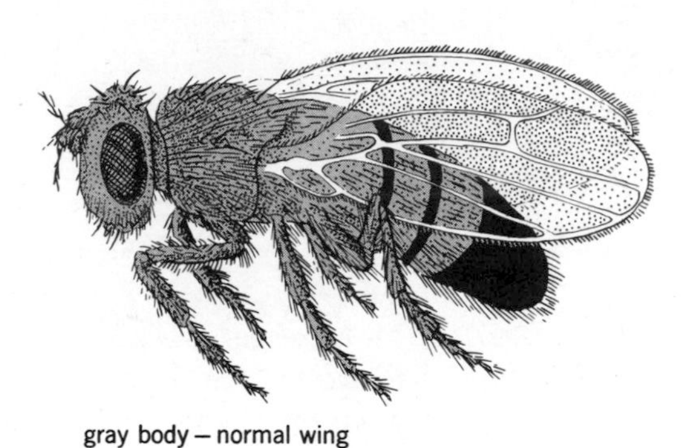

Figure 15.6. Flies showing the traits black body, curved wing (above) and grey body, normal wing (below).

dystrophy. Nearly sixty additional sex-linked traits, mostly pathological and recessive, are known. Figure 15.5 presents a pedigree chart for hemophilia. The chart illustrates some of the typical features of this type of inheritance: fathers never pass the trait to their sons, only to their daughters; sons inherit the defective gene from their mothers. Queen Victoria was a carrier of hemophilia; the appearance of the trait among her descendants has been extensively studied, and has had numerous political repercussions.

Muscular dystrophy, a destruction of muscle fibers, is carried by a recessive gene on the **X**-chromosome. It follows the typical inheritance pattern of sex-linked genes except that males usually die in their teens. Thus males rarely reproduce. Females, on the other hand, never have the disease but often reproduce and transmit the disease. No treatment is known since so many vital enzyme systems in the body are affected. The best course of action is genetic counseling. A carrier may not wish to have children if she knows there is one chance in four of having a son with this lethal disease.

Red-green color blindness is caused by a sex-linked recessive gene but does not lead to the serious consequences seen in hemophilia or muscular dystrophy (unless one misjudges the color of a stoplight!). Can you determine from its mode of inheritance why there are many more color-blind males than females?

Linkage, Crossover, and Chromosome Mapping

In discussing Mendel's work, we considered the mechanism of crosses where the genes were on nonhomologous chromosomes. What happens when different genes are located on the *same* chromosome? In *Drosophila*, it is known that genes for the recessive traits of black body (**b**) and curved wing (**c**) are on the same chromosome. The dominant alleles of these two are gray body (**B**) and normal wing (**C**). (See Figure 15.6.) An easy way to diagram a cross involving these traits is to indicate them on little "stick" chromosomes. Genes **B** and **C**, or their alleles, are indicated by dots on the stick chromosome, as follows:

B┃ **C**┃	or	┃**b** ┃**c**

A cross between a fly that is homozygous normal and one showing the recessive conditions would appear as:

Gray body, normal wing	**B**┃ ┃**B** **C**┃ ┃**C**	×	**b**┃ ┃**b** **c**┃ ┃**c**	Black body, curved wing
Gametes:	**B**┃ **C**┃	×	┃**b** ┃**c**	
F_1:		**B**┃ ┃**b** **C**┃ ┃**c**		Gray body, normal wing

Up to this point the phenotypic results (F_1) are identical to the type of dihybrid cross we described earlier. What happens if the heterozygous individual is crossed with a homozygous recessive one?

Gray body normal wing: B/C b/c × b/c b/c

Gametes: B/C and b/c × b/c and b/c

F_1: B/C b/c and b/c b/c

(gray, normal) (black, curved)

One half of the offspring are gray and normal winged and one half are black with curved wings. These results are not the same as those that would be obtained if these genes were on separate chromosomes. In that case there would be four phenotypes rather than two.

Actual performance of this cross would yield *mostly* gray, normal or black, curved offspring as shown; however, a *few* gray-bodied, curved-winged individuals and a *few* black-bodied, normal-winged members would also result. These two phenotypes are unexpected from the standpoint of the mechanisms we have been using and require a special explanation. Let us therefore refer to meiosis again. Recall that in prophase I, the homologous chromosomes pair and then pull apart. As they do this, some of the chromatids appear to stick together at various places; in reality, the chromatids of homologous chromosomes have exchanged homologous parts. In the cross involving body color and wing traits, this exchange can be diagrammed as follows where the open circles represent centromeres:

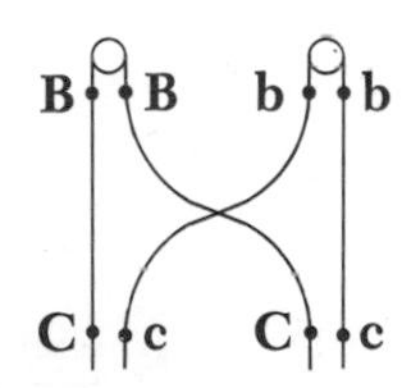

If the crossing chromatids broke and then rejoined at the crossover point, we would have

B B b b
C c C c

We would now have gametes

B/C B/c b/C b/c

If we repeated the cross of heterozygous gray, normal with black bodied, curved winged, our results would be as shown below.

Notice the percentages under each phenotype. Together, they indicate that 74 percent of the offspring result from the noncrossover gametes, and 26 percent from the crossover gametes. This is not an accidental occurrence, since a repetition of the cross will give the same proportion of offspring. Other linked genes give other characteristic values. T. H. Morgan and his student, A. H. Sturtevant, hypothesized that if genes were arranged in a linear manner along chromosomes, the genes closest together would undergo crossover less often than those farther apart. There would be a lesser likelihood and probability of breaks occurring between genes close together than when they are farther apart. Furthermore, the percent of crossover would be equivalent to the relative distance between two genes on a chromosome. Percent crossover becomes the number of crossover units between two genes. On the basis

P_1: B/C b/c × b/c b/c

Gametes: (noncrossover) B/C and b/c

\+ × b/c

(crossover) B/c b/C

Offspring: B/C b/c + b/c b/c

(gray, normal: 37%) (black, curved: 37%)

B/c b/c + b/C b/c

(gray, curved: 13%) (black, normal: 13%)

of this hypothesis, Morgan's group and other geneticists mapped the chromosomes of *Drosophila* and other suitable organisms. A number of gene locations are also known for human chromosomes.

Let us look briefly at this mapping technique. In the foregoing example, recall that body color and wing shape showed 26 percent crossover. Let us call this percentage 26 crossover units and map the two genes on a hypothetical chromosome as follows:

B C
|←---------26---------→|

Now let us add a third gene, cinnabar eye (**cn**) which breeding experiments show to be 9 crossover units from **B** and 17 units from **C.** It must be located as follows:

←---9---→ ←------17------→
B cn C
|←----------- 26------------→|

If brown eye (**bw**) were found to be 25 units from **C** and 42 units from **cn**, it would be positioned as

|←----------- 25 ------------→|
B cn C bw
|←--------------------42--------------------→|

By repeating this procedure many times, a virtually complete chromosome map may be constructed. Figure 15.7 shows a portion of a chromosome map for *Drosophila.*

Chromosome mapping in humans is difficult because the only source of data is pedigree charts. Using this technique, geneticists have mapped the loci of classic hemophilia, color blindness, glucose-6-phosphate dehydrogenase (G6PD) deficiency, and a blood group called Xg. These are linked genes on the **X**-chromosome. As you can see, these are unusual traits that lend themselves to pedigree analysis. Only a few linkage groups, mostly blood groups, are known for man's autosomal chromosomes.

Quantitative Inheritance

The traits considered previously are discontinuous: red or white flowers, white or red eye color, color blindness or normal vision, and so on. This is an either/or situation; there are no intermediates except in the few instances of incomplete dominance, and even there only one intermediate is possible.

Many traits, however, show numerous intermediate types and are also hereditary: for example, body size, height, weight. These traits are manifested by almost innumerable gradations, and the mechanisms of Mendelian-type heredity do not seem at first to apply to them. If we hypothesize that height is the result of simultaneous action of a whole series of genes, with a lack of dominance between alleles, then this continuous trait may be explained. Since each gene for tallness adds to the person's height, the accumulative effect of these genes determines the height of the individual. This phenomenon is known as *additive gene action.*

Assume that height results from five pairs of genes. By assigning capital letters to indicate genes for tallness and small letters to indicate genes for shortness, we represent an extremely tall person by **AA BB CC DD EE**, and an extremely short individual by **aa bb cc dd ee**. Different combinations of these genes produce different heights. For example, the combination **Aa Bb Cc Dd Ee** represents a genotype halfway between the two extremes. So would the genotype **AA BB Cc dd ee.**

This type of inheritance also explains another phenomenon commonly observed in nature. If we measure a number of live oak leaves, or weigh a hundred toads, or measure any continuous-type trait in a population, there will be a few of the extreme types and many intermediates. In live oak leaves, for example, a *frequency distribution* might look like this:

Leaf Length (mm)	Number of Leaves
50	1
51	0
52	3
53	88
54	15
55	5
56	4
57	1
58	2
59	2

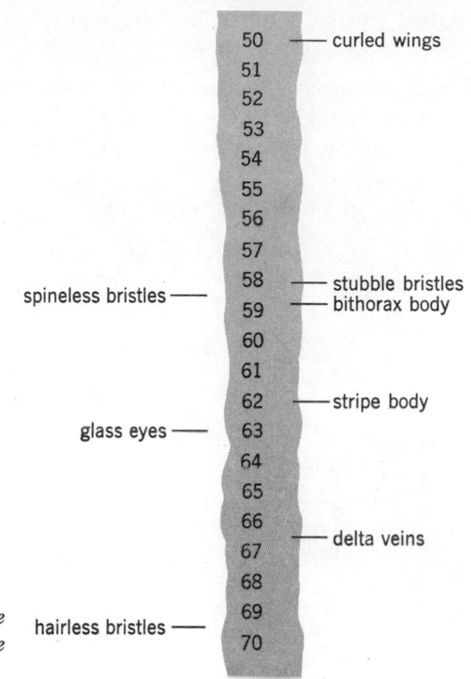

Figure 15.7. A linkage map for a central portion of one chromosome of Drosophila *showing the relative position of a few genes. The numbers are the crossover units.*

Generally, we attribute this kind of distribution to the additive actions of multiple genes. The role of environment must not be overlooked since it selects those phenotypes that function most efficiently for survival of the organism. The optimum leaf length for our live oak in its particular environment lies at about 53 mm. These leaf lengths function best under this set of environmental conditions. Practically all traits reflect this phenomenon.

The Mutation Concept

Even prior to 1900 biologists were aware that new traits occasionally appeared in plants and animals that had not been present in their ancestors. Eventually, the term *mutation* was applied to the sudden appearance of a new feature, which thereafter was passed from generation to generation. In other words, mutation denotes a hereditary change.

Geneticists had always been interested in this striking event, but the first real advance in understanding mutation took place in 1927 when the geneticist Herman J. Muller found that he could cause mutations in *Drosophila* flies by exposing them to X-rays. The X-rays in some way changed one or more genes in the reproductive cells of the organism. This technique provided a tool for accelerating the rate of mutation, which normally occurs infrequently. With this tool, biologists could better study mutations and their effects in many types of organisms. For his pioneer work, H. J. Muller received the Nobel prize in 1946.

In Chapter XVII the topic of mutation will be considered again with reference to the different kinds of mutations and their precise relationship to genetic materials.

The Genetics of a Haploid Organism: *Neurospora*

Life Cycle. As mentioned previously, most of our knowledge of the mechanics of genetics has derived from laboratory organisms like *Drosophila* and to a lesser extent from other common laboratory animals. All of these, however, have one major drawback: they are diploid. With the exception of sex-linked genes, every gene is represented at least twice, with all the complications of dominance-recessive relations and other interactions. An organism that is haploid during much of its life cycle would not present these problems and thus would facilitate several aspects of the study of genetics. In recent years organisms like the pink breadmold, *Neurospora*, have made this possible.

Neurospora is a common fungus, a pest in some instances, which grows by slender filamentlike extensions called *hyphae*. The hyphae contain nuclei that

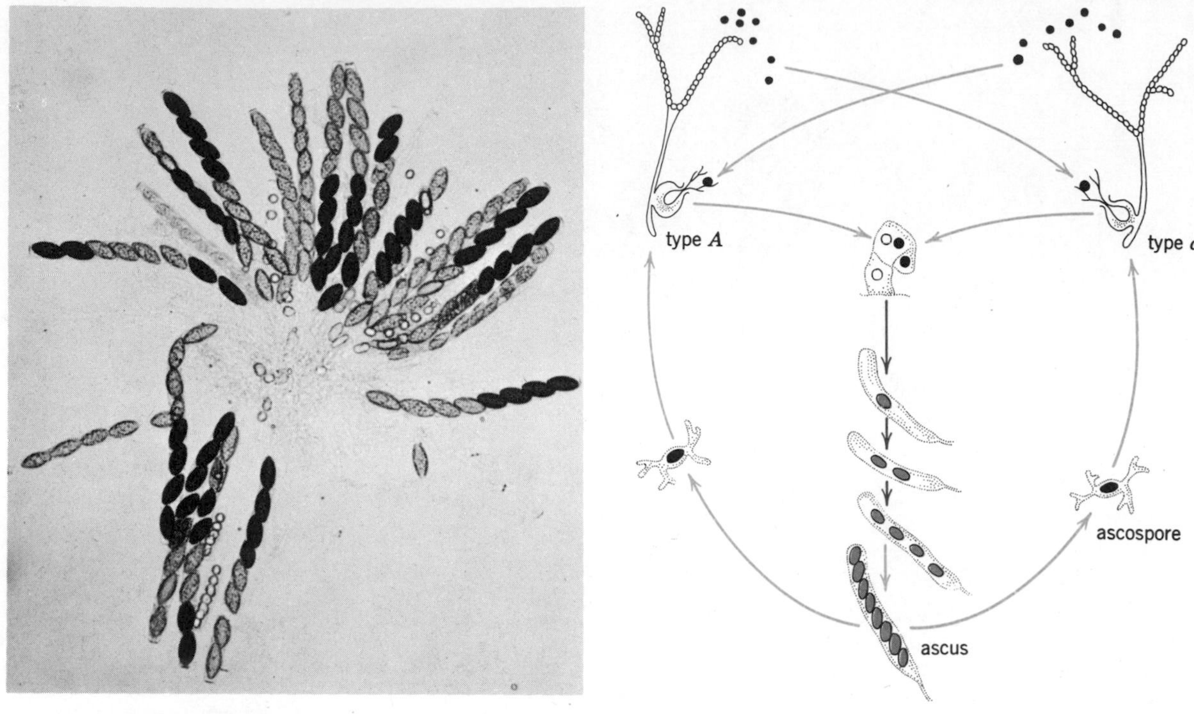

Figure 15.8. Life cycle of Neurospora. *The photograph shows the spores oriented within asci. The order of spores shows several patterns of segregation (Dr. D. R. Stadler).*

are haploid: this organism spends most of its life cycle in the haploid condition. *Neurospora* reproduces asexually by a series of continually spreading hyphae and by spores, which can be blown about to start new colonies. It also reproduces sexually, which is of importance here. Two mating strains, **A** and **a**, must be present before sexual reproduction will occur. Mating strain **A** will not cross with another **A**, neither will strain **a** cross with another **a**: only strain **A** crosses with strain **a** (Figure 15.8).

When the two strains are together, some of the hyphae of each form reproductive cells. No meiosis is involved since the nuclei are already haploid. Reproductive cells of opposite strains unite so that the equivalent of a fertilized egg, a diploid nucleus, is obtained. The specialized structure in which it is located becomes an *ascus*. The diploid cell in the ascus undergoes meiosis to form four haploid daughter cells, each of which enters one mitotic division to become two nuclei; now there is a total of eight haploid nuclei in the ascus. A spore wall forms around each nucleus, converting it into an ascospore. These eight spores will eventually be freed to form new hyphae. The reason that the ascus is so helpful to geneticists is that it is tubelike in form. The meiotic and mitotic products must line up side by side, in order of their formation, like peas in a pod, as shown in Figure 15.8.

Thus, the geneticist can cross two strains of *Neurospora* showing different characteristics, then remove each individual ascospore, germinate it, and observe the results. This is equivalent to being able to grow individual gametes or haploid spores into adult organisms in order to analyze each product of meiosis.

Neurospora contains seven chromosomes in each of its nuclei (haploid or diploid?). One of these chromosomes contains the gene for normal spreading-type growth, denoted by a + symbol. An allele that produces colonial growth is denoted *c*. Hence a cross between a strain showing normal growth with a strain showing colonial growth would be represented

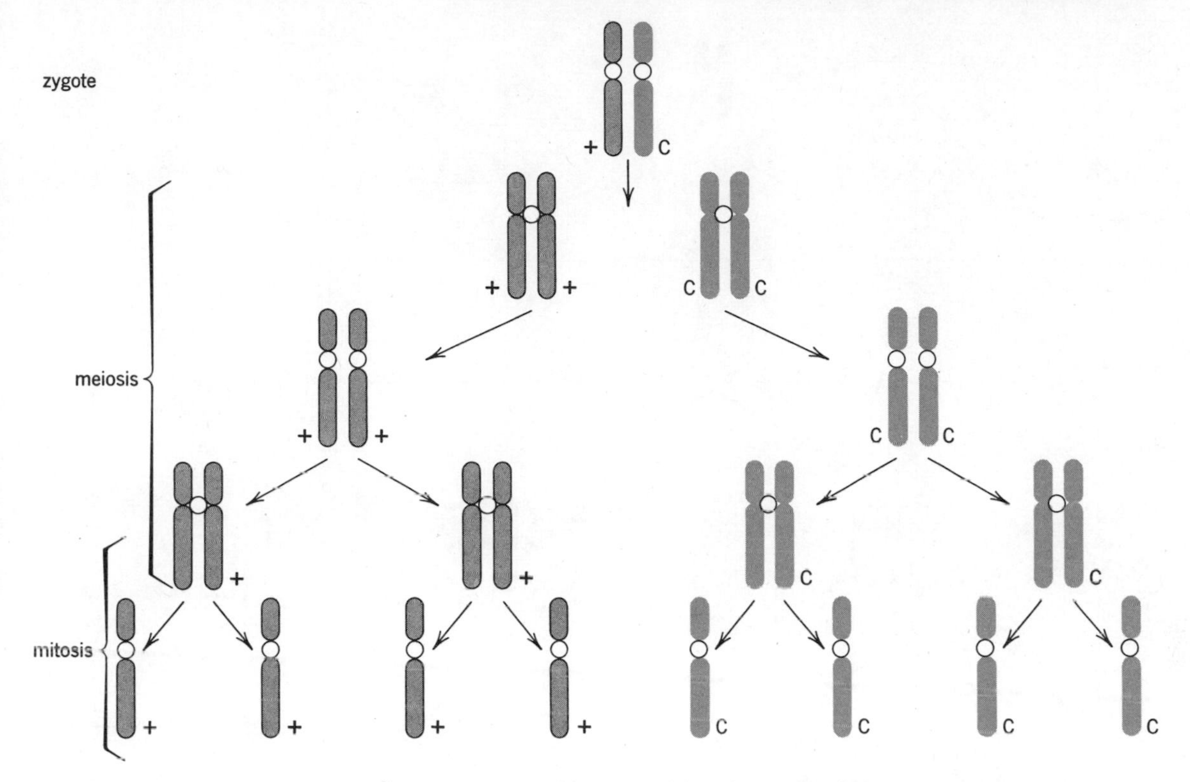

Figure 15.9. One pattern of segregation of the alleles for colonial (c) *and spreading* (+) *growth in* Neurospora.

as in Figure 15.9.

After the two nuclei fuse to form the zygote, meiosis immediately follows. Homologous chromosomes pair, then segregate. Meiosis terminates with the formation of four haploid daughter nuclei, two bearing the gene for spreading growth and two bearing the gene for colonial growth. Each undergoes a mitotic division to produce a total of eight nuclei as shown on the diagram. If these were separated from an ascus and germinated, four would show the normal growth pattern and four the colonial type.

Not only does *Neurospora* demonstrate segregation but it also shows independent assortment, linkage, and crossover. Figure 15.10 diagrams how crossover could occur in relation to the two traits discussed above. Note how it affects the order of the spores in the ascus.

The Work of Beadle and Tatum. Now let us examine a case of biochemical genetics that was demonstrated by G. W. Beadle and Edward L. Tatum in 1941, for which they received a Nobel prize. Prior to that time it had been postulated that genes controlled the production of enzymes that govern the numerous complex chemical activities in cells, which in turn bring about the phenotypic expressions of the genes. But how could this concept be demonstrated with living organisms? Beadle and Tatum used *Neurospora* for their experiments.

Normally, *Neurospora* can be grown on an artificial culture-medium containing only sugar, salts, and biotin (a vitamin). With these basic materials, called a *minimal culture-medium,* the fungus is able to synthesize all of its needed organic compounds such as amino acids, proteins, carbohydrates, fats, and nucleic acids—in short, all of the complex compounds found in living organisms. If these syntheses are controlled by enzymes, and enzymes are controlled by genes, then an altered gene, that is, a *mutation,* should disrupt some stage of this vital chain of events.

Beadle and Tatum proceeded to test this idea by

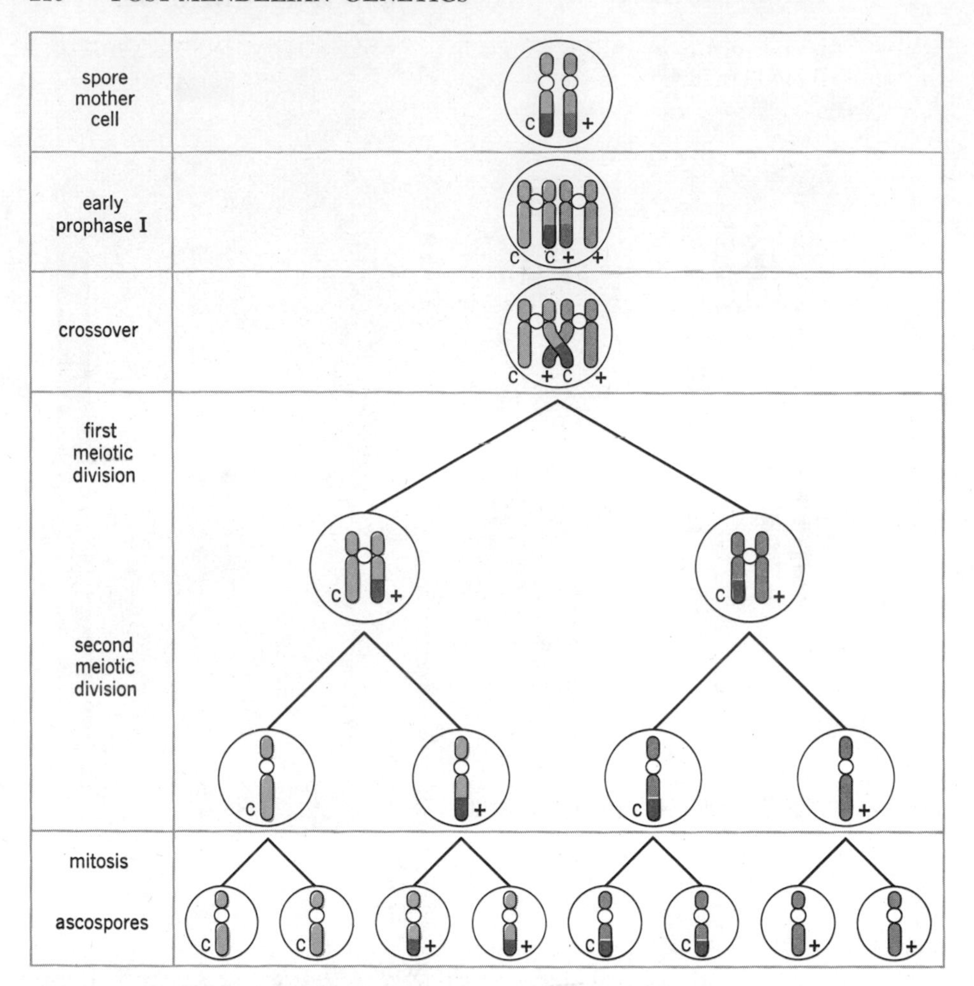

Figure 15.10. The pattern of segregation in Neurospora *showing crossover.*

exposing the spores of a normal *Neurospora* to mutagenic agents such as X-rays or ultraviolet light. The exposed spores were then grown into colonies on a culture medium containing all of the amino acids and vitamins essential to the growth of the fungus. These colonies provided sufficient additional material for carrying out the remainder of the experiment.

A portion of each colony was placed on the minimal culture-medium. If it grew normally, then the sample most likely had not undergone mutation. Those that did not grow had been altered in some way, probably by a mutation.

The altered samples (*mutants*) had to be tested to determine which gene or genes had mutated. This was a trial-and-error process in which the mutants were placed in a series of minimal culture-media, each containing a different amino acid. When a mutant survived and grew, it indicated which gene had been altered. For example, in one of the experiments performed by Beadle and Tatum, the mutant grew on a minimal culture-medium supplemented by the amino acid lysine. Evidently, this was a mutation affecting the gene that controlled the synthesis of lysine. To confirm this, the lysine mutant (if it really existed) should yield predictable results when crossed with normal *Neurospora*. Thus:

Parents:	Normal (+) crossed with lysine mutant (lys)
Zygote:	+ lys
Meiosis:	+ + lys lys
Mitosis: (ascospores)	+ + + + lys lys lys lys

When cultured, all spores should grow on the

minimal-plus-lysine medium but only half of them (+ strain) should grow on the minimal medium lacking lysine. This, in fact, happened and confirmed the *gene-enzyme-synthesis hypothesis*.

Beadle and Tatum found many other reactions in which there was a direct gene-enzyme relationship. They also found cases of chain-reaction syntheses where different genes controlled different steps in the reaction. A mutation of one of these genes disrupted the reaction unless the missing substance was added to the culture medium. In diagrammatic form, it might appear as

$$\text{Precursor material} \xrightarrow[\text{Enzyme 1}]{\text{Gene 1}} \text{Ornithine} \xrightarrow[\text{Enzyme 2}]{\text{Gene 2}} \text{Citrulline} \xrightarrow[\text{Enzyme 3}]{\text{Gene 3}} \text{Arginine}$$

Clearly then, a mutation of any of these three genes would interfere with this stepwise reaction. Experimentation also confirmed this concept.

Studies such as these carried out with *Neurospora*, *Chlamydomonas* (an alga), and bacteria have revealed information that would have been extremely difficult to obtain from more complex creatures like *Drosophila*. Not only did this type of investigation open up a new field of biochemical genetics but it also provided a frame of reference for some poorly understood hereditary events in higher animals. In human beings, sickle-cell anemia—a deficiency of hemoglobin in red blood cells—is known to result from an inherited defect in the hemoglobin molecule: the synthesis of hemoglobin is defective because of one "wrong" amino acid in the compound.

Bacterial Genetics

In recent years significant work in genetics has been accomplished through the use of bacteria, which, like *Neurospora*, are haploid. In addition, bacteria reproduce so rapidly that many individuals and generations can be produced in a short time.

Bacteria, however, pose a number of unique problems for the geneticist. First, the nuclear region is not separated from the rest of the cell by a nuclear membrane. The nuclear material consists of a single two-stranded DNA molecule compressed into a compact mass. When DNA is removed from a bacterium and is no longer compacted, the molecule assumes the form of a ring or circle. The DNA in bacteria never appears as a discrete chromosome even though it duplicates during reproduction. Fission is the most common form of reproduction, but nothing resembling mitosis occurs at any time.

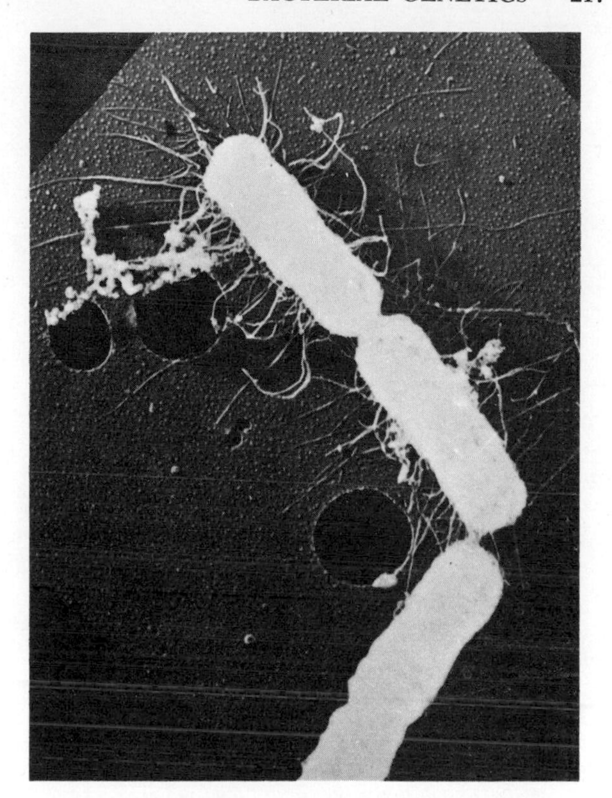

Figure 15.11. Bacteria conjugating (Dr. T. F. Anderson).

Despite these obstacles (or perhaps because of them) geneticists have found that bacteria often exchange genes with one another by three unusual methods: *conjugation*, *transduction*, and *transformation*.

Conjugation. Experimental evidence shows that the proper strains of certain bacteria temporarily unite (conjugate) and one, the donor, passes a portion of its DNA strand into the recipient (Figure 15.11). The recipient rarely receives a complete strand because the conjugating pairs usually separate before this happens. After a conjugating pair separates, the newly introduced DNA recombines with the recipient's DNA. The recipient eventually replicates this new combination of genes and passes them on to daughter cells by fission.

This method of transmitting hereditary substance evidently lacks the precision and orderliness of

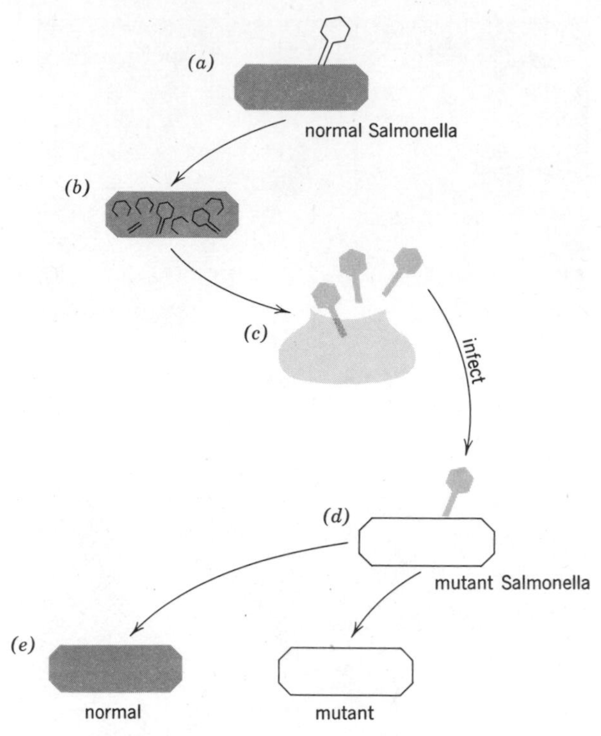

Figure 15.12. ***Transduction: the transport of genetic material from one bacterium to another by a virus.***

meiosis or mitosis. The two parents do not usually make equal genetic contributions to their progeny, all genes are linked since they are confined to one strand of DNA, and certain parental genes may never occur in offspring. Nevertheless, conjugation appears to function adequately for bacteria. They are highly adaptable organisms, as exemplified by the forms that have become resistant to antibiotics. At least part of this adaptiveness comes from the recombining of hereditary material in conjugation.

The classic studies that led to an understanding of conjugation were conducted by Joshua Lederberg and Tatum in 1946 using a bacterium called *Escherichia coli.* They worked out the minimal culture-medium on which normal (wild-type) *E. coli* would grow. They then utilized two mutant strains that would grow on the minimal medium only if certain amino acids and vitamins were added to it. One of these strains, called Y-10, required the addition of threonine, leucine, and thiamine; the other strain, Y-24, required phenylalanine, cystine, and biotin. If *mixed* cultures of the mutant forms, Y-10 and Y-24, are grown, some of the resulting colonies are able to grow on the minimal medium. A recombination of genes has evidently taken place so that at least some of the daughter cells received a recombined strand of nonmutant genes. This experiment led Lederberg and Tatum to conclude that transfer of genetic material occurred in bacteria. Other experiments and findings have verified this conclusion.

Recall the statement that a bacterial "chromosome" has a ring form. Presumably this ring opens during conjugation and passes in a linear manner through the cytoplasmic bridge between two conjugating bacteria. The passage of this chromosome into the female cell takes about two hours. If the mass of mating forms is vigorously stirred, the conjugants are separated and the chromosomal strand is broken. Part of the strand will be left in the female and part in the male. The piece received by the female will be passed on to daughter cells and function there. By separating bacteria at different times during mating, it is possible to vary the amount of the chromosome entering the female. By noting the traits of the daughter cells, we can map the chromosome in these bacteria.

Transduction. Another means of genetic transfer between bacteria involves use of a virus as the transmitting agent. Viruses called *bacteriophages* attach to bacteria, but only the DNA core of each virus particle actually enters the bacterium. The viral DNA directs the synthesis of more phage units until the bacterium is eventually destroyed.

Some types of phage, however, become associated with the genetic material in a bacterium and simply reside there without reproducing or harming the host. Strains of bacteria in which this occurs are called *lysogenic.* This is a hereditary trait. In lysogenic strains, the viral guest may be passed through many generations. Some of the virus forms occasionally start reproducing, an event that destroys (lyses) the host bacterium. In the process, the virus particles retain a copy of part of the bacterial chromosome. If one of these viruses infects another bacterium, the chromosomal fragment brought in from the old host may be recombined with the genes of the new one (Figure 15.12). Usually only a single gene is transmitted from donor to recipient bacteria, but even one gene may significantly change the genotype.

One experiment on *Salmonella* bacteria illustrates how transduction functions. One strain of *Salmonella* grows on a minimal medium only if an amino acid, tryptophan, is added; thus it is a "tryptophan mutant." If a phage is allowed to infect normal *Salmonella*—one that grows on a minimal medium—some of the cells will be destroyed, liberating new phage particles. If the mutant *Salmonella* is then exposed to these new phage forms, some of them become able to grow on the minimal medium without the addition of the amino acid. In other words, some of the tryptophan mutant forms have been transduced by the viruses into the normal or wild-type *Salmonella*. Could this be a mutation converting the tryptophan strain into the wild-type? Mutations do, of course, occur in bacteria, but in the case cited, the mutation rate would not be sufficiently high to account for the number of new forms that appear in the tryptophan population.

With reference to mutation in bacteria, we must note several points. Since most bacteria appear to be haploid, mutations appear immediately; that is, there is no dominant gene to inhibit the expression of the usually recessive mutant gene, as occurs in diploid organisms. Coupled with a rapid reproductive rate, this endows bacteria with tremendous adaptive potential. For example, if streptomycin is added to a culture of *E. coli*, most or all of the culture will die. There may, however, be a few mutant forms present that can resist the antibiotic, these forms survive and even reproduce. Note that the streptomycin did not *cause* the mutation but rather provided an environment in which the mutant form could express itself, becoming the normal or dominant population. The significance of this phenomenon in areas such as medicine is enormous. Many hospitals, for example, are now encountering *Staphylococcus* infections that resist all normal treatment methods, whereas this type of bacterium was easily controlled previously.

Transformation. A third type of hereditary transmission occurs when a strain of bacteria changes genetically after exposure to DNA that has been removed from another strain. In other words, the first strain absorbs and recombines with the isolated DNA from the second strain, thereby assuming some or all of the second strain's characteristics.

As an example, we can refer to experiments performed many years ago (1928) by an English bacteriologist, Fred Griffith, working with pneumococcus bacteria. When he injected mice with virulent strain S the mice died of pneumonia. When he injected mice with nonvirulent strain R or heat-killed strain S, the mice lived. All of this happened as predicted. Then he performed an experiment that had unexpected results: mice injected with a mixture of R strain and heat-killed S strain died of pneumonia! Griffith found, moreover, that these dead mice contained living virulent S strain pneumococci. Many repetitions of this type of experiment indicated that the nonvirulent strain had been transformed into the virulent type by coming into physical contact with the dead virulent strain. It became generally accepted that some kind of transforming principle was responsible for this dramatic change. In 1944, Avery, McCarty, and MacLeod, three biologists at the Rockefeller Institute, published a classical paper of their experiments demonstrating that the transforming principle was, in reality, DNA. Does transformation occur in nature or is it simply a laboratory phenomenon? Recent experiments show that it can occur between living strains of bacteria and thus may form a normal part of the hereditary mechanisms in bacteria.

Thus, in bacterial genetics, some new dimensions have been added to the study of heredity. Basic hereditary principles still apply, but their adaptations to the microbial world have many unusual aspects.

Principles

1. The behavior of genes and chromosomes during meiosis is parallel. When crossover occurs, the parallel behavior allows us to map the location of genes on chromosomes.
2. Organisms that are haploid in part of their life cycles are used to establish genetic control of cellular activity via control of enzyme production.
3. Studies of heredity in bacteria support the concept that recombination of DNA is the basis for hereditary variation.

Suggested Readings

Beadle, George W., "The Genes of Men and Molds," *Scientific American,* Vol. 179 (September, 1948). Offprint No. 1, W. H. Freeman and Co., San Francisco.

Cairns, John, "The Bacterial Chromosome," *Scientific American,* Vol. 214 (January, 1966). Offprint No. 1030, W. H. Freeman and Co., San Francisco.

Carlson, E. A., "A Decade of Progress in Modern Genetics," *American Biology Teacher,* Vol. 30, No. 9 (November, 1968).

Gordon, Manuel J., "The Control of Sex," *Scientific American,* Vol. 199 (November, 1958).

Jacob, Francois and Elie L. Wollman, "Viruses and Genes," *Scientific American,* Vol. 204 (June, 1961). Offprint No. 89, W. H. Freeman and Co., San Francisco.

Levine, R. P., *Genetics.* Second edition. Holt, Rinehart and Winston, New York, 1968, pp. 47–110, 127–147.

McKusick, Victor A., "The Royal Hemophilia," *Scientific American,* Vol. 213 (August, 1965).

Mittwoch, Ursula, "Sex Differences in Cells," *Scientific American,* Vol. 209 (July, 1963). Offprint No. 161, W. H. Freeman and Co., San Francisco.

Sutton, Walter S., "The Chromosomes in Heredity," *Biological Bulletin,* Vol. 4 (1903), pp. 231–251.

Zinder, Horton D., "Transduction in Bacteria," *Scientific American,* Vol. 199 (November, 1958). Offprint No. 106, W. H. Freeman and Co., San Francisco.

Questions

1. Explain why all the sons of a color-blind mother must also be color-blind. Why is it that a male cannot be heterozygous for color blindness?
2. How may cells from human males be distinguished from those derived from human females? Committees controlling athletic meets have used these differences. Is this an adequate criterion?
3. Why is it necessary to go back to meiosis to explain crossover?
4. What is the relation between Sutton's chromosome theory and the method used for mapping genes?
5. What is the relation of environment to gene expression? What mode of inheritance illustrates this point particularly well?
6. In what respect is the study of hereditary mechanisms simpler in haploid than in diploid organisms? For instance, why did Beadle and Tatum use *Neurospora* rather than *Drosophila* or white rats?
7. What is the hereditary advantage in being diploid rather than haploid?
8. On what basis could you argue that bacteria like *E. coli* reproduce sexually?
9. For what reasons are bacteria especially useful in studies on mutation?
10. How do bacteria illustrate the concept that the basis for hereditary variation lies in the recombination of DNA?
11. What is the difference between transformation and transduction?

CHAPTER XVI

Development

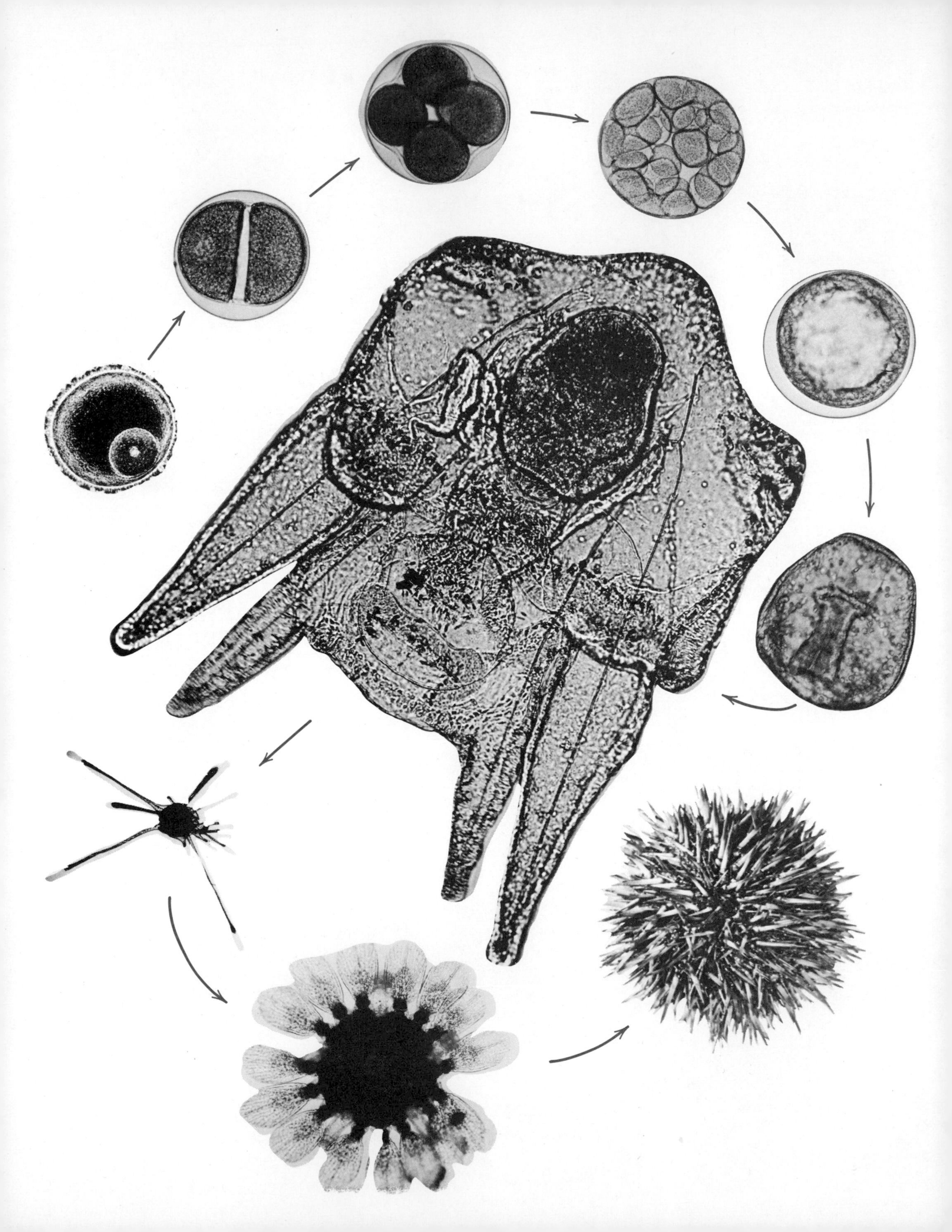

Stages in the development of a sea urchin, with an enlarged view of a larval stage at center.

Development

Stages of Development

In most organisms, reproduction involves more than the production of gametes and fertilization. Because it is small and single celled, the undifferentiated zygote resulting from fertilization usually bears little resemblance to the adult organism. For the reproductive process to be complete, the events of development must take place. Development involves three types of events: an increase in the number of cells (*cleavage*), the laying down of the outline of the form and structure of the organism (*morphogenesis*), and the specialization of cells and structures to perform their specific duties (*differentiation*).

Cleavage. Cleavage consists of the first divisions of the zygote, which are carried out with little intervening time. The number of cells greatly increases but the total amount of protoplasm remains the same as it was in the zygote. In animals, this process ends with the formation of the *blastula*, a hollow ball of thousands of small cells (Figure 16.1*f*). To this point no morphogenesis or differentiation has occurred, and there has been little growth. After blastulation, cell division continues at a generally lower rate, with an accompanying growth of cells, morphogenesis, and differentiation.

Morphogenesis. Morphogenesis, the origin of form and structure, begins in animals with the laying down of the general

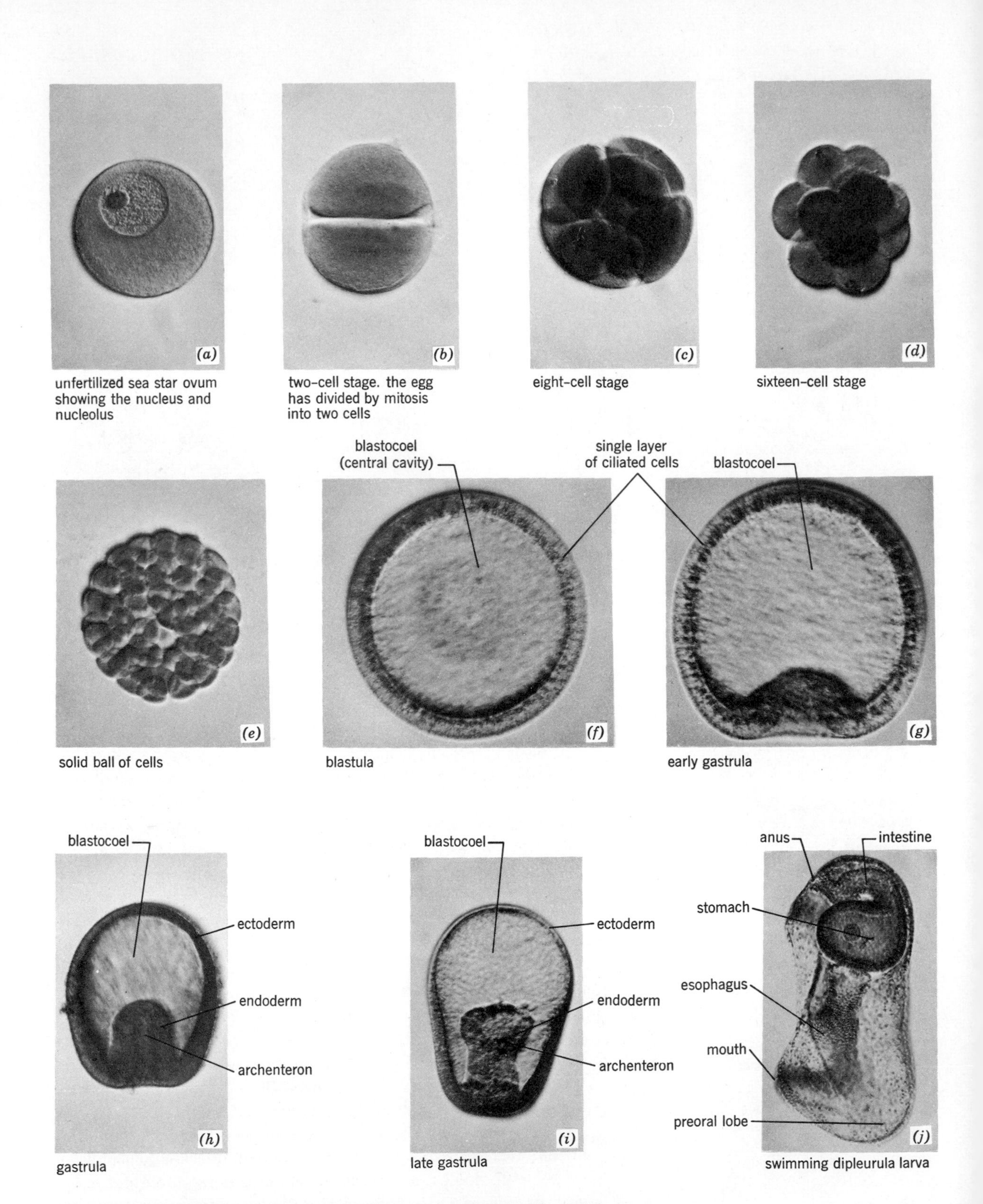

Figure 16.1. Early development of a sea star. The outer layers of the skin and the nervous system develop from the ectoderm whereas the primitive digestive tract (archenteron) is composed of endoderm. (CCM: General Biological, Inc., Chicago.)

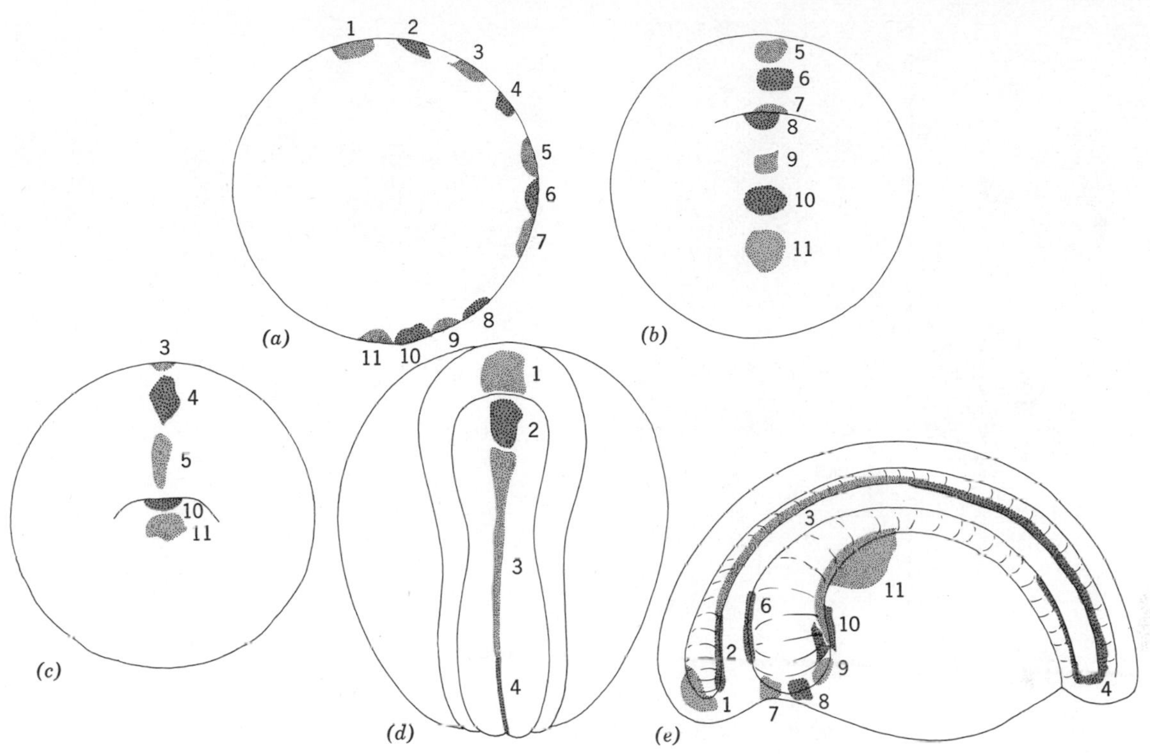

Figure 16.2. Tracing of cell movements using vital dyes. Each numbered area was dyed a distinct color (a). *Parts* (b) *and* (c) *show that some of the areas have moved into the interior of the embryo. Parts* (d) *and* (e) *show the positions of these areas later in development as viewed from the top and side of the embryo. (From Vogt, 1929; Balinsky,* An Introduction to Embryology, *second ed., W. B. Saunders, Co., Philadelphia, 1965.)*

pattern of the digestive tract. This event, termed *gastrulation*, occurs when a portion of the blastula wall grows inward as indicated in Figure 16.1g to 16.1*i*. During a later phase of gastrula development, certain cell groups are separated from the primitive gut and from the cells that will form the outer layers of the skin and the nervous system. These groups eventually form such structures as muscles, gonads, circulatory system, and excretory system (Figure 16.1*i* and 16.1*j*). The determination of which group of cells affects the formation of each animal structure is an arduous task. One method that has uncovered considerable information on this topic is the marking of areas in the blastula with dyes unharmful to the cells (vital dyes). (See Figure 16.2.) The movements of these marked cells and the structures they affect can thus be discerned.

Differentiation. Differentiation, the specialization of cells and tissues, occurs with morphogenesis. In fact, the two are parts of the same process of development and we separate them only for descriptive purposes. For example, the formation of the gut (digestive tract) and the specialization of cells to function in the gut take place concurrently. Differentiation is difficult to study because it often takes the form of physiological and biochemical specializations within cells. Two cells in a well-developed gastrula may look alike under the microscope in spite of the fact that one is differentiating into a part of the nervous system and the other is committed to a role in the muscular system. It is even more difficult to understand how one basic kind of cell, such as neural epithelium, forms a variety of specialized cells within the same (nervous) system. Differentiation is obviously one of the most significant events in development. We shall look at its possible controlling mechanism later in the chapter.

Variations in the Stages. Organisms carry out cleavage, morphogenesis, and differentiation in various ways. Even among the multicellular animals the

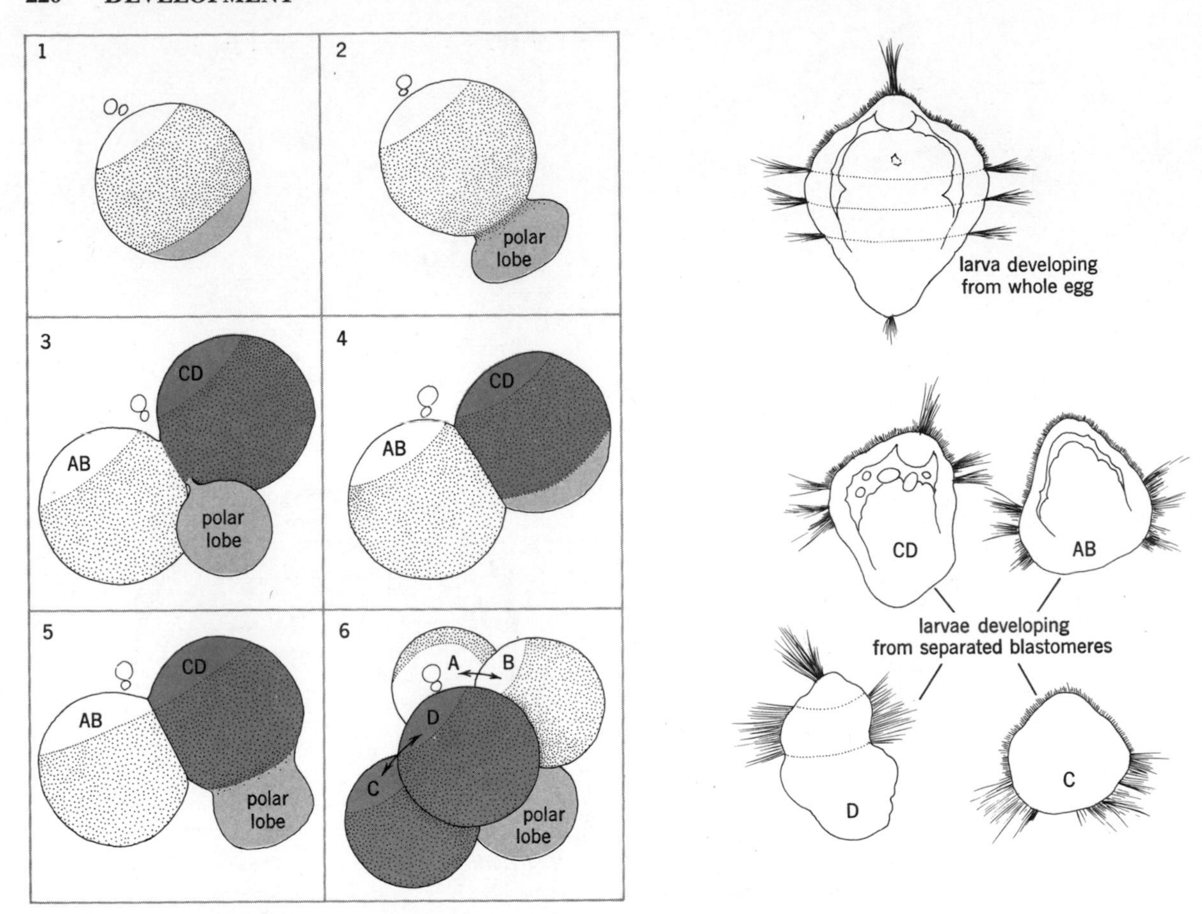

Figure 16.3. The polar lobe in cleavage of the mollusc **Dentalium.** *The materials (yellow) that comprise this lobe are only incorporated into one of the cells resulting from each division (part 4). Developmental abnormalities result if these substances are absent, as shown at lower right. The letters identify each cell of the four-cell stage. (From Fig. 1,* **Wilson, J. Exp. Zoology,** *Vol. 1, 1904.)*

processes are diverse. Cleavage in molluscs, for example, differs in fundamental ways from cleavage in the vertebrates. In the vertebrates, the new cells produced after the four-cell state lie directly above the old cells, whereas in the molluscs the new cells lie in the furrows between them. Moreover, in molluscs, prior to each of the early cleavages a bulge, the polar lobe, forms at the end of the cell farthest from the nucleus (Figure 16.3). This fuses with one of the resulting cells. No such structure exists in the vertebrates. If the cells of the early cleavage stages in a mollusc are separated, each forms a partial embryo. Even at this early stage, however, the cells have unequal developmental potentials, as shown in Figure 16.3. By contrast, in vertebrates and many other forms like sea urchins, separated early cleavage cells will each form an entire embryo.

Another variation of cleavage among the multicellular animals is related to the amount of yolk in the egg. When very little yolk is present, as in a sea star, the early cleavages form equal-sized cells. With moderate amounts of yolk, as in the eggs of frogs, the cells are noticeably smaller at one pole of the egg than at the other (Figure 16.4). If large amounts of yolk are contained in the egg, as in bird or shark eggs, only the upper portion of the egg cleaves and the embryo develops as a disk on the upper surface of the yolk.

In many animals, a free-living larval stage develops from the gastrula. This is a developmental stage that differs markedly in appearance from the adult. Some animals have no larval stage because

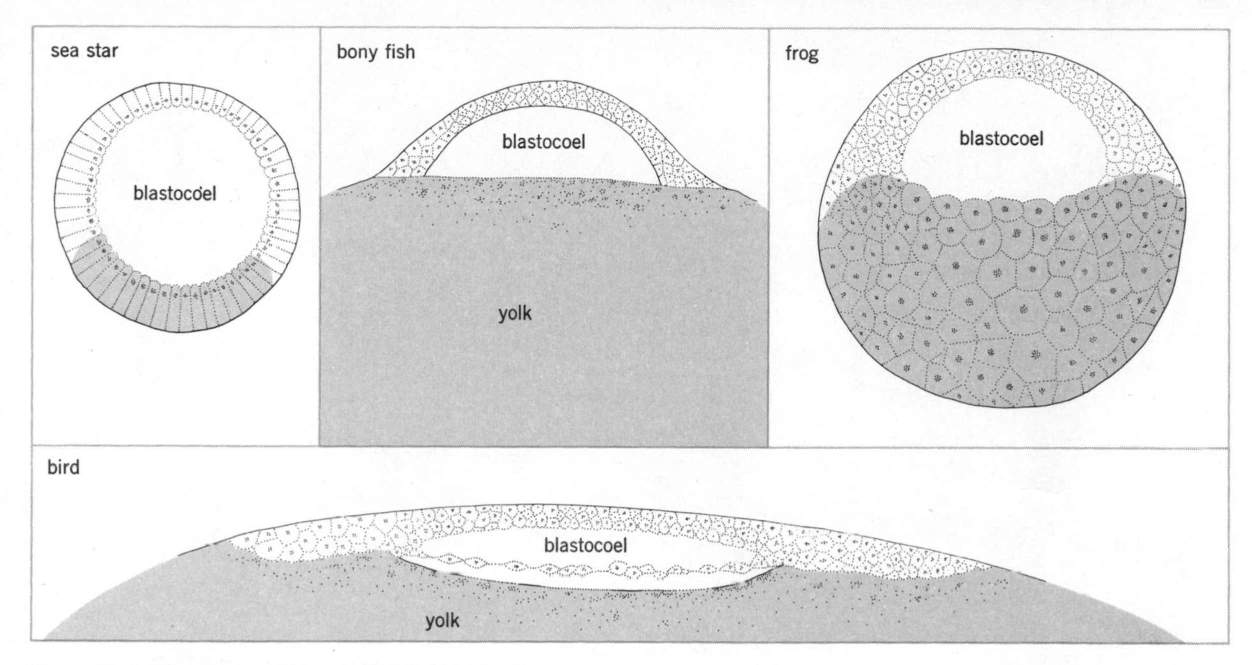

Figure 16.4. The effect of the amount of yolk on cleavage patterns in several animals.

of the plentiful supply of food available to the developing embryo. This supply comes from the mother either by way of a placenta, as in mammals, or by stored yolk deposited in the egg.

Like the green alga *Acetabularia,* the adults of many organisms are single celled. Obviously, no cleavage occurs in their development; but morphogenesis and differentiation do. In *Acetabularia* (Figure 16.5), the zygote becomes attached to the sea bottom by a group of rootlike structures called rhizoids. A slender stalk grows upward and an umbrella-shaped cap forms at the top. The nucleus is found at the base of the stalk. Morphogenesis and differentiation have both occurred.

Morphogenesis is easy to identify in *Acetabularia* as the three different parts of the zygote develop. Differentiation is difficult to see because various areas *within* the single cell are specialized for particular roles partly by their shape and location and partly by functional differences. Many single-celled plants and animals show rather similar patterns.

All multicellular organisms have developmental patterns that involve cleavage. In plants it is often difficult to distinguish cleavage, morphogenesis, and differentiation because all three occur at the same time in separate parts of the same organism. In the production of the embryo in flowering plants, however, we find that the first cell divisions occur without morphogenesis. Then the cotyledons (food-storing, leaflike structures) and the root and stem structures form (Figure 16.6). In appearance, the cells show no specialization from one structure to another. Later development produces the early stages of leaves on the stem and the first signs of the differentiation of cells.

After the seed germinates, cleavage, morphogenesis, and differentiation continue throughout the life of the plant, mainly at the tips of the stem and root. Thus, mature and embryonic tissues are present at all times in a mature, growing plant. In most multicellular animals, development does not continue throughout the life of the individual. Instead it ceases by the time the organism becomes mature.

Some Factors Controlling Development

The patterns of development we have discussed show that within an embryo, cells that appear similar migrate to various areas of the embryo and differentiate into distinct forms. If we look at different types of organisms, we find an even more confusing range of events. Since all organisms operate with essentially the same metabolic machinery, what accounts for the diversity of these events?

We can find our answer more easily if we ask two

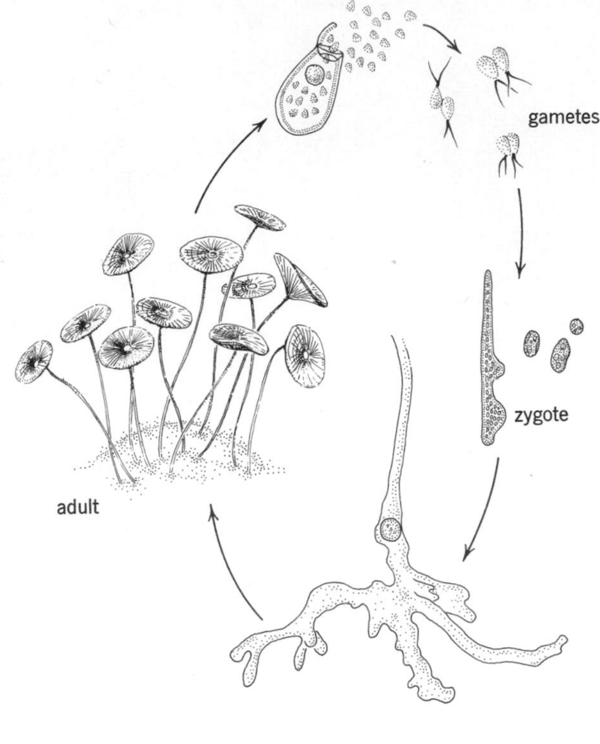

Figure 16.5. Life cycle of Acetabularia.

other questions. How can two cells that have the same hereditary information differentiate dissimilarly? How much impact will a different set of hereditary information have on the development of a cell?

Three kinds of experiments supply answers to the first of the subsidiary questions. Important factors are: (1) the position of the cell in relation to other cells; (2) the changes in the nucleus during development; (3) the way in which the cytoplasm of the egg is divided.

Cell Position. With certain plants it is possible to take a single cell and culture it on an artificial medium. The cell divides and produces a large mass of undifferentiated cells. At this time the cells look approximately alike. With the increase in the size of the cell mass, however, we find cells near the center specializing into root and stem tissues. Eventually, an entire plant will be formed from the culture.

If the experimenter continually separates the cells, preventing them from accumulating into a mass, no differentiation will occur. It is the relative position of the cells that causes some of the cells to differentiate when they are allowed to remain in contact with one another. Auxins are known to play a role in this differentiation in plants.

A similar principle operates in animal embryos. If an experimenter removes a piece of tissue from the developing brain region of a chick embryo and grows it in appropriate nutrients, the tissue forms nerve cells as it would normally do in the embryo. But if the tissue is cut into tiny pieces that are grown separately, differentiation does not take place. Presumably the numerous cells comprising a piece of tissue have a chemical effect upon one another that is lacking in small aggregates of cells.

Hans Spemann, a German embryologist, became interested in this problem many years ago. In 1924, after extensive experimentation, he discovered that one specific region of the amphibian embryo controlled (induced) the differentiation of the nervous

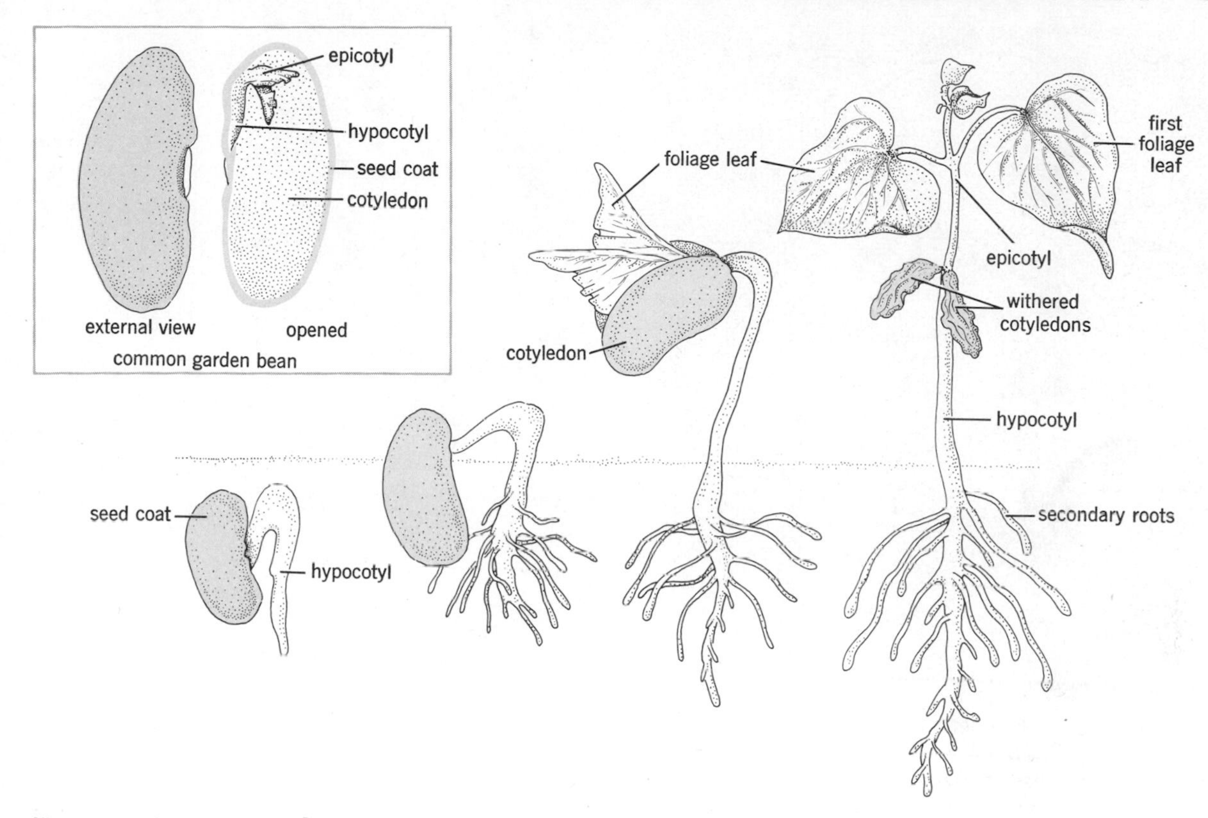

Figure 16.6. Germination and growth of the common garden bean.

system of the embryo. This region, the dorsal lip of the blastopore, was named the *organizer*. Even a piece of this tissue, transplanted to another part of the embryo, causes the formation of an additional nervous system. Since Spemann's time, other organizers have been discovered and studied. These establish the basic concept of one mass of cells influencing the development of another.

Cytoplasmic Variations. Unequal separation of various components of the cytoplasm occurs naturally in molluscs by the formation of the polar lobe and its fusing with one of the cells resulting from the cleavage. Since this lobe is so distinct from the rest of the embryo, it may be easily removed, thus removing any substance peculiar to the lobe. Such experiments should give some indication of the role of these substances in development and of the importance of the manner in which cytoplasmic materials are distributed to the cells. Similar experiments involve the separation of the cells at the two-cell stage, so that each cell develops separately, one with the substances found in the polar lobe, one without.

In general, the cells containing the polar lobe substances develop into nearly normal larvae. Those cells that do not contain polar lobe material develop into abnormal partial larvae that are missing major organ structures (Figure 16.3, right). Studies of the fate of individual cells in the normal embryo show that the missing portions normally develop from the cells that contain polar lobe substances.

We can conclude that one part of the cytoplasm contains a substance not found in another part of the cytoplasm. Usually the distribution of these substances plays an important role in determining which group of cells will form which structures in the embryo.

Nuclear Transplant Experiments. The most direct way to discover the role of the nucleus in differentiation is to pull the nucleus out of the cell and replace it with one from another cell or from the cell of another species. Work with the embryos of

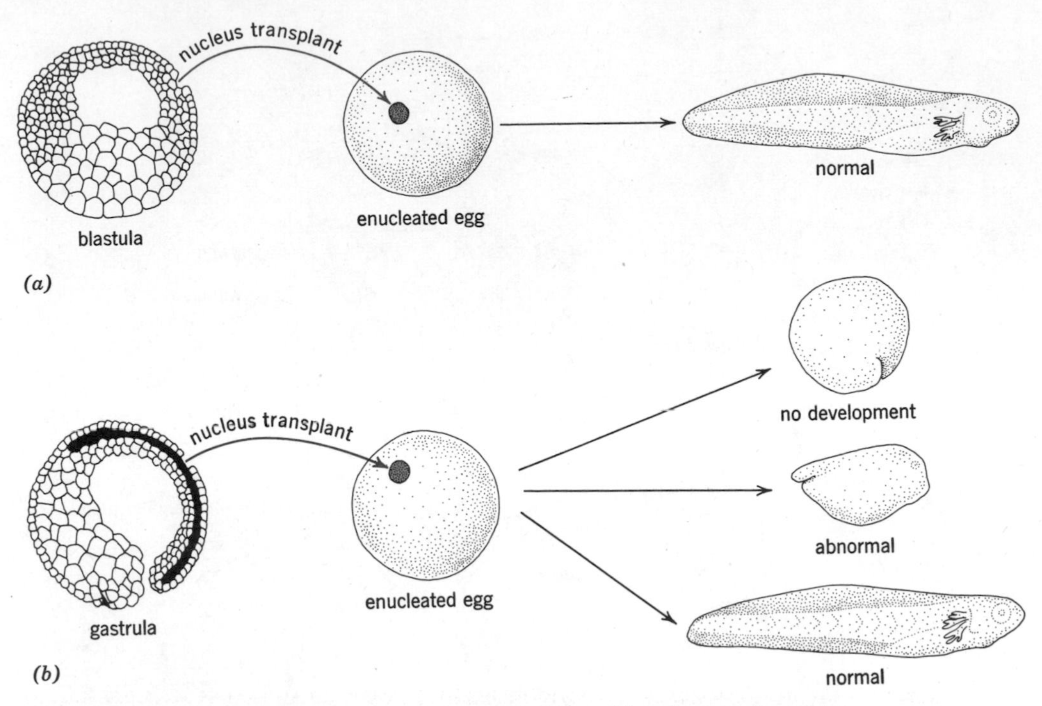

Figure 16.7. Nuclear transplant experiment. A nucleus from a blastula transplanted into an enucleated egg yields a normal larva (a). ***Enucleated eggs receiving nuclei from a gastrula yield varied developmental forms*** (b).

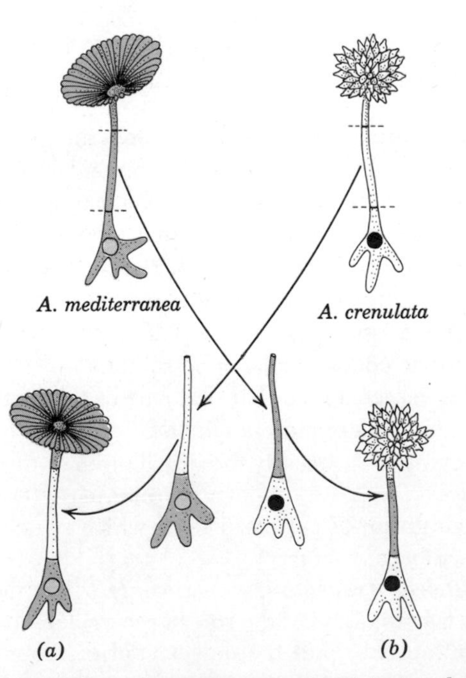

Figure 16.8. Reciprocal transplant experiments in **Acetabularia** ***showing nuclear control of regeneration.***

frogs has shown that the nuclei change with the age of the embryo. If the nuclei are removed from unfertilized eggs and replaced with nuclei from embryos that are in the blastula stage or younger, nearly all the resulting cells will produce normal embryos (Figure 16.7*a*). If the nuclei transplanted into enucleated eggs come from older embryos, the resulting embryos vary widely; some do not develop at all, some develop normally, and some develop into abnormal embryos (Figure 16.7*b*).

Even though the transplanted nuclei always contain the same hereditary information, their activity changes with age. The aging of the nuclei is not simply a matter of time but of time and interaction with the cytoplasm. Nuclei may even be transplanted from blastulae back to enucleated eggs. By repeating this mechanism, the nuclei may be prevented from losing their "youthful" characteristic (ability to control the normal development of an enucleated egg).

This type of experiment has also been conducted with *Acetabularia.* In these experiments the nucleus is not removed; instead the base of the plant that contains it is cut off and replaced with the base of another plant.

Two species of *Acetabularia*, *A. crenulata* (*cren*) and *A. mediterranea* (*med*), have frequently been used. The most obvious difference between the two is the shape of the cap. If the cap and stalk are removed from a member of either species, the plant regrows a new cap of its normal variety. If the capless stalk of a *cren* is grafted onto the base of a *med*, the plant regenerates an intermediate cap. If this cap is removed, all subsequent caps will be of the same kind as the nucleus (*med*). (See Figure 16.8.) A reciprocal experiment (*med* stalk, *cren* base), produces a *cren* cap. We can conclude that DNA, by controlling the production of new protein, regulates the type of cap regenerated regardless of the type of cytoplasm in the stalk.

How does the nucleus achieve this effect? If we cut the base and the cap from a plant, the piece of stalk regenerates a cap of the same type as the plant *one time only*. After this the plant dies. Obviously, some substance of short life span that also contains the information necessary to control regeneration is present in the cytoplasm. Messenger RNA fits these requirements.

During development in midges (a type of insect) all nuclei except those that will eventually form gametes in the adult undergo a special kind of division during which most of the chromosomes are lost. The nuclei that will form gametes migrate to one end of the developing embryo. Here, in an area having special granules in the cytoplasm (polar granules), they undergo normal mitosis. Thus, the gametes retain the normal chromosome number for the species. If the polar granule area is first exposed to ultraviolet radiation, the nuclei that migrate to this area also undergo the peculiar division in which chromosomes are lost. This means that the insect will be sterile because the cells that develop from this irradiated area will not contain the normal chromosome number for the gametes.

Since the area of polar granules has been shown to be rich in RNA, and since, in the cell, nucleic acids are the substances most sensitive to ultraviolet radiation, the evidence indicates that damage to RNA has destroyed this peculiar aspect of the control of development in midges.

Operon Theory of DNA-RNA Control. The effects described in the preceding sections can be explained by the operon theory for which two Frenchmen, François Jacob and Jacques Monod, were awarded the Nobel prize in 1965.

They hypothesized that two special types of genes are involved—operators and regulators. The *operator genes* function like switches that are turned on or off by the product of the *regulator genes*. According to this theory, at any one time not all of the DNA is directing the synthesis of messenger RNA; some of it is switched off.

A regulator gene may be located anywhere in the DNA and most likely acts, through a messenger RNA, to direct the production of a protein capable of affecting an operator gene. The form of this protein that is produced under the regulator gene's direction prevents (represses) the activity of an *operon*, a unit composed of an operator gene and the structural genes it controls.

Repression occurs when the protein combines with the operator gene and prevents the initiation of messenger RNA synthesis along these structural genes (Figure 16.9). None of the enzymes normally produced under the direction of the genes will be synthesized.

The repressor protein may be changed to an inactive form if a specific substance is present. This substance, which differs for different repressors, may be synthesized in the cytoplasm (proteins, for example) or derived from the environment (ions and amino acids, for example). In either case, the presence of this substance induces the production of specific proteins by turning on the operator gene or, in other words, by removing the repression of mRNA synthesis along this portion of DNA.

Some of the early experiments used by Jacob and Monod to support their theory involved genes in *E. coli* that direct the production of enzymes required for the utilization of galactose, a six-carbon sugar. Under most conditions the repressor protein, synthesized under the direction of mRNA from the regulator gene, is in a form that interacts with the operator gene. This consequently prevents mRNA synthesis by any of the structural genes in that operon. However, when galactose is present under suitable conditions, the form of the repressor is changed so that it can no longer interact with the operator gene. As a result, the synthesis of mRNA is no longer repressed along the structural genes of the operon, and thus the enzymes are produced that are necessary for utilization of galactose.

In this case the repression of genes is removed by

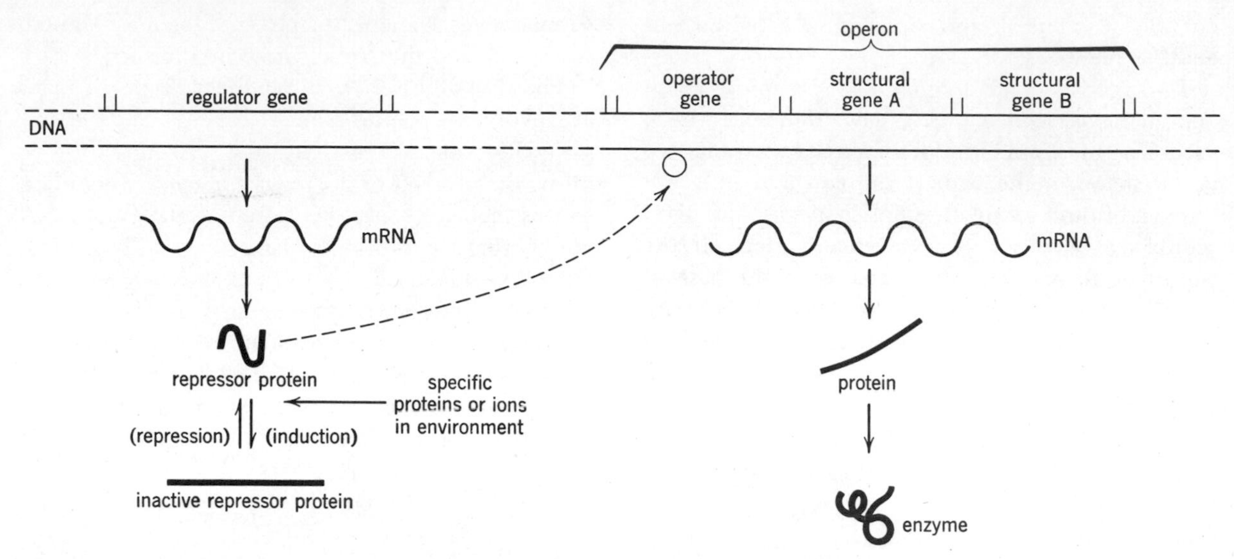

Figure 16.9. A model of the operon theory. The regulator gene controls segments of DNA by means of a repressor protein. This protein affects the functioning of the operator gene or operon by either repressing its activity or allowing it to function. (Modified from Hartman and Suskind, Gene Action, *Prentice-Hall, Englewood Cliffs, N.J., 1965*).

the presence of a suitable substrate for the activity of these genes. Another example from *E. coli* shows how repression may be stopped by the *removal* of a substance that makes the activity of the genes unnecessary. For example, typical environments contain sufficient amounts of the amino acid tryptophan to sustain bacterial growth. If tryptophan is absent, the bacteria are capable of synthesizing the enzymes required for deriving tryptophan from another amino acid, acetylornithine. Here, the protein produced by the regulator gene is in a repressor form unless tryptophan is absent and acetylornithine is present.

Most of the evidence supporting the operon theory comes from experiments with bacteria. Indications are strong, however, that the general outlines proposed by the theory also operate in higher organisms. In these, the DNA of the chromosomes is surrounded by a protein coat. Removal of this protein seems to be an essential part of inducing mRNA synthesis.

Two types of evidence support this idea. We mentioned in Chapter VIII the puffs that appear on giant chromosomes in insects. This may represent the uncovering of DNA so that mRNA synthesis can occur. It is known that RNA occurs in higher concentrations near the puffs than elsewhere in the vicinity of the chromosomes.

The second type of evidence comes from work done mainly by Barth and Barth using different ions on developing frog embryos. They found that ions could bring about differentiation of cells in the embryo and that various ions differed in their ability to induce these effects. More importantly, the ions that were the most effective were the same ions that best prevented the formation of the protein coat-DNA complex.

Viewing the operon theory from a higher level of organization, the total organism, we can hypothesize that each of the cells in the developing embryo may be in contact with cytoplasm which contains slightly dissimilar substances. These various substances will induce activity in different portions of the DNA by affecting diverse repressor proteins. Consequently, the makeup of the cells continues along divergent pathways of specialization.

Principles

1. Development comprises cleavage, morphogenesis, and differentiation.
2. Differential gene action at various stages of the life cycle is important in development.
3. Differentiation is controlled and directed by multiple interacting factors that include cell position, cytoplasmic materials, and nuclear control.

Suggested Readings

Butler, J. A. V., "How Genes Are Controlled," *Science Journal*, Vol. 2, No. 3 (March, 1966).

Ephrussi, Boris, and Mary C. Weiss, "Hybrid Somatic Cells," *Scientific American*, Vol. 220 (April, 1969).

Fischberg, Michail and Antonie W. Blackler, "How Cells Specialize," *Scientific American*, Vol. 205 (September, 1961). Offprint No. 94, W. H. Freeman and Co., San Francisco.

Galston, Arthur W., *The Life of the Green Plant.* Second edition. Prentice-Hall, Englewood Cliffs, N.J., 1964, pp. 81–105.

Gray, George W., "The Organizer," *Scientific American*, Vol. 197 (November, 1957). Offprint No. 103, W. H. Freeman and Co., San Francisco.

Gurdon, J. B., "Transplanted Nuclei and Cell Differentiation," *Scientific American*, Vol. 219 (December, 1968). Offprint No. 1128, W. H. Freeman and Co., San Francisco.

Hadorn, Ernst, "Transdetermination in Cells," *Scientific American*, Vol. 219 (November, 1968). Offprint No. 1127, W. H. Freeman and Co., San Francisco.

Konigsberg, Irwin R., "The Embryological Origin of Muscle," *Scientific American*, Vol. 211 (August, 1964). Offprint No. 191, W. H. Freeman and Co., San Francisco.

Spratt, Nelson T., Jr., *Introduction to Cell Differentiation.* Reinhold Publishing Co., New York, 1964.

Steward, F. C., "The Control of Growth in Plant Cells," *Scientific American*, Vol. 209 (October, 1963). Offprint No. 167, W. H. Freeman and Co., San Francisco.

Waddington, C. H., "How Do Cells Differentiate?" *Scientific American*, Vol. 189 (September, 1953). Offprint No. 45, W. H. Freeman and Co., San Francisco.

Questions

1. Would you expect to find cleavage stages in the development of a single-celled organism? How about morphogenesis and differentiation?
2. Differentiation is almost entirely a physiological phenomenon. Is this true of cleavage?
3. During what cleavage stage does differentiation evidently begin?
4. How does the quantity of yolk in an egg affect the type of cleavage it undergoes?
5. Describe an experiment that shows that the differentiation of a cell depends partly on its association with other cells.
6. Does the nucleus of a cell change in some way during development? Cite an experiment to support your answer.
7. Can the directions from the genetic material of *E. coli* to the rest of the cell change if the cell's environment changes? List the steps involved.
8. What features of *Acetabularia* make it especially suitable for experimentation involving the role of the nucleus in differentiation?
9. What is the significance of the experiment in which gametes of midges were exposed to radiation?
10. Why would it be difficult to argue that heredity is more important than cellular environments during development?

CHAPTER XVII

Chemistry of the Genetic Material

Chromosomes damaged by radiation. Notice the fragments of chromosomes. (Dr. Arnold H. Sparrow, Brookhaven National Laboratory.)

Chemistry of the Genetic Material

As more and more information became available about inheritance, the chemical identity of the genetic material became an increasingly intriguing problem. Early geneticists knew that the chemical substance had to have a mechanism for duplicating itself as exactly as genetic material had been observed to do. This substance had to have enough different forms to account for all variations in organisms. It needed some way to control the activities of the cell. The Watson-Crick model of the structure of DNA satisfied the first two criteria. Recent studies of RNA control of protein synthesis have satisfied the last criterion.

DNA: The Hereditary Material

Early Evidence. Our knowledge of the structure and function of DNA is very recent, although DNA was discovered a century ago (in 1869) by Friedrich Miescher, a German physiologist. He isolated a substance from the nuclei of pus cells and called it nuclein. Although he studied this substance for many years, its function eluded him. In the 1880s a number of biologists proposed that nuclein was associated with the transmission of hereditary traits. This advanced idea was not accepted, even by Miescher, and lay dormant for over fifty years.

DNA was first considered a possible genetic material because its characteristics seemed to match those known for such material. The evidence to support this early suggestion now comes

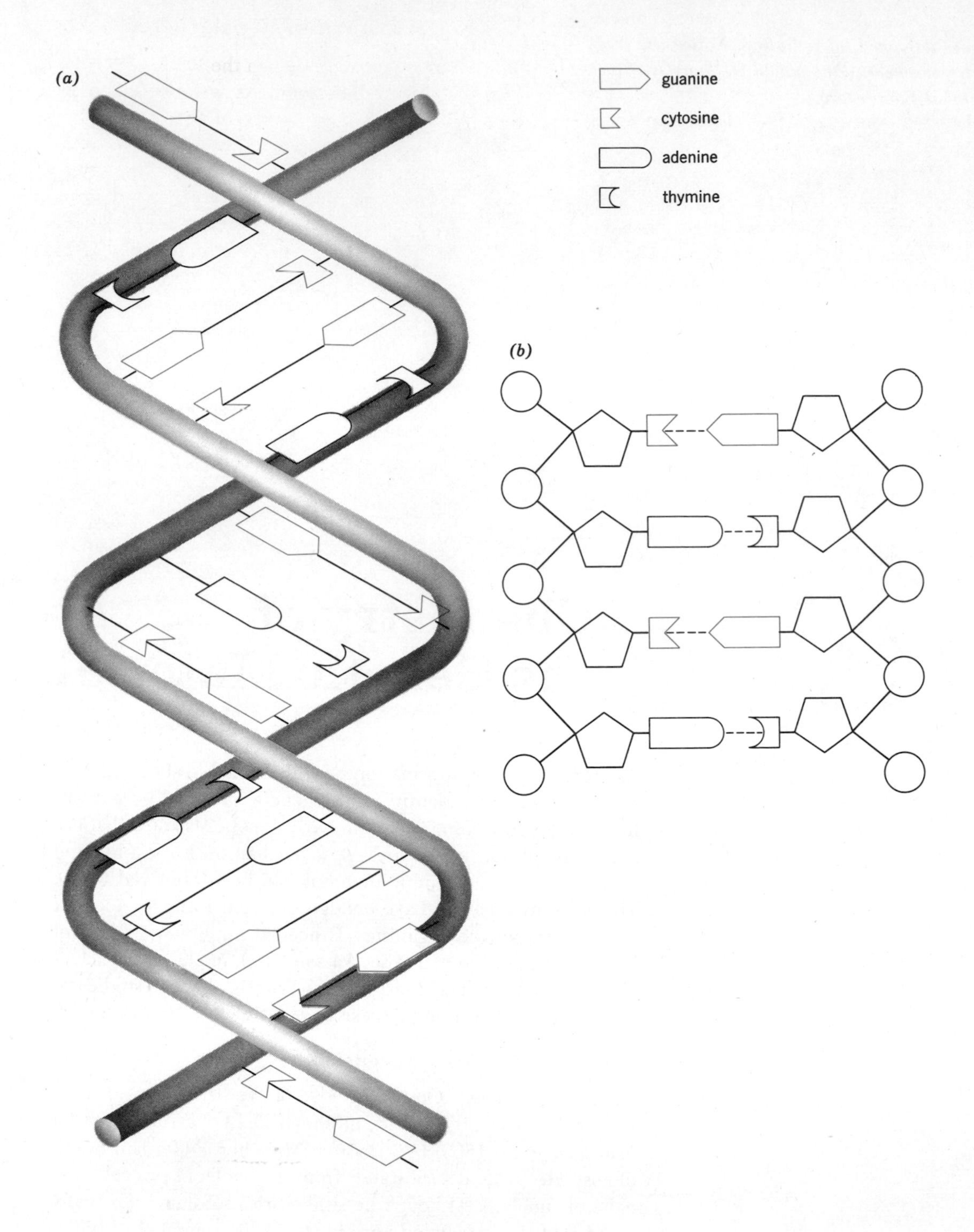
(a)
guanine
cytosine
adenine
thymine
(b)

from several different sources. Early experiments showed that the nucleus had a large concentration of DNA. Later it was shown that the chromosomes were composed of DNA and protein. However, these findings did not constitute really good evidence for DNA being the genetic material. It is known that each cell in an organism has the same genetic information; thus, there should be the same amount of genetic material in each. Finally, optical techniques for weighing DNA produced the information that the amount of DNA present in the cells of an organism was constant.

In these optical weighing techniques, a dye is used that is colorless until it combines with specific groups found solely in DNA. Here a reddish-purple color develops. The dye content is determined by measuring the percentage of a green light that is able to pass through the nucleus. The greater the dye content, the less light that passes through.

Because the degree of staining by the dye is nearly proportional to the quantity of DNA in the nucleus, DNA content may be calculated after measuring the size of each nucleus (if we assume the nucleus is a sphere). Obviously, this calculation provides only an approximate value, but nevertheless it shows that DNA content of nuclei is nearly constant in an organism.

Nucleic acids had been shown to absorb ultraviolet radiation (wavelength of 260 $m\mu$) strongly. This information was the basis for later evidence that a nucleic acid was the genetic material. First, ultraviolet light of this wavelength brought about mutations; that is, it affected the genetic material. Second, in some species many of the processes of development could be abolished by treating the cells with ultraviolet radiation, suggesting that interference with genetic control of development had occurred.

As described in Chapter XV, studies of bacteria during the 1920s showed that freshly killed bacteria of one strain could cause a change in the genetic characteristics of another strain. In the 1940s further work in the same field indicated that only the DNA portion of the bacteria could bring about this effect called *transformation.*

A crucial finding was that digested DNA (free nucleotides) did not cause transformation, whereas intact pieces of DNA did. This showed that the important factor was not merely the presence of the nucleotides but rather something about the relation of these units to one another in the DNA molecule.

Structure and Function. As we mentioned in Chapter VII, DNA is composed of two long chains of nucleotides that spiral around each other to form a double helix (Figure 17.1). These chains are structurally interdependent. Since, of all the parts, the bases of the nucleotides approach each other closest, the fact that the bases will pair in only one way is very important to the structure of the DNA helix. This specific pairing of bases also enables us to predict the sequence of bases in one chain if we know the sequence in the other one. The order of bases in the chain is the genetic code itself, the genetic information.

How does this chain relate to the structure we can see—the chromosome? Except in bacteria and some algae, the long DNA molecule forms the core of the chromosome but not its whole. A coat of protein surrounds the DNA. Although the protein plays no part in the genetic information carried by the chromosome, it is important in repressing the activity of the chromosome (see Chapter XVI).

The exact relationship of the nucleotide units to the visible chromosome is difficult to understand. We do not know how many nucleotide pairs make up a chromosome. Only shrewd guesses can be made. When we talk of nucleotides, we are talking of the biochemical structural units of the chromosome. These are neither the visible cytological unit, the chromosome, nor the functional unit that we discussed as the gene in Chapters XIV and XV.

Even the term gene becomes imprecise when we consider the many viewpoints from which the genetic material may be considered. One definition of this material emphasizes the functional aspects of DNA, and applies the term gene to the segment of the DNA that controls the production of a protein; another definition, at the other end of the spectrum, emphasizes the unit that may change or mutate. This unit, of course, may be only one nucleotide in length.

We have already described the way in which three bases form the genetic unit that determines which amino acid will be located at a particular point in a particular protein. The translation of a triplet-code unit of DNA into an amino acid in a protein requires the presence of both transfer and messenger RNA. Messenger RNA is a relatively long molecule that carries the directions for the production of only one kind of protein. Each triplet determines the location

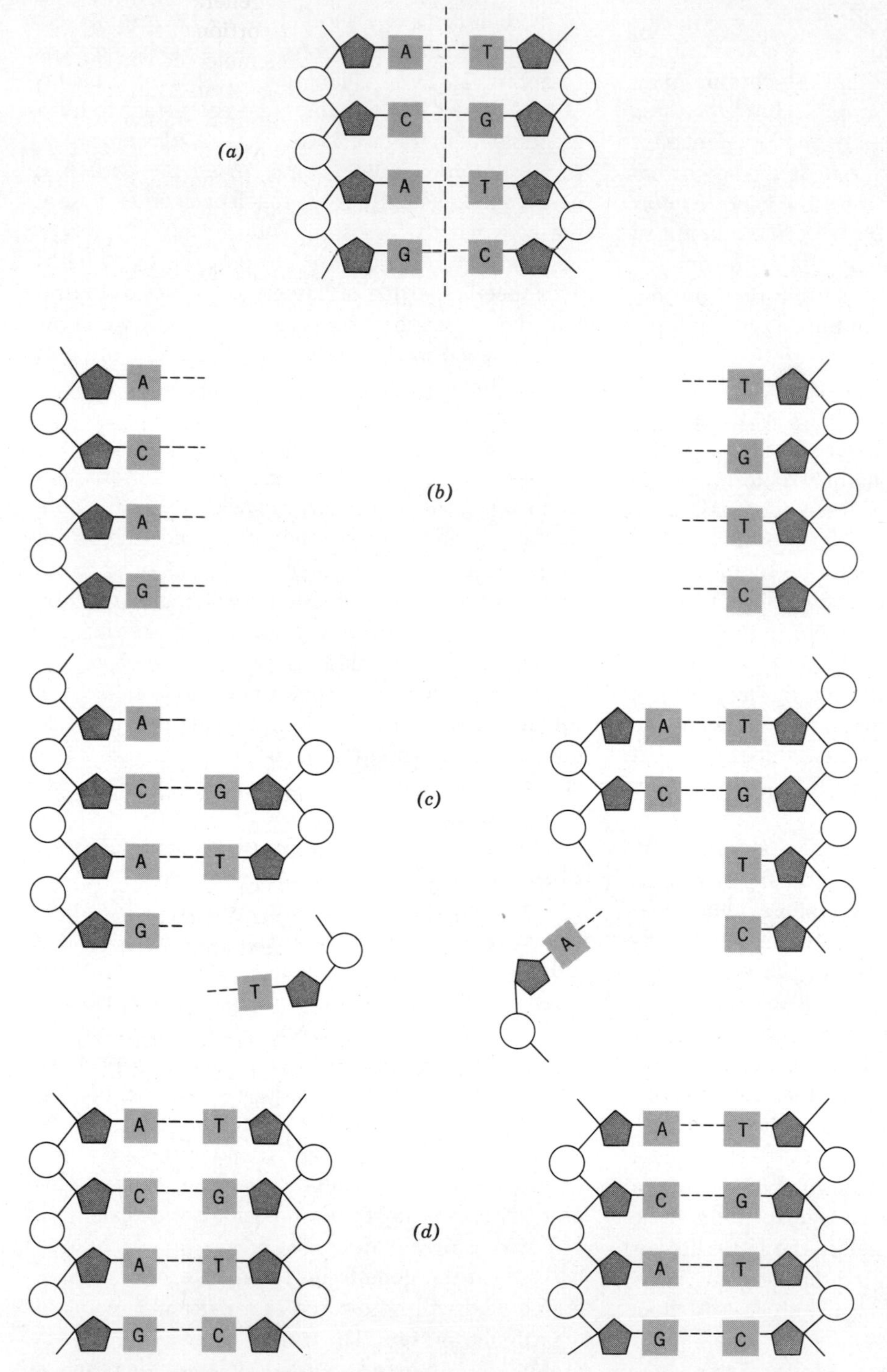

Figure 17.2. Duplication of DNA. (a) *and* (b) *Separation of the two strands.* (c) *Free nucleotides joining the strands.* (d)· *Resulting duplicate strands.* (*Reprinted with permission, Copyright © 1957 by Scientific American, Inc. All rights reserved.*)

of one amino acid. Thus, this RNA represents the genetic information translated into a slightly different form.

Transfer RNAs are evidently produced by short segments of DNA that may be nearly alike in many different species. One kind of transfer RNA may be used in the synthesis of many different proteins. An amino acid becomes attached to a specific transfer RNA, which pairs with bases at the appropriate sites along messenger RNA (see Chapter VII). The formation of peptide bonds completes the synthetic process.

Duplication. One other aspect of DNA structure must be emphasized. If DNA is the basic genetic material, then it must be duplicated when the chromosomes are duplicated. The method by which this is done must have a high degree of precision and reproducibility.

An explanation for the duplication of DNA is indeed available (Figure 17.2) based on the Watson-Crick model for its structure. At the time of duplication, the hydrogen bonds must be broken so that the two intertwined nucleotide chains can unwind and separate. The bases of each chain then become associated with the bases of free nucleotides. The kind of base with which each base will pair is, of course, determined by the specific pattern of base pairing—adenine and thymine, cytosine and guanine. As the free nucleotides become associated with those of an old strand, they are joined to each other by sugar-to-phosphate bonds, thereby forming a new strand. Notice that each of the old nucleotide chains has served as a pattern by which a new complementary chain is produced. This results in two copies of the original double helix. The copies are then enveloped by protein coats to become chromosomes, which in turn may be incorporated into the daughter cell products of cell division.

How can we be sure that DNA replicates in the manner described above? Perhaps a copy is made of the intact helix, or the DNA is broken into separate nucleotides and copied. Proof was provided by using an isotope of nitrogen, N^{15}. Bacteria were grown on a medium containing N^{15} until all of the nitrogen in the bacterial DNA was of this type. These bacteria were then placed on a medium containing ordinary nitrogen, N^{14}. Samples of bacteria were removed as they began to reproduce, so that their DNA could be analyzed. The investigators found *hybrid* DNA in the first-generation bacteria—the DNA contained a 1 : 1 proportion of N^{15} and N^{14}. Presumably one strand of each molecule was the N^{15} variety and its complementary strand was made of N^{14}. These results, and additional similar experiments, appear to confirm the hypothesis stated previously that DNA replicates by untwining its strands and synthesizing complements to match them.

Mutation. In Chapter XV mutation was described briefly as the appearance in an organism of a new feature that was thereafter inherited. Now we can examine this process in greater detail.

One class of mutation involves minor changes in the structure of the DNA molecule. These minor changes, called *point mutations,* occur during DNA duplication and involve such errors as loss or repetition of nucleotide units, misbonding between units, and replacement of one base in the DNA chain by another. None of these changes alters the appearance of the chromosome but they are potentially capable of effecting a change in the mRNA, thereby changing the enzymes present and producing a different phenotype. Examples of this type of mutation in *Neurospora* were described in Chapter XV. Additional examples that we have already mentioned include many of the different eye color mutants in *Drosophila,* albinism in man, taillessness in cats, and white flower color in pea plants.

A second major class of mutations involves visible changes in chromosome structure; thus they are called *chromosomal mutations.* When chromosomes are undergoing movements and changes, as during meiosis, abnormalities may occur. Thus, a portion of a chromosome may be broken off and lost (a deletion), a segment may be duplicated so that its genes are represented twice, or a chromosomal segment may be inverted (Figure 17.3). The most severe of all mutations is the loss of all or a major part of a chromosome. The removal of this much DNA from an organism's genotype often causes death.

Examples of chromosomal mutations are known mostly from observations on giant chromosomes in insects. These chromosomes show identifiable patterns of dark bands, so that any change in their arrangement is detectable (Figure 17.4).

The cause of mutations in natural populations is unknown despite much interest and many studies by biologists. In 1927, H. J. Muller showed that X-rays could induce mutations in *Drosophila,* and since his

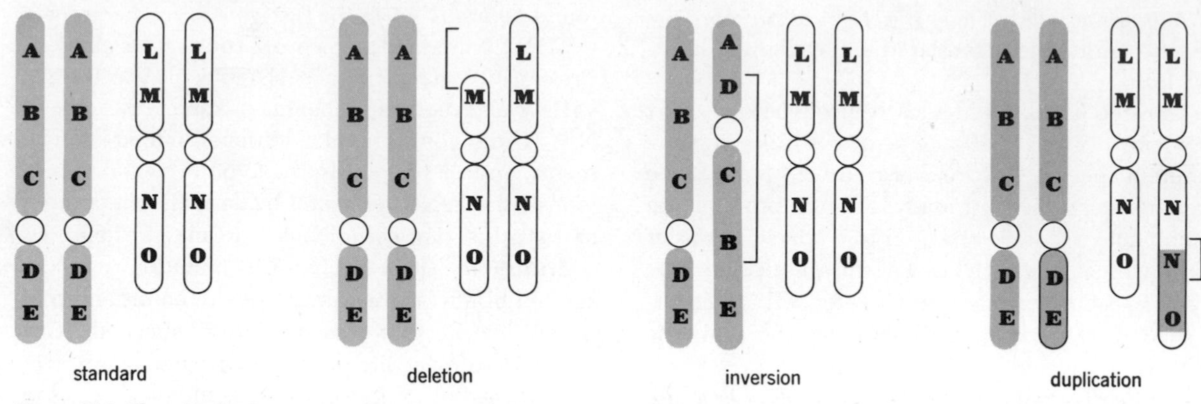

Figure 17.3. Three kinds of mutations that produce visible changes in chromosome structure.

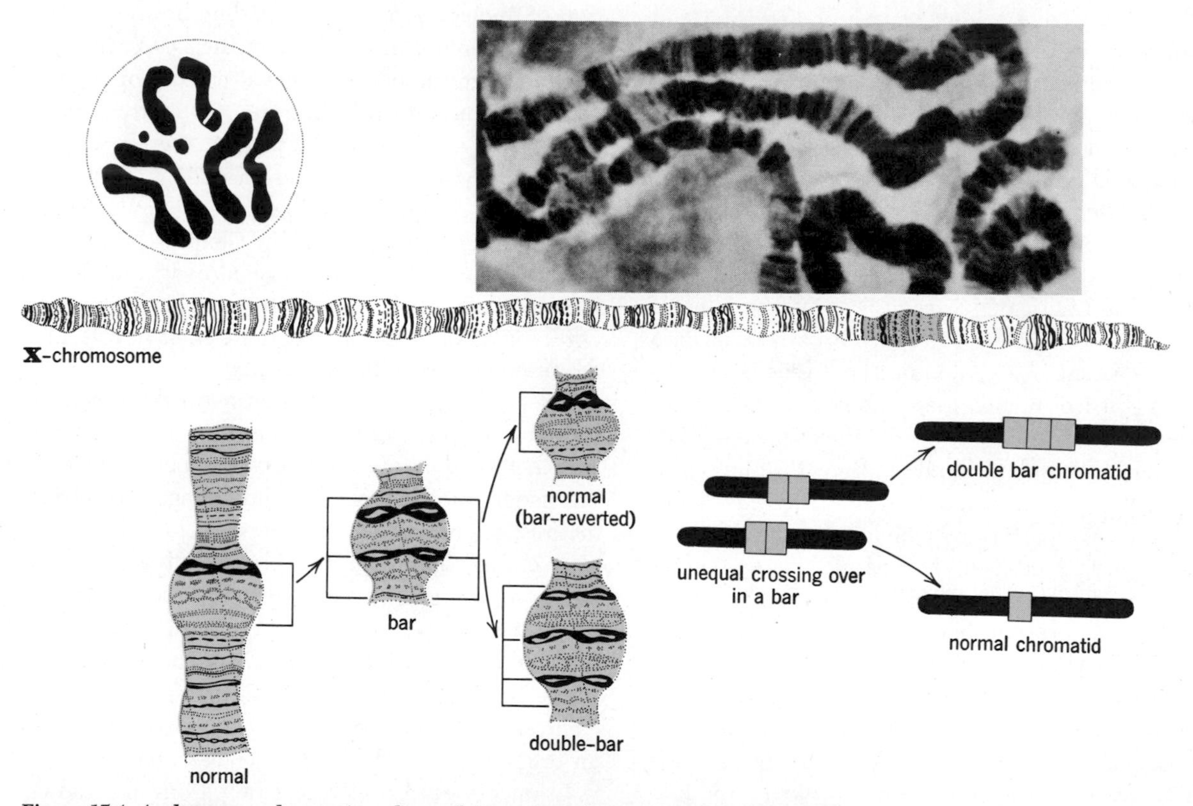

Figure 17.4. A chromosomal mutation observed in a giant chromosome from Drosophila. *The location of the segment of the chromosome involved is shown in a full complement of chromosomes (upper left) and in a detailed drawing of the affected chromosome (center). The marked banding of the giant chromosomes (photo) makes identification of these mutations possible. The appearance of the normal form and of a duplication (called bar) are shown. An additional form (double bar) may result from unequal crossover of two bar regions.*

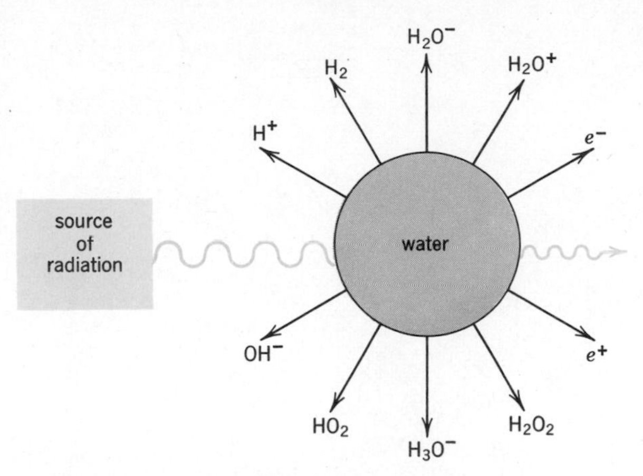

Figure 17.5. Highly reactive substances formed by the irradiation of water. (Adapted from **Atomic Radiation**, *RCA Service Co., 1957.)*

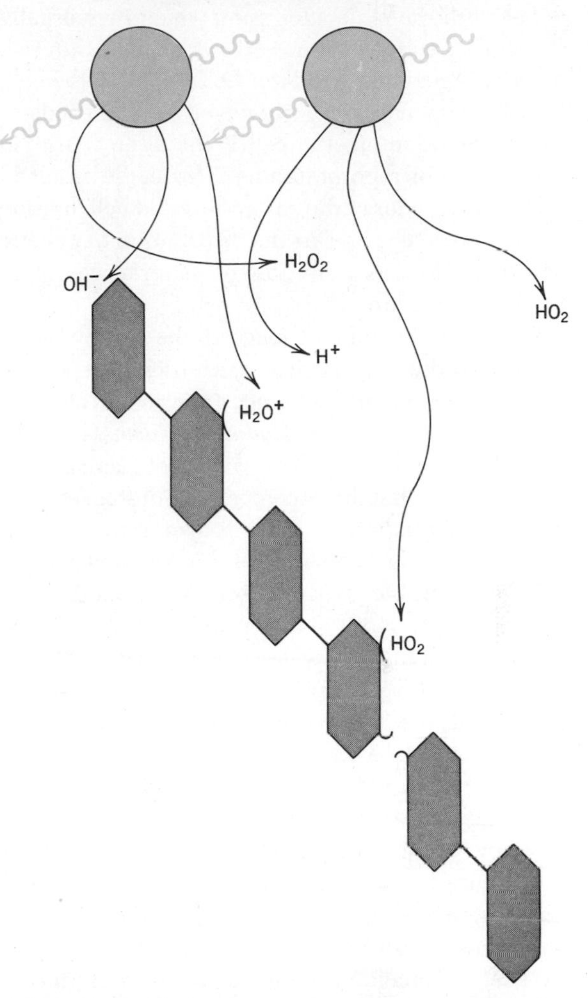

Figure 17.6. Structural modifications in a large molecule caused by ions from irradiated water. (Adapted from **Atomic Radiation**, *RCA Service Co., 1957.)*

day many other mutagens (such as ultraviolet light, and certain dyes and chemicals) have been discovered.

Mutation and Ionizing Radiation

One of the most potent mutagenic agents employed in the laboratory is *ionizing radiation.* Ionizing radiation consists of particles or waveform energy transmissions which, as they pass through matter, cause ions to form. The most damaging to living matter are such forms as X-rays and gamma radiation. These exceedingly penetrating, high-energy, transmissions strike atoms in the living tissue and cause the loss of electrons. These, in turn, may cause extensive ionization in molecules of any size, from water to proteins and nucleic acids.

If small molecules such as water are struck by the radiation, highly reactive substances such as peroxides or free hydrogen atoms may be formed (Figure 17.5). These substances do not remain in this form for long but react with other substances in the cell. They may react with many compounds in the cell, but we shall consider only DNA, RNA, and proteins.

These reactions may cause two types of damage. The molecules may be broken or the bonding pattern in the molecule may be changed (Figure 17.6). If either happens to DNA, then one of the types of mutations we have listed can occur. Such changes can kill the cell by interfering with mitosis or the cell's metabolism. If the cell lives, these changes will be duplicated in each succeeding cell division. Consequently, such damage to the cell is permanent. If RNA is affected, temporary changes can result in the protein produced. As the DNA produces more RNA, however, a new supply of normal RNA replaces the defective RNA. The same changes in DNA, RNA, or protein result directly if the radiation strikes them (Figure 17.7).

Two other important kinds of effects on the cell may be brought about if its proteins are modified. If the proteins that are enzymes (or parts of them) are changed in shape, they may no longer fit together with the molecules upon which they usually act, and therefore their activity will be altered. Since new enzymes will later be formed by the cell, this change is temporary. However, permanent damage may be done to the cell if the altered protein is a portion of a chromosome. This could result in modified behavior of the chromosome at cell division and in drastic changes in the distribution of genetic material. Of course, any loss of material would be permanent (Figure 17.8).

The possibilities for damage to the cell by radiation are so diverse that it is easy to understand the many derangements of homeostatic mechanisms in an irradiated organism. *Radiation sickness* results from the cumulative effects of direct damage and the abnormal but highly reactive substances produced by the radiation. The first symptoms are nausea and vomiting, reddening of the skin, and general fatigue. Later, bleeding (for example, from gums or nose), congestion of the lungs, and ulceration of the intestine with salt and water loss may develop. Low immunity to disease and anemia are also common. Active cells are more susceptible to radiation than relatively inactive ones. It may be noticed that the cells affected are dividing fairly rapidly.

An animal may recover from these symptoms yet die or be seriously affected by the long-term effects of exposure to radiation. Cancer or leukemia, continued anemia, cataracts, sterility, and nerve damage may appear after considerable time has passed.

Genetic damage in the gamete-producing tissues will not appear until the next generation. Larger numbers of mutations, mostly to deleterious alleles, and increased fetal or young animal deaths result. This kind of damage is the most dangerous to the species in the long run and is the most difficult to detect. Because low doses of radiation over a long period of time can cause these genetic effects, this form of damage is also the most insidious.

Since radiation sources such as X-rays, radioactive fallout, radioactive substances on watch dials, and natural radiation are so common, the problem of the effects of radiation on biological systems is an important one. What can be done? At present, facts are being gathered but there are still no satisfactory answers. Even if there were, the final answer would be dependent on public acceptance of the danger and of the solutions. The problem is not only scientific but also psychological and sociological.

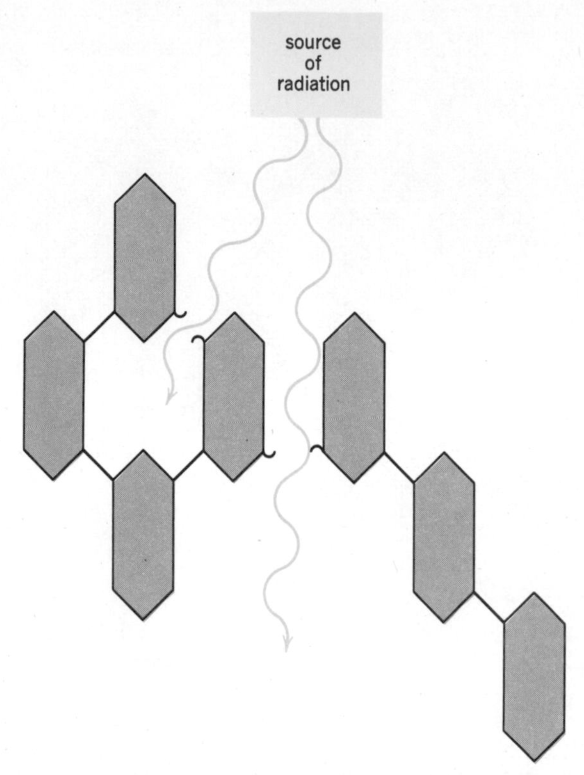

Figure 17.7. Disruption of bonds in a large molecule by direct radiation. (*Adapted from* Atomic Radiation, *RCA Service Co., 1957.*)

Principles

1. DNA has been established as the genetic material because the Watson-Crick model provides an adequate explanation for duplication, mutation, and control of cellular activity.

2. Ionizing radiation alters the structure of DNA as well as the structure of other molecules such as RNA, protein, and water.

Suggested Readings

Alexander, Peter, *Atomic Radiation and Life.* Penguin Books, Baltimore, 1965.

Auerbach, C., "The Chemical Production of Mutations," *Science,* Vol. 158, No. 3805 (December 1, 1967).

Beadle, George W., "Structure of the Genetic Material and the Concept of the Gene," in *This Is Life,* edited by W. H. Johnson and W. C. Steere. Holt, Rinehart and Winston, New York, 1962, pp. 185–211.

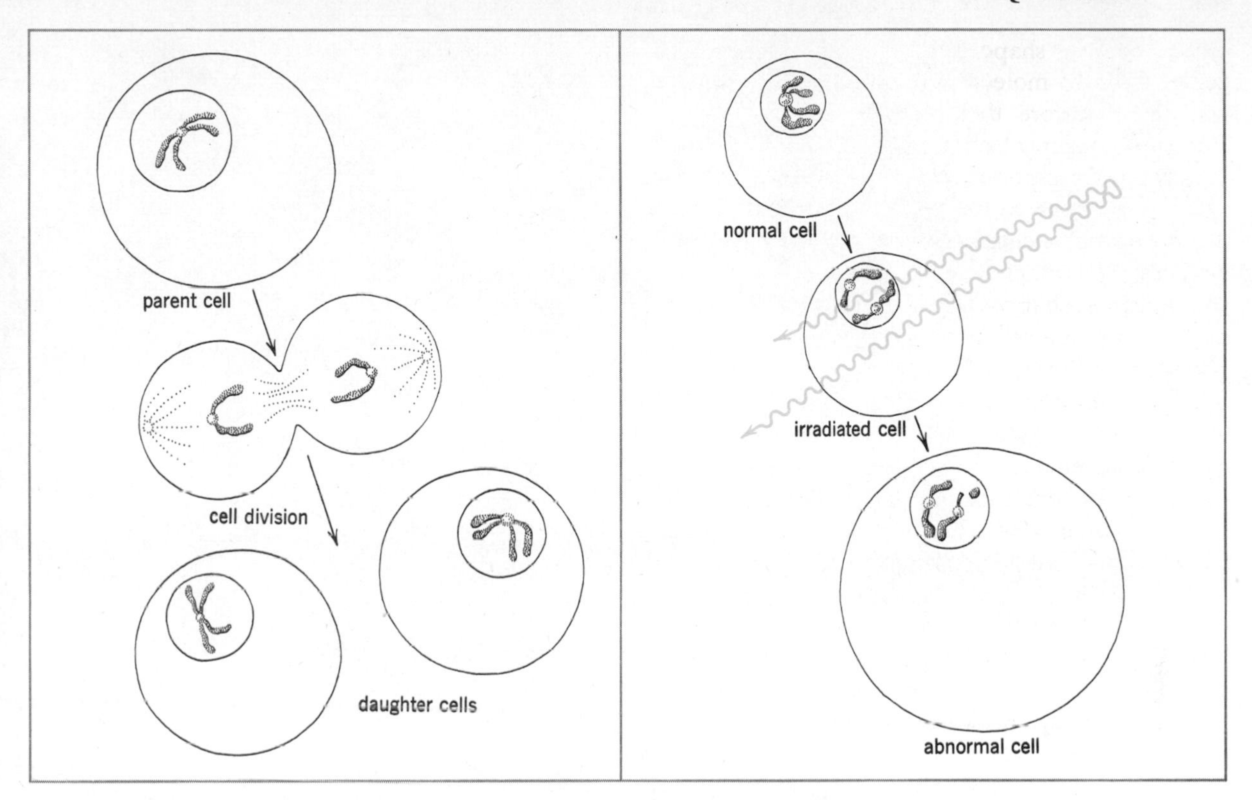

Figure 17.8. Modification of chromosomes by radiation. Normal mitosis (left) may be stopped and the chromosome number doubled (right). Pieces of chromosome will be lost in subsequent mitosis. (Adapted from Atomic Radiation, *RCA Service Co., 1957.)*

Beerman, Wolfgang and Ulrich Clever, "Chromosome Puffs," *Scientific American,* Vol. 210 (April, 1964). Offprint No. 180, W. H. Freeman and Co., San Francisco.

Hollaender, Alexander and George E. Stapleton, "Ionizing Radiation and the Cell," *Scientific American,* Vol. 200 (September, 1959). Offprint No. 57, W. H. Freeman and Co., San Francisco.

Levine, R. P., *Genetics.* Second edition. Holt, Rinehart and Winston, New York, 1968, pp. 18–28, 151–172.

Lindell, Bo and R. Lowry Dobson, *Ionizing Radiation and Health.* World Health Organization, 1961.

Loutit, John F., "Ionizing Radiation and the Whole Animal," *Scientific American,* Vol. 201 (September, 1959).

Muller, H. J., "Radiation and Human Mutation," *Scientific American,* Vol. 192 (November, 1955). Offprint No. 29, W. H. Freeman and Co., San Francisco.

Platzman, Robert L., "What Is Ionizing Radiation?" *Scientific American,* Vol. 201 (September, 1959).

Spiegelman, S., "Hybrid Nucleic Acids," *Scientific American,* Vol. 210 (May, 1964). Offprint No. 183, W. H. Freeman and Co., San Francisco.

Questions

1. What early evidence indicated that DNA was the basic hereditary material?
2. What did Watson and Crick contribute to this idea? Why was this important?
3. What is a chromosome? What is the relationship between DNA and chromosomes?
4. If the bases in one chain of DNA are known, can we predict the sequence of bases in the other chain? How?
5. Trace the steps involved in the translation of a DNA code in the nucleus into an enzyme out in the cytoplasm.
6. Name the ways in which mutations occur. Can man deliberately cause any of these mutations? How?
7. Explain how ionizing radiation can produce changes in cells.
8. Does radioactive fallout material constitute a source of ionizing radiation? Explain.

CHAPTER XVIII

Population Genetics: An Introduction to Evolution

Monarch butterflies. Dense populations occur in areas where they spend the winter.

Population Genetics: An Introduction to Evolution

In attempting to learn how traits are inherited we mate two individuals that differ in one or more respects and observe the characteristics in their offspring. Although this technique has been useful in discovering the basic laws of heredity and the mechanism for inheritance of traits, it does not provide a genetic basis for understanding evolution.

Because an individual's genetic makeup does not change during his life span, we must look at the unit that does change over time. This unit, which is composed of all individuals that can interbreed and their offspring, is a reproductive unit called a *Mendelian population.* The unit is characterized by a changing genetic composition throughout an existence that may span millions of years.

This change in genetic composition and the consequent modification in the phenotypes constitute evolution. The population, not the individual, evolves. Therefore, if we wish to understand the genetic basis of evolution, we must understand the genetics of populations.

In previous chapters the Punnett Square was used to combine gametes from two parents to show all possible types of offspring. The same technique can be applied to population genetics if we use, in proper number, all the gamete types from all parents in the population. In other words, we shall select gametes from a pool containing all the genes in the population. We call this the *gene pool* of the population.

The Hardy-Weinberg Law: Populations in Equilibrium

Let us imagine a population of hamsters composed of 100 males and 100 females. There are 49 homozygous gray, 42 heterozygous gray, and 9 homozygous black animals of each sex. Gray coat color (**B**) is dominant over black (**b**). If each individual produced ten gametes, the gametes produced by the *males* should appear as in Table 18.1.

Table 18.1.

Genotypes of the Males	Gametes B	Gametes b	Totals
49 **BB**	490	0	490
42 **Bb**	210	210	420
9 **bb**	0	90	90
	700	300	1000
	or	or	or
	70% (0.70)	30% (0.30)	100% (1.00)

The heterozygous individuals should produce as many gametes carrying recessive genes as they do gametes carrying dominant genes, since the genes are present in equal numbers in these individuals. The homozygous individuals will produce only one kind of gamete in each case. The figures 0.30 and 0.70 in the table are gene frequencies (or gamete frequencies, since they are equal). They tell us that 30 percent of the genes in this population are recessive genes while 70 percent are dominant. The females in the population should produce the same number of the same types of gametes since they contain the same genes in identical proportions. A Punnett Square can also be constructed to show how often fusion occurs among the various combinations of gametes.

		Gametes from the Males		Summary of Resulting Genotypes
		0.7 **B**	0.3 **b**	0.49 (49%) **BB**
Gametes from the Females	0.7 **B**	0.49 **BB**	0.21 **Bb**	0.42 (42%) **Bb**
	0.3 **b**	0.21 **Bb**	0.09 **bb**	0.09 (9%) **bb**
				1.00 (100%)

In the previous chapters, we assumed that the number of each type of gamete produced by each individual would be equal. Here, on the other hand, the homozygous gray individuals far outnumber the homozygous black individuals, so this population will *not* produce equal numbers of the two kinds of gametes. We have, therefore, inserted in the Punnett Square the frequency of each kind of gamete along with its symbol.

Since the probability of the union of any two types of gametes depends on their frequency, the percentages of resulting genotypes are computed by multiplying the frequency of the gamete from one parent by that of the gamete from the other parent. The results are summarized at the right of the Punnett Square.

If we examine the offspring in this population, we find 49 percent homozygous gray, 42 percent heterozygous gray, and 9 percent black. These are the same percentages observed in the parental generation. In fact, the gene frequency would be the same in each succeeding generation. Two men, G. H. Hardy, a British mathematician, and W. Weinberg, a German physician, noticed this fact independently and stated what is known as the *Hardy-Weinberg law: under certain conditions, gene frequencies and genotype frequencies remain the same from one generation to the next in sexually reproducing populations.*

Obviously, most populations cannot be controlled in a manner that will result in each individual contributing only a given number of genes to the next generation. The same kind of reasoning can be used, however, to set up a formula that will apply to a general population if we substitute p and q for the frequency of the dominant and recessive genes, respectively.

		Gametes from the Males		Summary
		p(**B**)	q(**b**)	p^2 = freq. **BB**
Gametes from the Females	p(**B**)	p^2(**BB**)	pq(**Bb**)	$2pq$ = freq. **Bb**
	q(**b**)	pq(**Bb**)	q^2(**bb**)	q^2 = freq. **bb**
				1.00 (100%)

In this Punnett Square, p equals the frequency of the dominant gene (0.7 in the last example) and q

equals the frequency of the recessive gene (0.3 in the last example). We now see that there will be p^2 homozygous gray animals, $2pq$ heterozygous animals, and q^2 black animals. (Verify this by substituting 0.7 for p and 0.3 for q and checking against the results in the table.) We may use this method of calculation as long as each individual has an equal chance of reproducing, and the offspring have an equal chance of surviving.

A second look at the substitution shows that since there are only two kinds of genes (**B** and **b**), then $p + q = 1$. This can be restated as: the frequency of the dominant genes plus the frequency of the recessive genes equals all the genes in the population. Similarly, since there are only three possible genotypes (**BB**, **Bb**, and **bb**), $p^2 + 2pq + q^2 = 1$. These two equations are useful in calculating the gene and genotype frequencies for a population.

For example, if we are studying a simple recessive trait like vestigial wing (very short wing) in *Drosophila,* which is present in 4 percent (0.04) of a laboratory population, we can complete all the genotype and gene frequencies for the population, provided the conditions for a Hardy-Weinberg equilibrium are met. We define the letters from the two equations in the usual fashion.

p = the frequency of the dominant gene
q = the frequency of the recessive gene
p^2 = the frequency of the homozygous dominant genotype
$2pq$ = the frequency of the heterozygous genotype
q^2 = the frequency of the homozygous recessive genotype

Since vestigial wing is a recessive trait, we know that all individuals showing it must have a genotype **vv**.

The frequency of this group is q^2 (by definition) and is equal to 0.04. Since

$$q^2 = 0.04$$
$$q = \sqrt{.04} = 0.2 \text{ or } 20\%$$
$$p + q = 1$$
$$p + 0.2 = 1$$
$$p = 1.0 - 0.2 = 0.8 \text{ or } 80\%$$
$$p^2 = (0.8)\,(0.8) = 0.64 \text{ or } 64\%$$
$$2pq = 2(0.8)\,(0.2) = 0.32 \text{ or } 32\%$$

In summary, 80 percent of the genes are the dominant (**V**) for normal wings, while 20 percent are recessive (**v**) for vestigial wings. Sixty-four percent of the flies are homozygous dominant (**VV**), 32 percent are heterozygous (**Vv**), and 4 percent are homozygous recessive (**vv**). These results are summarized in the form of a Punnett Square in Figure 18.1.

Let us now suppose that during a study of 2500 people we find that 1600 have freckles. What is the frequency of the gene for freckles in this population? First, we must know that freckles are inherited as a simple Mendelian dominant. Because 900 of the 2500 do not have freckles, q^2 equals $^9/_{25}$ if we follow the usual symbols in the Hardy-Weinberg formula. Therefore, the frequency of this gene for the absence of freckles is

$$q = \sqrt{^9/_{25}} \text{ or } ^3/_5 \text{ or } 60\%$$
$$p + q = 1$$
$$p + 0.6 = 1$$
$$p = 0.4$$

The frequency of the dominant gene for freckles is 40 percent in this population.

Forces Changing Populations

In the statement of the Hardy-Weinberg law there is an important qualification—"under certain conditions." All of our calculations have assumed that:

1. Natural selection was not occurring.
2. Migration into or out of the population did not occur.
3. No mutations were occurring (or that mutations in the two directions were equal).
4. The population was large enough not to be affected by random changes in gene frequencies.

These assumptions had to be made since the absence of any one of these conditions would cause a shift in the gene frequencies. Since a change in the gene pool is what constitutes evolution, we shall examine these four conditions separately.

Natural Selection. Selection for the best-adapted individuals is the most potent force that changes the characteristics of populations. The best-adapted organisms are those that leave the most offspring, that is, contribute the most genes to the gene pool. This is the basic driving force of evolution. We can divide selection into three types on the basis of the phase of the life cycle at which it occurs: before the organism reaches reproductive age, during mate selection and actual mating, or during fertilization and embryonic life. Probably the type of natural selection easiest to envision is that which occurs prior to reproduction. It is obvious that if the organism does not survive to reproduce, it will not pass genes to the next generation.

Selection during mating can occur only if the organism survives to reproductive age. The mechanism of this type of selection varies; the females may select one type of mate in preference to others, the pollen may not be carried as far, the flowers may not be as attractive to pollinators, or most of the population may have already completed their breeding before this individual is ready. In any of these cases, the effect may vary from complete prevention of breeding to an only slightly lessened chance.

The third type of selection operates on the gametes or embryos. Gametes are often infertile or at a disadvantage in reaching or being reached by other gametes (Figures 18.2 and 18.3). Even if fertilization occurs, death of the zygote or embryo from genetic causes may occur at any time.

To illustrate, let us return to the population of hamsters (p. 250), starting with the same basic ratios of individuals. This time, rather than assuming equal reproductive success of all genotypes, let us place the homozygous **BB** hamsters at a disadvantage by allowing only five gametes from each to be involved in the production of the succeeding generations. This is half the number contributed by each **BB** individual in the calculations in Table 18.1. The other two groups continue to contribute the same number of gametes. (See Table 18.2.)

Table 18.2.

Genotypes of the Animals	Gametes B	Gametes b	Total
49 **BB**	245	0	245
42 **Bb**	210	210	420
9 **bb**	0	90	90
	455	300	755
	or approximately 60% (0.60)	or approximately 40% (0.40)	or 100%

	p(**V**) 80%	q(**v**) 20%
p(**V**) 80%	p^2(**VV**) 64%	pq(**Vv**) 16%
q(**v**) 20%	pq(**Vv**) 16%	q^2(**vv**) 4%

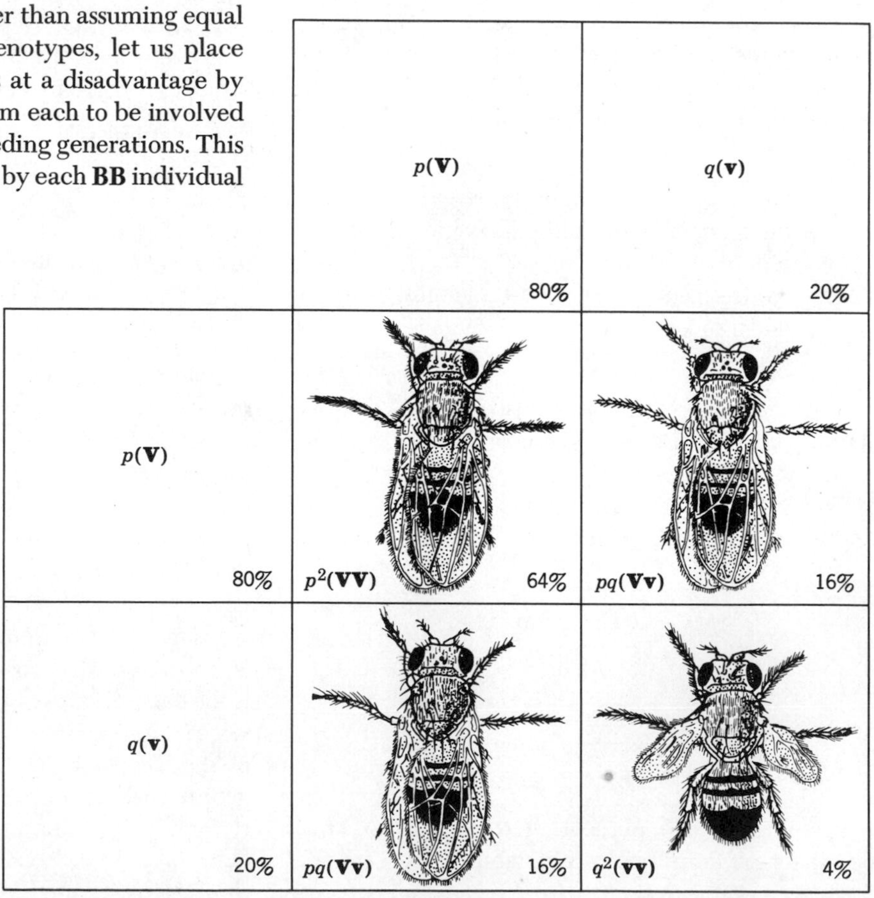

Figure 18.1. Frequency of vestigial wing in a population of **Drosophila.**

Figure 18.2. Sea anemones spawning. The released gametes appear as two oval dense white masses to the right of the anemones. Gametes released into the water have a small chance of surviving.

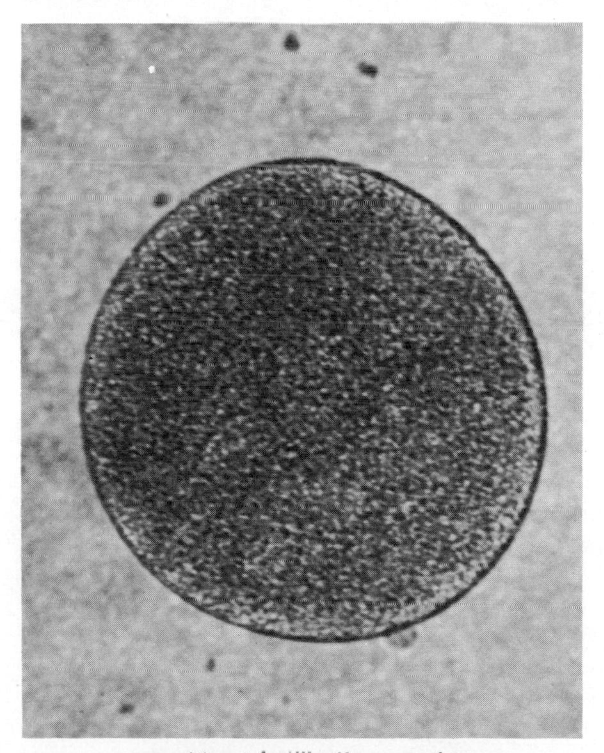

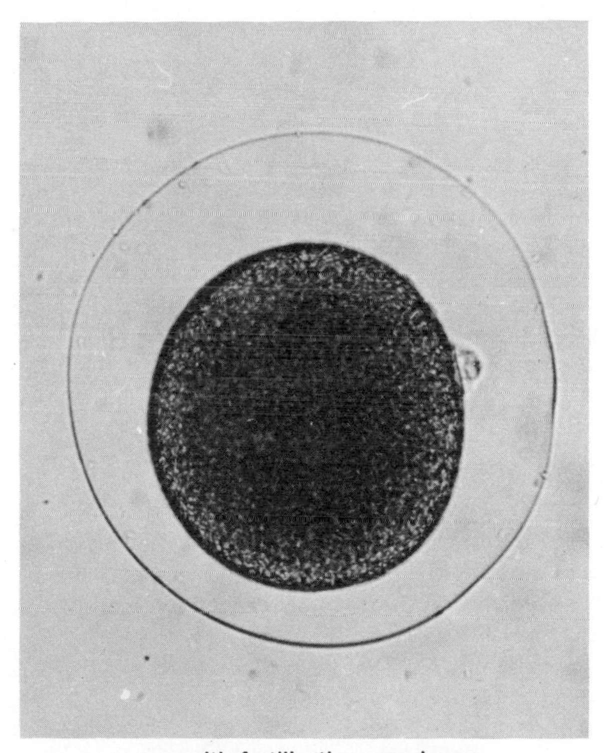

Figure 18.3. Starfish eggs. The egg on the right has been fertilized and is beginning normal development whereas the one on the left has failed to develop despite fertilization.

The result of our interference is as follows:

		Gametes from the Males		Summary
		0.6 **B**	0.4 **b**	0.36 (36%) **BB**
Gametes from the Females	0.6 B	0.36 BB	0.24 Bb	0.48 (48%) **Bb**
	0.4 b	0.24 Bb	0.16 bb	0.16 (16%) bb

In just one generation, the frequency of the genes has shifted appreciably owing to a 0.5 selection factor against the homozygous dominant individuals. This is rigorous selection but by no means extreme. On some occasions selection may completely eliminate all carriers of a particular gene.

Migration. The second factor that modifies gene frequencies is immigration from a population in which the gene frequencies are different, or selective emigration of one genotype from its population. In most widely distributed species the semi-isolated populations are kept as integral parts of the species by the frequent migration from one group to another of individuals or gametes and the genes they contain. Such exchange of genes prevents the differentiation of the gene pools of the populations and their division into distinct species.

Mutation. Mutation is the third factor modifying the genetic characteristics of the population. By itself, mutation is usually unimportant, because it commonly occurs only once in 100,000 to once in 100,000,000 gene duplications. Even if mutations were common, the mutation of the new gene back to the original allele would tend to moderate the effect on gene frequencies. But mutation combined with selection is a potent force in modifying populations because mutation supplies the only new source of variation upon which selection may act.

Population Size. Most populations are large enough so that the failure of some members to reproduce does not significantly change gene frequencies in the gene pool. If a population should decrease drastically to perhaps a few dozen members, a phenomenon called *genetic drift* may take place. In this small population, the chance failure of one or a few members to contribute to the gene pool may lead to the total loss of certain genes or alleles. Imagine a population of eight mice on a small island, of which only two mice contain a dominant gene for long tail. If one of these two is killed by some catastrophe prior to reproduction, the frequency of this dominant in the population will change drastically. The population may "drift" down adaptive pathways that would be unlikely in a large population. Of the four main forces that change populations, genetic drift is considered the least important.

Some Applications

If so many factors in nature modify the gene frequencies predicted by the Hardy-Weinberg law, what good is this concept? It has many different uses, all stemming from the fact that the law provides a model with which to compare populations found in nature. If these natural populations do not conform to the Hardy-Weinberg predictions, one or more of the conditions stated in the law has not been fulfilled.

Calculations based on the Hardy-Weinberg law are also of interest in studying human populations. Couples that have an inherited disease somewhere in their families often want to know the likelihood of one of their children being affected. Even couples that do not have any family history of undesirable traits often want to know the chances of their having children with one of these characteristics. Questions are often asked about albinism, because the complete lack of pigment in the skin, hair, and eyes causes the person to differ markedly in appearance from most of the population.

In order to predict the chance that a child in a given family may be an albino, it is essential to know the prevalence of the trait in the general population, the history of the trait, if any, in the two families, and the method of inheritance of the trait.

This simple recessive trait is present in about one person out of 5000 in the population. This means that $q^2 = \frac{1}{5000}$ or 0.0002 if the letters are defined in the usual way. The frequency of the recessive gene (**a**) would be 0.014.

$$p + q = 1.0$$
$$p + 0.014 = 1.0$$
$p = 0.986$ or 98.6% of the genes are **A** (for normal trait)

$p^2 = (0.986)\ (0.986) =$ approx. 0.972 or 97.2% of the population are **AA**

$2pq = 2(0.986)\ (0.014) =$ approx. 0.028 or 2.8% of the population are **Aa**

If the prospective parents were both normally pigmented and had no known history of albinism in their families, we could tell them that the chance that both of them were heterozygous would be (0.028) (0.028) or about 8 chances in 10,000. Even if they were both heterozygous, only one child in four should be albino. As in all cases where no family history of a relatively rare recessive trait occurs, the chance of this couple having a child with the recessive trait is very low.

Of more interest to us for its theoretical implications is the study of the frequencies of traits in different populations. Many human groups in the past were quite isolated from one another for long periods of time. This isolation allowed differences in gene frequencies to develop between groups. We are familiar with differences such as skin color which, according to the existing evidence, is an adaptation to the amount of sunlight in the ancestral environment. But other marked differences occur in the frequencies of traits, such as the frequencies of the A–B–O blood groups, for which we know only indirect adaptive significance.

The nationalities from central Europe all show fairly similar blood group frequencies. We would expect this because of relatively free intermarriage. A sample of Belgians and a sample of Germans form the most widely divergent groups in this area but are still not very different (see Table 18.3). On the other hand, two tribes of Indians from Montana show extreme divergence. Evidently little intermarriage has occurred between these tribes. This information helps our knowledge of evolution by demonstrating that differences in relatively nonadaptive traits can arise between isolated populations and by giving us information on the degree of isolation that these human groups have experienced.

Many people are interested in controlling or eliminating the reproduction of individuals possessing so-called undesirable characteristics. Reproductive control unfortunately does not provide a quick answer to the problem of eliminating undesirable recessive traits. If we modify the Hardy-Weinberg equations to include the effects of such selection on gene frequencies, we can use the basic idea of population genetics to show how long selection would take to eliminate or reduce the trait to any particular level. In Japan, a recessive gene was found that caused the homozygous individual to be deaf and mute. The frequency of this gene in the population was 0.009 (9 recessive genes out of every 1000 genes). If *all* the individuals showing the trait were prevented from marrying and having children, the frequency of the recessive gene in the population would be reduced at the rate shown in Table 18.4. These calculations indicate that reduction of the frequency of a recessive trait is very slow. Is control worth all the problems it entails? Society must make that judgment.

Table 18.3. Frequencies of the A-B-O Blood Groups in Various Populations

Modified from W. C. Boyd, *Genetics and the Races of Man*, Little, Brown and Co., Boston, 1956.

Population	Place	Percentage of O	A	B	AB
American Indians					
Utes	Montana	97.4	2.6	0	0
Blackfeet	Montana	23.5	76.5	0	0
Navaho	New Mexico	77.7	22.5	0	0
Flatheads	Montana	51.5	42.2	4.7	1.6
Belgians	Liège	46.7	41.9	8.3	3.1
Spaniards	Spain	41.5	46.5	9.2	2.2
Frenchmen	Paris	39.8	42.3	11.8	6.1
Germans	Berlin	36.5	42.5	14.5	6.5
Greeks	Athens	42.0	39.6	14.2	3.7

Table 18.4. Frequency of One Gene for Deaf-Mute in Japan

Generations of Selection	Frequency of the Gene	Frequency of Homozygous Individuals
0	0.009	0.000081
14	0.008	0.000064
32	0.007	0.000049
139	0.004	0.000016

Principles

1. In sexually reproducing populations the gene frequencies and genotype frequencies remain the same from one generation to the next if certain conditions prevail.
2. This genetic equilibrium may be altered by selection, migration, mutation, and genetic drift.

Suggested Readings

Boyd, William C., "Genetics and the Human Race," *Science,* Vol. 140, No. 3571 (June 7, 1963).

Dobzhansky, Theodosius, "Changing Man," *Science,* Vol. 155, No. 3761 (January 27, 1967).

Dobzhansky, Theodosius, "Evolutionary and Population Genetics," *Science,* Vol. 142, No. 3596 (November 29, 1963).

Huxley, Sir Julian, "Eugenics in Evolutionary Perspective," *Perspectives in Biology and Medicine,* Vol. VI, No. 2 (1963).

Petrokis, Nicholas L., Kathryn T. Molohon, and David J. Tepper, "Cerumen in American Indians: Genetic Implications of Sticky and Dry Types," *Science,* Vol. 158, No. 3805 (December 1, 1967).

Questions

1. State the Hardy-Weinberg law.
2. Is this only a useful theoretical model, or a realistic statement of conditions as they occur in nature? Defend your viewpoint.
3. Is a *gene pool* an abstract concept or a reality? Explain.
4. If provided with the frequency of a certain gene in a population, how do you determine the frequencies of the genotypes in which it may occur?
5. List the major forces that act to change gene frequencies in populations.
6. Which of these forces plays the dominant role in directing the course of evolution? Explain how it works.
7. Which of these forces introduces entirely new genes into the population?
8. Using the Hardy-Weinberg concept, explain why eliminating an undesirable recessive phenotype from a population reduces the frequency of the gene only slightly.
9. Find out whether the state in which you live has any "eugenic" laws aimed at reducing the frequencies of undesirable genes. If so, are the laws really effective from a biological viewpoint?

CHAPTER XIX

Adaptation and Speciation

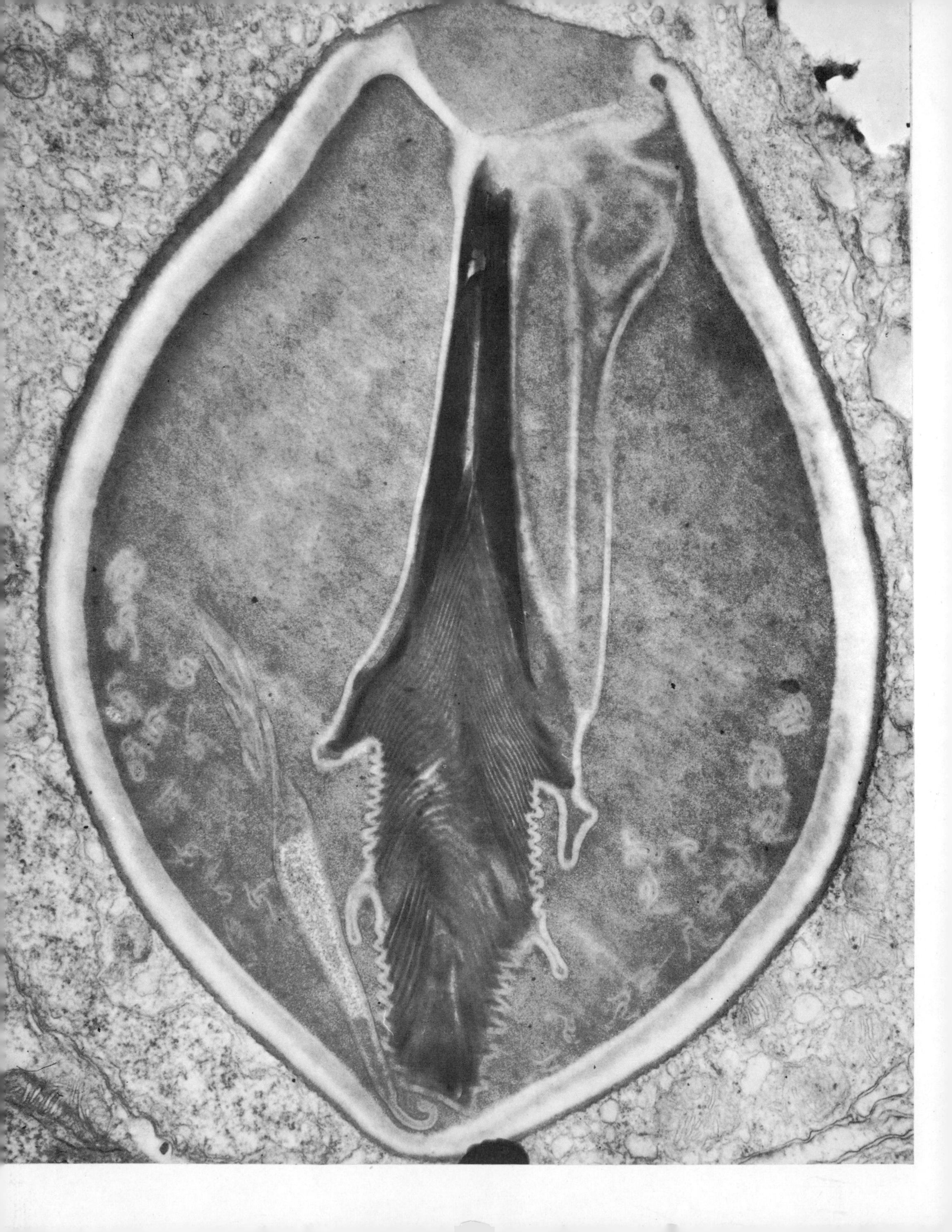

Longitudinal section of a nematocyst (stinging cell) from Hydra. *These structures are used for protection and for food-gathering* (×17,000) *(Dr. G. B. Chapman).*

Adaptation and Speciation

Although the gene pool of a Mendelian population tends to remain the same, natural selection, mutation, and migration are continually causing shifts in the gene frequencies. Since selection is the force that gives direction to these changes, the organisms in the population become progressively better adapted to their environment.

In what ways do organisms become better adapted? We can classify adaptations into three general categories: (1) morphological, (2) physiological, and (3) behavioral. At first it seems that examples should fall neatly into one or another of these categories. But careful examination confirms what our knowledge of the interrelations of parts of organisms leads us to suspect—that adaptations in the structure, function, or activity of an organism are generally accompanied by related adaptations in the rest of the organism.

Types of Adaptations

Morphological Adaptations. Because adaptations in form and structure are the most noticeable type in animals and plants, they have long fascinated scientists as well as laymen. The animals in a zoo, for example, provide an excellent variety of morphological adaptations. Here one can easily see the relation between teeth and diet, limbs and locomotion, coat color and

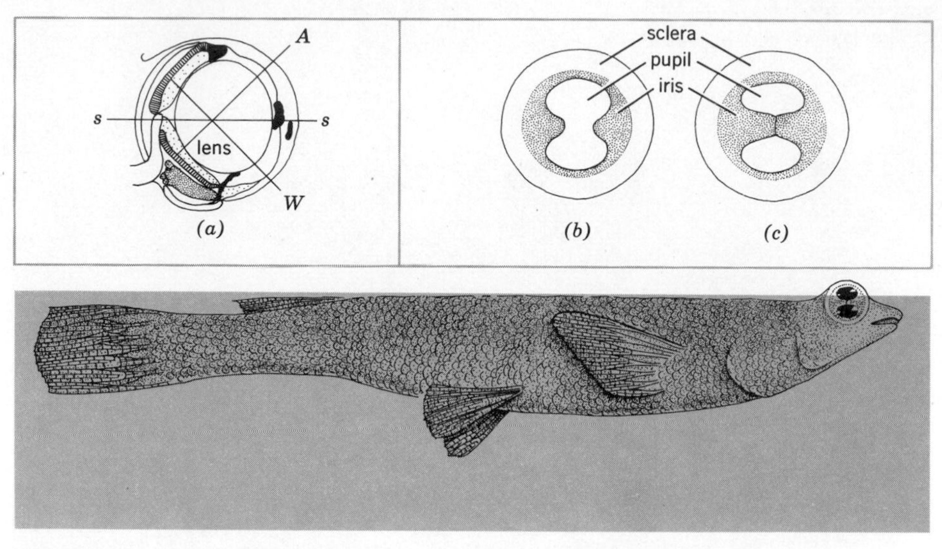

Figure 19.1. The "four-eyed" fish. The upper diagrams show development of the two pupils in an eye. The two lenses allow the fish to see simultaneously in and above the water. In (a), *the line* ss *represents water surface;* A *is line of sight upward into air and* W *is line of sight downward into water* [(a)-*Adapted from A. Putter, Handbuch der gesamten Augenheilkunde, 1912. Aufl., Bd. 2:* (b) *and* (c)-*Schneider and von Oreilli,* Mitt. d. naturt. Ges. in Bern., *1907.*]

native habitat. Some adaptations may seem unusual to us, such as the adaptation in fish shown in Figure 19.1. The two lenses allow the fish to watch simultaneously for insects skimming on or near the water, and for any food source or predator beneath the surface.

Plants also exhibit numerous morphological adaptations, including the basic form of the plant itself, the seeds (Figure 19.2), and leaves (Figures 19.3 and 19.4). Many interesting modifications are found in flower structure. These are primarily related to ways of bringing about pollination. For example, wind-pollinated flowers are small and drab in color, whereas insect-pollinated ones are frequently colorful and large. Tropical flowers are especially noted for their sometimes bizarre forms. An Australian orchid, for example, resembles a female bee. The male attempts to mate with the flower and, in the process, spreads pollen from one flower to another. Some plants live and flower under water and obviously require many special adaptations in leaf structure and flowers.

Physiological Adaptations. Physiological adaptations involve *functional* adjustments that help to insure survival. Improvements in the metabolic pathways involved in respiration, digestion, and other body functions fall in this category. An example is the Red Scale, an insect that damages citrus. At one time, treatment consisted of cyanide gas fumigation of the trees. After some years of this treatment, however, a strain of scale appeared that was resistant to the cyanide spray. Stronger concentrations of the spray killed the scale but also damaged the citrus trees. How did the scale become resistant to cyanide? The most plausible explanation involves mutation and selection modifying the Hardy-Weinberg equation, as in the last example in Chapter XVIII.

The desert ground squirrel of the southwestern United States is able to extract water, by chemical means, from the dry seeds that it lives on. In captivity these squirrels exist entirely on dried seeds and never drink water.

The production of specific pheromones by many animals is a physiological adaptation, although its use is obviously behavioral.

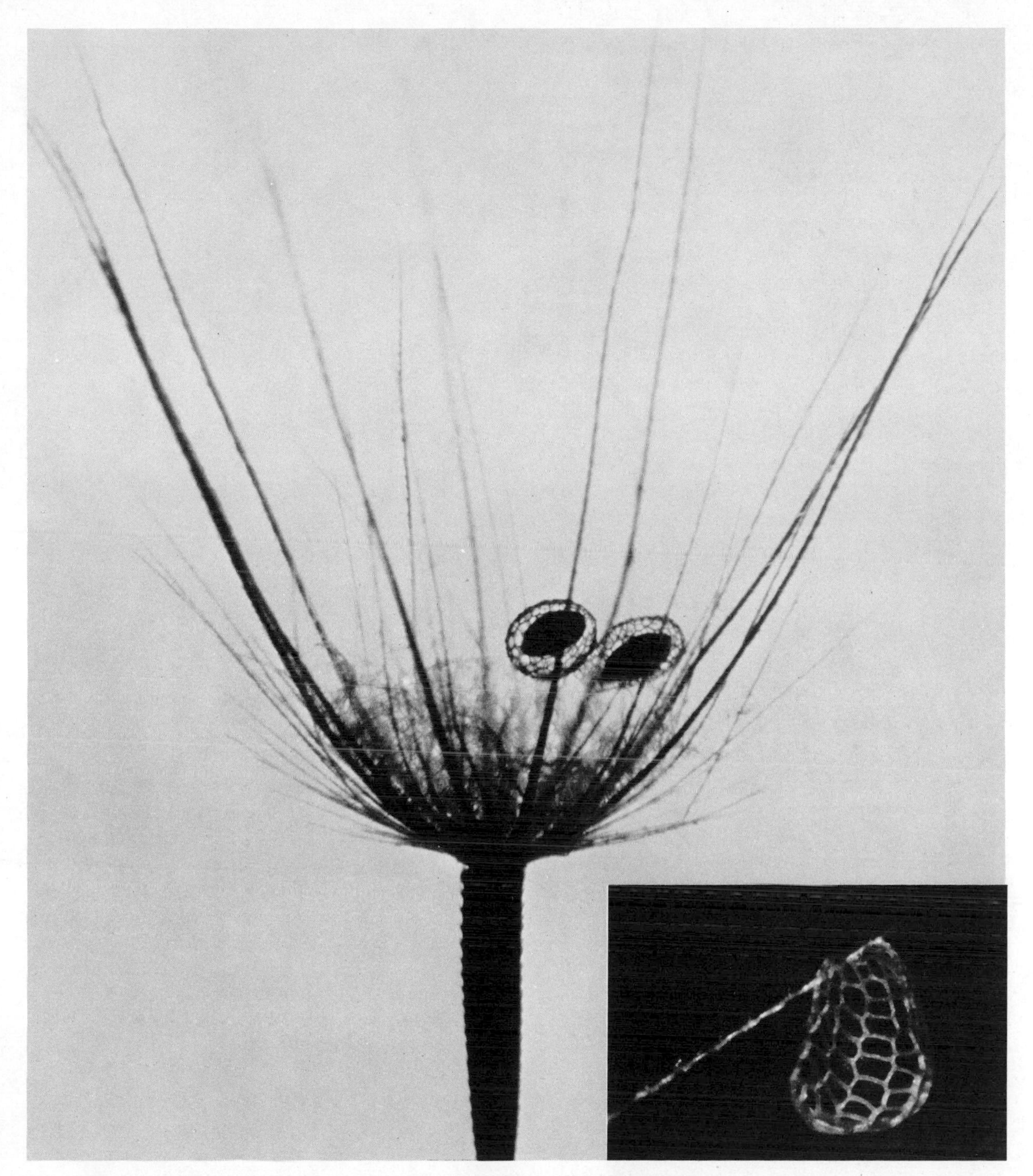

Figure 19.2. Coordinated seed dispersal in a parasite and its host. The netlike structure surrounding the seed of the parasite often catches on a bristle of the host's seed (insert) (Dr. P. R. Atsatt).

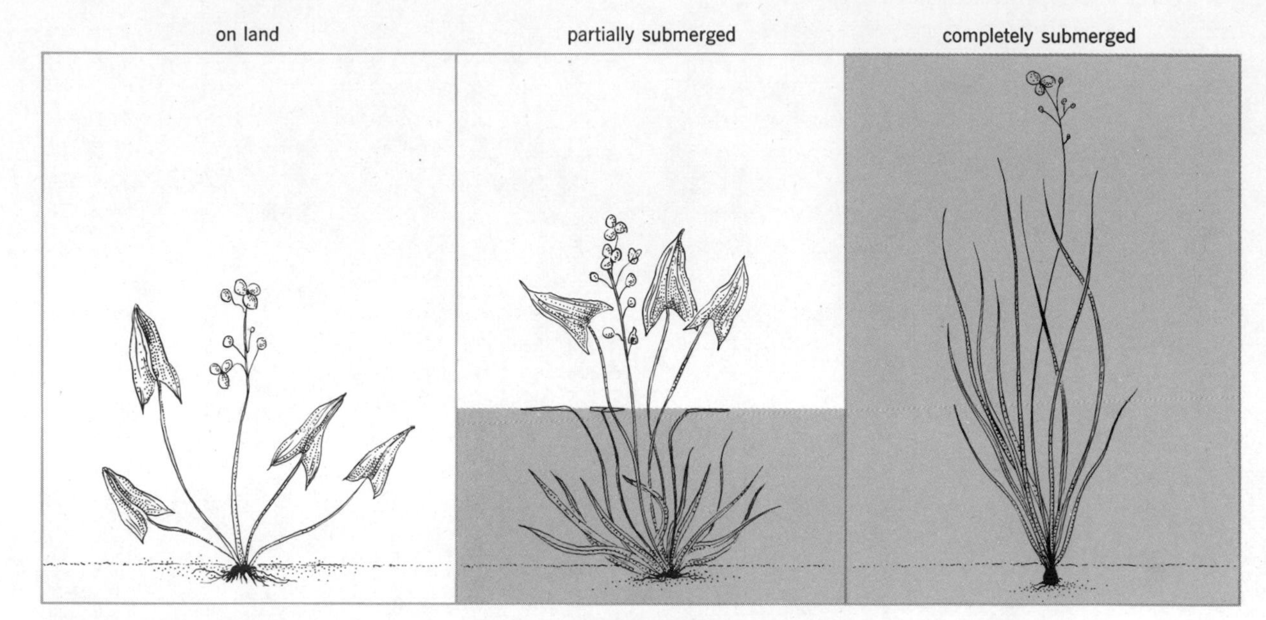

Figure 19.3. Divergent leaf shapes in the arrowleaf plant in different environments. (Redrawn from Bruce Wallace and Adrian M. Srb, Adaptation, *second ed., 1964, Prentice-Hall, Englewood Cliffs, N.J.)*

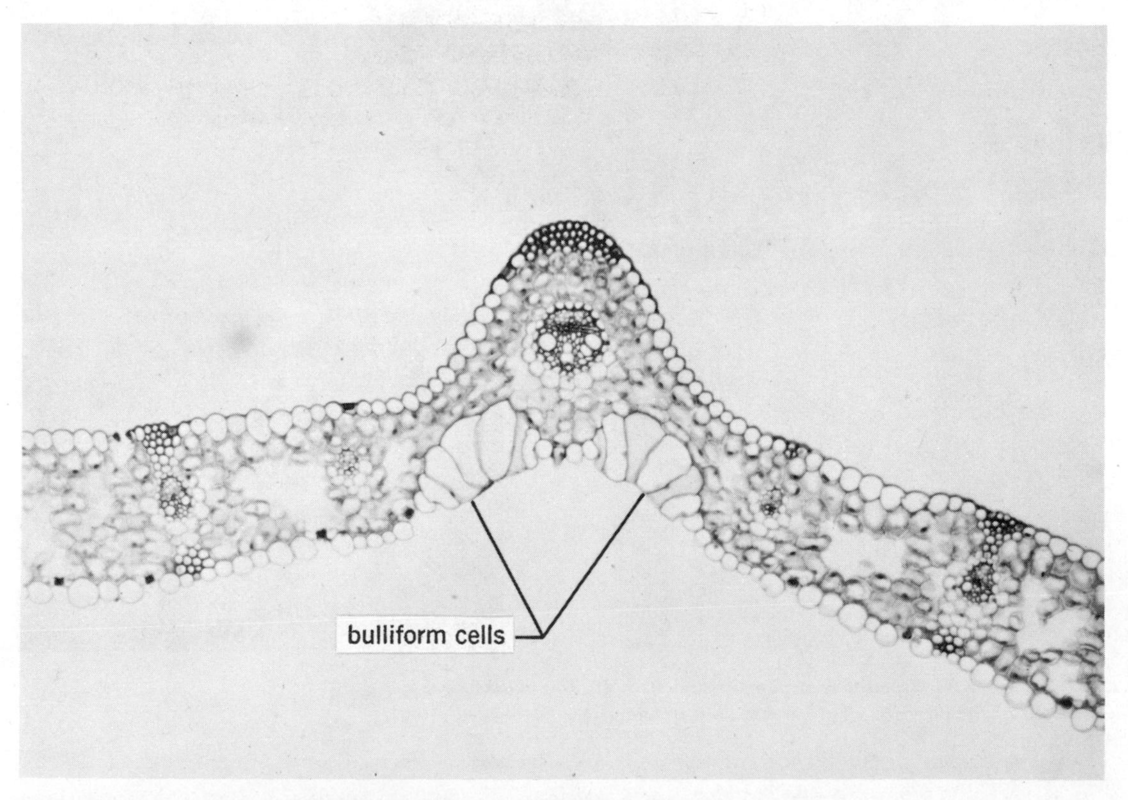

Figure 19.4. A cross section of a grass leaf (Poa) *showing bulliform cells. Under dry conditions these lose water and allow the leaf to fold.*

Figure 19.5 illustrates a physiological adaptation in plants. Other examples include the odor produced by insect-pollinated flowers, and the chemicals produced from the leaves or roots of some plants like the brittlebrush that inhibit other plants from growing nearby.

Behavioral Adaptations. Since the study of behavior is new in biology, it is more difficult to find simple examples of *behavioral adaptations.* Many of the best examples involve subtle modifications. The release of a black fluid, called "ink," by a sea hare (Figure 19.6) or by a squid is a clear example. When threatened by a predator, the squid shoots its ink into the water, usually becomes paler in color, and swims away. The ink assumes a roughly squidlike configuration in the water. Not only does the paler color of the squid makes it less noticeable, but the black ink tends to draw the attention of the predator away from the squid and helps it to escape.

One animal that shows all three types of adaptations in one related set is the skunk. Its distinct coloration, the ability to produce scent, and the pawing of the ground and ejecting of the scent are all important adaptations in the skunk's mechanisms for self-protection. Which adaptation should have arisen first? Why would natural selection, that is, selection by nature, not by man, then favor the other adaptations?

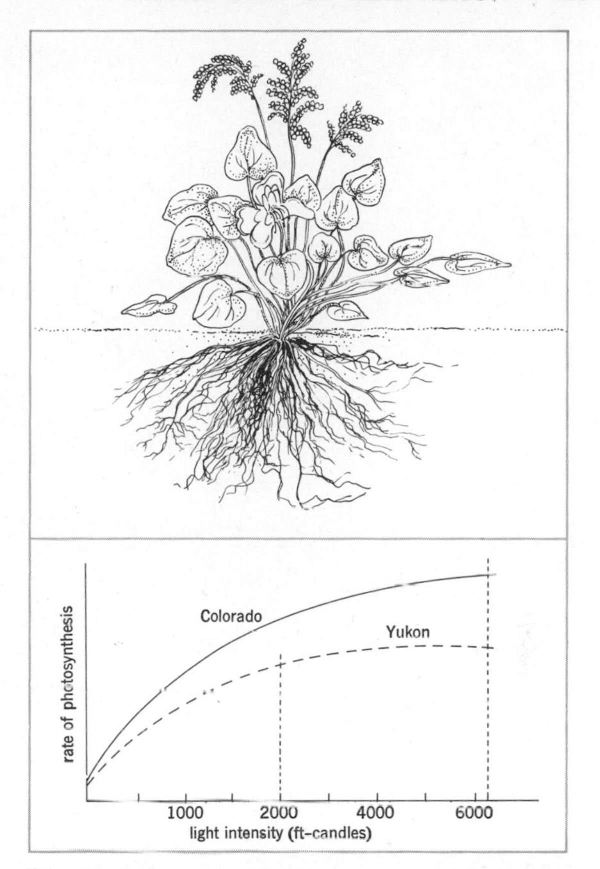

Figure 19.5. The rate of photosynthesis in Oxyria *is adapted to the plant's environment. Can you think of a hypothesis that might account for the difference in the rate of photosynthesis between the state of Colorado and the Yukon territory?* (*Dr. H. A. Mooney.*)

Mechanisms of Adaptation

Variations in Environment. Even though we often think of an organism's environment as being relatively uniform, its entire geographic range usually spans a variety of environmental conditions. In addition, natural selection generally emphasizes different traits in the different areas of an organism's range. Consequently, a species nearly always consists of a group of interbreeding Mendelian populations whose gene pools differ slightly. As you might predict, a species that is distributed over many markedly different environments will consist of populations with many slightly different adaptations (see book frontispiece). The slightly differing gene pools or Mendelian populations that constitute a species are known by a variety of terms such as local populations, demes, races, varieties, or subspecies.

An example of how gene pools change adaptively from one population to another is illustrated by the coloration of the beach deer mouse (*Peromyscus polionotus*) that inhabits northern Florida. If these animals are to survive, their color must be closely matched to the color of the soil. A coat coloring that is distinct from the soil color would allow predators to spot the animals much more easily.

In northern Florida the soil color varies from reddish near the Atlantic coast, to dark in the central part, to very light sand along the Gulf coast. The color of the mice follows this same pattern. Since the populations of mice from all these areas can interbreed and are much alike except for color, we assume that they came from ancestors that all looked very much alike. The observable differences among populations have arisen because the differences in the environments of these areas have selected individuals best adapted to each. We are able, therefore, to identify the populations to which various mice belong.

Figure 19.6. The sea hare (a) *releases a black fluid* (b) *when disturbed. This presumably helps the sea hare to escape predators by confusing or repelling them.*

Certainly more differences than color occur among these populations. If we were to measure the mice and the environments on the east and Gulf coasts, we could probably enumerate a good number of distinctions in each. Under existing conditions, genes from the population on one coast can be exchanged with genes from the population on the other coast through the interbreeding of mice from each population with individuals from the population in the central part of the state.

Variations in the environment also lead to adaptive changes in the gene pools of plant species. An example is found in a common European roadside weed (*Camelina sativa*). The form of the plant is usually low and thickly branched, with small round seeds. Farmers easily remove it from their flax fields because flax plants are tall and nonbushy. In addition, the seeds of the two are easily separated at harvesting time since flax has large flat seeds. In portions of Europe, however, some populations of the weed have adapted to growing among the flax plants. These populations are tall, like flax plants, and have large flat seeds. Natural selection has adapted *Camelina* to compete successfully in the flax plant environment.

Not every species, of course, shows such variation in its gene pool. The range of the Eastern Mockingbird extends from Canada into southern Florida, but there are no demonstrably different populations within this range. Even if we examine mockingbirds from all over North America, there are only two recognizable populations in the species, those from eastern North America and western North America. Mockingbirds are evidently so mobile and nonspecialized in their living requirements that they inhabit nearly all of North America.

Barriers. Consider again the beach deer mice in Florida. What would happen if the central population did not exist? Then the only possible way in which exchange could occur between the gene pools of the two coastal populations would be the migration of individuals from one coast to the other. The distance is too long for this to occur often. Consequently, *reproductive isolation* (lack of gene exchange) would result; the distance would form a barrier between the populations. With no exchange of genes between the two populations, the mutations occurring in one could not reach the gene pool of the other.

The tendency of natural selection to cause the two groups to become dissimilar would continue unabated. The counterforce of migration and gene exchange would no longer slow down the divergence of the two populations. They would become progressively more dissimilar. It is impossible for this to occur unless a barrier to gene exchange has arisen.

In most cases the first barrier that prevents gene exchange is a *geographical barrier*. The organisms are prohibited from reaching one another by some feature of the environment between them. It may be a river, ocean, mountain, desert, forest, or any unsuitable habitat. As the two groups become more dissimilar through time, other barriers to reproduction may arise. Examples of these are temporal, ecological, behavioral, and hybrid inviability barriers. All these barriers prevent effective gene exchange, but they function at different times in the life cycle or in different ways.

Temporal and ecological barriers prevent the two organisms from even meeting during the breeding times. One group may breed earlier than the other (*temporal*) or in a different type of habitat (*ecological*). An interesting example of ecological isolation is found in some small birds living in Mexico (Figure 19.7). Two related species, the Red-eyed Towhee and the Green Towhee, live very close to each other in some areas. In one area in southeastern Mexico, they even live on the same mountain without any apparent interbreeding. The Red-eyed Towhee lives in the brushy growth in oak and oak-pine forests at lower elevations, whereas the Green Towhee lives in the brushy plants in fir and oak pine forests at higher elevations. Even though they live near each other, they do not live in the same environment, and therefore do not interbreed.

Farther north in Mexico, the two species again come into contact. However, in this area man has changed the environment a great deal by cutting the virgin forests. The brushy regrowth is an optimum habitat for both species. Consequently, both nest in these areas and interbreeding occurs. Here, the population is quite variable and intermediate between the normal appearance of the two species. Obviously, the two species can interbreed but differences in habitat have usually kept them apart.

Behavioral barriers often prevent the interbreeding of individuals of different species that do meet during the breeding season. Some important factor

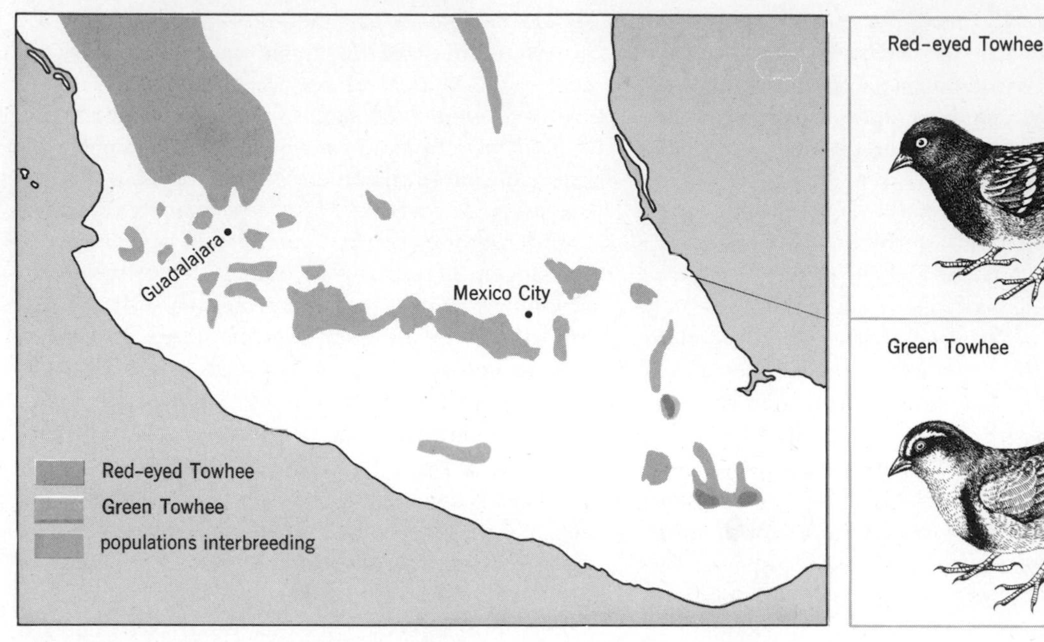

Figure 19.7. The distribution of towhees in Mexico. Interbreeding of these towhees occurs over a broad range where man has modified the environment.

in the courtship differs between the two. Consequently, the displays necessary to initiate mating are not all present and mating does not occur.

The Eastern and Western Meadowlarks look almost identical. In the Midwest, where their ranges overlap, the male of one species may court the females of the other. All may go well with the courtship until just before the male mounts the female. As he approaches the female, he gives a call note that is different in the two species. The female, hearing the wrong note, will not allow courtship to proceed.

Hybrid inviability does not prevent interbreeding. It only prevents interbreeding from generating gene exchange. The hybrid may die at any stage of development up to the onset of reproduction. In fact, even if the hybrid lives but is incapable of reproducing because of sterility or its unattractiveness to other members of the species, the result will be the same.

An instance involving flax provides an example. If two species are crossed, the hybrid seeds fail to germinate. However, perfectly normal germination occurs and luxuriant fertile plants are produced if the seedcoats are first removed from the embryos. Here, the seedcoat that comes from the tissue of the maternal plant interacts with the hybrid embryo in some way that prevents germination.

In nature, these hybrids would not grow. Any plants that interbred with members of the other species would leave no offspring. Therefore, natural selection would favor plants that did not interbreed and more barriers would be introduced to reinforce this one of hybrid inviability.

Speciation: The Results of Adaptation

Let us now assume that the two groups of Florida deer mice are again able to reach each other. What will happen? Will the groups be able to interbreed again or not? The result will depend on the extent to which barriers have been built up. If, when the geographic barrier breaks down, other barriers are sufficiently developed to prevent gene exchange, the two groups will remain distinct and continue to be affected only by their own selective pressures. The changes in one gene pool that have been brought about by selection and mutation will not be transmitted to the other gene pool and will not affect its evolutionary path.

If, on the other hand, the two groups do interbreed, genes from the two pools will become intermingled and the differences between the groups will become less marked than before. Distinctions between some individuals may be even greater because of recombinations of the greater variety of genes, but most individuals will resemble some intermediate form.

In reality, we have been discussing the method by which two species may develop from one. This process is called *speciation.* When the two populations make contact again, a lack of interbreeding indicates that each is now, in fact, a distinct species. This lack of interbreeding is a meaningful test to show distinct species in sexually reproducing organisms because two populations that cannot interbreed, cannot affect the course of each other's evolution. The most favored formal definition of a species includes this idea of lack of gene exchange: *a species is an interbreeding or Mendelian population or group of populations that is reproductively isolated from other such populations.*

This definition of a species is a useful one but biologists generally agree that it cannot be rigidly applied in all situations. The following example illustrates the dilemma biologists not uncommonly encounter. Two species of *Anolis* lizards inhabit adjacent islands in the Lesser Antilles. One (*Anolis trinitatis*) is bright green and has 36 chromosomes. The other (*A. aeneus*) is gray and mottled and has 34 chromosomes. They are geographically and reproductively isolated since they live on separate islands. Both, however, were apparently introduced onto the island of Trinidad by man, became established, and are now hybridizing. The hybrids are intermediate in color and have 35 chromosomes. Should we continue to recognize these lizards as separate species or not? A similar situation may arise when wild animals are associated in captivity. Lions and tigers, for example, do not hybridize in nature but have produced "ligers" and "tiglons" in captivity. Forms that reproduce mainly by asexual means obviously present another point of difficulty with our definition of species. Because bacteria, parasitic worms, and many fungi fall into this class, the problem is an important one.

Speciation can occur in other ways; one other deserves mention. Closely related species of plants often interbreed to some extent in nature. These hybrids are sterile, or nearly so, if the chromosomes derived from the two parents do not pair normally in meiosis. This results in the chaotic distribution of chromosomes to the gametes and generally the gametes die. On occasion, however, some of the hybrid plants show a doubling of chromosome number as a result of abnormal meiosis. When this occurs, all chromosomes can pair at meiosis and viable gametes result. Note that these gametes contain twice the number of chromosomes found in gametes from the parent plants (see Figure 19.8). These *tetraploid* (four sets of chromosomes) plants may now be fertile and reproduce readily, but they will be incapable of interbreeding with the parents. Usually the new species resembles the parent species closely, yet such a population fits our definition of a new species since it would be reproductively isolated.

This means of speciation, known as *polyploidy,* is extremely rapid since the chromosome doubling takes place within one generation. Although it is rare in animals, it must be a common means of evolution in flowering plants since numerous species are known to be polyploids. Examples include cotton, wheat, tobacco, apples, poplars, and numerous wild flowers. In fact, most of man's domesticated plants were developed by hybridizing different varieties, usually producing new polyploid types.

Adaptive Radiation

Let us now expand the model of speciation from two isolated populations to several reproductively isolated groups. Earlier we stated that the environment was certain to differ from place to place throughout the range of an organism. Each of the isolated units is therefore exposed to its own selection pressures and different adaptations arise. If the isolation persists long enough, several distinct species result. The differences in structure among the various groups would be related to the adaptation of each group to its own environment. Since all of these groups come from a common ancestor, becoming different as they adapt to new environments, this process is called *adaptive radiation.*

Because of their respective adaptations, each group would probably be doing something unique in its environment, for example, feeding in a slightly different place, nesting in a special spot, growing on another soil, etc. Since the groups exploit different parts of the environment, they are said to occupy separate *niches.* As soon as these niches become distinctive enough to avoid undue overlap, the

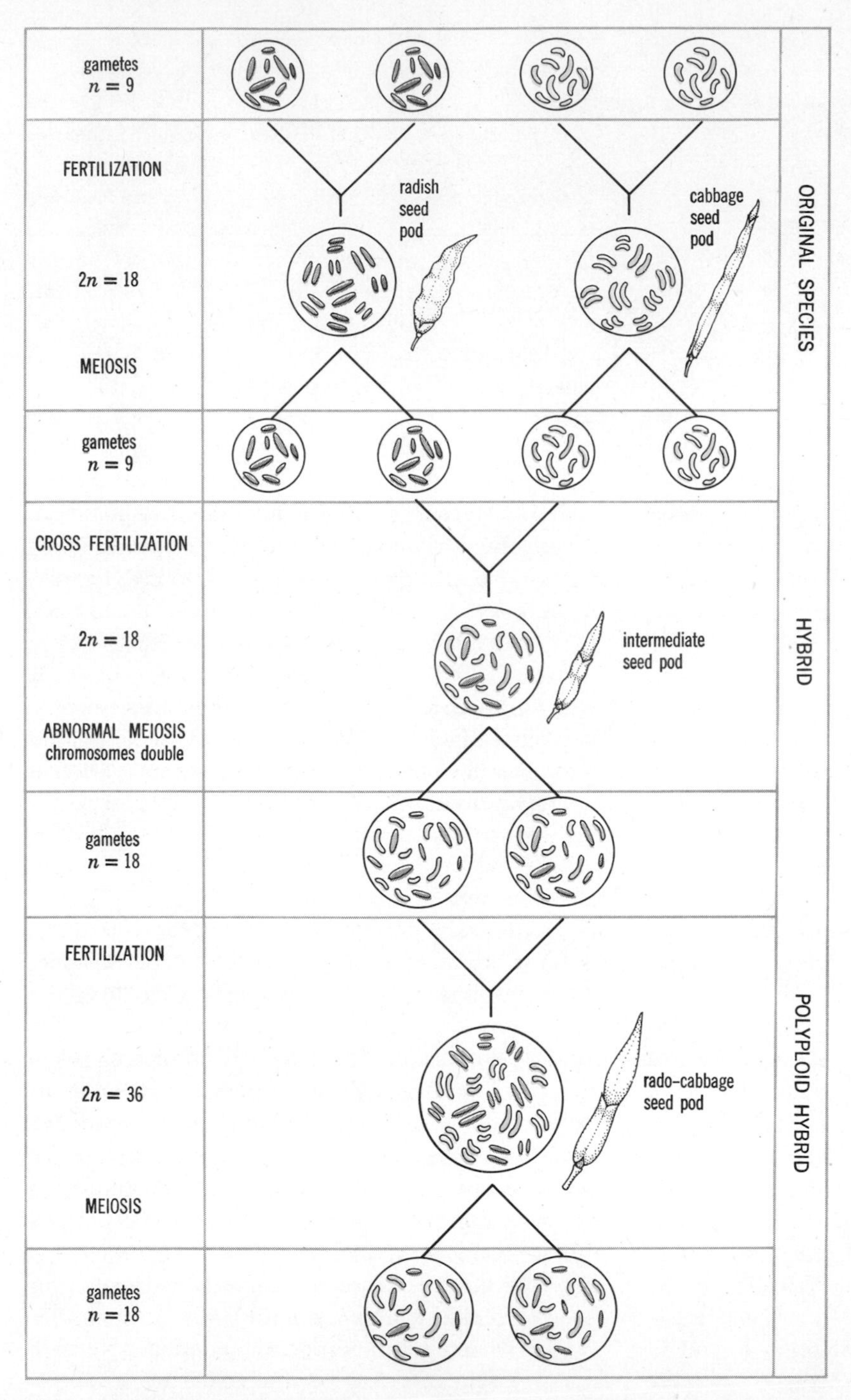

Figure 19.8. The chromosome number and appearance of seed pods in a fertile polyploid hybrid of radish and cabbage. The hybrid is not fertile until doubling of the chromosome number occurs.

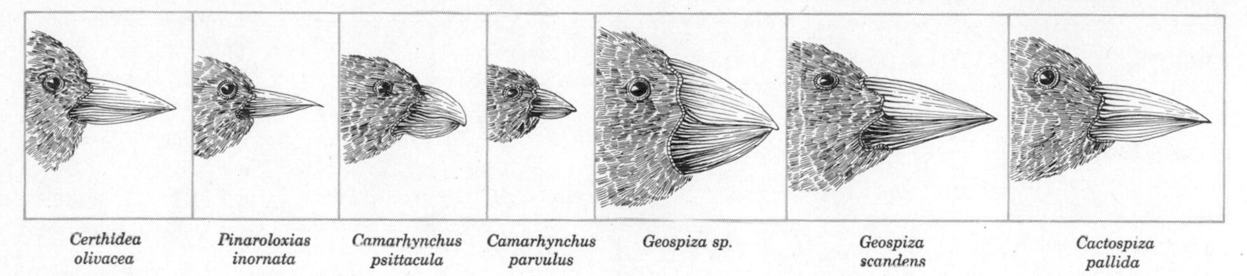

Figure 19.9. Shape of beaks in Darwin's finches. A group of small finches reached the Galapagos Islands where no other land birds lived. The radiation to feed in the available habitats has resulted in the marked differences in beak shapes. Insect eaters (left) and seed eaters (Geospiza) are two examples.

groups can come together without interbreeding or severe competition.

The events leading to adaptive radiation are identical with the process described earlier as speciation. However, adaptive radiation refers specifically to the *diversity* that comes about during speciation. Studies of groups of living organisms and study of the fossil record emphasize adaptive radiation as one of the major events in the evolutionary process. Again and again evidence has shown that when a group of plants or animals entered a new environment with many unoccupied niches, it evolved in many directions and became adapted to the new habitats. Such adaptation has usually resulted in organisms that may appear different superficially but whose basic structural similarity indicates their close relationship.

A classic example of adaptive radiation in birds has been worked out in great detail for Darwin's finches, which inhabit the Galapagos Islands. The ancestors of these birds arrived on the islands long ago, perhaps blown there from South America in a severe storm. The islands were unoccupied by other land birds and offered numerous new habitats. As time passed, the birds spread over the islands, became isolated on some of them, and evolved many specialized feeding habits. Their radiation in feeding habits resulted in the marked diversity in beak shapes (Figure 19.9) observed in the present-day descendants of the original finches. All have the basic finch-type bill, but natural selection has altered it for seed crushing, bark removing, insect eating, and other special jobs. If we think in terms of longer periods of evolutionary time, adaptive radiation probably accounts for the diversity of life in all major groups of organisms. For example, the diverse types of mammals we see today—those that run, fly, swim, burrow, climb trees, and so on—arose from an ancient ancestral group that underwent extensive radiation over a vast period of time.

Evolutionary Convergence. Many times during the course of adaptive radiation relatively unrelated organisms have become adapted by natural selection to similar environments. Obvious examples are birds and bats, or whales and fishes. Superficial similarities may be marked but the basic structure of the organisms betrays their distinct ancestries.

Whales and fishes live in the same general environment and their general body form is much the same. Their respiratory systems contain marked differences, however. When a lung-breather like the whale becomes adapted to life in the water, a system such as gills that allows gas exchange to occur under water would be most useful. This acquisition was not available to whales, because no mutation for the development of gills occurred. Instead, mutations occurred that modified the way the whale utilized its oxygen. Lungs are still the means of gas exchange and these can accommodate rather large supplies of air. The real adaptations are found in the operation of the circulatory system and in the ability of certain cells to tolerate low oxygen supplies—but not in the respiratory system. When the animal dives, the heart rate drops and most of the blood is shunted to the nervous system, with little going to the extremities. The sensitive nervous system receives sufficient oxygen to prevent damage and to allow respiratory movements to be suspended. Cells in the extremities receive little oxygen and must tolerate a low oxygen supply.

This type of adaptation to life in the water does not appear to be as advantageous to the organism as the development of gills. However, lungs were already present and gills were not. Obviously, natural selection must work with available materials and structures or with those that happen to develop because of mutation or genetic recombination.

Natural selection operating on the variability present in the genotypes of populations can cause better adaptation of organisms to their environment. Coupled with reproductive isolation, these adaptations bring about speciation. To determine the relationships of species to one another, however, we need to know the types of evidence for evolution and their possible meanings.

Principles

1. Natural selection adapts organisms to the environment by means of morphological, physiological, and behavioral modifications.
2. Populations adapted to different environments may undergo speciation if they are isolated from one another for a sufficient period of time.
3. The diversity of plant and animal life results from adaptive radiation.

Suggested Readings

Ehrlich, Paul R. and Peter H. Raven, "Butterflies and Plants," *Scientific American*, Vol. 216 (June, 1967). Offprint No. 1067, W. H. Freeman and Co., San Francisco.

Irving, Laurence, "Adaptations to Cold," *Scientific American*, Vol. 214 (January, 1966). Offprint No. 1032, W. H. Freeman and Co., San Francisco.

Lewis, Harlan, "Speciation in Flowering Plants," *Science*, Vol. 152, No. 3719 (April 8, 1966).

Romer, Alfred S., "Major Steps in Vertebrate Evolution," *Science*, Vol. 158, No. 3809 (December 29, 1967).

Savage, Jay M., *Evolution*. Holt, Rinehart and Winston, New York, 1963, pp. 66–93.

Sheppard, P. M., *Natural Selection and Heredity*. Harper and Row, New York, 1960.

Stebbins, G. Ledyard, *Processes of Organic Evolution*. Prentice-Hall, Englewood Cliffs, N.J., 1966, pp. 1–127.

Wallace, Bruce and Adrian M. Srb, *Adaptation*. Second edition. Prentice-Hall, Englewood Cliffs, N.J., 1964.

Wecker, Stanley C., "Habitat Selection," *Scientific American*, Vol. 211 (October, 1964). Offprint No. 195, W. H. Freeman and Co., San Francisco.

Questions

1. In light of the previous chapters, does heredity act as a conservative force in evolution or as an agent of change? Why?
2. Name three categories of adaptation and provide an example of each. Can you think of some examples in addition to those in the chapter?
3. Why is it incorrect to say that the environment *causes* organisms to change (adapt) in order that they may better survive? Can you think of an experiment to test this idea?
4. List some of the ways in which reproductive isolation might occur. Does this always lead to speciation? Why?
5. What is the best evidence that two populations have become two species?
6. If two animals breed in captivity, does it prove that they are not separate species? Explain your answer.
7. How can polyploidy bring about speciation in only one generation?
8. What is the relation between speciation and adaptive radiation?
9. Why do superficial similarities not necessarily reflect evolutionary relationships?

CHAPTER XX

Evolution: Evidences and Theories

Fossil remains of coccolithophorids (microscopic organisms) from the tropical Indian Ocean (×4,900) (Courtesy Dr. Annika Sanfilippo).

Evolution: Evidences and Theories

Evolution in its simplest and broadest sense means changes in gene frequency within a population over a period of time. Natural selection guides these changes: only those persist that will better adapt organisms to their environment. The origin of *new* characteristics lies in mutation. In mutation there is also a change in gene frequency (from zero to some small percentage), and the survival of these changes is likewise subject to the test of adaptability.

Over long periods the accumulation of changes may be sufficient to separate once similar populations into distinct groups. In the course of evolutionary history this divergence has apparently led to different classes (mammals, birds, fish, etc.), different phyla (insects and corals, for example), and even different kingdoms (plants and animals).

Evidences of Evolution

How extensive has evolution been? Which groups are closely related and which are distantly related? The same evidence can be used to answer both questions.

The evidences of evolution may be divided into two classes: direct or fossil evidence, and indirect evidence such as that gained from the distributions of organisms and from their comparative structure, development, and function. If we assume that evolution has occurred, these many facts present a coherent and intelligible picture. If we assume that evolution has not occurred, these facts become a confusing array of disconnected observations. As evolution enables us to explain more facts about organisms in terms of known phenomena, our explanation of the history of life in evolutionary terms becomes more certain. Since the fossil evidence is the most important to our knowledge of the history of life, we shall discuss it first.

Fossils. A fossil is any evidence of preexistent life (Figure 20.1). Many people think of fossils as organisms turned to rock. Such fossils are relatively rare. More often, the organism has been slowly replaced by sand, mud, or a mineral. In a few cases, part of the organism itself is still present. In others, only an impression of the organism remains. Regardless of the method of preservation, the important fact is that preservation has occurred. Consequently, we have information about an organism that existed in the past. The information varies, but allows us to say that a particular kind of organism did exist at some definite time. For this reason we consider fossils direct evidence that evolution has taken place.

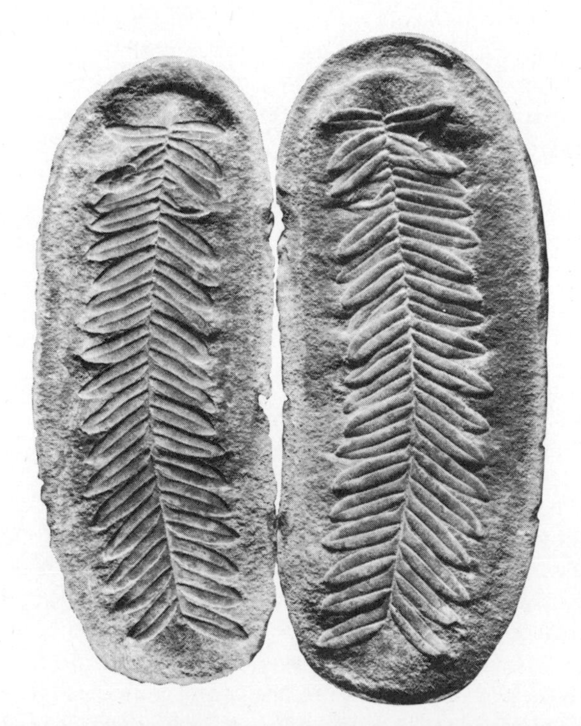

Figure 20.1. A fossil fern leaf. (John H. Gerard from Monkmeyer Press Photo Service.)

If we can discover the approximate age of the fossils, we can arrange them in order of their appearance on the earth. By carefully noting the details of the structure of fossils and the time at which they existed, it is possible to construct a history for many groups of organisms. This history tells us when they existed, in what forms, and the relationships among members of the group.

Dating techniques that utilize radioactive substances are the most widely known. We shall illustrate the process for carbon, which is used for dating recent materials. Other radioactive substances used for dating older remains include uranium and radioactive potassium. The process is essentially the same for any radioactive substance—a substance that emits energy, particles, or both, all of which are measurable. When the carbon 14 method is used, a sample of the fossil to be dated is analyzed for two carbon isotopes, C^{14} (carbon 14 or radioactive carbon) and C^{12} (carbon 12 or ordinary carbon). If the organism being dated has died recently, the ratio of C^{14} to C^{12} will be the same as it is in the atmosphere today. The earlier in the earth's history the organism died, the greater will be the amount of C^{14} that will have decayed to C^{12} through the loss of energy and particles (Figure 20.2). Thus, the ratio of C^{14} to C^{12} will necessarily be lower. By using the rate of decay for C^{14} to C^{12}, which is already known, we can calculate how old the fossil must be. The age that we obtain is not completely accurate, but the range of error is small enough to make this technique a valuable tool.

Since not all fossils can be readily dated by radioactive methods, we must use some other methods as well. In studying fossil materials it has been found that certain fossils occur only in rocks of one particular age. These fossils are called *indicator species.* After several rock formations bearing these species have been dated by the radioactive methods, the presence of these indicator species may be used to identify other rock formations of the same age.

It is important to emphasize the detail sometimes found in the fossil record. Some plant fossils have even shown detail within the cells, and in a few cells, the stage of mitosis has been identified by the appearance of the chromosomes. The changes that occurred as mammals arose from reptiles are also recorded in fine detail in fossil form (Figure 20.3). There are fossils that show almost every minor change in skull structure in this sequence of animals.

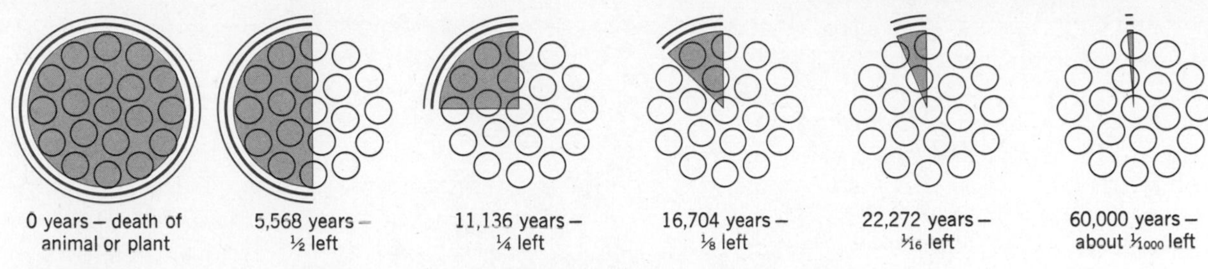

Figure 20.2. The proportion of carbon 14 remaining in fossils of different ages is represented by the blue segments. (Adapted from an illustration by Adolph E. Brotman for Time-Life Books, *1962, Time Inc.)*

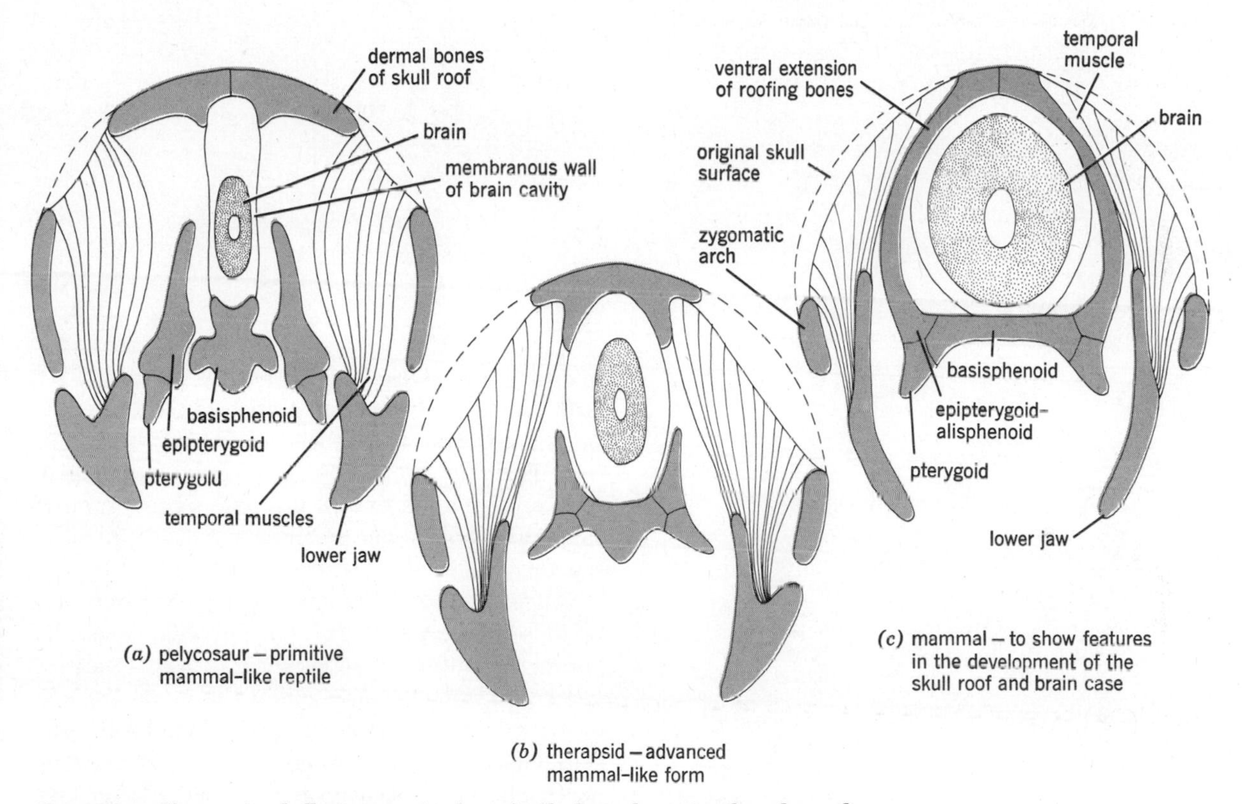

Figure 20.3. Changes in skull structure in three fossils from the series that shows the evolution of reptiles to mammals. Even in this abbreviated series, it is easy to follow the changes in the bones (blue). (After Alfred S. Romer, The Vertebrate Body, *third ed., 1962, W. B. Saunders Co., Philadelphia.)*

If all groups of organisms showed such fine fossil records, many of the problems concerning the history of life would be settled. Just as the record is rich in some areas, it is noticeably impoverished in others. The organisms that do not have hard parts (for example, a skeleton of some kind) such as most worms, algae, and jellyfish, do not often appear as fossils. The events that must occur for an organism to become a fossil account for this comparative rarity.

For an organism to become a fossil, it must be buried before it has decayed. Thus, the chance of fossilization is increased for an organism with hard parts, since these portions decay less rapidly. In

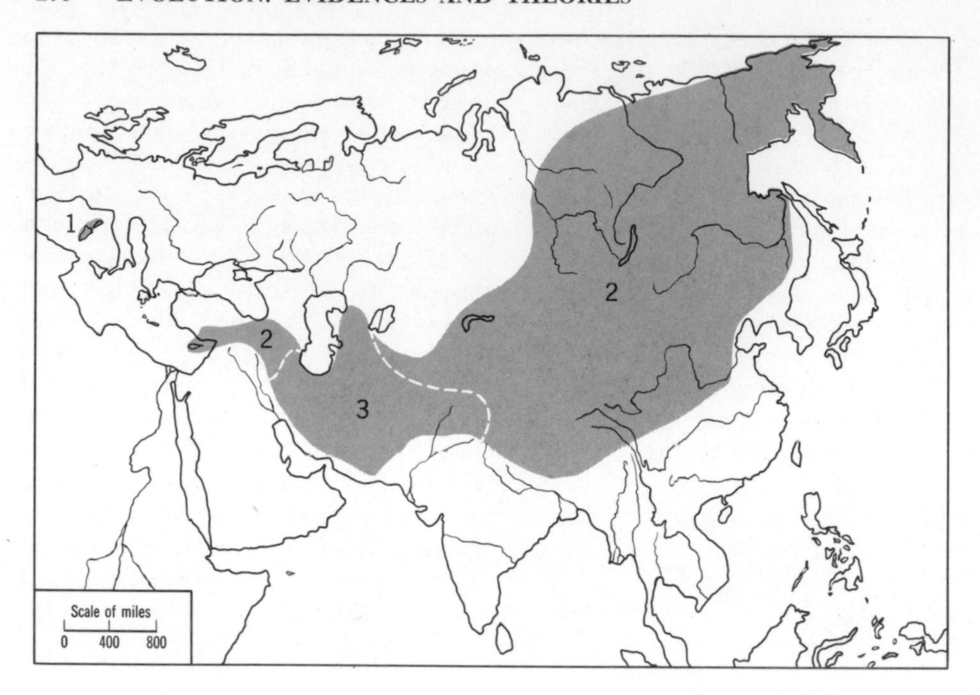

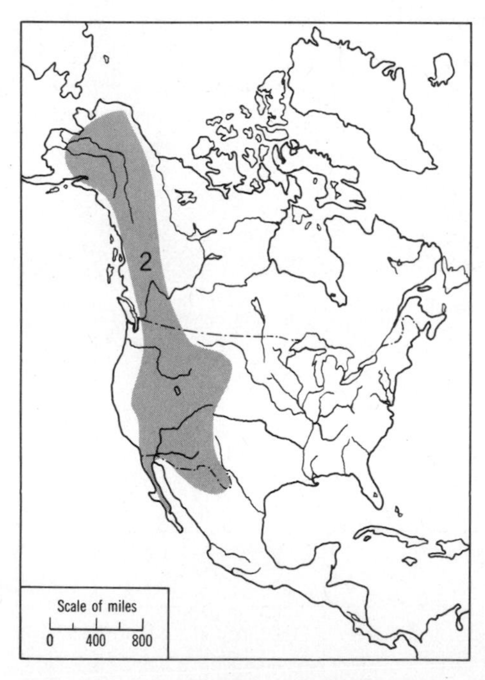

Figure 20.4. Present distribution of bighorn sheep. The numbers indicate how many species of bighorns inhabit each area.

addition, after burial, hard parts are better able to withstand the pressure and heat in the rock. Soft parts are frequently reduced to a thin layer of carbon. The chance of finding this thin layer is small. In many instances, even if it is found, little information is available about the organism the layer represents.

Biogeography. The study of the distribution of organisms over the earth is known as biogeography. Some distributions show patterns that are exactly as we would expect them to be if the various species of the group had originated from a common ancestor through a long series of minor changes. For example, nearly all species of sheep are represented in Central Asia, which seems to be the place of their origin. The area of origin is often the place where the majority of the species of any group live, because the longer the group has lived there, the more opportunity there has been for speciation. In addition, the older fossils of sheep and their relatives are found in Central Asia. As sheep spread north and east through Siberia during recent glacial times, they found a route to North America by way of the land bridge across the Bering Strait. The bighorn sheep of the Rocky Mountains is more closely related to the sheep in northeastern Siberia than to any of the other nine species of sheep. This is the distribution we would expect if the sheep, when isolated, differentiated into distinct species. (See Figure 20.4.)

Some patterns of distribution of closely related groups seem to argue against evolution from a common ancestor. However, as we examine all the evidence, including fossil remains and the habitat requirements of the species, these examples also show the exact distribution we would expect if evolution had occurred.

The distribution of magnolias presents a problem that is difficult to solve. As Figure 20.5 shows, closely related species are found in widely separated localities with no relatives anywhere between. How can this be? The fossil record gives us the key to this distribution. Earlier in the earth's history the magnolias were widely spread over Europe, Asia, and North America. During the glacial periods, the ice sheets forced the plants into small refuges in southeastern North America and Asia. In Europe, the Mediterranean Sea and the mountains of Asia Minor prevented the plants from moving farther south to areas in which they could have survived. After the glaciers retreated, the magnolias spread back to the north in Asia and North America but could not reach the areas of Europe they had occupied before because intervening areas were too dry or too high.

The plants and animals of extreme southern Florida also show an interesting and apparently contradictory pattern. Almost all of the plants and marine fish of southern Florida are related to forms found in the West Indies, whereas the birds, mammals, amphibians, and reptiles are nearly all related to forms found over more northerly parts of the United States. To explain this, we assume that the water between the West Indies and Florida was a *barrier* (preventing organisms from crossing) or a *corridor* (a means of distribution) depending on whether the organism could tolerate exposure to salt water. To further complicate the situation, some birds and many insects have come from the West

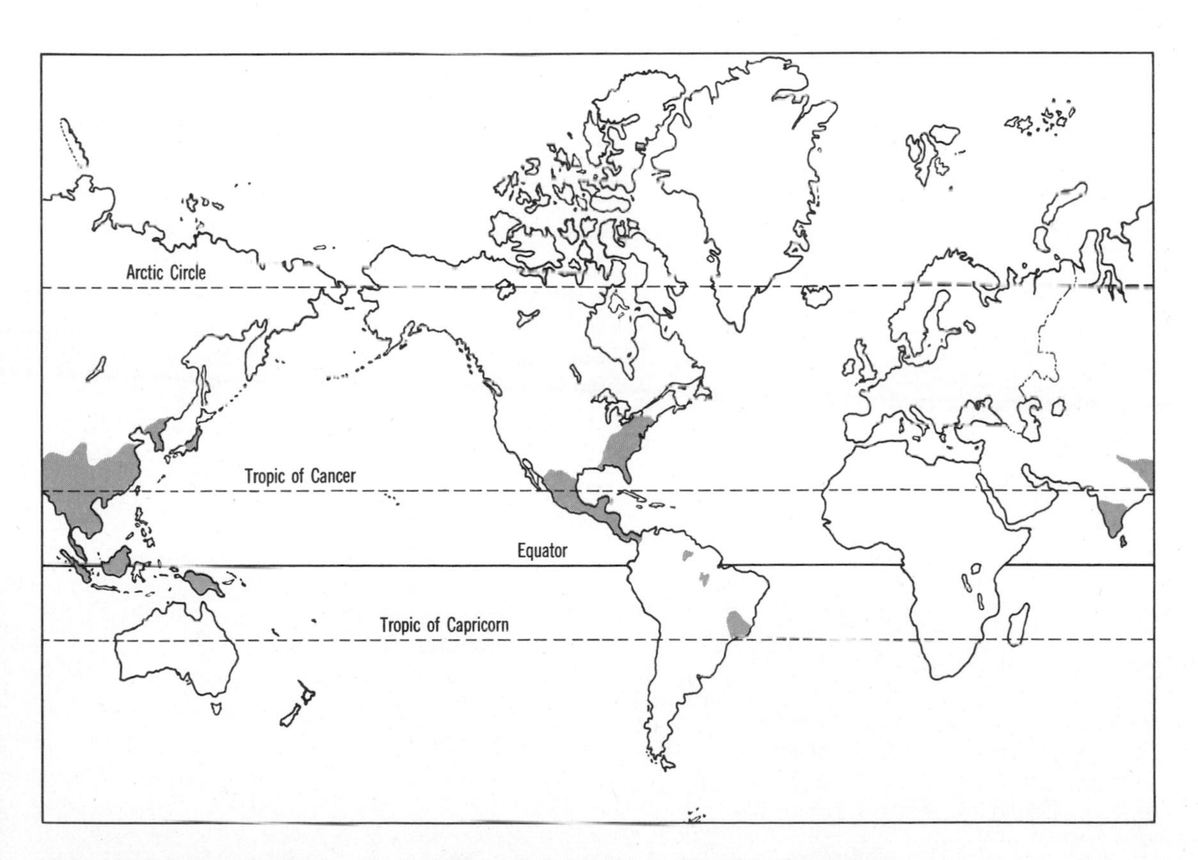

Figure 20.5. Present distribution of magnolias and their relatives. Note that the groups are widely separated.

Indies to southern Florida, many of them being blown across by hurricanes.

Comparative Structure and Development. Comparisons of the structure and development of organisms provide considerable information for use in deciding the degree of relationship among groups. This information is particularly helpful in working with groups that are not well represented in the fossil record and with relationships among major groups (phyla) where fossils are few.

Most comparative information is obtained from studying the structures of embryonic and adult organisms. We assume that if evolution has occurred, those groups that have developed more recently from a common ancestor should resemble one another more closely in their development and adult structure than those that have been distinct groups longer.

A careful examination of the anatomy of any group of related animals shows that they are always organized around a basic body plan of some sort. Among the vertebrates for example, the basic architecture of a limb is the same, although the limb is often highly modified for a special function. Thus the forelimbs of a frog, alligator, bird, elephant, bat, and man are all based on a five-digit appendage containing a humerus, radius and ulna, wrist bones and hand bones (Figure 20.6). The same general musculature, blood vessels, and nerve innervations can also be traced in these animals.

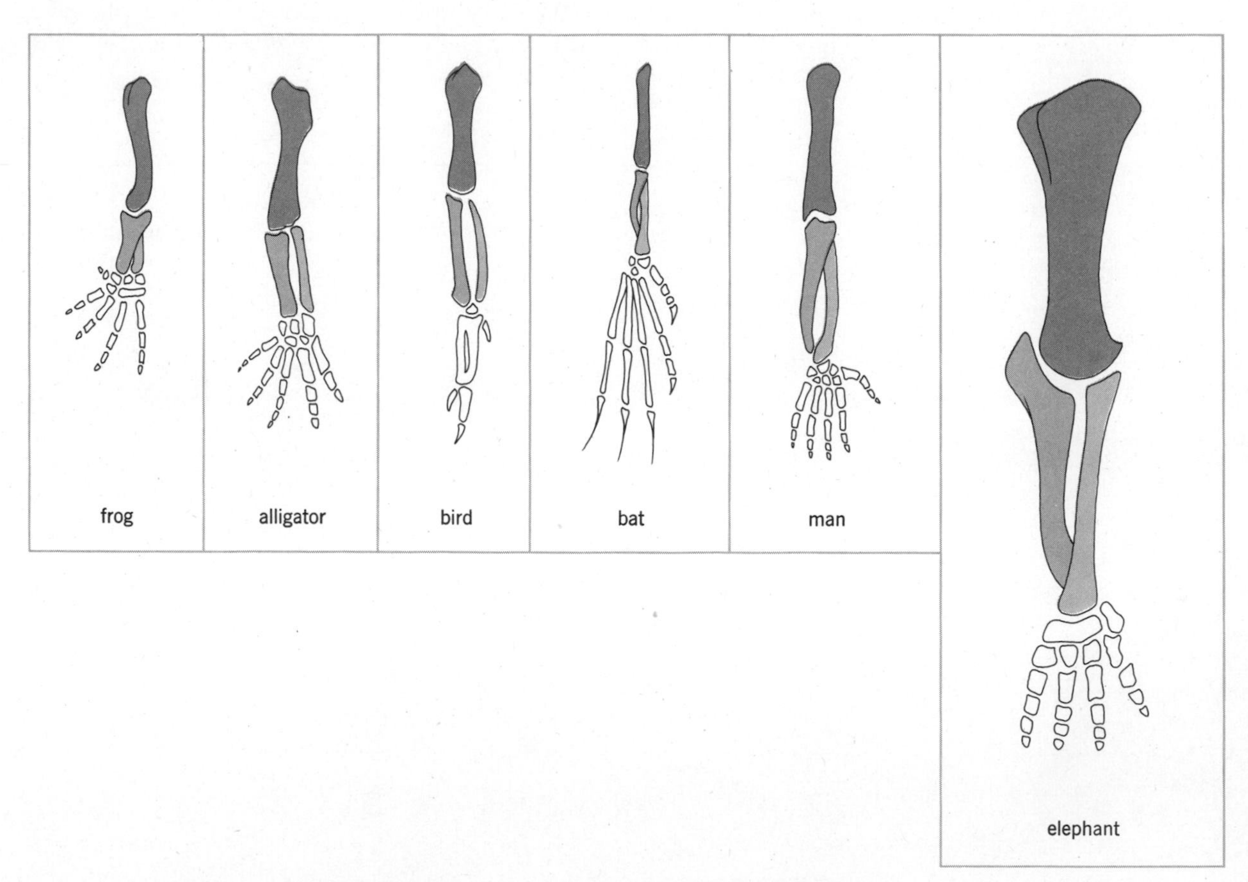

Figure 20.6. The skeletal plan of the forelimbs of a series of vertebrates. Note that the basic plan is the same for the different representatives; however, each limb is modified for a specific function. In each diagram, the darkest blue element represents the humerus, the lightest the radius, and the intermediate shade the ulna.

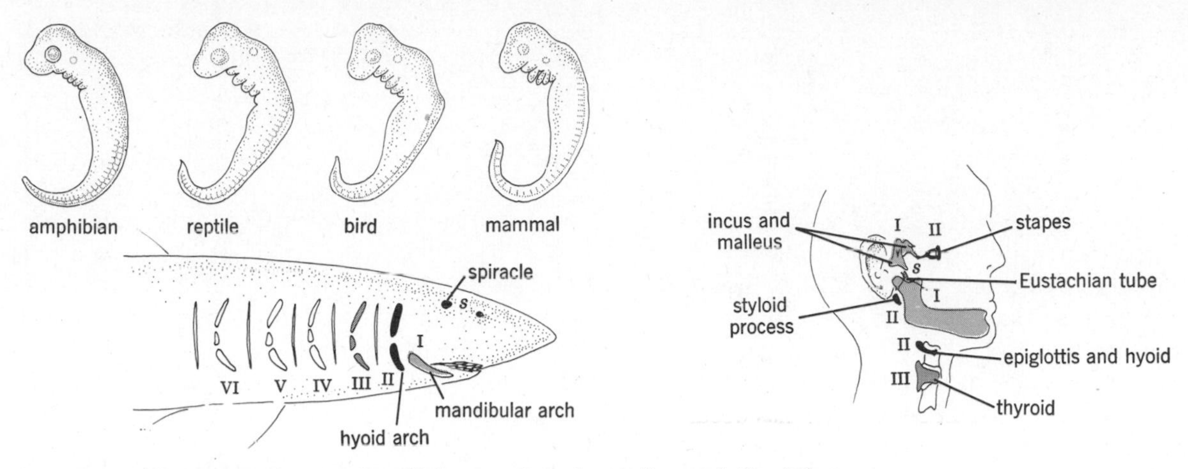

Figure 20.7. All vertebrates have similar gill slits as embryos (upper diagrams). The different fates of the structures associated with the gill slits in a shark and in man are illustrated in the lower diagrams.

This basic limb plan is not clearly evident in a highly specialized animal like a horse until we examine the animal's early developmental stages. At this time many structural features appear in generalized form before the animal develops the special adaptations that characterize it as a horse. For this reason, embryology has been one of the most useful means of unraveling confusing relationships among some animals.

Structures that exhibit a similar basic form and have a common origin in the embryo are termed *homologies*. The vertebrate forelimbs listed earlier are all homologous structures. If we examine the basic limb plan of insects, we find no resemblance, anatomically or embryologically, to the vertebrate limb. The term *analogous* is applied to structures that are used for similar functions but that are not basically alike and have not arisen from the same embryonic source. Biologists interpret analogous structures to signify an absence of close kinship between two groups. This interpretation is reinforced when other anatomical systems and developmental patterns are compared in the two groups.

Some seemingly nonadaptive structures occur in the embryos of higher organisms, but through modification become distinct adaptive structures in the adults of different groups. The gill slits of vertebrate embryos are a good example (Figure 20.7). In fish, most of the gill-slit structure develops into gills and gill slits in the adults. The embryos of amphibia, reptiles, birds, and mammals also have gill slits, but these do *not* develop into gills (except in some of the amphibians). In later development, parts of these structures become modified into parts of the ear and parathyroid and thymus glands. Not only does this evidence indicate a close evolutionary relationship among the many vertebrate groups, but it also seems nearly inexplicable on any other grounds.

Relationships among plants are also studied by examining their structure and development. Some plants of the nightshade family superficially appear to be quite different from others. Yet tomatoes, potatoes, tobacco, nightshade, and jimsonweed (*Datura*) all belong to this family. The flowers and fruits of these groups of plants are homologous (Figure 20.8). These and other similarities support the idea that these plants are all closely related and should be included in the same family.

One caution must be repeated about the interpretation of similarities. The two preceding examples illustrated similarities in basic structures that are present because the organisms had a common ancestor. Distinctions among the groups have arisen through natural selection in different environments, a pattern of evolution we call *divergence*. Plants, like animals, sometimes show convergence. For example, two groups of desert-adapted plants—the cacti and the euphorbs—both have spiny, water-storing,

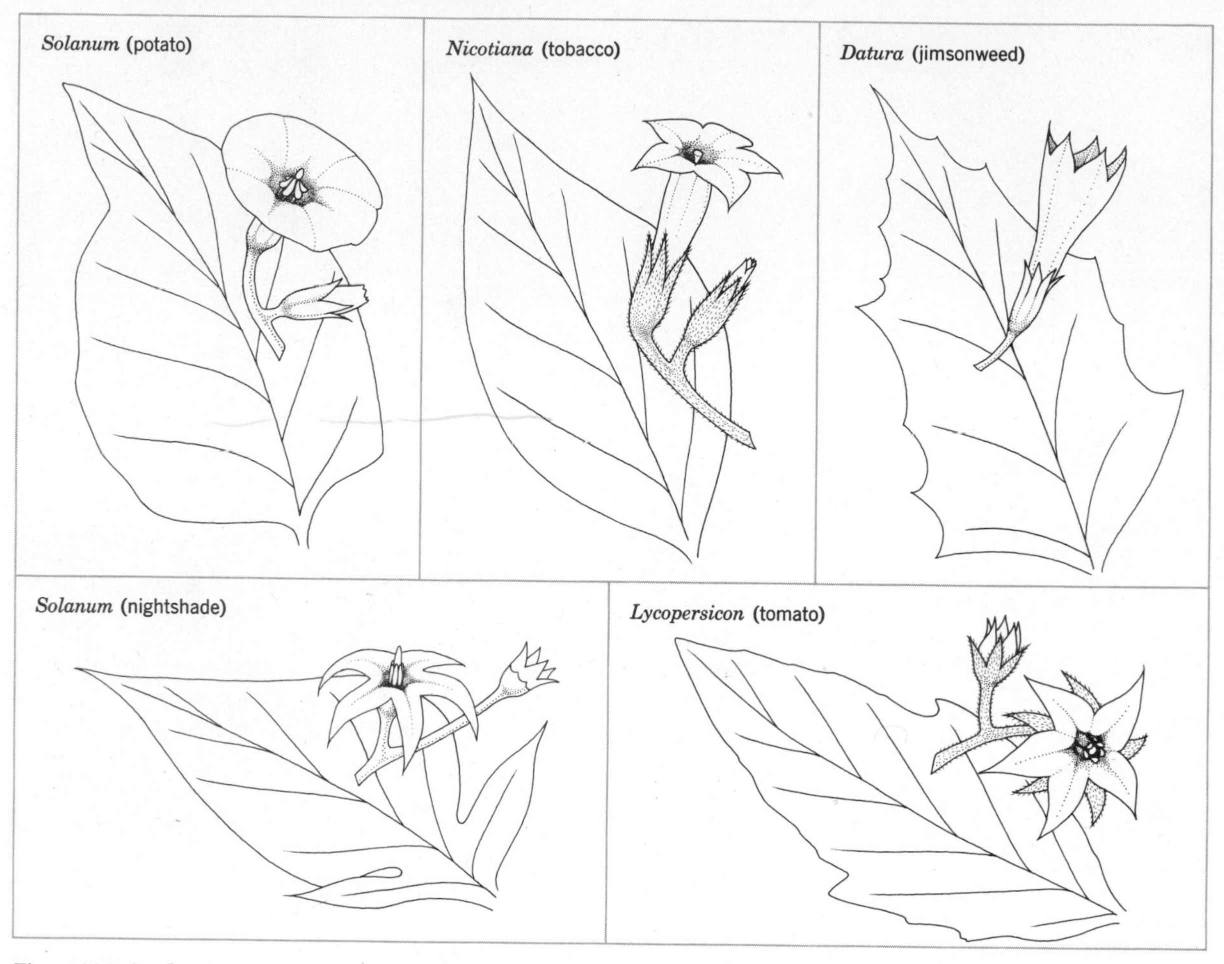

Figure 20.8. Similarities in flower structure show that these plants are all members of the same family.

leafless stems (Figure 20.9). Examination of the flowers shows that the two groups are unrelated (Figure 20.10); they have converged through selection by similar environments although they exist on different continents and spring from different kinds of ancestors. The similarity of the two groups results from selection for the harsh desert environment.

Within the cacti and euphorbs at least two different kinds of spines are found. In the euphorbs, the spine is a reduced leaf, whereas at least some of the cacti have spines that are modified bud scales and therefore not as closely related to foliage leaves. The tissue of the stem that stores water is also of two different types. In euphorbs, it is derived from the cortex, a tissue that would in most plants be only a thin layer within the structure we call bark. In cacti, this water-storage tissue is derived from the pith in the center of the stem and other tissues nearby.

Comparative Biochemistry. In recent years a number of biochemical tests have been devised to help study the degree of relationship among groups of organisms. If blood serum from a human is properly injected into a laboratory animal such as a rabbit, the rabbit's blood system builds antibodies against the proteins that were in the human blood. When this rabbit's serum and human serum are mixed in a test tube, a precipitate forms. If serum from an animal closely related to humans, like a chimpanzee, is tested against the rabbit serum, a

Figure 20.9. A cactus (left) and a euphorb (right) that show marked similarities, the result of convergent evolution.

Figure 20.10. The flowers of euphorbs (a) *and cacti* (b) *are markedly dissimilar. This is evidence that the two groups are not closely related. (Photo of euphorbs courtesy CCM: General Biological, Inc., Chicago.)*

precipitate again forms but not as strongly. If serum from a very distantly related form, such as a cat, is tested, no visible precipitate occurs. By refining this antigen-antibody reaction and employing optical measuring devices, this test has been used to work out confusing animal relationships. For example, this test showed that the nearest relatives of whales are the even-toed hoofed mammals such as cattle, deer, and pigs. This kinship was not known from any other evidence.

When protoplasm from any source is analyzed for its chemical constituents, certain findings are always the same. Carbon, hydrogen, oxygen, and nitrogen comprise 99 percent of the elements; carbohydrates, lipids, proteins, and nucleic acids are the organic constituents; and water is the most abundant compound. In life forms from *E. coli* to humans, DNA not only directs control of cellular activities via enzyme synthesis, but appears to involve the same code of nitrogen bases in all organisms. Metabolic utilization of energy through the steps of cellular respiration is alike in nearly all living entities. Biologists interpret all of the similarities described here as indicating a fundamental kinship of life forms.

A consideration of the biochemistry of living matter often leads to speculation about the origin of life itself. Many scientists feel that life evolved from nonliving matter back in dim eons of time when the environment of the earth differed greatly from what it is today. Geological evidence indicates that the atmosphere once consisted of ammonia, methane, water vapor, and hydrogen. The presence of energy, in the form of heat, ultraviolet light, and lightning, caused various chemical reactions in this mixture. Some of these produced simple organic substances, such as carbohydrates and amino acids, to form in pools of water. Eventually more complex molecules resulted until self-reproducing entities occurred. These crude life forms, perhaps consisting of nucleic acids, used simpler organic materials for energy. Natural selection would presumably operate to make this population of molecular life increasingly better able to reproduce and survive. The ability to synthesize one's own energy supply became the next requirement for continued life since the supply of organic matter in the environment would eventually be depleted. A simple photosynthetic apparatus evolved and from then on the expansion of life was almost explosive. This rapidly increasing mass of life evidently altered the atmosphere greatly by adding oxygen to it as a byproduct of photosynthesis.

This hypothesis for the chemical origin of life is highly speculative since the event occurred so long ago and left no fossil record. Credit for this concept goes mostly to a Russian biochemist, A. I. Oparin. When his book, *The Origin of Life on Earth,* was translated into English in 1938, it aroused much interest as well as numerous experiments to test Oparin's idea that organic substances could be formed in an environment containing no life. The first success was obtained by Stanley Miller in 1953 at the University of Chicago. Miller confined a mixture of gases, like those thought to exist on the primitive earth, in a container and discharged electrical sparks through it to provide an energy source. Amino acids appeared in the chamber after a period of time. Since the 1950s, experiments similar to this have been performed many times by such widely known biologists as Melvin Calvin and Sidney Fox. A large variety of organic substances will form in this manner, including the purine and pyrimidine units necessary for constructing DNA and RNA. These experiments, of course, do not prove that such events happened billions of years ago or that they led to the evolution of living matter. On the other hand, this general concept is the most widely accepted hypothesis that biologists offer at the present time concerning the origin of life.

Classification. Over the hundreds of millions of years that life has existed on the earth, natural selection acting on available variability has brought about a tremendous number of different kinds of organisms. By the eighteenth century many scientists in Europe had given much thought to naming and grouping plants and animals by schemes that attempted to place similar organisms together. Most of their efforts failed. In the middle 1700s, however, the Swedish naturalist Carolus Linnaeus introduced a useful system for classifying organisms. He gave each organism two Latinized names—the first representing its genus and the second its species; for example, *Homo sapiens,* the scientific name for man. Linnaeus then ordered these names into a sequence of increasingly larger categories (see Table 20.1), each containing a greater variety of organisms than the preceding one. This system of branching categories allowed Linnaeus to indicate relative degrees of similarity. Thus, two organisms, such as the coyote and the red fox, placed in the same family are more

Table 20.1. Classification of Some Organisms

	Common Names				
	Domestic Dog	**Coyote**	**Red Fox**	**White-tailed Deer**	**Gray Squirrel**
Kingdom	Animalia	Animalia	Animalia	Animalia	Animalia
Phylum	Chordata	Chordata	Chordata	Chordata	Chordata
Class	Mammalia	Mammalia	Mammalia	Mammalia	Mammalia
Order	Carnivora	Carnivora	Carnivora	Artiodactyla	Rodentia
Family	Canidae	Canidae	Canidae	Cervidae	Sciuridae
Genus	*Canis*	*Canis*	*Vulpes*	*Odocoileus*	*Sciurus*
Species	*familiaris*	*latrans*	*fulva*	*virginianus*	*carolinensis*
	Leopard Frog	**House Fly**	**White Oak**	**White Pine**	**Pine Moss**
Kingdom	Animalia	Animalia	Plantae	Plantae	Plantae
Phylum	Chordata	Arthropoda	Tracheophyta	Tracheophyta	Bryophyta
Class	Amphibia	Insecta	Angiospermae	Gymnospermae	Musci
Order	Salientia	Diptera	Fagales	Coniferales	Polytrichales
Family	Ranidae	Muscidae	Fagaceae	Pinaceae	Polytrichaceae
Genus	*Rana*	*Musca*	*Quercus*	*Pinus*	*Polytrichum*
Species	*pipiens*	*domestica*	*alba*	*strobus*	*commune*

like each other than they are like any other organism that is placed in a different family. A third organism, the gray squirrel—dissimilar enough from the first two to be placed in a different family and a different order—still has enough similar characteristics to be placed in the same class with them.

When Linnaeus first proposed his systems, he classified the organisms on the basis of characteristics that seemed fundamental to him. The concept of evolution played no part in his choices. Since the process of evolution by natural selection was discovered, biologists have tried to select those characteristics that indicate a common ancestry for the organisms for use in classification. This work on classification modified Linnaeus' work but did not replace it. A modern version is presented in the Appendix. Because of the accumulation of knowledge about plants and animals, Linnaeus, without knowing anything about evolution, was able to select characters that indicated evolutionary relationships. This is one evidence for evolution that is obtained from classification.

As we compare the various kinds of evidence about any group of organisms, we find that almost all of them point to the same general kind of relationship. This coincidence is another strong argument for evolution.

Major Trends and Features of Evolution

Detailed presentation of any one theory of the evolutionary relationships among organisms is beyond the purpose and scope of this book. However, we shall discuss certain general characteristics of evolution that are found throughout the history of life.

Change. The most obvious trend in the evolution of life is that of change through time. Since environments are always changing, the species that inhabit

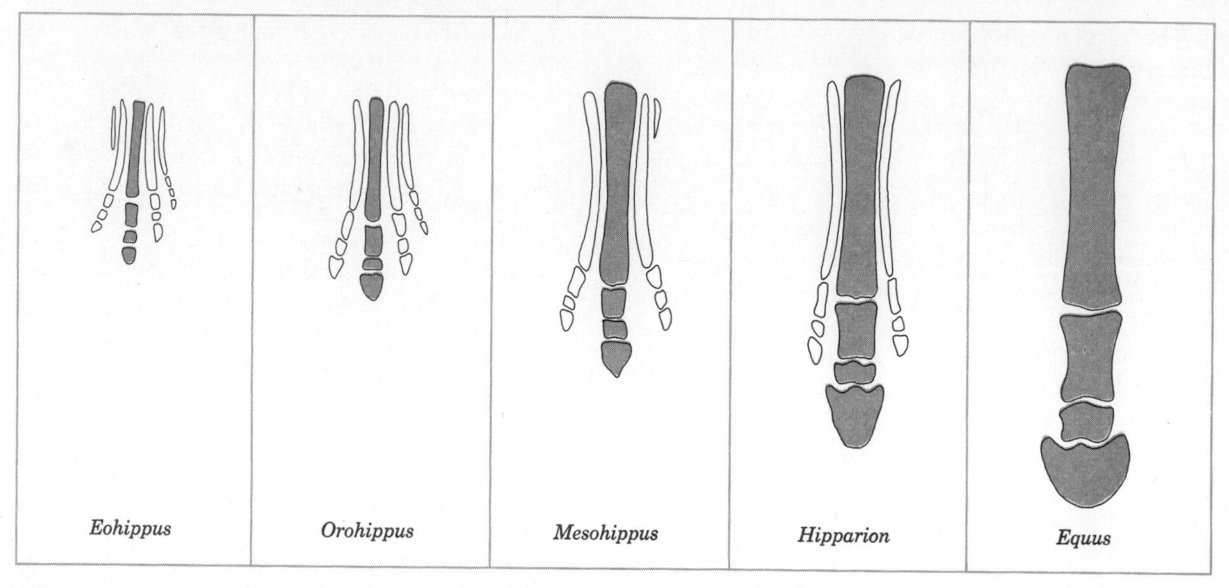

Figure 20.11. Diagrams illustrating stages in the evolution of the forefoot of the modern horse. Note the changes that have taken place from the five-toed ancestor Eohippus *to the contemporary species* Equus.

them must adapt to the changes in order to continue their existence. As we saw in the previous chapter, adaptations may take the form of striking morphological changes or subtle physiological adjustments.

Fossils sometimes provide a history of the changes that have taken place in a group of plants or animals. One of the best documented of these is the horse. Its evolution has been traced from a tiny, five-toed ancestor to the modern highly specialized creature we now term a horse (Figure 20.11).

Change occurs from genetic recombinations and the spontaneous, random occurrence of mutations. Natural selection determines whether these changes are beneficial or detrimental, since there is no assurance that change will occur when the species needs it.

Invasion of New Environments. A study of fossils and biogeography shows that plants and animals have continually spread into new environments throughout evolutionary history. Some of these invasions were enormously important, such as the emergence of the ancient group of fishes that gave rise to land animals. Invasions of new environments have often led to adaptive radiation, as described in Chapter XIX.

Increasing Complexity. With the invasion of additional habitats and progressive adaptation to them, new and increasingly complex structures appeared. As complexity increased, many organisms became specialized to utilize only a small part of their original environment. This allowed other species to utilize the remaining portions. Bats, the only group of mammals capable of feeding in the air, represent this type of specialization and complexity. An exception to the evolutionary trend to complexity is found in some parasites. These sometimes become progressively simplified in structure. Tapeworms, for example, lack a digestive tract, muscular system, and nervous system. On the other hand, internal parasites always have complex life cycles and highly specialized reproductive organs.

Expansion. The three trends we have described resulted in an enormous increase in the amount of life on the earth. As more environments were inhabited and organisms became increasingly adapted to various parts of them, the total number of species and individuals increased. This combination of trends allowed living things to utilize a growing percentage of the earth's total energy input. This energy input is the final barrier to expansion.

The total number of living things has increased through time but many groups of organisms first expanded only to become extinct. At one time the dinosaurs were the dominant type of animal life. They have been extinct for seventy-five million years. At about the time the dinosaurs began to decline, the number of mammals began to increase. Increases in groups like this have offset the decrease in other groups and maintained the expansion of life.

Evolutionary Time. A final major feature of evolution concerns the time scale that is involved. Evolution guided by natural selection can only occur if vast spans of time are available. This is necessary since most mutations do not benefit a population's gene pool. Only rarely does a mutation take place when it will facilitate an adaptive change in a group of organisms. Vast numbers of detrimental mutations occur before chance links an appropriate one with a needed adaptive change. Such a trial-and-error process obviously required a large time scale to produce the diversity of life we see today.

Has a sufficiently large span of time been available for evolution by natural selection? Various dating techniques place the age of the earth at four and one-half billion years or more. Life has probably been present for at least half of this time period, and most students of evolution feel that this is sufficiently long.

Theories of Evolution

All the evidence that supports evolution has not by itself yielded any explanations of the actual mechanism that governs the process. Men looked at the evidence available and constructed explanations to account for it. As new evidence accumulated and was evaluated by other scientists, the old theories were either reinforced or else discarded and replaced by others.

The success of each of the various theories of evolution has depended partly on the amount of evidence available to the man propounding a theory and partly on his attitude and those of the people judging his theory.

The germinal concepts of the theory of evolution were present in the minds of many men before 1800. These ideas did not attract much attention because the evidence to support them was sparse and the prevailing ideas of the time were against them. Nevertheless, these theories did cause some men to look for evidence and to think about the problem of the history of life.

Lamarck's Theory. Shortly after 1800, Jean Baptiste Lamarck attempted to explain evolution by his theory of the *inheritance of acquired characteristics*. Briefly summarized this theory states the following:

1. The environment introduces a *need* for some structure in the organism.
2. The organism attempts to meet this need.
3. In response to its efforts the structure of the organism is changed.
4. The change in the structure of this organism is passed on to its offspring.

The classic example used to explain this theory is the development of a long neck in giraffes. Short-necked ancestors of the giraffe needed to reach foliage in the trees in order to obtain food. To meet this need they stretched their necks as far as possible. Stretching the neck resulted in longer necks in the offspring. The continuation of this process through many generations resulted in the long-necked giraffe as we know it today. Although this is untenable in view of our current understanding of heredity, no information on heredity was available to contradict the theory when Lamarck propounded it.

The idea that environment causes changes in organisms is an attractive one. It was a popular and widespread concept in Lamarck's day, even influencing some of the writings of Charles Darwin. Even today, many uninformed people turn to this idea to explain heredity and adaptation.

Darwin's Theory. By the time that Charles Darwin completed his investigations in 1858, two important works in other fields had provided clues to another solution. To Charles Lyell, the origin of the geological formations of the earth did not need to be considered mysterious but rather as something that could be understood in terms of the natural processes that affect these same forms today. This means that the changes in the earth's surface have been brought about by uniform physical forces acting throughout the history of the earth. No cataclysmic events need to be postulated to account for the earth's surface features. In addition to explaining the appearance of the surface of the earth, Lyell's theory implied that the earth was considerably older than had been supposed up to that time.

Thomas R. Malthus, an economist, published an essay on the relationship of human populations to the resources supporting them. He stated that since resources increased arithmetically and populations increased geometrically, some forces obviously operated to check population growth. He listed war, famine, and pestilence. In this, Darwin found the idea of competition between organisms for the available resources.

In the 1830s, Darwin spent five years as the naturalist on a round-the-world expedition. During this trip he noticed variation among organisms and the adaptability resulting from some of the variations. Over a number of years of further study he developed a theory of the origin of species by *natural selection.* This theory contained four major points.

1. Within a population, individuals show considerable variation. For instance, not all seeds of a species of plant are exactly the same color; some are a little lighter than the average, others a little darker.

2. Populations tend to produce more offspring than can possibly survive, because the available resources are limited.

3. The offspring must compete for the resources that are available.

4. Of the offspring, those individuals that are best fitted will survive. Because the selection of the individuals that are to survive is made by the environment, Darwin called this natural selection.

Modern Modifications. Darwin, like Lamarck, had little understanding of the basis of variation even though Mendel was his contemporary. With the rediscovery of Mendel's work in about 1900, the physical basis for this point of the theory was supplied. It is now possible to discuss sources of variations and the means of selection in a meaningful manner, as we have done in Chapters XVIII and XIX.

Although Darwin's four points are still used as a general statement of the means by which organisms evolve, the meaning of the phrases is somewhat different from what it was in Darwin's time. By survival of the fittest, Darwin meant the survival of those specimens that were largest and strongest. Consequently, the discussion of competition tended to be limited to physical strife.

We now think of survival of the fittest in the sense of organisms best suited to leave the most offspring. It is not merely the ability to survive but the ability to survive and reproduce that is important. The combination of genes that enables its bearer to produce the highest number of offspring will be found most frequently in subsequent generations.

Competition is no longer thought of in the narrow sense of direct physical combat. It involves the utilization of necessary resources and acquisition of reproductive partners. The competing organisms may never see each other, or, like plants, may be quite incapable of combat. Such competition is less bloody but no less real and effective.

Why do most biologists believe that this explanation of evolution is the correct one? On the basis of the evidence, it seems to offer the most reasonable theory to account for what is at hand. It will indeed be modified by new evidence in the future but probably will not be completely discarded. Too much available evidence fits well with this theory for us to suppose that complete rejection will be required.

Principles

1. Charles Darwin proposed a theory of evolution based on variation, competition, and consequent natural selection.

2. The basic mechanism of evolution is now known to be changes in gene frequencies of populations through time, guided by natural selection.

3. Evidence for the historical development of life is derived from fossils and comparative studies of organisms.

Suggested Readings

Cain, A. J., *Animal Species and Their Evolution.* Harper and Row, New York, 1960.

Deevey, Edward S., Jr., "Radiocarbon Dating," *Scientific American,* Vol. 186 (February, 1952). Offprint No. 811, W. H. Freeman and Co., San Francisco.

Eglinton, Geoffrey, and Melvin Calvin, "Chemical Fossils," *Scientific American,* Vol. 216 (January, 1967). Offprint No. 308, W. H. Freeman and Co., San Francisco.

Eiseley, Loren C., "Charles Darwin," *Scientific American,* Vol. 194 (February, 1956). Offprint No. 108, W. H. Freeman and Co., San Francisco.

Fox, S. W., and R. J. McCowley, "Could Life Originate Now?" *Natural History,* Vol. LXXVII, No. 7 (August–September, 1968).

Harris, W. J., "The Origin of Life—A Master Molecule?" *Advancement of Science,* Vol. 24, No. 121 (March, 1968).

Hurley, P. M., "Radioactivity and Time," *Scientific American,* Vol. 181 (August, 1949). Offprint No. 220, W. H. Freeman and Co., San Francisco.

Keosian, John, *The Origin of Life.* Second edition. Reinhold Publishing Corporation, New York, 1968.

Kettlewell, H. B. D., "Darwin's Missing Evidence," *Scientific American,* Vol. 200 (March, 1959). Offprint No. 842, W. H. Freeman and Co., San Francisco.

Kurten, Björn, "Continental Drift and Evolution," *Scientific American,* Vol. 220, No. 3 (March, 1969).

Lorenz, Konrad Z., "The Evolution of Behavior," *Scientific American,* Vol. 199 (December, 1958). Offprint No. 412, W. H. Freeman and Co., San Francisco.

Newell, Norman D., "Crisis in the History of Life," *Scientific American,* Vol. 208 (February, 1963). Offprint No. 867, W. H. Freeman and Co., San Francisco.

Seilacher, Adolf, "Fossil Behavior," *Scientific American,* Vol. 217 (August, 1967). Offprint No. 872, W. H. Freeman and Co., San Francisco.

Simpson, George Gaylord, *Life of the Past, an Introduction to Paleontology.* Yale University Press, New Haven, Conn., 1964.

Stebbins, G. Ledyard, *Processes of Organic Evolution.* Prentice-Hall, Englewood Cliffs, N.J., 1966, pp. 132–175.

Questions

1. Are footprints or leafprints from the past considered fossils? Why?
2. Why are indicator species useful for dating rock formations?
3. Fossilization is often called a rare event, yet fossils are abundant in some areas. How do you explain this?
4. How does biogeography provide evidence to support evolution?
5. What is the basis for the idea that similar structures in organisms indicate relationship?
6. Give an example of convergence among plants.
7. Dehydration synthesis and hydrolysis are universal chemical events in the living world. In what category of evidence would you place this?
8. Why are discussions on the origin of life more speculative than those on other areas of evolution?
9. List the major trends in the course of evolution.
10. Explain the meaning of "Lamarckian evolution."
11. How did Darwin describe natural selection? Describe natural selection as viewed by modern biologists.
12. The establishment of evolution as a biological principle is said to be the major unifying concept for all areas in biology. Can you defend this statement?

Ecosystems: Physical Aspects and Structure

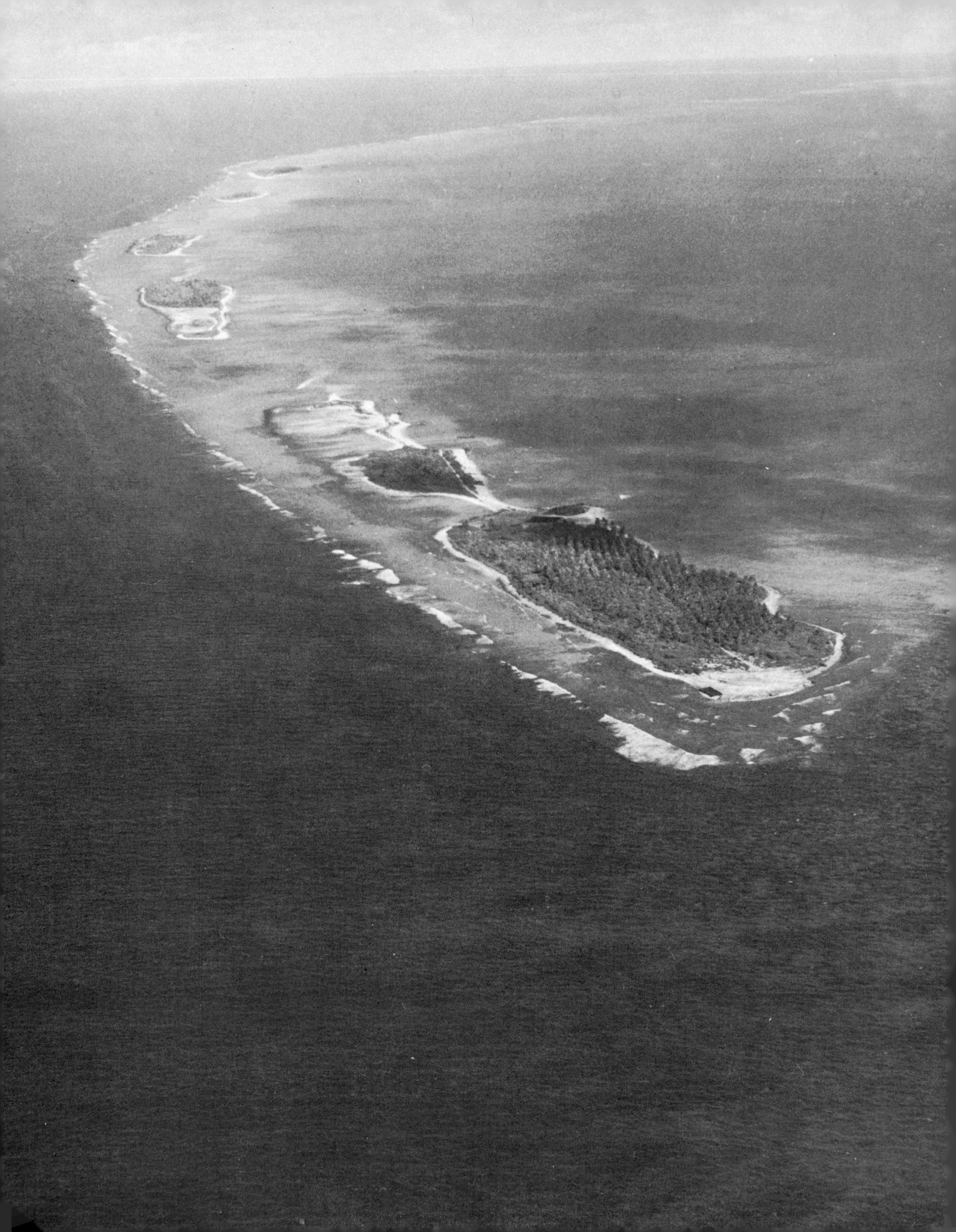

Aerial view of coral atoll.

CHAPTER XXI

Ecosystems: Physical Aspects and Structure

The final five chapters of this book, beginning with this one, present some of the major concepts derived from studies that view organisms and their environment as a biological unit. This area of biology is known as *ecology*. In a sense, we introduced the ecological approach in the chapters on population genetics and evolution, since they dealt with populations interacting with their environments through natural selection. These topics provide the historical basis for understanding many of the organismic-environmental relationships we see at the present time.

Most people feel at home with the topic of environment, since it is part of their everyday experience. Everyone has observed a number of different environments, from woodlands to bodies of water. In fact, we can step into our backyard and immediately observe a community of grass, shrubs, and trees reasonably well adapted at that particular moment to its physical and biotic surroundings. It even seems easy to predict what will happen in this artificial community if environmental conditions such as temperature and rainfall change drastically. Studying the environment may seem to be simply a matter of measuring all of these factors and noting their influence on the organisms in it. This is a deceptive assumption, as biologists have demonstrated many times. For example, the success or failure of a backyard community may depend on subtle environmental fac-

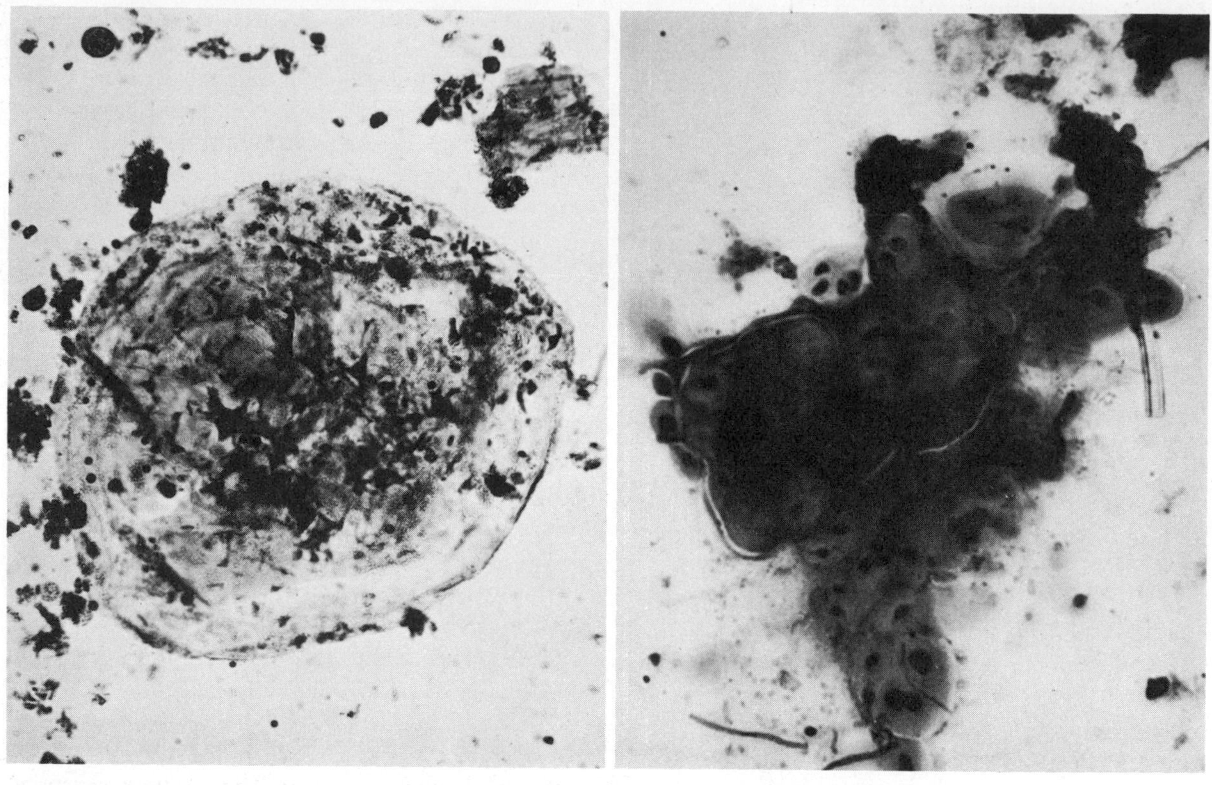

Figure 21.1. Cysts of a protozoan (left) and a blue-green alga (right). These cells have thick protective walls that enable them to withstand extreme temperatures.

tors like the presence or absence of certain trace elements in the soil, the pH (acid-base relations) of the soil, the amount of humus present, and the kinds of insect populations that may appear, to name only a few. In addition, organisms themselves change the environment in many ways, for example, by increasing the moisture content of the soil or by fixing nitrogen (changing atmospheric nitrogen into nitrates that can be utilized by plants). Understanding the backyard community requires an investigation of not only the influence of environment on organisms but also the influence of organisms on environment. Furthermore, the reciprocal interactions of communities of organisms and their environment results in varying degrees of self-regulation. This mechanism enables a community to function smoothly despite fluctuation in the physical environment caused by weather, seasons, etc.

It is especially important for us to realize that man is not only a manipulator of his environment but also a functional *part* of the environment, since he depends on biological processes for food, water purification, oxygen, and many other of his vital needs. To achieve short-term gains by modifying natural environments, he must become aware of the long-term effects of his power to bring about radical changes in the living world. He must learn to interact properly with the environment or suffer drastic consequences. Our heritage of dust bowls in the plains, of polluted rivers, and extensive soil erosion are examples of man's failure to consider the intricacies of natural communities. One of the objectives of this section is to help you make intelligent decisions about ecological matters that affect all citizens: air pollution, sewage disposal, drainage projects, and the like.

Ecology has been traditionally defined as the study of relationships between organisms and their environment. A modern viewpoint, enunciated by a contemporary ecologist, Eugene Odum, defines ecology as "the study of the structure and function of nature." This definition attempts to avoid the distinction between organisms and environment by treating them as an integral unit. This is the viewpoint we propose to develop. First, we need to present some background material on the major physical aspects of the environment (*abiotic factors*) so that we can better develop the modern concept.

Physical Aspects of the Environment

Heat and Light. *Solar radiation* is the most important of the abiotic factors because it supplies the energy that supports all life. It reaches the earth as sunlight, a mixture of short and long wavelengths partially screened or filtered by the atmosphere. The ecologically important wavelengths are infrared (heat), light, and ultraviolet.

Heat, recorded in terms of temperature, is obviously important in the activities and distribution of many living things. Generally speaking, all living things operate most efficiently at temperatures somewhere between the freezing and boiling points of water. Thus, a considerable portion of the earth, including all the oceans, presents hospitable temperatures for plants and animals. Most forms of life lack an internal temperature-controlling system; they are *poikilothermal.* These forms often compensate for temperature changes with behavioral activities and other adaptations. Generally, poikilothermal forms function best at temperatures between 41° and 93° Fahrenheit, a fact that limits their distribution. A few forms of life exhibit extreme heat hardiness (tolerance), such as the blue-green alga which lives in hot spring water of 162°F in Yellowstone National Park. A few also have extreme cold tolerance, particularly certain plant seeds, spores, and protozoan cysts which survive temperatures far below the freezing point. (See Figure 21.1.)

Only birds and mammals are *homoiothermal,* that is, have a relatively uniform body temperature. They are less restricted in their distribution by temperature extremes; however, their metabolisms require more energy, and this may become a limiting factor. A few groups like bats are able to exploit the advantages of both poikilothermy and homoiothermy by alternating these states to adjust to different temperatures or different seasons.

Forms of life that inhabit areas subject to drastic changes of season frequently show adaptations for temporarily avoiding detrimental extremes. Many plants become *dormant* during cold or very dry seasons by shedding their leaves. Examples include maples, oaks, hickories, and other broad-leafed trees of our temperate forests. In many grassland areas of the world, the broad-leafed trees often shed their leaves during the dry seasons. In northern regions the amphibians, reptiles, and a few mammals such as ground squirrels and woodchucks *hibernate,* that is, become physiologically inactive during the winter. In hot, arid areas the insects, spiders, and reptiles enter a similar inactive state termed *estivation* during the hottest seasons.

Organisms display many additional adaptations to heat or cold, including body size, length of extremities, coloration, number of young, and other physiological manifestations (Figure 21.2). Many years ago naturalists observed that many species of mammals and birds averaged smaller body sizes in southern populations than in northern ones. A good example is the familiar White-tailed Deer. In the northern United States, mature bucks may weigh over 300 pounds, whereas in the Carolinas they average about 200 pounds. The smallest are found in the Florida Keys, where males seldom exceed 80 pounds. Moreover, exposed body parts such as ears and tails tend to be shorter in cold climates, and coloration tends to be lighter.

Figure 21.2. The Arctic fox maintains a constant body temperature despite the cold temperatures of its environment. The short ears and snout help reduce heat loss (© Walt Disney Productions).

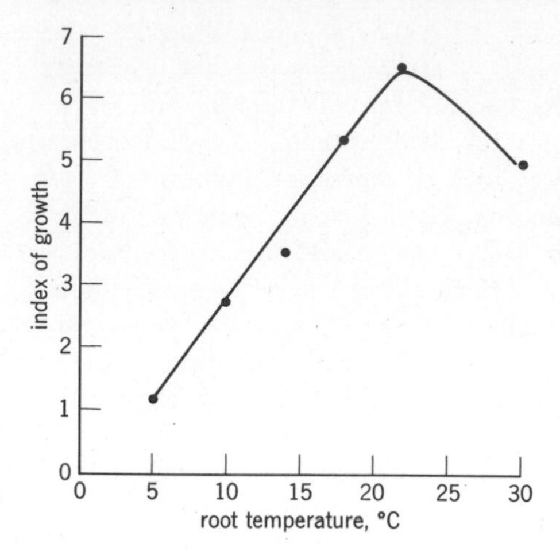

Figure 21.3. Effects of temperature on rate of development of some roots. An increase in temperature results in faster growth up to about 22°C. (Data from A. H. Gibson, Aust. Jour. Biol. Sciences, 19(2), 1966.)

We might expect metabolic reactions to be considerably influenced by temperature. This is indeed the case. For example, the rate of development of eggs or the growth of plant seedlings usually shows a definite relation to temperature, as indicated in Figure 21.3.

Another important aspect of solar radiation is *light,* a factor with complex ecological effects. Probably the most obvious and most important role of light is in relation to photosynthesis, described in Chapter IV. Another indication of the importance of light is the fact that virtually all forms of animal life have evolved light receptors of some sort. In terms of their response to light, animals can be divided into *nocturnal* (night-active), *diurnal* (day-active), and even *crepuscular* (twilight-active) types. A majority of mammals are nocturnal, the majority of birds are diurnal, and bats, many moths, and a few birds are crepuscular. The color patterns throughout the animal kingdom are another indication of the importance of the receptor-light relationship and are related to the functions of protection and mating (Figure 21.4). Even plants, although

Figure 21.4. Protective coloration in an insect (Dr. E. S. Ross).

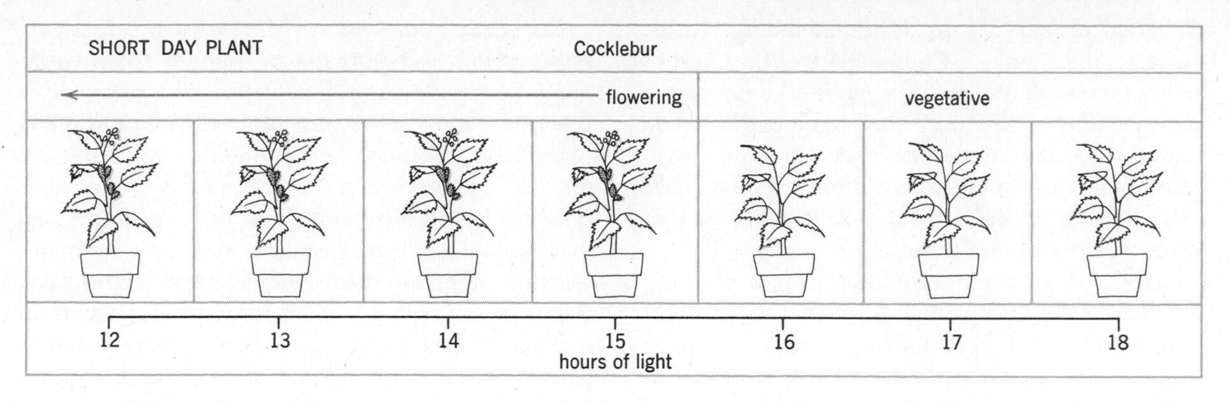

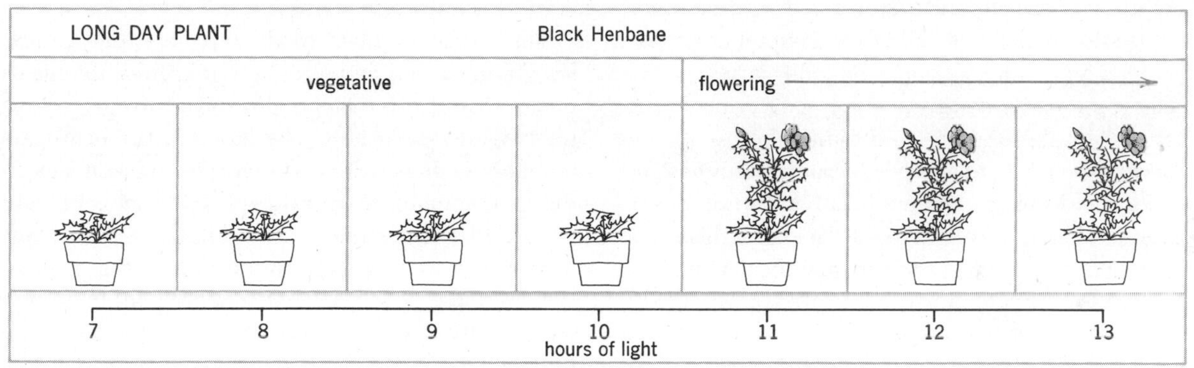

Figure 21.5. The photoperiodic response of a short-day and a long-day plant to varying periods of light per twenty-four hour day. The Cocklebur will flower only if the hours of light do not exceed fifteen. The Black Henbane, on the other hand, requires at least eleven hours.

they lack specific light receptors, respond to light by growth movements.

In another type of light reaction, many plants and animals exhibit seasonal changes in response to the number of daylight hours in a twenty-four-hour cycle; this response is called photoperiodism. In plants, it often determines the flowering and fruiting time (Figure 21.5). In animals the breeding cycle often functions in relation to seasonal day length. This is well illustrated in birds, since their gonads are extremely small during winter months and then enlarge greatly during the spring of the year (see p. 119). Experiments have related such biological events in birds to day length (Figure 21.6). This interaction sounds simple until we consider that it involves hormonal changes occurring in response to day length as perceived through the eyes of the bird. In other words, this is the sequence: eye → hypothalamus → pituitary → gonads.

Ultraviolet light is important in two respects. First,

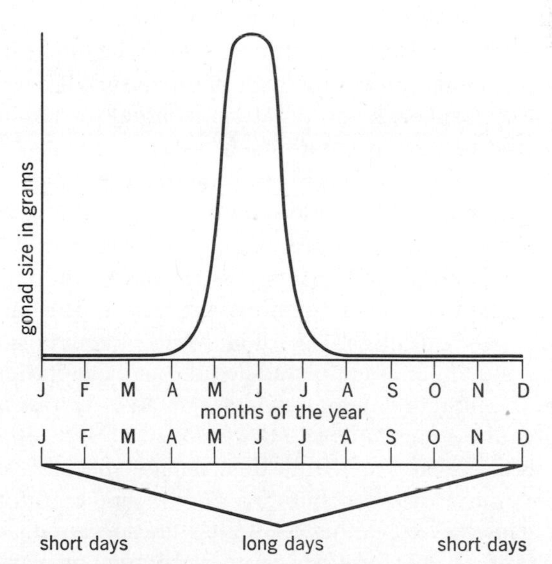

Figure 21.6. In many species of birds, long day lengths act through neuro-hormonal pathways to stimulate a growth in gonad size.

it is lethal to many forms of life, including viruses, bacteria, and fungi. Consequently, the earth's shielding layer of atmospheric gases is an important ecological factor. Secondly, the wavelengths that do reach the earth's surface are necessary to the synthesis of vitamin D in the skin of birds and mammals. In the absence of this vitamin, rickets, a disease that causes defective bone formation, occurs.

Gases. The atmospheric gases *oxygen* and *carbon dioxide* are vital components of an organism's surroundings. Animals and plants constantly remove oxygen, while photosynthesis restores it as fast as they use it. For soil- and water-dwelling organisms, the situation is different. These environments contain far less oxygen and its concentration varies.

Carbon dioxide constitutes only a fractional part of the atmosphere, yet is vital to the photosynthetic process as well as to many physiological reactions in plant and animal tissues. It is readily soluble in water, where it forms carbonic acid. In this form it influences the pH of the surroundings or may be combined with calcium to help form bones, shells, and limestone formations such as coral reefs. Carbon dioxide reaches the environment from the respiration of living things. Moreover, industrial processes release large amounts of it into the atmosphere. Here we encounter a consequence of human activity that could change the environment of an entire planet. The smog encountered in London and California provides an uncomfortable example of the kinds of atmospheric changes that do take place.

Water. Water constitutes a major factor in the environment and in the lives of organisms. It enters the atmosphere by evaporation, eventually returning to the terrestrial environment as rain or snow. As lakes, rivers, and oceans, it forms major habitats for living things. As ice and snow, or as moisture in the atmosphere, it influences many aspects of plant and animal ecology. Of course, water is also the most abundant compound within living tissues. The basic anatomy and physiology of all forms of aquatic and marine life are highly modified through adaptations to the liquid medium in which they live. Terrestrial organisms must adapt to the problem of preventing excess loss of water from their bodies; special body coverings, various respiratory devices, and excretory organs may serve this need. Dehydration is a major limiting factor for life in terrestrial communities just as the absence of oxygen limits life in aquatic environments.

Chemical Substances. Chemical substances, both organic and inorganic, are important environmental factors in soil and water. One category of these substances, dissolved salts, is vital for sustaining life. These substances are termed *biogenic* salts or nutrient materials and include nitrogen and phosphorus salts as well as potassium, calcium, sulfur, and magnesium. They are primarily necessary for maintenance of green plants which provide the basic source of nutrition for other forms of life. The term *nutrient* is misleading, since these salts are used in various parts of the photosynthetic process rather than as a source of energy.

The so-called "plant food" (fertilizer) that can be bought in garden shops contains the three chemicals that are most often lacking in soil—nitrogen, phosphorus, and potassium. The label on the bag bears numbers such as 6–8–4, which refer in sequence to the percentages of nitrate, phosphoric acid, and potash in the fertilizer. This knowledge of plant nutrients is tremendously important in the production of food crops, since soils that are poor in certain elements can be fortified by the application of the proper fertilizer. These nutrients function in the same way for freshwater and marine plants. When ordinary commercial fertilizer is applied to a small lake, the microscopic plant life soon becomes so abundant that the water turns cloudy. If fertilizer is added seasonally, the yield of animal life (fish) increases greatly. This has been done in salt water bodies along the coast of Scotland to increase the numbers of flounder.

This presentation has barely touched on the vast topic of chemicals in the environment, but perhaps now the complexity and importance of these materials in nature becomes clear. In Chapter XXII we shall consider the cyclical movement of chemicals through the environment.

Structure of the Ecosystem

Let us return to Odum's modern definition of ecology—*the study of the structure and function of nature.* In this definition, nature includes all of the populations of an area functioning with their nonliving environment. Nature is thus perceived as an ecological system, in other words, an *ecosystem.* This is an extremely useful concept since it is feasible to study the structure and function of an ecosystem in a quantitative manner.

The term ecosystem usually refers to a relatively

large unit in nature such as a forest, grassland, desert, or a lake that is self-sufficient when supplied with sunlight. One way to describe the features of such a system is to take a small pond as an example (Figure 21.7). Four major components of an ecosystem can be distinguished.

1. The *nonliving* or *abiotic* components. The minerals, biogenic salts, bottom sediment, water, and remains of organisms make up this entity. The chemical substances are continually cycling between the environment and the life forms in the ecosystem. Some of these chemical cycles are described in the next chapter. Water, of course, has numerous functions in any ecosystem.

2. The *producer* organisms. These are always photosynthesizers and, in aquatic habitats, consist mostly of microscopic algae of many kinds. Producers convert radiant energy (light) into organic materials and thus provide the basic energy supply for all life in the pond. Energy, unlike other abiotic substances, cannot cycle; hence a continuous supply must enter the ecosystem.

3. The *consumer* organisms. These include the animal life of the pond from microscopic protozoans to the largest animals present. Feeders on plant life are termed *primary consumers* or *herbivores*. *Secondary consumers* feed on herbivores and each other, thus are often called carnivores or predators. Still others like catfish serve a scavenging role and eat

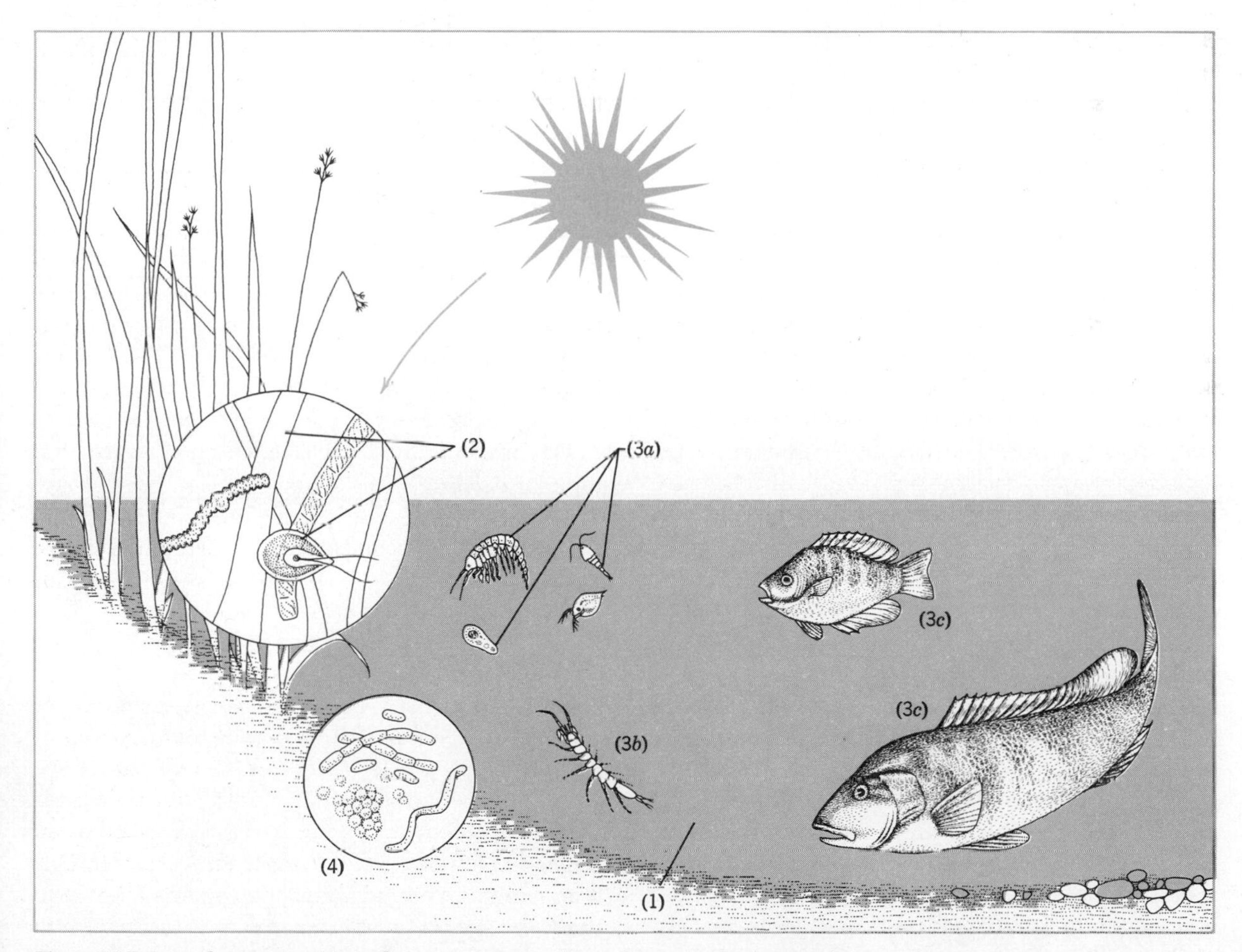

Figure 21.7. A pond ecosystem. (1) *nonliving components,* (2) *producers,* (3a, b, c) *successive consumer levels,* (4) *decomposers.*

any available organic matter. There may be a series of consumer levels depending on the complexity of the ecosystem.

4. The *decomposer* organisms. These include bacteria, fungi, and other organisms that break down the complex compounds of "dead" protoplasm to release simpler substances into the environment. This important act is necessary to continue the cycling of basic chemical materials in the system. Some consumers, like the catfish mentioned above, are also decomposers.

The four basic structural elements described for a pond also apply to terrestrial ecosystems. The producer, consumer, and decay organisms are different but the same basic energy flow prevails. For example, a forest ecosystem contains trees and shrubs as producers; insects, squirrels, and rabbits as herbivore consumers; foxes, raccoons, and hawks as carnivores; and bacteria, fungi, and insects as decomposers. Apply these ideas to an ecosystem with which you are familiar.

Food Chains and Ecological Pyramids

The producer-consumer relations in an ecosystem are often expressed as a food chain. For a grasslands community, a food chain might look as follows:

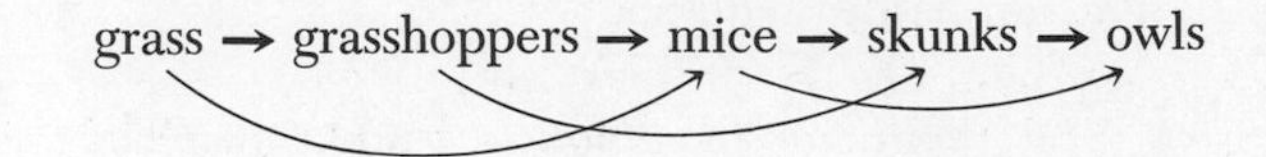

The straight arrows represent major feeding relationships, and the curving arrows are additional ones that probably exist. Of course, other food chains exist in the community involving other animals and other trophic levels (types of nutrition) such as decomposers and parasites. If all of the chains were combined, as they are in a real community, the result would be a weblike arrangement rather than a chain. Figure 21.8 shows a food web for a woodland community. It expresses a far more complex set of conditions than does a food chain.

Food chains and webs do not express the quantitative aspects of ecosystem trophic conditions. One

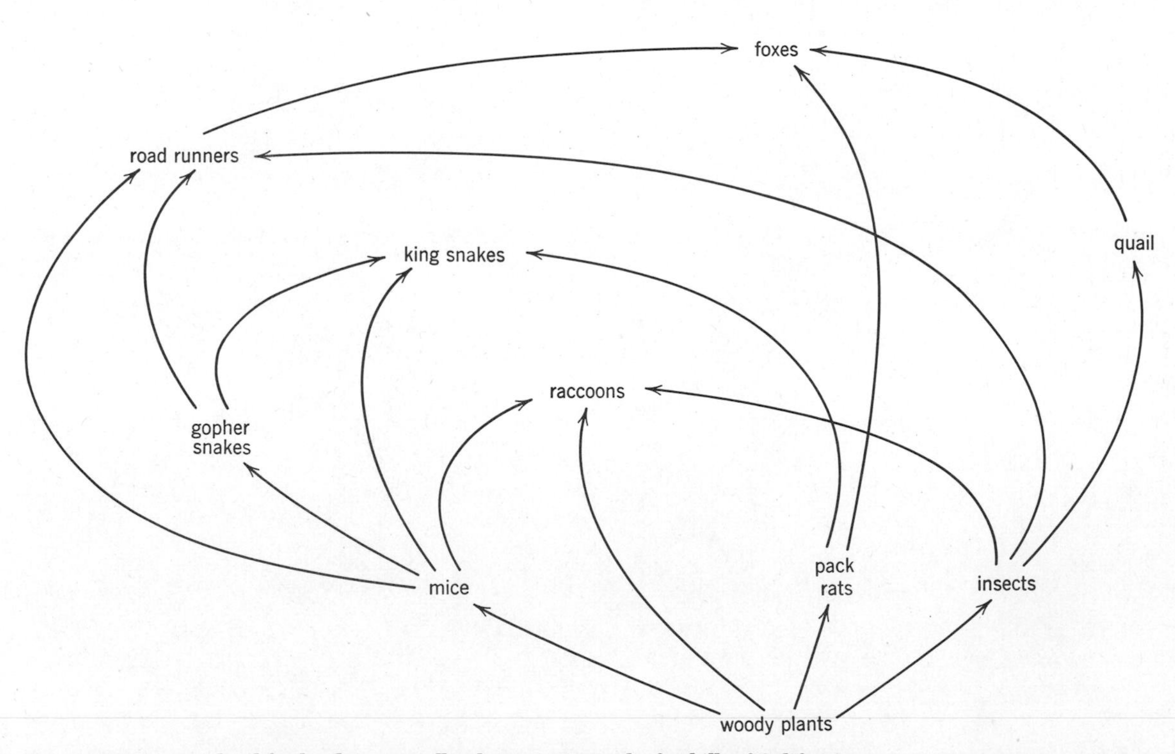

Figure 21.8. A generalized food web in a woodland community in the foothills of California. The arrows represent major feeding relationships and show the direction of energy flow.

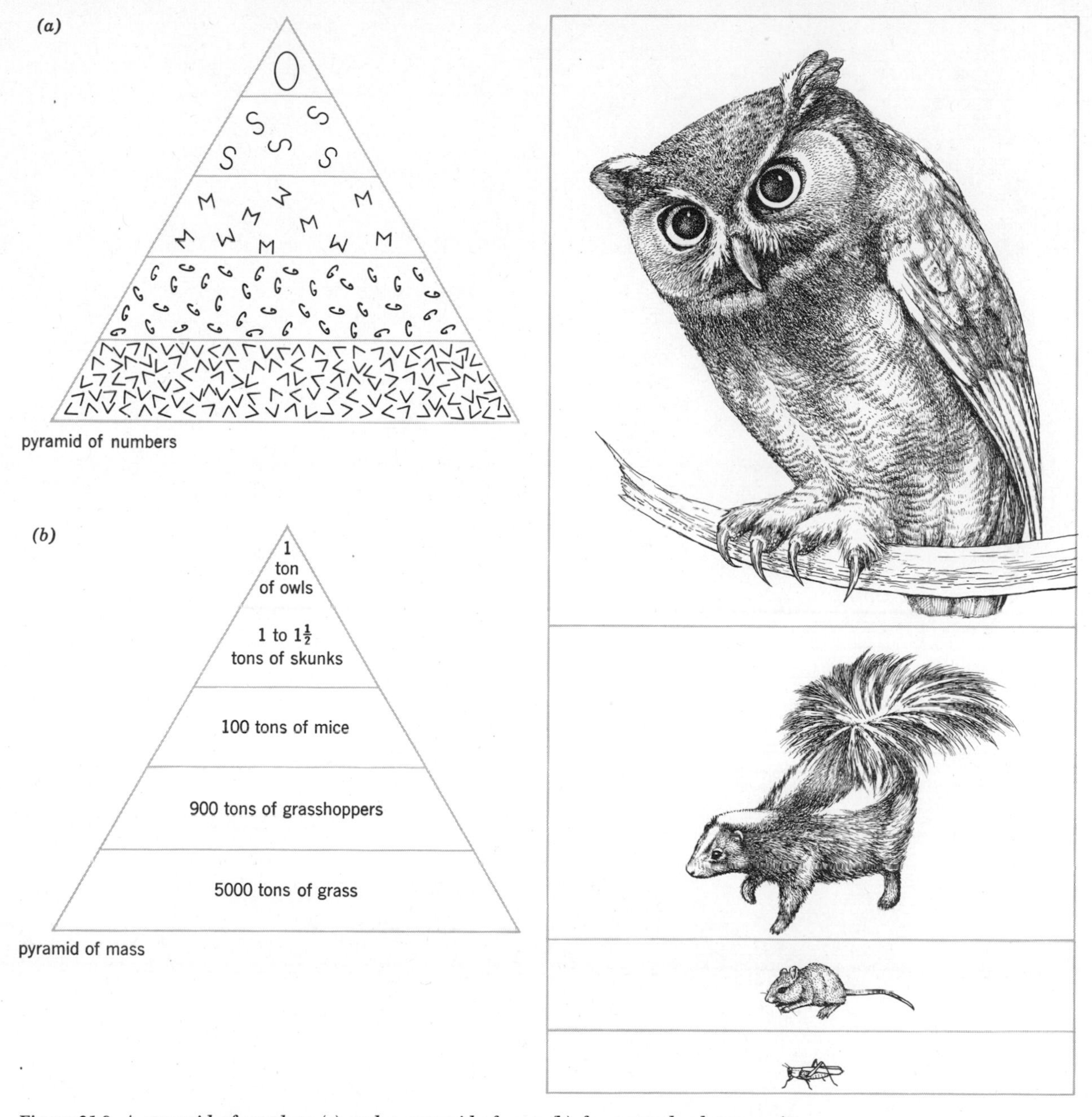

Figure 21.9. A pyramid of numbers (a) *and a pyramid of mass* (b) *for a grassland community.*

method of doing this is to use ecological pyramids. For example, if a count were made of grass plants, grasshoppers, mice, skunks, and owls per unit area in the grasslands, it would be possible to construct a *pyramid of numbers* as shown in Figure 21.9*a*. Each successive layer contains fewer organisms than the one below it.

This type of pyramid has limited value since it contains no information about the size or energy content of each level in the food chain. In addition, some types of food chains would form inverted pyramids of numbers. One oak tree may support hundreds of insects that in turn may contain thousands of internal parasites.

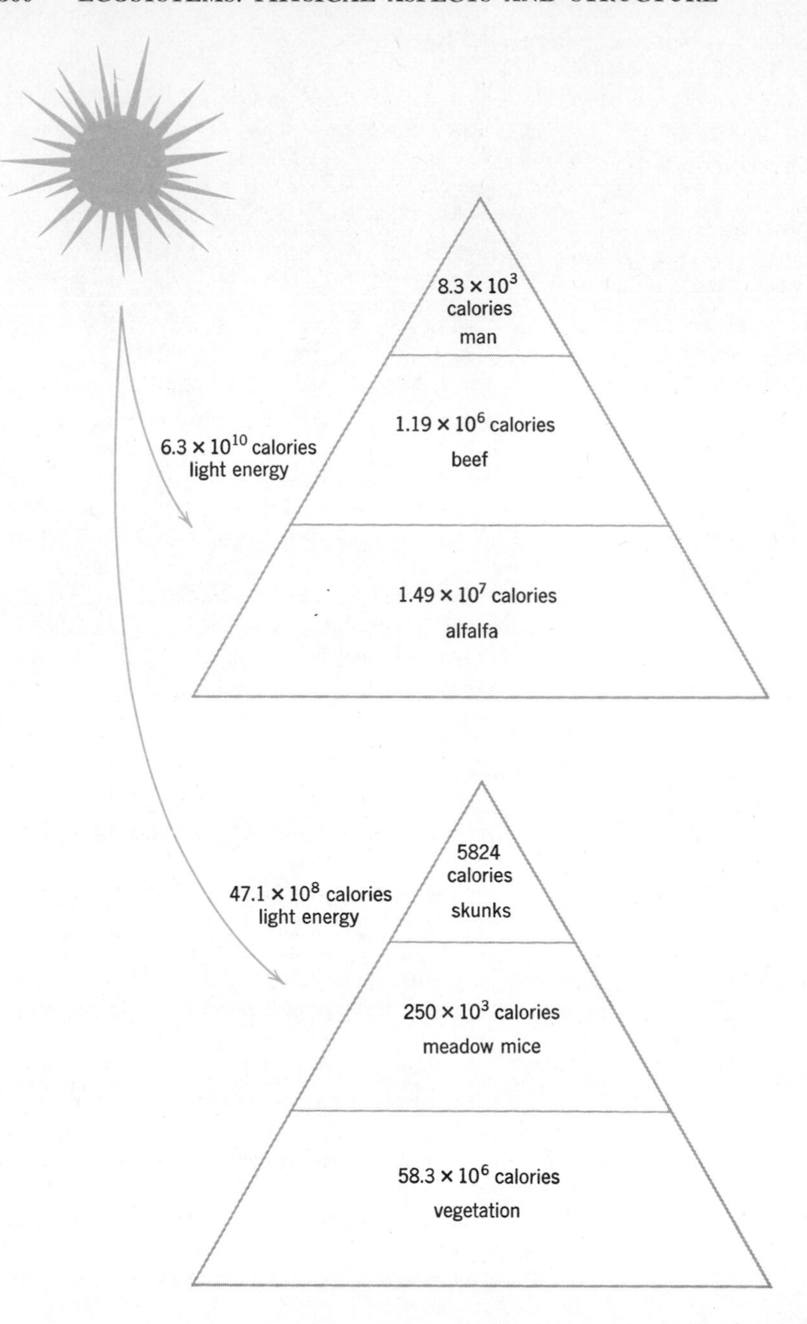

Figure 21.10. Pyramids of energy for an alfalfa community (top), and a grasslands community (bottom). Note the decreasing amounts of energy in each successive level for both pyramids.

To overcome these problems, ecologists sometimes construct a *pyramid of mass* by weighing the producers and consumers from a measured area. Figure 21.9*b* shows a pyramid based on data for a grasslands community.

This pyramid begins to give us a better idea of the energy requirements in the community. Thus, it indicates that an enormous mass of grass is required to support a relatively small mass of owls; each stage in the pyramid has far less mass than the preceding one; and an imbalance in any part of the pyramid would necessitate adjustments in the rest of it.

A third kind of pyramid, a *pyramid of energy,*

perhaps best illustrates the actual food relations among a community of organisms. To make this pyramid, the ecologist must determine the caloric value of each level in the food chain. Figure 21.10 presents energy pyramids for two different kinds of communities.

These pyramids emphasize the point that energy transfer or conversion from one level to the next is not highly efficient. The inefficiency occurs because every organism uses energy in maintaining its own life processes, and because every energy conversion between organisms is unavoidably wasteful (as explained in the next chapter). Consequently these pyramids can never be inverted.

Man is obviously a part of ecosystems and ecological pyramids. His influence and control over nature is so great that he must learn to interact properly with ecosystems. His failure to do this has caused catastrophes of huge proportions.

Principles

1. An ecosystem is a large unit in nature made up of the populations in an area interacting with their nonliving environment and with one another.
2. The major structural features of an ecosystem consist of abiotic materials and populations of producers, consumers, and decomposers.
3. Man or any other organism must interact properly with the rest of his ecosystem if he is to survive, because he is an integral part of that system.

Suggested Readings

Anderson, A. J. and E. J. Underwood, "Trace-element Deserts," *Scientific American,* Vol. 200 (January, 1959).

Brock, Thomas D., "Life at High Temperatures," *Science,* Vol. 158, No. 3804 (November 24, 1967).

Kelsey, Paul M., "Hibernation and Winter Withdrawal," *The Conservationist,* Vol. 23, No. 2 (October–November, 1968).

Odum, Eugene P., *Ecology.* Holt, Rinehart and Winston, New York, 1963, pp. 1–36.

Pratt, Christopher J., "Chemical Fertilizers," *Scientific American,* Vol. 212, No. 6 (June, 1965).

Storer, J. H., *The Web of Life.* New American Library of World Literature, New York, 1956.

Wald, George, "Life and Light," *Scientific American,* Vol. 201 (October, 1959). Offprint No. 61, W. H. Freeman and Co., San Francisco.

Woodwell, George M., "The Ecological Effects of Radiation," *Scientific American,* Vol. 208 (June, 1963). Offprint No. 159, W. H. Freeman and Co., San Francisco.

Questions

1. What is an ecosystem? Name two that are present in the area where you live.
2. Name the major abiotic factors functioning in ecosystems.
3. In what respect is light the most fundamental abiotic factor in an ecosystem?
4. Can you think of a large community of living things that is perpetually dark? How does it obtain its energy supply?
5. What are the four major *structural* elements of an ecosystem?
6. How would you classify man in this structural arrangement?
7. Provide an example of human activity damaging an ecosystem in your area.
8. Provide an example of how man has probably improved an ecosystem.

CHAPTER XXII

Ecosystems: Functional Aspects

Some organisms that occur in a coral-reef ecosystem.

Ecosystems: Functional Aspects

Two major functional events of importance to ecosystems are the energy relations within it, and the cyclical passage of materials through it.

Energy flow and utilization constitute a major topic in ecosystems. Energy is defined as the ability to do work. It may manifest itself in a variety of forms: light, heat, chemical, mechanical, and electrical. All may function in ecological systems although some, such as light and heat, are more important than others. The behavior of energy is described by two laws.

Energy Flow in the Ecosystem

The *First Law of Thermodynamics* holds that energy may be transformed from one type to another but that it can be neither created nor destroyed. It is frequently transformed into an unusable form like heat, and is thus lost to the ecosystem, but the energy itself has not been destroyed. (See Figure 22.1.)

The *Second Law of Thermodynamics* says that an energy transformation is never entirely efficient. When energy is changed from one form to another—for example, light energy to chemical energy in photosynthesis—some of the energy is dispersed into unusable heat. Another facet of the second law pertains to the energy required to maintain all structural organizations. Without a constant input of energy, all entities degrade into simpler states.

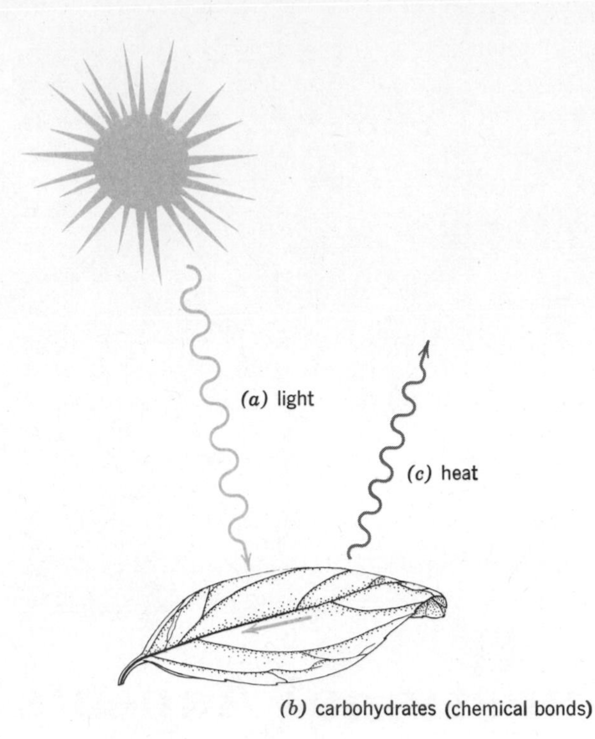

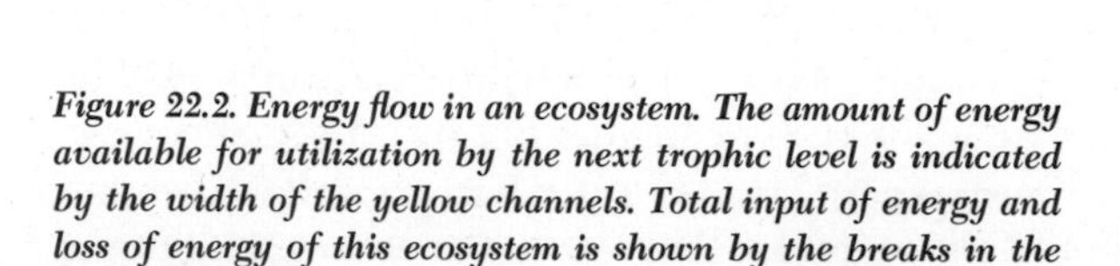

Figure 22.1. The two laws of thermodynamics state that energy may be transformed from one type (sunlight, a) *to another (chemical bonds,* b). *In each transformation some energy becomes heat* (c), *an unusable form for organisms.*

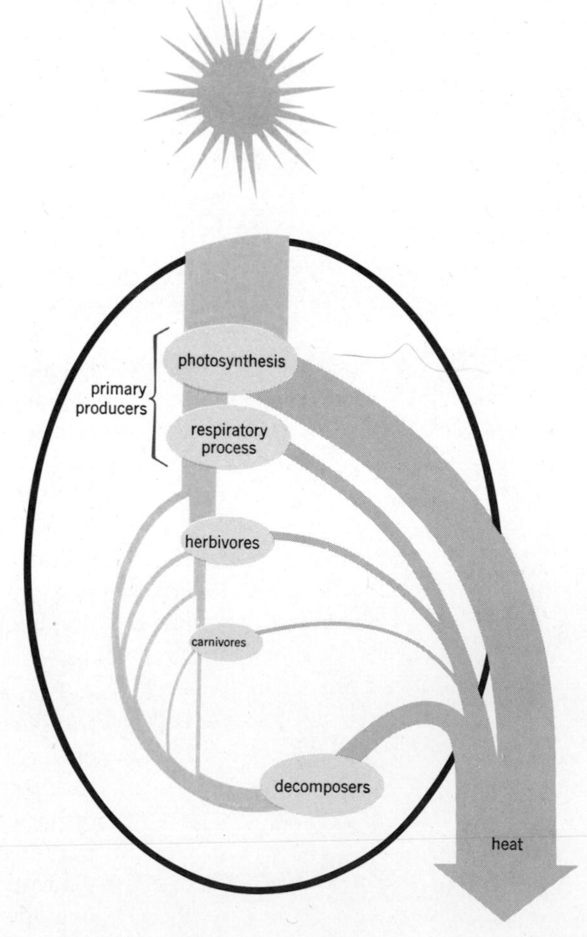

Figure 22.2. Energy flow in an ecosystem. The amount of energy available for utilization by the next trophic level is indicated by the width of the yellow channels. Total input of energy and loss of energy of this ecosystem is shown by the breaks in the heavy black line.

These laws of energy function in ecosystems as they do throughout the physical world. The ecological pyramids discussed in Chapter XXI result from the functioning of the second energy law.

Ecologists sometimes speak of the *energy flow* through communities, which also illustrates the second law of energy. The heavy black line in Figure 22.2 represents the community or ecosystem. Energy enters the system as sunlight to be converted by the producers, through photosynthesis, into potential energy. Observe that a considerable amount of this energy is lost to the system as heat and is thus shown leaving. Cellular respiration also accounts for some of the energy loss. The potential energy that remains is then utilized by the primary consumers (herbivores). Again, a large portion of it is lost to the system, and so it continues to the final consumers, the carnivores. The striking feature of the energy-flow concept is that tremendous quantities of energy are necessary to support the vital activities of the community, since during energy flow large quantities are converted into an unusable form.

If the energy demands of the community become greater than the supply, the consequences are evident. If carnivores become more abundant than their food supply, the dynamics of the community will be disturbed. Likewise, the disturbances become severe if all the carnivores are removed. This actually happened some years ago when all large predators were removed by man from the Kaibab plateau in Arizona. Within a short time, the deer (primary consumers) became too abundant to be supported by the vegetation (primary producers). The entire area took on the appearance of an overgrazed pasture and thousands of deer eventually starved. Unfortunately, this is common consequence of man's attempts to reorganize ecosystems to suit his whims.

Long food chains function less efficiently than short ones because of the great loss of energy between consumer levels. Man takes advantage of this factor by creating short chains with his food crops. When he lengthens the chain by growing edible animals such as cattle, the process becomes much less productive (Figure 21.10, top). This is the major reason why the poorer peoples of the world are grain eaters and frequently suffer protein deficiencies.

Ecosystems have complex energy relations that have evolved over long periods of time. Man's ignorance of this aspect of nature has led to great damage in many parts of the world. Poor farming or grazing practices have often destroyed large parts of the natural producer segment of ecosystems. Consumers, of course, declined or disappeared as their energy source declined. This kind of destruction resulted during biblical times from overgrazing by sheep and goats in many of the areas around the Mediterranean Sea, and has continued into the present century in other parts of the world. In the last century our own western grasslands were damaged by agricultural practices not greatly unlike those that ruined many Mediterranean coastal areas.

Chemical Cycles in the Ecosystem

Between thirty and forty chemical elements are required by living things. Some, like carbon, hydrogen, oxygen, and nitrogen, are needed in sizable amounts, while others are necessary only in small quantities. In whatever amount, most of the elements circulate in characteristic paths from environment to organism and back to environment. These pathways are known as inorganic-organic cycles, or *biogeochemical* cycles. A description of the phosphorus cycle demonstrates some of their functional aspects.

Recall from the chapters on metabolism that phosphorus played a key role in the formation of ATP. In addition, phosphorus is an important structural component of tooth and bones. The movement of phosphorus in and out of the living world is diagrammed in Figure 22.3.

Let us begin the cycle with the large reservoir of phosphorus found in rock deposits formed over long periods of time. Central Florida, for example, contains one of the largest known deposits of this type. By erosion and weathering phosphates are released, some to be utilized by plants and then by animals. Phosphatizing bacteria serve in an important decomposing role to keep dissolved phosphates circulating in ecosystems. A considerable amount of phosphate material is washed into the sea by rivers to form marine deposits. This material becomes an important component in marine communities since phosphates, along with nitrates, appear to be the major nutrients for marine plant life. Marine birds return some of the phosphorus to terrestrial communities, as indicated in the diagram. In addition,

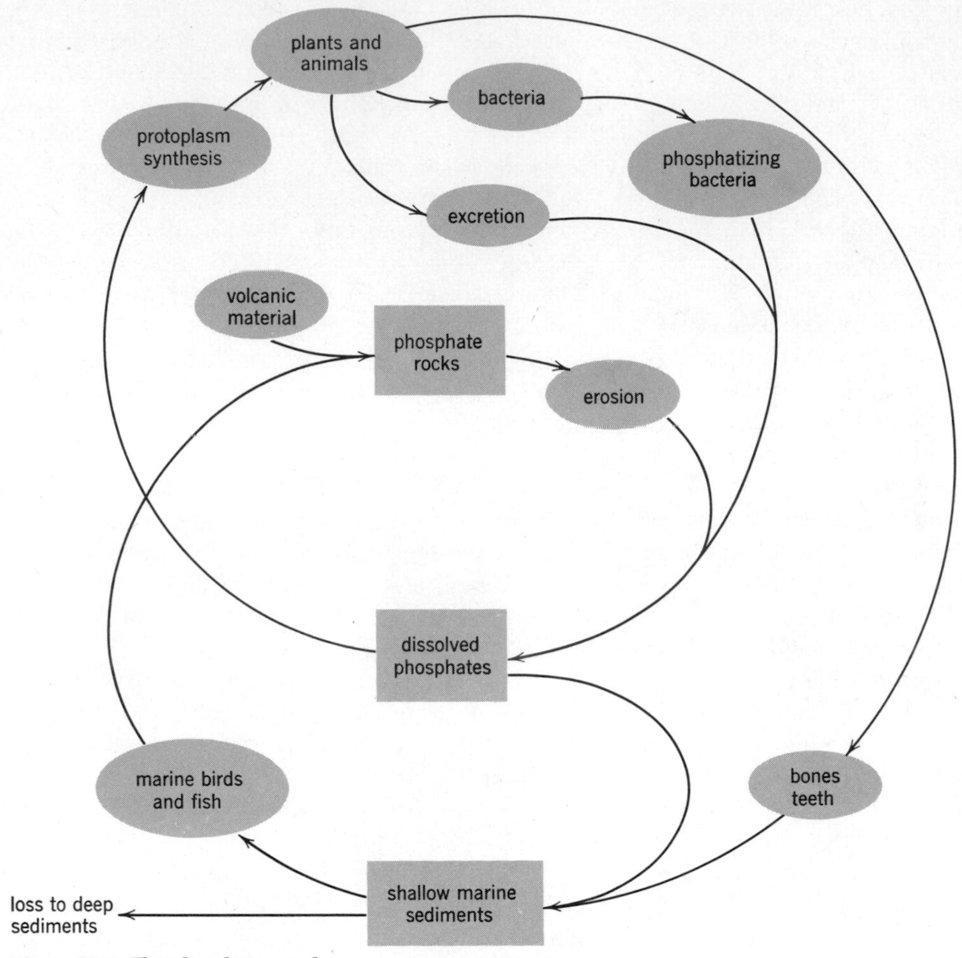

Figure 22.3. The phosphorus cycle.

man utilizes various marine creatures and recycles small amounts in this way. Much of the phosphate, however, ends up in marine sediments where it remains out of circulation until a geological upheaval converts the sea bottom into a land mass. Then, weathering and erosion continue the cycle. At present it appears that more phosphorus is being "bottlenecked" in marine deposits than is reentering the cycle, but this is difficult to judge on a short-term basis. Large amounts of phosphate are mined at present for agricultural and industrial use.

A second vital biogeochemical cycle is the nitrogen cycle shown in Figure 22.4. Nitrogen (N_2) comprises 78 percent of the atmosphere but is biologically inert in this form. Nitrogen-fixing and nitrifying bacteria in the soil convert N_2 into nitrates that in turn are utilized by plants. From plants, nitrogenous compounds enter animals to be used for synthesizing vital compounds like proteins and nucleic acids. The decomposition of animal and plant bodies releases nitrogen compounds that can again be used by plants. Another group of microorganisms, the denitrifying bacteria, degrade nitrogenous materials into gaseous nitrogen that enters the atmosphere.

Biogeochemical cycles have several features in common. In all mineral cycles, for example, specialized microorganisms, such as the phosphatizing bacteria, play essential functions. Plants are also

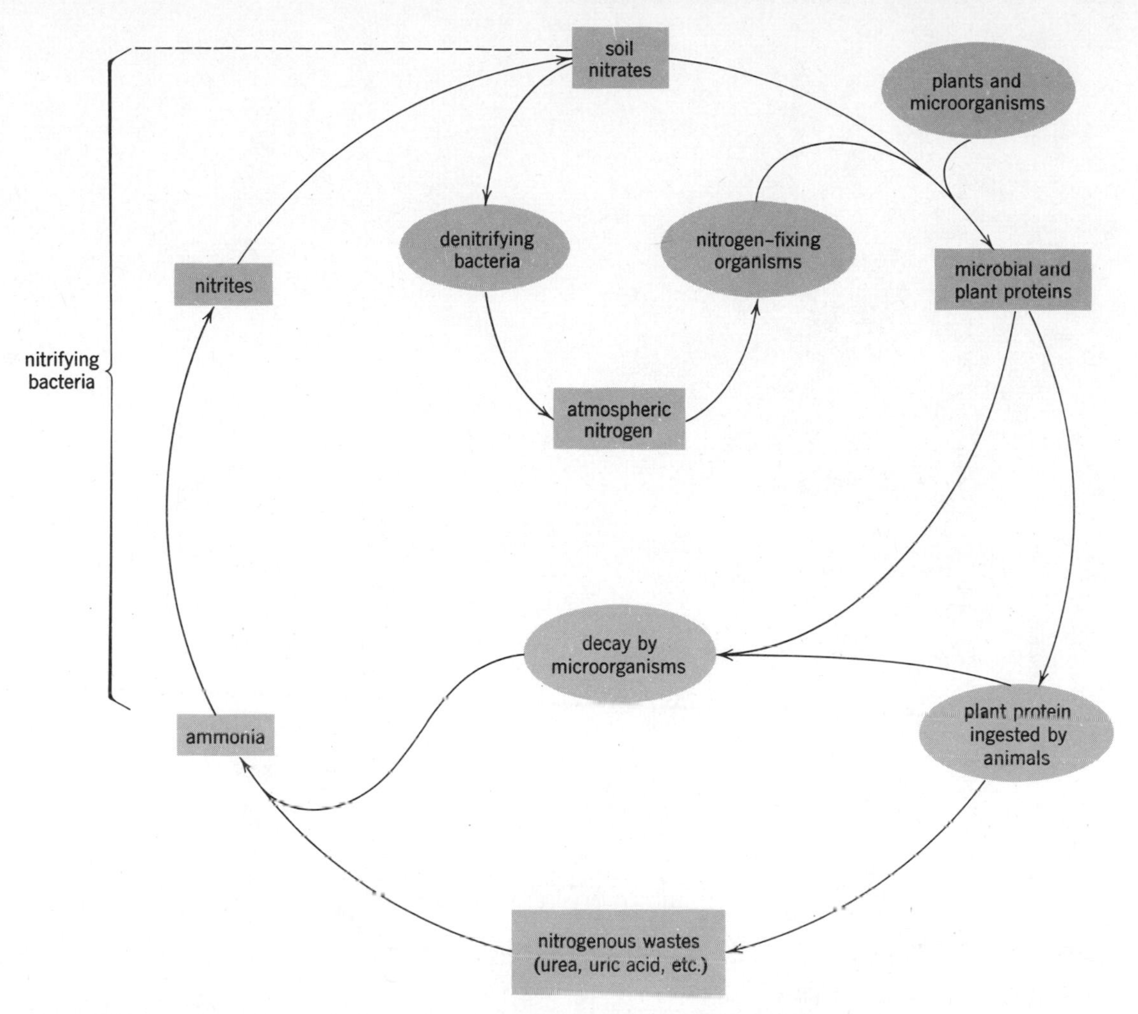

Figure 22.4. The nitrogen cycle. (*After A. Nason,* **Textbook of Modern Biology,** *1965, John Wiley and Sons, New York.*)

basic units and are generally responsible for converting the mineral into a form that animals can utilize. In fact, most cycles could probably function without animals.

Mineral cycles have become much more clearly understood since the introduction of techniques for producing and detecting radioactive isotopes. These techniques have enabled ecologists to introduce a "tagged" chemical into the environment and follow its fate. In several instances, radioactive phosphorus has been put in lakes and then traced as it entered the phosphorus cycle. It was observed in some instances to pass from organisms into mud sediments and then back into organisms again. Other data such as rate of uptake, which could not have been determined by any other technique, were also obtained.

Changing Communities: Succession

Ecosystems and the communities within them do not remain static; they change through time. The orderly and frequently predictable changes a community undergoes are termed *succession.* This process continues until the community reaches the *climax* stage, a fairly stable stage. Everyone has observed succession in a backyard that is no longer being maintained, in an abandoned field, or in a stagnant pond. All of these are changing communities. Ecologists have studied the successional stages of many

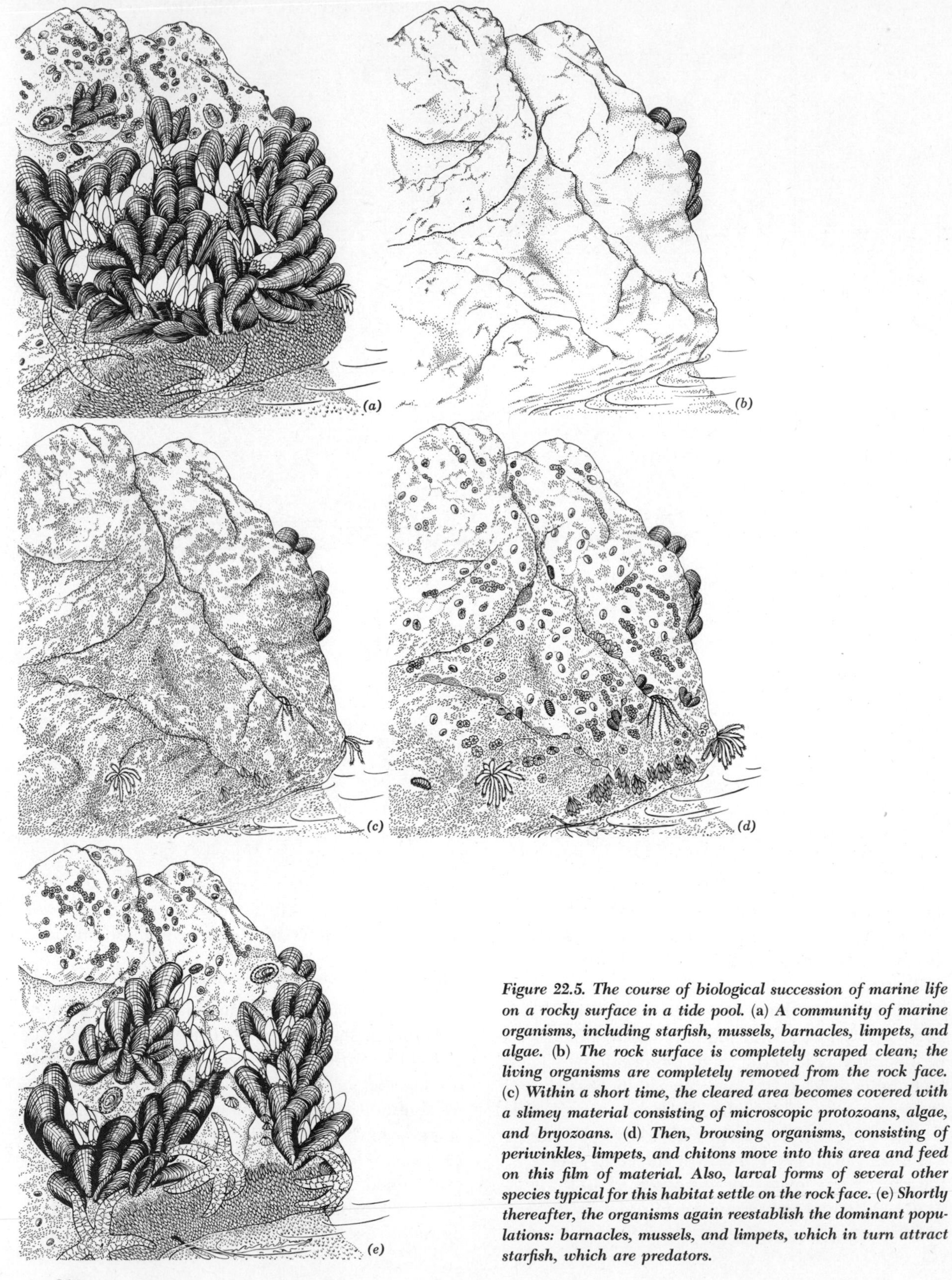

Figure 22.5. The course of biological succession of marine life on a rocky surface in a tide pool. (a) *A community of marine organisms, including starfish, mussels, barnacles, limpets, and algae.* (b) *The rock surface is completely scraped clean; the living organisms are completely removed from the rock face.* (c) *Within a short time, the cleared area becomes covered with a slimey material consisting of microscopic protozoans, algae, and bryozoans.* (d) *Then, browsing organisms, consisting of periwinkles, limpets, and chitons move into this area and feed on this film of material. Also, larval forms of several other species typical for this habitat settle on the rock face.* (e) *Shortly thereafter, the organisms again reestablish the dominant populations: barnacles, mussels, and limpets, which in turn attract starfish, which are predators.*

Figure 22.6. Red mangrove. Notice the branched roots that may catch soil and debris carried by currents and wave activity.

types of communities (Figure 22.5). Their findings indicate that communities go through a sequential period of development, attain some sort of maturity, and eventually may be replaced by another set of producers and consumers.

Many people are familiar with the bare rock → lichens → fungi and ferns → herbs → shrubs → trees idea of succession. This is a long-term development beyond the experience of a human lifetime, but there are examples of much shorter duration. Along the coast of Florida and in other areas, channels are frequently dredged to facilitate the passage of large boats along the coastline. The mud from the dredging operation is deposited in mounds at the side of the channel. This creates small barren islands. Eventually the seeds of salt-tolerant grasses and herbs find their way to these islets and germinate. This sparse cover will likely be joined shortly by mangrove plants whose seeds are dispersed by tides and currents. These plants grow into a dense, shrubby border around the islands. Their peculiar root system holds the mud in place and may even enlarge the perimeter of the island (Figure 22.6). If the island has sufficient elevation to be permanently above water, the grasses and herbs may be gradually replaced by shrubs such as sea grape, groundsel, and even trees, especially pines. But if the island is frequently flooded by salt water, mangroves cover the entire space and form a climax community. This type of succession takes place over a relatively few years, but can disappear in a matter of hours in a severe storm.

Most of the detailed studies on succession have concerned larger communities such as the broadleaf forests of the eastern United States, the grasslands of the midwest, and aquatic environments. In the eastern United States, the climax community consists of various types of broadleaf forests such as the oak-hickory or beech-maple forest. In the southeast, succession creates a problem for the pulpwood industry. Pines are more desirable than other trees for making paper, but when pines are harvested from an area, broadleaf trees may succeed or replace them so that a former pinewood forest becomes a hardwood forest. Only by restraining the young broadleafs can the timber grower assure another crop of pines; in other words, he must interfere with succession.

An interesting ecological aspect of this type of community is that forest fires are necessary for a pine forest to maintain itself. If protected from fire, the pines are often succeeded by broadleafed trees. Forestry studies in the 1920s indicated how fire affects longleaf pines. The seedling of this pine is covered by a bushy canopy of needles and is always in danger of being shaded by rapidly growing broadleafed competitors. In addition, the seedling is susceptible to a fungus growth on its needles. If fire passes through the area, it usually destroys the competing hardwoods as well as the fungus growing on the pine seedlings. The dense covering of needles protects the bud of the seedling so that the plant survives. The young pine grows extremely rapidly for two or three years, finally becoming tall enough to escape damage from the next ground fire. Thus, with periodic burning, a pine flatwoods or forest can maintain itself.

This situation is not peculiar to southern pine forests. The jack pine of Michigan, Wisconsin, and Minnesota and the knobcone pine of California apparently require heat from a fire to open tightly closed cones. Some weeks after the fire the seeds fall into the bed of ashes, which are ideal for germination. The seeds of competing species have been destroyed. Here we see an adaptation dependent on the presence of periodic fires! Specialists in forest management have utilized this concept and initiate controlled burning in certain situations.

What principles can be derived from studies of succession? First, that organisms change the physical and chemical aspects of the environment in which they live. Mice burrowing in the soil, or leaves accumulating and decaying are examples. As the environment changes, so do the kinds of plants and animals inhabiting it. This is the most noticeable feature of the process; hence the term succession. Accompanying the succession of changing forms is an increase in the diversity of species and an increase in the amount of organic matter in the ecosystem. The excess organic material (decaying plant and animal remains) plays a dominant role in changing the physical and chemical nature of an area. Finally, as a community becomes stable, there is a decrease in the net production of living matter. In other words, as succession slows in a community, a balance is obtained between the utilization of energy (respiration) and the formation of new organic matter. For this reason, man uses *early* succession stages for the production of food crops or for certain crops like the pines just described.

Man has experienced a mixture of success and failure in his attempts to modify ecosystems for his long-term benefit. The misuse of soil, excessive timbering and grazing practices, and indiscriminate use of insecticides represent failures. Success has been attained in a few areas like wildlife and forestry management. Rice cultivation is an example of the successful manipulation of the marsh ecosystem. Unfortunately, misuses from overexploitation and pollution are increasing at an alarming rate as the result of man's greater power to change the face of the earth and his rapid increase in numbers.

As biologists have attained a better understanding of ecosystems, they have attempted to integrate man's activities with those of nature. Problems associated with water pollution, erosion, overgrazing, and mining are receiving serious attention. The effects of the Second Law of Thermodynamics are inevitable, and perhaps man is now ready to delay rather than hasten the consequences.

Population Dynamics

We have used the term *population* in various places throughout this book, assuming that the meaning was correctly understood. Thus, a genetic population referred to a self-perpetuating group of organisms, whereas in ecology a population is defined as all of the members of a species inhabiting a particular area. A population, however considered, is an important biological unit and in this section we discuss a few of its major features.

A population consists of individuals, and thus many of its features are a reflection of the characteristics of its members. A population also has a number of unique traits that are an outgrowth of its being a numerical concept or entity. We shall consider these quantitative aspects first.

Density. Density refers to the number of individuals per unit of area (or volume). Five meadowlarks per acre, two thousand diatoms per liter of pond water, 10,000 people per square mile are all expressions of density. This is an important aspect of population study in view of the prior discussion of energy relations and structure in the ecosystem.

Natality. This term refers to the production of new individuals, the ability of a population to in-

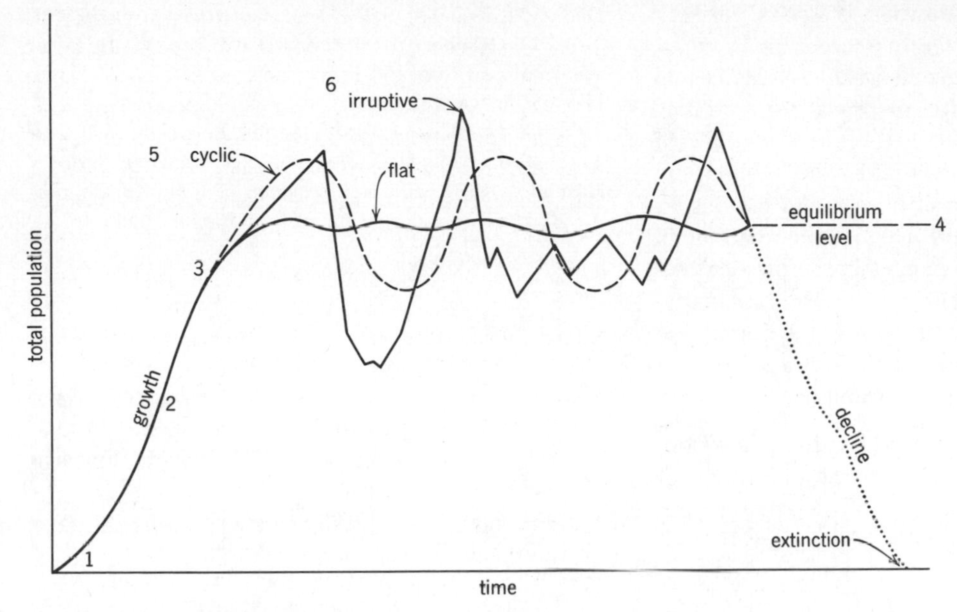

Figure 22.7. Some possible growth curves. For stage numbers (the arabic numerals), see the text. (After G. Clarke, Elements of Ecology, *1954, John Wiley and Sons, New York.)*

crease. The natality rate (birth rate) observed in a population living under natural conditions is usually not the maximum rate possible under ideal conditions.

Mortality. The inevitable fate of all organisms in a population is death, which can be expressed in terms of a mortality rate. Mortality is obviously a major factor in population control, but its measurement in wild populations is often difficult unless the organisms have a characteristic feature from which it is possible to tell the age of the individual. For example, the scales of some fish have growth rings, and mountain sheep have horns from which age can be determined. But unfortunately these instances are rare.

Age Distribution. If we study birth and death rates, we find a pattern of age groups. A new, growing population consists of many young members and few old ones. In a mature population, the number of older members may become greater than the young ones.

Age is important in relation to both natality (in terms of reproductive age) and mortality. Again, the ecologist encounters difficulty in determining age classes in natural populations unless the structure of the individual organism provides a clue.

Population Growth Rate and Form. At this point, we can summarize these brief comments about populations in a growth curve, as shown in Figure 22.7. This curve is also termed a *sigmoid curve*, owing to its S-shape, or a *logarithmic growth curve* because of the rapid increase in numbers during stage 2. Curves of this type indicate what happens to a population with the passage of time. Although the curve shown is theoretical, similar curves have been plotted for many laboratory and natural populations. The following statements refer to the numbered stages on the diagram.

1. Stage *one* is the establishment of the population and involves adaptations to the environment, survival, and initial reproduction. Since relatively few individuals are involved, the rate of increase is slow.
2. Stage *two* is obviously a time of rapid growth. Here the maximum theoretical reproductive rate may be attained, or at least closely approximated.
3. In stage *three*, the population encounters environmental resistance of various sorts and thus levels off.
4. Stage *four* is a theoretical equilibrium position around which the population may oscillate or fluctuate. This is the mean or average size of the population if measured at various intervals of time.
5. Stages *five* and *six* illustrate two kinds of de-

partures from the equilibrium stage. Relatively symmetrical changes (stage five) are called *oscillations* and result from events such as changes in natality and mortality rates or predator-prey interactions. Those oscillations are often the result of density-dependent factors within the population, since they are a function of numbers of organisms. Stage six shows sharp and irregular fluctuations in numbers resulting from changes in the physical environment. These often result from *density-independent* factors operating on the population from without. The increase in a mosquito population during the rainy season is an example.

Poor planning and hasty modification of environments by man tend to increase the amplitude of fluctuation, creating "boom or bust" situations instead of the more desirable equilibrium levels. Sudden, drastic changes in an environment often produce explosive eruptions of insects or other organisms to make pests out of species that had previously been rare. Finally, of course, the population may decline and die.

Principles

1. The basic function of an ecosystem is to capture and utilize energy.
2. The laws of energy apply to ecosystems as they do throughout the physical world, as illustrated by energy flow diagrams or pyramids of mass and energy.
3. Chemicals circulate through ecosystems in characteristic pathways termed biogeochemical cycles.
4. Communities of organisms in an ecosystem go through a series of changes termed succession, until a fairly stable stage, the climax, is reached.
5. A population like an individual shows growth, maturity, and senescence.

Suggested Readings

Arnold, James R. and E. D. Martell, "The Circulation of Radioactive Isotopes," *Scientific American,* Vol. 201 (September, 1959).

Bormann, F. H. and G. E. Likens, "Nutrient Cycling," *Science,* Vol. 155, No. 3761 (January 27, 1967).

Comfort, Alex, "The Life Span of Animals," *Scientific American,* Vol. 205 (August, 1961).

Cooper, Charles, "The Ecology of Fire," *Scientific American,* Vol. 204 (April, 1961).

Deevey, Edward S., Jr., "Bogs," *Scientific American,* Vol. 199 (October, 1958). Offprint No. 840, W. H. Freeman and Co., San Francisco.

Deevey, Edward S., Jr., "The Human Population," *Scientific American,* Vol. 203 (September, 1960). Offprint No. 608, W. H. Freeman and Co., San Francisco.

Harcourt, D. G., and E. J. Leroux, "Population Regulation in Insects and Man," *American Scientist,* Vol. 55, No. 4 (1967).

Leopold, A. S., "Too Many Deer," *Scientific American,* Vol. 193 (November, 1955).

Odum, Eugene P., *Ecology.* Holt, Rinehart and Winston, New York, 1963, pp. 37–64, 77–93.

Powers, Charles F. and Andrew Robertson, "The Aging Great Lakes," *Scientific American,* Vol. 215 (November, 1966). Offprint No. 1056, W. H. Freeman and Co., San Francisco.

Stewart, W. D. P., "Nitrogen-Fixing Plants," *Science,* Vol. 158, No. 3807 (December 15, 1967).

Woodwell, George M., "Toxic Substances and Ecological Cycles," *Scientific American,* Vol. 217 (March, 1967). Offprint No. 1066, W. H. Freeman and Co., San Francisco.

Wynne-Edwards, V. C., "Population Control in Animals," *Scientific American,* Vol. 211 (August, 1964). Offprint No. 192, W. H. Freeman and Co., San Francisco.

Questions

1. What is the relation between the Second Law of Thermodynamics and the ecological pyramids described in Chapter XXI?
2. What is the ultimate fate of energy that is converted into unusable forms such as heat?
3. Can you relate the idea of energy flow through communities to the concepts of energy production in cellular respiration as presented in Chapter V?
4. In the geological past, a large quantity of carbon was "bottlenecked" in one place in the carbon cycle. In what form do we find this today?
5. Is carbon still being trapped this way? Why?
6. What are some examples of ecological succession in the locality where you live?
7. Does a field of corn represent an early or a late stage in succession? What concept about energy production at different succession stages does this illustrate?
8. What characteristics do populations and individuals share?
9. What characteristics do populations have that an individual cannot possess?
10. If population growth curves have a characteristic S-shape, is it possible to predict at which point a rapidly growing population (as humans, for example) will level off? Why?

CHAPTER XXIII

Interaction in Ecosystems

Interaction between two adult baboons showing grooming behavior while offspring nurses (Dr. I. De Vore).

Interaction in Ecosystems

". . .the structure of every organic being is related in the most essential yet often hidden manner to that of all other organic beings, with which it comes into competition for food or residence, or from which it has to escape, or on which it preys. This is obvious in the structure of the teeth and talons of the tiger; and in that of the legs and claws of the parasite which clings to the hair on the tiger's body."

Charles Darwin, *The Origin of Species*

This statement by Darwin illustrates the theme of the chapter, since we wish to explore some of the ways in which organisms interact and react with one another and with their physical environment. These interactions have their beginnings in evolution, since they represent adaptations that have enabled populations to survive through time. As discussed in earlier chapters, the mechanism of this adaptation is natural selection, and the outcome is a population adapted to interact more effectively with its biotic and physical environment.

The interactions between organisms and their environment are recognized by many people. As a mental exercise, consider a familiar organism and then think of ways in which its structure is adapted to the environment: locomotion, nutrition, protection, and so on. Here are a few examples to illustrate the point.

Water, because of its density, presents problems of locomotion to organisms that live in it. It is no coincidence that the

Figure 23.1. Convergent evolution of body form shown by an extinct reptile (a), *a shark* (b), *and a mammal* (c). *All are adapted for an aquatic mode of life.*

problem was at least partially solved by the evolution of streamlined bodies, even in diverse groups, as illustrated in Figure 23.1. In the ocean, a shark, a reptile, and a mammal evolved similar body forms as an adaptation for movement through a dense medium.

On an even larger scale, we can consider the consequences of the evolutionary event known as adaptive radiation, illustrated in Figure 23.2. In this process, interactions between organisms and their environment lead to divergent body forms rather than similar ones.

One of the most interesting aspects of biology is the many ways in which organisms live in relation to other organisms, ranging from beneficial to detrimental associations. As we saw in the preceding chapter, interactions between producer and consumer populations form a vital part of the ecosystem. Now we want to consider various types of organismic interactions in greater detail.

Adaptation to Competition

Competition. Competition is the outcome of organisms requiring the same materials from the environment. For plants, these are sunlight, water, minerals, and growing space. For animals, it involves food, nesting sites, and mates. In most instances, there is a subtle interaction: the two competitors may never even see each other. For example, grasshoppers, mice, rabbits, sheep, and antelope may all act as primary consumers in the same community. Several carnivores may utilize the same prey.

One result of studying competition was Cause's *competition-exclusion principle.* In essence, this principle states that two species with identical requirements cannot live in the same niche; one will be driven away (excluded) or will change its requirements. In reality, related species with similar requirements often inhabit the same community. Various conditions make this possible. Frequently, they utilize slightly different food sources or, in the case of plants, require different nutrients from the soil. Many animals may utilize the same food but at different times of the day. Butterflies and moths are an example of this. Related organisms often occupy different parts of the same community as in a woodland where some organisms forage and live in the tree tops, while their relatives live on or near the ground.

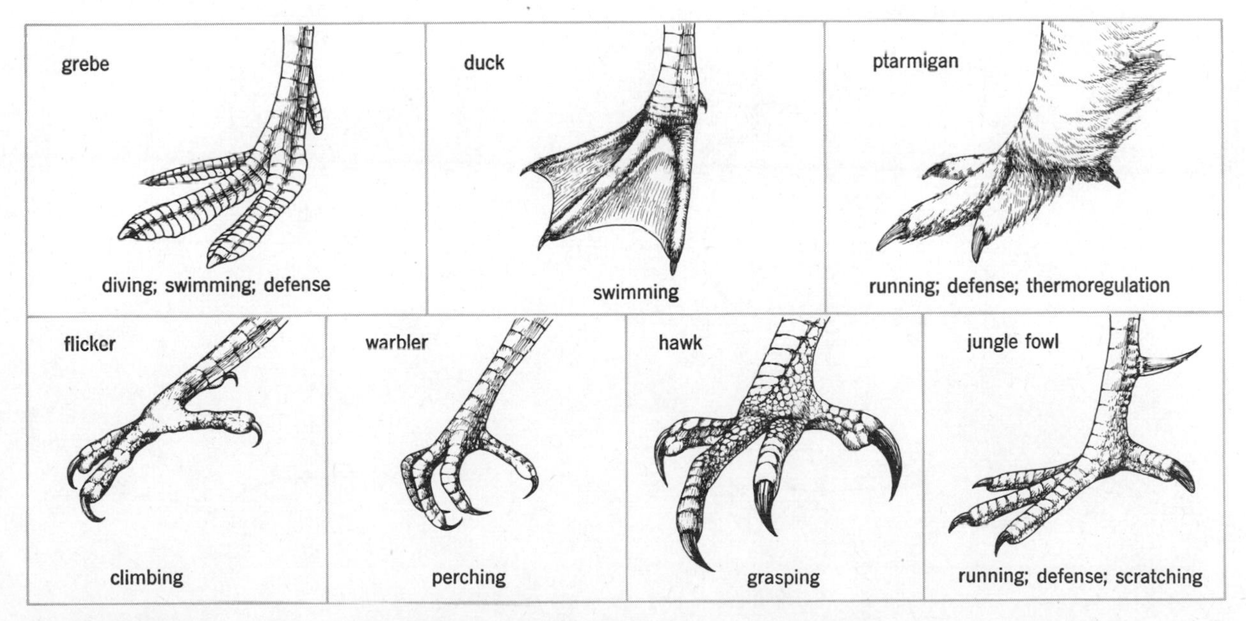

Figure 23.2. Divergent shapes of feet in birds. The structure of each is an adaptation for the functions listed.

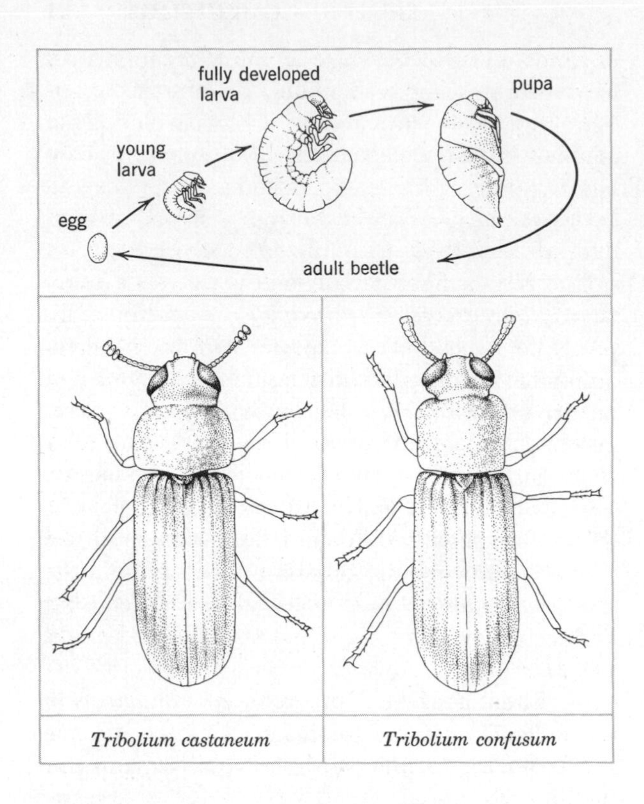

Figure 23.3. Two species of the beetle Tribolium *and their life cycle. These two species have been used in studies of interspecific competition.*

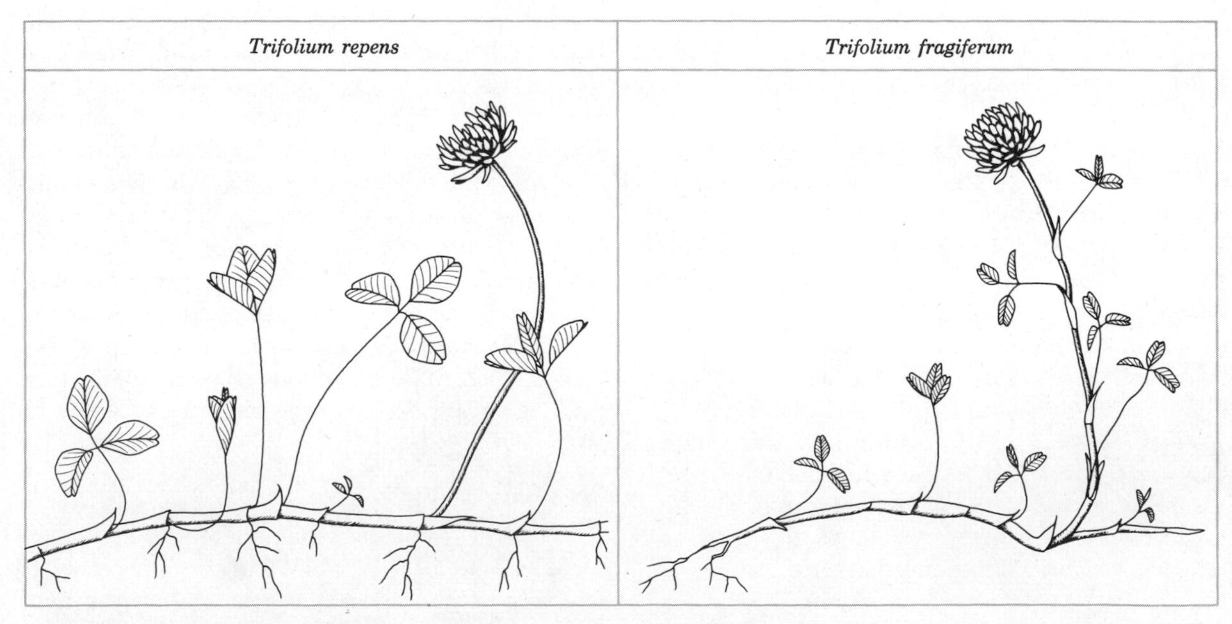

Figure 23.4. In plants, similar studies on competition have been made using the genus Trifolium. *Even though the two species shown here resemble one another, their forms are adaptations to slightly different conditions.*

Since manifestations of competition are largely historical—we assume that a present-day distribution pattern is the consequence of competition in the past—investigators work with laboratory models in order to observe the event while it is taking place. Some of the most rewarding models involve species of the flour beetle, *Tribolium* (Figure 23.3). Experiments with them were conducted by Dr. Thomas Park at the University of Chicago. These small beetles will live, feed, and breed in a jar of flour. Two species can be started in a jar and their progress followed. Moreover, the physical features of their habitat can be regulated. One of the outcomes of the experiments using two species of *Tribolium* was that one species eventually died out. It was an unsuccessful competitor. When Park manipulated the climate of the flour habitat in terms of temperature and humidity, he found that he could determine which species would be successful and which would fail. For example, when *Tribolium castaneum* lived with *T. confusum*, it was found that a hot-wet flour environment always led to survival by *T. castaneum*. A cool-dry one always led to survival by the other species, *T. confusum*. Each species had an optimum climate in which its population dominated the environment. Note that this was a subtle action; neither species directly attacked the other, yet both were competitors.

Experiments also have been made using plants. In one, seeds of two competing clovers, *Trifolium repens* and *Trifolium fragiferum*, were germinated and studied under controlled conditions (Figure 23.4). Subtle differences were found in their development, even though the mature plants appeared similar. *T. fragiferum* had larger seeds than *T. repens*, an advantage in nourishing germinating seedlings. However, the seedlings of *T. repens* grew their foliage leaves more quickly. Seeds of *T. repens* germinated more rapidly at temperatures of 25°C and below, but seeds of *T. fragiferum* were capable of more rapid germination at 35°C. In later growth, *T. fragiferum* continued to form new leaves later than *T. repens* but bore its leaves higher on the stems. Also, the stem of *T. repens* became prostrate early in development while the stem of *T. fragiferum* remained vertical. We assume that these features aid the plants as they compete in nature.

We have been considering cases of competition between different species. Members of the *same* species also compete. They constitute a population in which all members have identical requirements. Obviously this competition must be carefully regulated in various ways so that it will not harm the population. Let us examine a few examples of how this is done.

Territoriality. Territoriality is a behavioral activity exhibited by many animals (Figure 23.5). In general, the animal, usually a male, defends a circumscribed area from other males, then eventually courts the female there, and perhaps raises young. This behavior is particularly noticeable in birds and can be observed in any backyard in the springtime. Usually the male bird uses a singing perch, his song identifying the area to other males and females of his species. In an aquarium, one can observe male fish defending certain areas of the tank against the other fish. Even domesticated animals like dogs show territorial behavior as exemplified by a small dog chasing a much larger one out of its yard—but no farther!

Territorial behavior apparently serves a variety of functions depending on the animal. It avoids conflict between members of the population since the auditory and visual signals serve as forms of communication. In some cases, territoriality limits population size in accordance with the ability of the area to supply food and nesting sites.

Peck Order. Peck order is another behavioral adaptation that functions to decrease intragroup strife. It is most commonly observed in social aggregates such as flocks and herds where members of the group become arranged in social hierarchies. Usually the strongest, largest, or maturest individual dominates the other members, who, in turn, dominate individuals below them in the hierarchy, and so on. In a flock of chickens, for example, the dominant ones eat first at the feed trough, get first choice of roosting sites, etc. After the hierarchy is established, the flock functions with a minimum of internal strife. Most gregarious animals exhibit this pecking order and it is even tempting to apply it to human beings!

Symbiosis

The term symbiosis, which literally means living together, may be familiar. We use it in this context to apply to any intimate association between two forms, regardless of the harmful or beneficial outcomes of the relationship.

Commensalism. Symbiotic interactions in which

Figure 23.5. Elephant seals, during the breeding season, establish territorial boundaries in which any male intruder will be met with a fight (a). *In part* (b) *a subordinate male is approaching the territory of a dominant male who, in turn, is exhibiting a threat posture.*

only one of the organisms benefits and neither is harmed is termed commensalism (Figures 23.6 and 23.7). We shall describe only a few examples. Interested students can find additional ones in most ecology textbooks.

A classic example of commensalism is the remora fish, commonly called a "shark sucker." These fish have a suction disk on top of the head by which they attach themselves temporarily to much larger forms like sharks, whales, or sea turtles. This does not damage the host yet allows the remora to share tidbits left over from the host's meals. So far as is known, the host does not gain any advantage from this relationship.

Many plants have adapted to an epiphytic way of life; that is, they grow upon the trunks or limbs of other plants deriving only support but not nourishment from them. Residents of the southeastern coastal areas are familiar with the gray, streamerlike, Spanish moss that drapes so many trees. This epiphyte, a flowering plant rather than a moss, does no damage to its host unless it becomes thick enough to shade its leaves.

In many of the sandy ridge areas of Florida, a

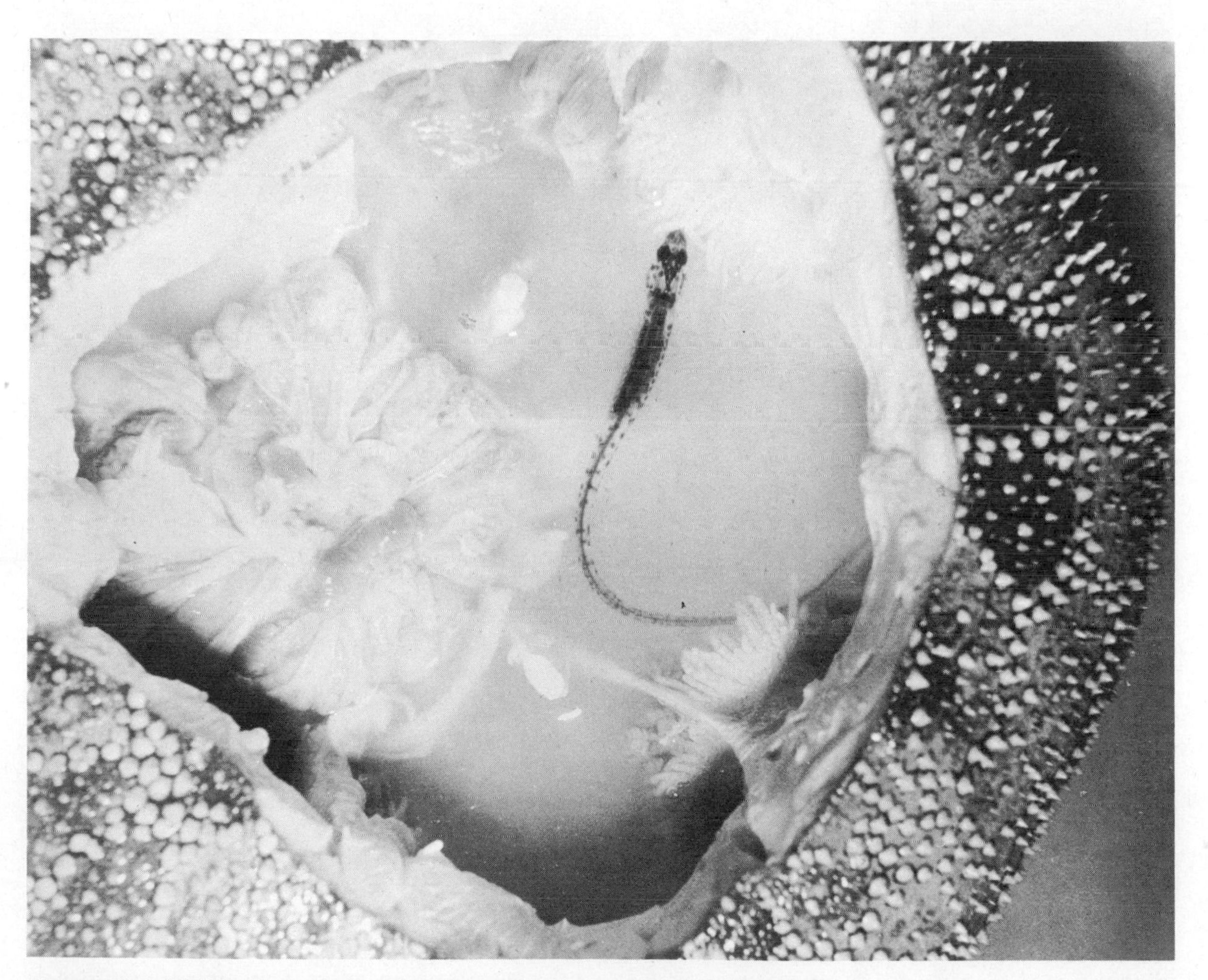

Figure 23.6. A fish living in the body cavity of a starfish. A portion of the body wall of the starfish has been removed.

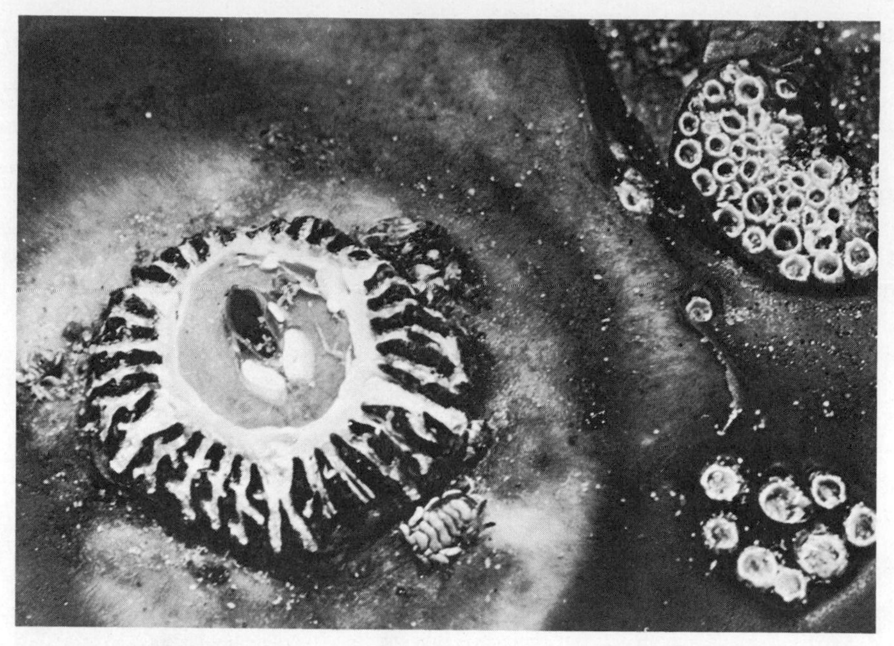

Figure 23.7. Barnacles living on a whale's back.

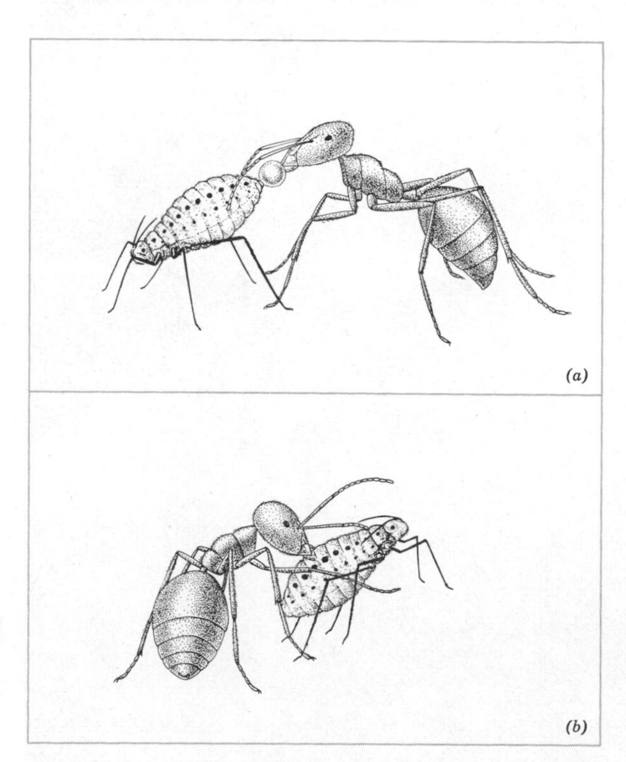

Figure 23.8. Mutualism between ants and aphids. (a) *Ant feeding on secretion produced by the aphid.* (b) *Ant carrying an aphid.*

small deer mouse, *Peromyscus polionotus,* lives in burrows that it digs in the sand. A lizard utilizes the opening of the mouse burrow from which it digs a short side tunnel slanting almost to the surface of the soil. When disturbed, the lizard can burst up through the soil to escape a predator. The mouse is neither harmed nor helped by this arrangement, but the lizard's shelter conditions improve.

These few examples illustrate that commensalism occurs among diverse organisms and may involve feeding arrangements, shelter, or support.

Mutualism. It would not be difficult to imagine a commensal relationship evolving into a mutually beneficial one. An interaction in which both populations benefit is termed mutualism. As in all of these relations, there is a wide range of interactions from voluntary mutualism, as in a flock of birds, to inseparable types like the combination of an alga and a fungus known as the *lichen.*

Lichens consist of a mass of fungal cells that contain cells of a green alga. The fungus holds moisture and provides minerals for both members; the algal cells make carbohydrates by photosynthesis, which provides energy to both members. Neither member exists alone in nature.

Many trees including pines, oaks, hickories, and maples have a mutualistic arrangement with a fungus in the soil. The fungus forms a covering (mycorrhiza) around the roots and facilitates the absorption of water and nutrient salts. The fungi are nourished by organic material from their host. Young pine seedlings seem particularly dependent on this mycorrhiza for normal growth since it helps them extract mineral nutrients from the soil.

Many species of ants have evolved a mutualistic arrangement with aphids. Ants feed upon a material that exudes from the body of this small insect (Figure 23.8*a*). Some species of ants keep aphids in their colonies, feeding and maintaining them while utilizing their exudation. Other species carry the aphids to plants during the day (aphids feed by sucking plant juices) and return them to the ant hill at night (Figure 23.8*b*). Both ants and aphids benefit by these arrangements although both can live quite well alone. This is not an obligatory mutualism as is the case with lichens.

In marine habitats, a number of small creatures are involved in a "cleaning symbiosis." For example, at least six species of small shrimps, frequently brightly colored, crawl over fish, picking off parasites and cleaning injured areas. This is not an accidental occurrence, since fish are observed to congregate around these shrimp and stay motionless while being "inspected." The shrimp may even forage among the gills and mouth cavity of the fish. Several species of small fish are also cleaners, and nearly all of these show adaptations for this way of life, namely, long snouts, tweezerlike teeth, and bright coloration. This type of mutualism involves a number of species.

The dependency of many plants on insect pollination (Figure 23.9) and the association of nitrogen-fixing bacteria with the roots of leguminous plants indicate that mutualism is a basic biological principle. Even members of the same species which aggregate for purposes of food gathering, protection, or reproduction are practicing a form of mutualism.

It is unfortunate that interactions such as competition have received so much attention in the past because this has tended to overshadow the biological importance of mutualism. Both are vital in the evolution of living things.

Parasitism. It may have occurred to the reader that intimate associations like commensalism and mutualism could become parasitic if one of the

Figure 23.9. An insect visiting a flower. (© *Walt Disney Productions.*)

members began to utilize the other as a food source. Perhaps this is how parasitism evolved. Parasitism is a relationship in which an organism spends part or all of its life cycle on or within another organism and uses the food or the tissues of the host for nourishment. Thousands of species of wasps deposit their eggs in the bodies of other insects. The larvae use the internal tissues of the host for food. One large group of wasps even uses plants as the host tissue.

There are many degrees of parasitism and some barely fit the definition. Thus, in "social" parasitism the female Brown-headed Cowbird does not build a nest, but rather, she deposits her eggs in the nests of other birds. The young cowbird is cared for by the host bird, sometimes to the detriment of the host's own progeny.

Parasites can be divided into *ectoparasites,* which live on the exterior of the host, and *endoparasites,* which dwell inside. Examples of ectoparasites are fleas and lice in animals (Figure 23.10) and dodder in plants (Figure 23.11). The highly specialized parasitic way of life requires many structural adaptations. Thus, fleas have laterally flattened bodies for crawling through their host's fur, and legs adapted for clinging to hair (Figure 23.10). Internal parasites are even more modified. Parasitic worms, for example, frequently show a drastic reduction in all body systems except the reproductive apparatus (Figure 23.12). This latter system is usually complex and extremely fecund. Not infrequently the life cycles of internal parasites include a stage of asexual reproduction. Since the evolution of internal parasitism was probably long and precarious, internal parasites are often highly specific in choice of hosts. Diagrammatic life cycles for two other common parasites are shown in Figures 12.7 and 12.8.

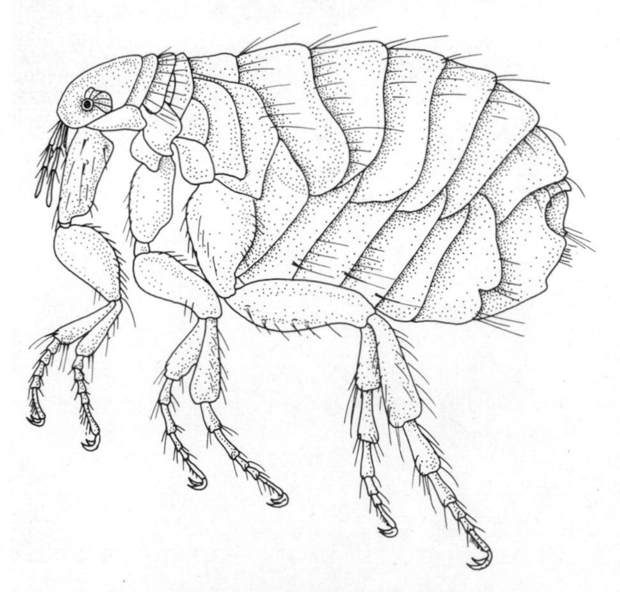

Figure 23.10. A common ectoparasite, the flea.

Parasitism is evidently a successful way of life judging by the large number of forms in both plant and animal kingdoms that utilize it to some degree. The term "successful" as used here means that the parasite must not damage its host too severely or its way of life would become untenable—if the host dies, so does the parasite. There are even instances of parasitism within the same species. In several of the abyssal (deep-sea) fishes the male becomes permanently attached to the much larger female. Do you see the adaptive advantage of this arrangement in the lightless depths of the ocean?

Predation

Predation is another type of exploitation of one species by another. In a way it is a form of parasitism, except that the predator does not use its host (prey) as a habitat. Commonly, the term predator refers to animals that feed on other animals. Predation has been studied extensively in natural environments and in the laboratory, so there is abundant literature on the subject.

In Chapter XXI we presented the food chain

grass → grasshoppers → mice → skunks → owls

This is also a chain of predators. A number of studies indicate a close correlation between the abundance of a predator and the abundance of its prey. A study by D. A. MacLulich many years ago showed a correlation between population fluctuations in hare and lynx over a number of years. As the hare population rose, the lynx population also increased. As the hare population declined, the lynx population did likewise. An example of the same kind of activity is shown in Figure 23.13, which graphs the effect of predators (other insects) on mealybugs that live in citrus trees. Here again the predator-prey populations are evidently tied to each other, as can be observed from the graph. It is tempting to conclude that the predators *cause* the decline in prey. This

Figure 23.11. Dodder, a plant ectoparasite, grows into an entangled, orange-colored mass on its host.

cannot be assumed in every instance because there are many factors other than predation that bring about periodic population fluctuations. For example, one of the mealybug's predators was handicapped by cold weather and thus was less abundant during the winter months. This predator's population changed in response to seasons rather than to abundance of its prey.

Predation is a natural and necessary occurrence in a community's energy flow. It benefits the prey population when acting as a control on population numbers. Moreover, if the victims of predation are the sick, old, or less well-adapted members of the prey population, their loss in the long run may benefit the gene pool of the species.

There are examples of predation in the plant kingdom, such as the familiar Venus fly trap. Other animal-catching plants include the vaselike pitcher

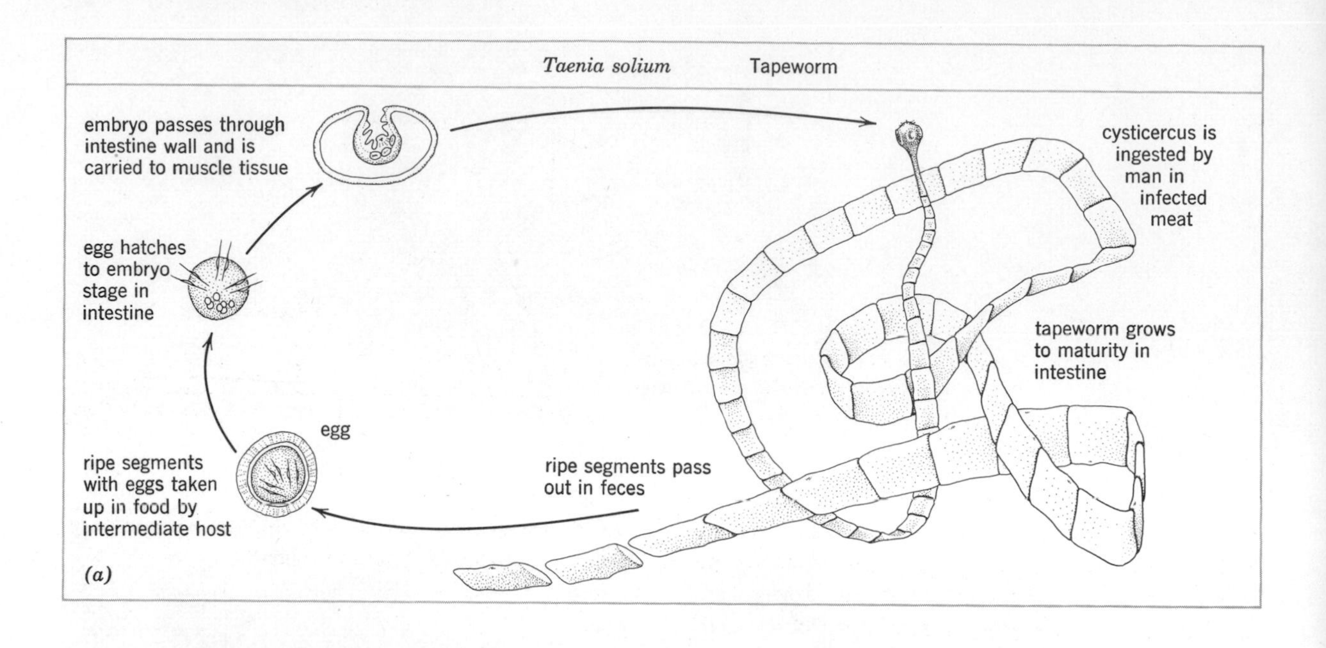

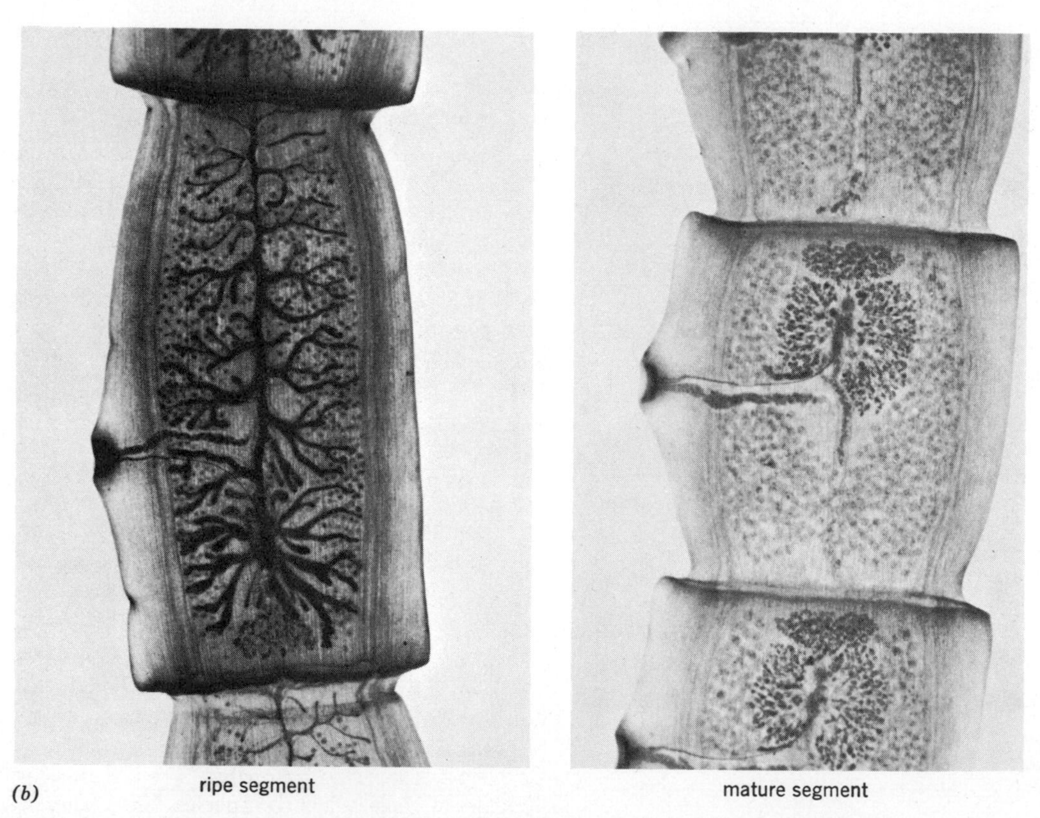

Figure 23.12. (a) *Life cycle of the tapeworm.* (b) *The photomicrographs show that nearly all of each segment is composed of reproductive structures (dark granules and branched structures).*

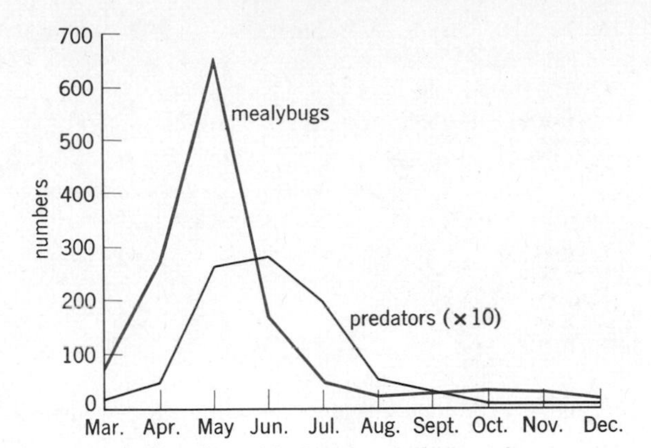

Figure 23.13. Population changes in mealybugs and their predators (Dr. P. DeBach).

plant into which insects fall and then are digested, and certain soil fungi that entrap nematode worms. In all of these examples the predators are thought to derive certain nutrient materials, primarily nitrogen compounds, from their prey.

Finally, we might consider briefly the role of man as a predator. First, he is the most voracious and wasteful of all predators. Second, he has frequently attempted to interfere with other predator-prey relationships, especially with respect to game animals. Seldom have these attempts led to any beneficial ends, as shown earlier with the example of the deer on the Kaibab plateau. Rachel Carson's *Silent Spring* presents cases of the catastrophes that follow inadvertent poisoning of insect predators and parasites when insecticides are used to control the prey species; upon removal of predators, prey often increase to higher levels than existed prior to the control attempt.

We hope the reader realizes that all interactions are solutions to the problem of survival. All presumably arose by natural selection as adaptations for obtaining nutrition, protection, and mates. As man continues to utilize natural communities and ecosystems, he should recognize the importance of *all* interaction and avoid manipulations that suit only his preconceptions of what is "good" and "bad" in these associations.

Principles

1. Two common forms of interaction among organisms are competition and predation. Forms of interaction that lessen competition, such as territoriality and symbiosis, have been favored by natural selection.

Suggested Readings

Argo, Virgil N., "Insect-trapping Plants," *Natural History*, Vol. LXXIII (March, 1964).

Argo, Virgil, "The Bull-Horn Acacia," *Natural History*, Vol. LXXIV, No. 1 (January, 1965).

Batra, Suzanne W. T. and Lekh R. Batra, "The Fungus Gardens of Insects," *Scientific American*, Vol. 216 (November, 1967). Offprint No. 1086, W. H. Freeman and Co., San Francisco.

Brower, Lincoln P. and Jane Van Zandt, "Investigations into Mimicry," *Natural History*, Vol. LXXI, No. 4 (April, 1962).

Carson, Rachel, *Silent Spring*. Houghton Mifflin, Boston, 1962.

Debach, Paul, "Population Studies of the Long-tailed Mealybug and Its Natural Enemics on Citrus Trees in Southern California," *Ecology*, Vol. 30 (1949), pp. 14–25.

Freudenthal, Hugo, John Lee, and John McLaughlin, "Some Symbionts of the Sea," *Natural History*, Vol. LXXV, No. 9 (November, 1966).

Guhl, A. M., "The Social Order of Chickens," *Scientific American*, Vol. 194 (February, 1956). Offprint No. 471, W. H. Freeman and Co., San Francisco.

Hall, Edward T., "Territorial Needs and Limits," *Natural History*, Vol. LXXIV, No. 10 (December, 1965).

Harper, John L. and J. N. Clatworthy, "The Comparative Biology of Closely Related Species. VI. Analysis of the Growth of *Trifolium repens* and *T. fragiferum* in Pure and Mixed Populations," *Journal of Experimental Botany*, Vol. 14 (February, 1963), pp. 172–190.

Limbaugh, Conrad, "Cleaning Symbiosis," *Scientific American,* Vol. 205 (August, 1961). Offprint No. 135, W. H. Freeman and Co., San Francisco.

Mykytowycz, Roman, "Territorial Marking by Rabbits," *Scientific American,* Vol. 220 (May, 1968). Offprint No. 1108, W. H. Freeman and Co., San Francisco.

Odum, Eugene P., *Ecology.* Holt, Rinehart and Winston, New York, 1963, pp. 93–109.

Questions

1. Why is it difficult to observe competition occurring between two different species in nature, that is, under natural conditions?
2. Name two types of *intraspecific* competition. Why is it easier to observe this type of competition than the interspecific type?
3. What evidence could you provide to support the idea that territoriality and peck order are present in human societies?
4. Can you provide examples of mutualism, commensalism, and parasitism not mentioned in the chapter?
5. Name some ways in which man is a predator in the ecosystems. How could he be a *beneficial* predator in an ecosystem?

CHAPTER XXIV

Ecological Geography of Terrestrial Environments

A juniper tree in the high Sierras.

Ecological Geography of Terrestrial Environments

Communities of plants and animals vary over the surface of the earth, largely in relation to climatic differences. These communities are not haphazard assemblages; the same general types of producers and consumers appear together in similar climatic zones. In this chapter we examine the distribution of these major ecosystems over the earth, their general composition, and some of the principles that influence their distribution.

Terrestrial ecosystems are classified on the basis of their biotic components; thus they are often called *biomes*. They are named according to their dominant plant forms; tundra, coniferous forest, temperate deciduous forest, grasslands, desert, tropical forest, and temperate rain forest. Some authorities subdivide these thus making a longer list. However, recognition of these seven will serve our purpose. The world distribution of these biomes is shown in Figure 24.1.

Tundra

The tundra is a vast, treeless zone bordering the Arctic Ocean in North America, Europe, and Asia (Figure 24.2). It generally extends from the treeline to the areas perpetually covered with ice and snow. Obviously, this area has a cold climate: the ground remains permanently frozen to within a few inches of the surface, and the growing season is only about sixty days. There is also an alpine tundra on the peaks and high slopes of moun-

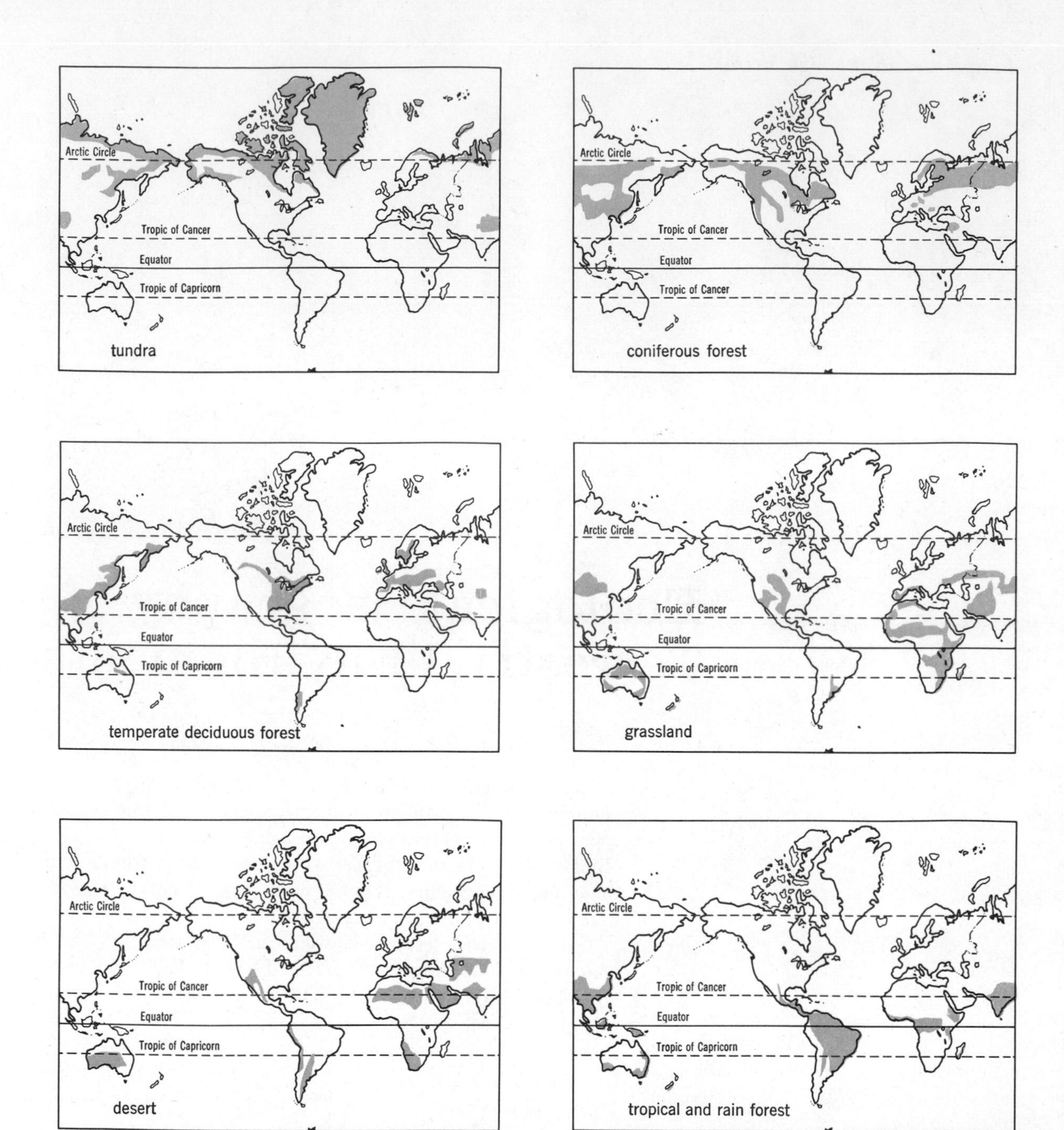

Figure 24.1. Distribution of major biomes.

Figure 24.2. The arctic tundra. The absence or small size of trees (foreground) is characteristic.

tains, as in the western Rocky Mountains, the Alps, and the Himalayan Mountains.

In general, tundra vegetation consists of lichens, mosses, grasses, and dwarfed woody plants. The composition of the vegetation in any particular portion of the tundra varies with the thickness and fertility of the soil. Grasses and sedges compose the alpine tundra. In all cases, numerous adaptations for survival in this extreme environment are found: dwarfism, small hairy leaves for water conservation, ability to survive in the frozen state even when flowering, and means of vegetative reproduction.

We might suppose that animal life would be sparse under these conditions, but this is not true throughout the year. During the summer great numbers of waterfowl nest in the tundra, and several species of insects—especially mosquitoes and black flies—are abundant. Permanent residents are few. Small rodents including the well-known lemmings are the most abundant mammals. Other characteristic forms of animal life are caribou (reindeer in Eurasia), arctic hare, arctic fox, gray wolf, grizzly bear, polar bear, and the snowy owl. These forms show adaptation for winter survival, including white coloration, ability to hibernate for periods of time, and means for burrowing under the snow.

Northern Coniferous Forest

The northern coniferous forest is also called boreal forest or taiga. It consists of a vast belt of evergreen forest that crosses North America, Europe, and Asia south of the tundra. Large "fingers" of this biome stretch southward in North America and Eurasia. In this area the winter climate is cold and severe, similar to that of the tundra; however, the growing season is much longer—three to five months.

Figure 24.3. The northern coniferous forest.

Figure 24.5. The crossbill, a common resident of northern coniferous forests. The bill is adapted for removing seeds from pine cones.

Figure 24.4. Snowshoe hare. Like many animals of the northern coniferous forest, this hare has a different coat color in summer than in winter. Notice the streaks of dark fur showing on the head. (© Walt Disney Productions.)

Vegetation in this biome is composed mostly of needle or scale-leaved evergreen trees: spruces, firs, pines, and cedars (Figure 24.3). Their leaf shapes prevent excessive evaporation of water and freezing. Even the flexible branches and conical shape of the trees aid them in avoiding snow damage during the winter. Their dense growth habit produces a deep shade that inhibits the growth of herbs and shrubs on the forest floor. Not uncommon are thickets of broadleafed plants like birch and alder. The heavy concentration of plants in this biome presents an abundant food supply of leaves, wood, and seeds. Consequently, it supports a considerable assemblage of consumers.

Along the west coast of North America from Alaska to central California is a luxuriant, humid, coniferous forest with a warmer climate than the remainder of the biome. Hemlock, cedar, fir, and arborvitae compose much of this forest, with the spectacular redwood trees present in restricted areas of northern California. Coastal fogs frequently drift through these forests and the resulting moisture functions as precipitation. These forests are sometimes called temperate rain forests.

Animal forms of the northern coniferous forest include moose (called "elk" in Eurasia), snowshoe hare (Figure 24.4), rodents, wolf, fox, lynx, porcupine, and birds such as crossbills, siskins, and evening grosbeaks. Crossbills are uniquely adapted for this ecosystem; the tips of the bill cross, rather than meet, as a device for removing seeds from cones (Figure 24.5). This biome is of considerable economic importance because of its extensive pulpwood, lumber, and fur resources.

Temperate Deciduous Forest

The term *deciduous* refers to plants that lose all their leaves for a part of the year, a trait common to many of the trees and shrubs in this biome (Figure 24.6). These trees are sometimes spoken of (in the vernacular of the lumber industry) as broadleafs or hardwoods in contrast to coniferous, narrow-leafed softwoods.

Deciduous forest communities originally formed a continuous band across eastern North America, the British Isles, Central Europe, large portions of China, and southeastern Siberia. A few small areas are found in the Southern Hemisphere. Today much of this area is occupied and utilized by human beings so that the biome no longer exists in its original form. Its initial boundaries should be kept in mind, however, since they form the basis for the present distribution of many organisms.

Figure 24.6. The deciduous forest.

The term *temperate* very generally describes the climate of the biome, although this is misleading to some extent. In many instances temperate forests have a distinct seasonal pattern with cold winters and hot summers, but on the average, the climate is moderate. A significant aspect of the climate is the moderate rainfall (30–60 inches per year), which is evenly distributed over the seasons. Among other things, this accounts for the broadleaf forest climax that is typical of the ecosystem.

This biome consists of a series of community types, each with a different set of dominant plants. Thus an oak-hickory forest is most abundantly populated by species of oaks and hickories. A beech-maple community is dominated by beech and maple trees. All of these communities share the striking seasonal contrasts of being leafless during the winter and

Figure 24.7. A pine flatwoods community. Saw palmetto (palm-like leaves) is an abundant member of this community.

Figure 24.8. The tropical rain forest.

heavily foliaged during the summer. This trait alone has a considerable influence on the animals that inhabit these communities. Common trees include beech, tulip tree, sycamore, maple, oak, hickory, elm, birch, basswood, and a variety of conifers (Figure 24.7).

Larger animals typical of the biome include the familiar white-tailed deer in North America and other species elsewhere, a number of large cats such as mountain lions and wildcats, foxes, and black bear. Many species of rodents, including squirrels, chipmunks, and mice, inhabit the biome in addition to familiar forms like raccoons and opossums. The birds of temperate forests are largely arboreal and tree-nesting. In North America, warblers, flycatchers, woodpeckers, vireos, and wrens are common, although these are but a few of the long list of species. Tree frogs, salamanders, and numerous snakes are typical. All told, these communities have a rich fauna. Tree dwelling is facilitated by the sharp claws of squirrels and woodpeckers, movable scales of some snakes, and various nesting habits adapted to trees. In the colder portions of the biome, many forms either migrate during winter or else "den up" under the ground or in dead logs to escape the cold. Woodchucks, bats, chipmunks, and numerous insects spend the winter in hibernation. Others, like the black bear, undergo semi-hibernation.

The southeastern United States is usually included in the temperate deciduous forest ecosystem even though it contains two large vegetation units that are not deciduous. The magnolia-oak forest is one of these, although it also contains some deciduous species. Some ecologists term this forest a broad-leafed, evergreen, subtropical forest, although the local designation for these woodlands is *hammock.*

Extensive areas of the southeastern coastal plains are occupied by pine forests known as pine flatwoods (Figure 24.7). These frequently occur on poorly drained flat areas and are dominated by longleaf, slash, or loblolly pine. A visitor to this forest gets the impression of a scattered array of pines and a sharply delimited shrub layer consisting predominantly of saw palmettos. This ground layer, however, may contain a large variety of shrubby and herbaceous plants.

Tropical Forest Biomes

Contrary to the impression we are likely to acquire from movies and television, the tropics are not uniformly enveloped by steamy, impenetrable jungles. Instead we find a variety of ecosystems such as deserts, scrub lands, deciduous forests, rain forests, and cloud forests. In this section we shall consider rain forests in particular, as they constitute a sizable belt of vegetation surrounding the earth in equatorial latitudes: the Amazon and Orinoco basins in South America, a large segment of Central America, a band in central and western Africa, and much of the Indo-Malayan region.

In terms of climate, the outstanding features are year-round temperatures that are consistently high, uniform lengths of day and night, and rainfall exceeding eighty inches a year. In regions where the rainfall is seasonal, such as monsoons alternating with dry periods, rain forests give way to a tropical deciduous forest. Notice that the deciduous adaptation here is in relation to rainfall rather than temperature.

An outstanding vegetational feature of rain forests is that virtually all of the plants are woody and grow as trees. Bamboos, which are grasses, exemplify this. Thick-stemmed woody vines (*lianas*) and *epiphytes* abound as one of the most typical features of the biome. Some of the climbers are "stranglers": they gradually envelope the host plant, kill it, and then remain as free-standing plants. In contrast to temperate forests, tropical forests contain a large number of species and relatively few individuals of the same kind. The dominant trees average 100–180 feet in height and are slender and unbranched except at the crown or top. Palms and tree ferns are often abundant. The thick crown presented by the trees casts a heavy shadow so that the forest floor is frequently dark, humid, and open. Dense, junglelike growth is found only in openings, where succession is occurring, or along river and stream banks where sunlight can penetrate (Figure 24.8). Many of the earlier accounts described tropical forests as being junglelike, possibly because the observers always traveled in these areas by boat.

Rain forests are well stratified in the sense that there are distinctive layers of vegetation and animals at various heights in the forest. The thick canopy formed by the tops of the taller trees constitutes one layer, and is inhabited by a characteristic assemblage of arboreally adapted organisms. Another layer, in contrast to the canopy, is the forest floor where an entirely different set of physical conditions prevail. Intermediate strata exist between these and have their own specialized forms of plant and animal life.

Table 24.1

Birds		Snakes	
Region	Number of Species	Region	Number of Species
Labrador	81	Canada	22
New York	195	United States	126
Panama	1100	Mexico	293
Colombia	1395	Brazil	210

In the American tropics, an epiphyte group called *bromeliads* are notable because they hold a gallon or so of rainwater in their nested leaves. In these small pools live many insects, frogs, and other tiny organisms.

Animal life in this biome is exceedingly varied in number of species. Table 24.1 shows the richness of two types of fauna as compared with those of other regions. As with plants, there are fewer *individuals* of each species than would be found in temperate areas. This diversity of species makes it impractical to list even typical forms; the following discussion merely points out a few noteworthy adaptations.

A visitor to a rain forest might be disappointed in the apparent absence of animal life, because many of the forms are either nocturnal or dwell in the treetop canopies. Many animals are adapted for fruit eating and nectar feeding. Many of the canopy dwellers build hanging nests, possibly as a solution to the competition for nesting sites or as protection against ants and other marauders. Most mammals and birds inhabiting rain forests tend to be smaller than their temperate-zone relatives. Reptiles and arthropods (insects, spiders, etc.), however, reach their largest size in tropical areas.

The soil in the tropics is not well suited to farming or grazing for domestic animals and is quickly damaged by these activities. Ecologists and students of animal husbandry have urged Africans to treat their wildlife as a crop to be managed and harvested with care. For example, one elephant is equivalent to about eighty sheep as a protein source, can be "harvested" without great difficulty, and does far less

Figure 24.9. Protein equivalents of African wildlife as compared to domestic sheep.

Figure 24.10. Grassland. Evidence of burrowing forms is common in such areas.

damage to the ecosystem than cattle or sheep (Figure 24.9). So far, these suggestions have been ignored and increasing attempts are being made to practice agriculture in the conventional way. P. W. Richards, an authority on tropical forests, believes that these forests will disappear in less than a hundred years unless the present rate of their destruction by man is reduced.

Grasslands

Grasslands occur on all continents and are frequently similar to one another in climate, physical features, and fauna (Figure 24.10). Grasslands have localized names wherever they occur: *prairie* for the tall grasslands of western North America and *plains* for the short grasses of that region; *pampas* and *savanna* in South America; *veld* in South Africa; and *steppe* in Russia. In tropical regions the term savanna is also applied to grasslands containing scattered trees.

A major climatic feature leading to the formation of grasslands rather than forests is the uneven seasonal distribution of rainfall. Grassland areas may have as much rain as forty to sixty inches a year, but it falls either erratically or, as in the tropics, interspersed with long dry periods. Trees cannot survive these droughts and the frequent fires that accompany them.

A large proportion of the animals of this biome are adapted for running, leaping, or burrowing. In North America, these include the jack rabbit, antelope, bison, fox, coyote, ground squirrel, and pocket gopher. Many ecological equivalents are noted over the world: Australian kangaroo and North American antelope; Australian marsupial mole and North American pocket gopher; North American wolf and Asiatic cheetah; and many more. Each member of these "pairs" performs a similar role in its respective grasslands area.

Man has made extensive use of this community. The richer grasslands of North America and Europe are major cereal-producing centers. The poorer grasslands over the world are frequently overused as grazing lands, often to the point of exhausting the land. In all parts of the world, grasslands have served as vast pasturelands for nomadic people with their cattle, sheep, goats, camels, and horses.

Deserts

As moisture becomes less available, grasslands give way to deserts (Figure 24.11). About one fifth of the surface of the earth consists of desert, and all major

Figure 24.11. Desert. These plants show adaptations for reduction of water loss.

continents contain a biome of this type. The Old World deserts include the Sahara and those found in Asia Minor, India, Tibet, China, and Mongolia. Much of the interior of Australia is a desert. Prominent deserts also occur in southern Africa, as a strip along the west coast of South America, and, of course, in our own southwest.

The major controlling climatic factor in this biome is low annual rainfall, generally less than ten inches, unevenly distributed during the year. Intense sunlight, hot days, and cold nights (an 80°F range in some cases), high evaporation rate, and constant winds are characteristic of this generally inhospitable environment. Winter months are much milder than summer months and may even be cold in the northern deserts. The highest temperature ever recorded (134°F) by a standard observation station was in Death Valley, California, which is probably the hottest desert on earth.

When rain occurs in the desert it usually takes the form of a cloudburst, often resulting in destructive flash floods and sheet erosion. At other times, sand and dust storms constitute hazards for desert-dwelling organisms.

Plants adapted for survival under extremely arid conditions are known as *xerophytes,* and as we describe some of the typical desert vegetation, we shall review a few of their adaptations. Eugene Odum has concisely summarized the forms adapted to deserts. These are annual plants that grow only when adequate moisture is present, succulent plants—such as cacti—which store water, and shrubs with numerous basal branches and small, thick leaves that are shed during extremely dry periods. The seeds of the annuals lie dormant in the soil for long periods of time, and then during a rainy period quickly germinate, grow, flower, and produce new seeds. The seeds, in other words, provide the organism's surviving link between two widely separated rainy periods. There is nearly always wide spacing among desert plants, which, in some instances, is known to be caused by a chemical released from the roots. This substance inhibits other plants from growing nearby (see p. 124). In addition, some desert plants release inhibiting substances from fallen leaves. In North American deserts, familiar plants include the creosote bush, sagebrush, bur sage, giant Saguaro cactus in some areas, and a number of grasses.

Reptiles, rodents, and scorpions are the most characteristic animal life of the deserts (Figure 24.12). Many of the animals are active at night and spend the daylight hours either underground or in shaded nooks. Many forms are burrowers. Most desert dwellers are lighter in color than their relatives in moister climates. This protective coloration camouflages them on the desert floor. Birds, insects, and many reptiles excrete nitrogenous wastes in the form of concentrated uric salts which further aid them in conserving water. An interesting physiological adaptation is the ability of some of the rodents to exist solely on water from their internal metabolic processes.

When infrequent rain pools are formed, frogs and toads, aquatic insects, and crustacea pass quickly through their breeding cycles. Estivation enables some forms to survive long periods of drought, and hibernation occurs during the winter among those forms of life found in more northerly deserts.

Ecological Principles and Biome Distribution

All biomes consist of assemblages of producers, consumers, and decay organisms; in other words, they possess the ecosystem structure described in Chapter XXI. Their energy flow diagrams and energy pyramids may vary somewhat, and the organisms composing them may be different, but all function in a similar manner. One example is given by the many kinds of ecological equivalents that exist among similar but widely separated ecosystems.

Energy flow is the most important functional feature in all communities and, of course, plants are the photosynthetic or energy-capturing agents in

Figure 24.12. A scorpion feeding on a spider. (© Walt Disney Productions.)

every case. As noted previously, biomes are based on typical assemblages of plants because they accurately reflect the broad, climatic features of the earth. Since the success of vegetational units depends on a combination of solar radiation, precipitation, and temperature, these factors are major ones in determining the distributional patterns of major ecosystems. The plants characteristic of each biome, being a product of natural selection and evolution, show many adaptations related to the special conditions of their specific environment. Conifers in the boreal forest are one example. Vegetation also modifies the climate in an environment. This reciprocal reaction is important in succession and in providing suitable living conditions for animals.

Since animals ultimately depend on plants for energy, the nature of the fauna of a biome is strongly influenced by the vegetation found there. Many of the adaptations described in this and other chapters involved features that enable animals to function efficiently with the vegetational complex in which they live. In a biome such as the grasslands, animals can be responsible for many of its major characteristics. In this instance burrowing, grazing, and even trampling strongly influence the kinds of plants that live in the community.

Principles

1. The major terrestrial ecosystems (biomes) are categorized by a distinctive biotic composition and are usually named according to their dominant plant forms.
2. The distributional pattern of biomes is determined largely by a combination of the factors of solar radiation, precipitation, and temperature.

Suggested Readings

Amos, William H., "The Life of a Sand Dune," *Scientific American*, Vol. 201 (July, 1959).

Barnett, L. and editors of *Life* Magazine, *The World We Live In*. Time, New York, 1955.

Cooper, C. F., "The Ecology of Fire," *Scientific American*, Vol. 204 (April, 1961).

Darling, Frazer F., "Wildlife Husbandry in Africa," *Scientific American*, Vol. 203 (November, 1960).

Jaeger, E. C., *The North American Deserts*. Stanford University Press, Palo Alto, California, 1957.

Llano, George A., "The Terrestrial Life of the Antarctic," *Scientific American*, Vol. 207 (September, 1962). Offprint No. 865, W. H. Freeman and Co., San Francisco.

McNeil, Mary, "Lateritic Soils," *Scientific American*, Vol. 211, No. 5 (November, 1964).

Odum, Eugene P., *Ecology*. Holt, Rinehart and Winston, New York, 1963, pp. 123–135.

Questions

1. Make a list of the biomes described in the chapter and name one or more indicator plants and animals for each.
2. Why are biomes named according to their dominant plant forms rather than their major animal forms?
3. In what biome type do you live? What are some of the community types in the biome where you live?
4. Can you state some ecological principles that apply to both desert and tropical forest biomes? Can these principles be applied to *all* biomes?
5. List the major factors that determine where the various biomes are located on the earth. Which of these factors, if any, plays a dominant role in your area?

Ecological Geography of Freshwater and Marine Environments

The intertidal region of the rocky shores during low tide. Note the zonation on the rock at center.

Ecological Geography of Freshwater and Marine Environments

This final chapter describes the two major habitats, freshwater and marine, in which water dominates the environment. Aqueous environments present some features not shown by terrestrial communities, although the same basic ecological concepts of ecosystem structure and function apply to both.

Some Major Features of Aqueous Environments

Heat. Thermal energy (heat) has important effects on lakes, seas, and other bodies of water owing to the high specific heat of water. That is, water absorbs tremendous amounts of heat and releases it slowly. Bodies of water accumulate heat during warm seasons and release it during cold ones. This process is vital to organisms living in water, for it means that extremes of temperature rarely occur. Water dwellers, in fact, live in a narrower range of temperature than land inhabitants. Moreover, land areas adjacent to large bodies of water undergo less extremes of temperature than other land areas. This is illustrated by citrus trees in mid-Florida. Following a severe freeze, the citrus trees bordering large lakes are not damaged, although the remainder of the grove suffers. On a larger scale, oceanic currents like the Gulf Stream influence the temperature and entire climate of continental coastlines.

Sunlight. As in terrestrial habitats, sunlight is the energy source for photosynthesis but water filters light waves as they pass through it. For example, red, orange, and ultraviolet waves are absorbed first, leaving only green, yellow, and blue to penetrate deeper. Even in the clearest water with the sun directly overhead, 300 feet appears to be the limit of effective photosynthesis. Usually photosynthetic zones are shallower. Consequently, deep lakes and all oceans contain only a small zone that will support producer organisms. Nevertheless, the entire biotic economy depends on these producers as do land areas.

The laws of energy and the concept of energy flow apply to freshwater and marine communities in the same way as they do to the land ecosystems.

Dissolved Gases. Oxygen is present in much smaller amounts in water than in the air. Air has the equivalent of 210 cubic centimeters (cc) of oxygen per liter (irrespective of temperature); at 15°C freshwater holds only 7.2 cc per liter, and sea water 5.8 cc per liter at saturated conditions; often the amount is less. Animals that breathe in water exist on these relatively small amounts of dissolved oxygen by virtue of their reduced metabolic rates and specially adapted breathing structures. Carbon dioxide dissolves readily in water and hence is present in greater amounts than in air, although in various forms: dissolved carbon dioxide, carbonic acid, carbonate, and bicarbonate. It is important in photosynthesis, regulation of pH, and other reactions.

Chemical Substances. Sea water contains over forty chemical elements, largely in the form of ions. This "ionic soup" provides an excellent chemical environment for life. Chloride and sodium ions are the most abundant. Other principle ions are calcium, magnesium, potassium, carbonate, sulfate, and bromine. Many of these also occur in fresh water but in smaller amounts. The dissolved materials most important to organisms are nitrogen compounds, phosphates, calcium salts, and silicates. Nitrates and phosphates are valuable nutrients for phytoplankton, and calcium is required by shellfish, corals, and other marine life. The use of nutrient salts by plants was discussed in Chapter XXI.

In both fresh and salt water, organisms face osmotic problems in retaining or excreting ions from their cells. In general, they have evolved osmoregulatory organs that enable them to live only in salt water or fresh water. Very few can survive in both.

In discussing marine habitats the term *salinity* is often used. Salinity refers to the number of grams of dry salts per 1000 grams of water (thus, parts per thousand, ‰). This measure, in theory, includes a number of salts, but in practice only the amount of chloride ions is determined. A mathematical factor is then applied to convert this amount to salinity. The salinity of oceanic waters approximates 35‰, whereas freshwater bodies generally have less than 1‰.

Major geochemical cycles frequently involve freshwater and marine environments, as illustrated by the phosphorus cycle described in Chapter XXII.

Freshwater Environments

Streams, rivers, ponds, lakes, and marshes are all freshwater habitats. These are sometimes divided into flowing-water and standing-water communities. The general structure of a pond ecosystem was presented in Chapter XXI.

Rivers have been utilized by man for travel, commerce, irrigation, and waste disposal. This last use has been abused to the point that many major rivers are now unsuitable for recreation or as sources of drinking water. Pollution, in fact, is a major ecological problem today. Water pollution usually results from domestic sewage (organic matter) or toxic chemical wastes from industrial sources. Organic matter in moderate amounts can be disposed of in streams and rivers by organisms of decay. In large amounts organic wastes cause a severe depletion of oxygen because the population of decay organisms becomes so large. The eventual result is that only anaerobic forms can survive. Toxic industrial wastes act either as oxygen depleters or as poisons to all forms of life. Insecticides from farm lands are known to cause severe pollution in some areas. Soil erosion also produces pollution in the form of silt in streams and rivers. Lakes are not often used for waste disposal, but some areas of the Great Lakes are now badly polluted by industrial and municipal wastes. All these forms of pollution can be eliminated or controlled, although legislation is usually necessary to bring this about. Unfortunately, such legislation invariably waits until an informed and alarmed public demands it.

In terms of geological time, lakes have short life spans. In fact, they provide classic examples of succession as they fill with sediments and encroaching plant life, eventually becoming marshes or bogs.

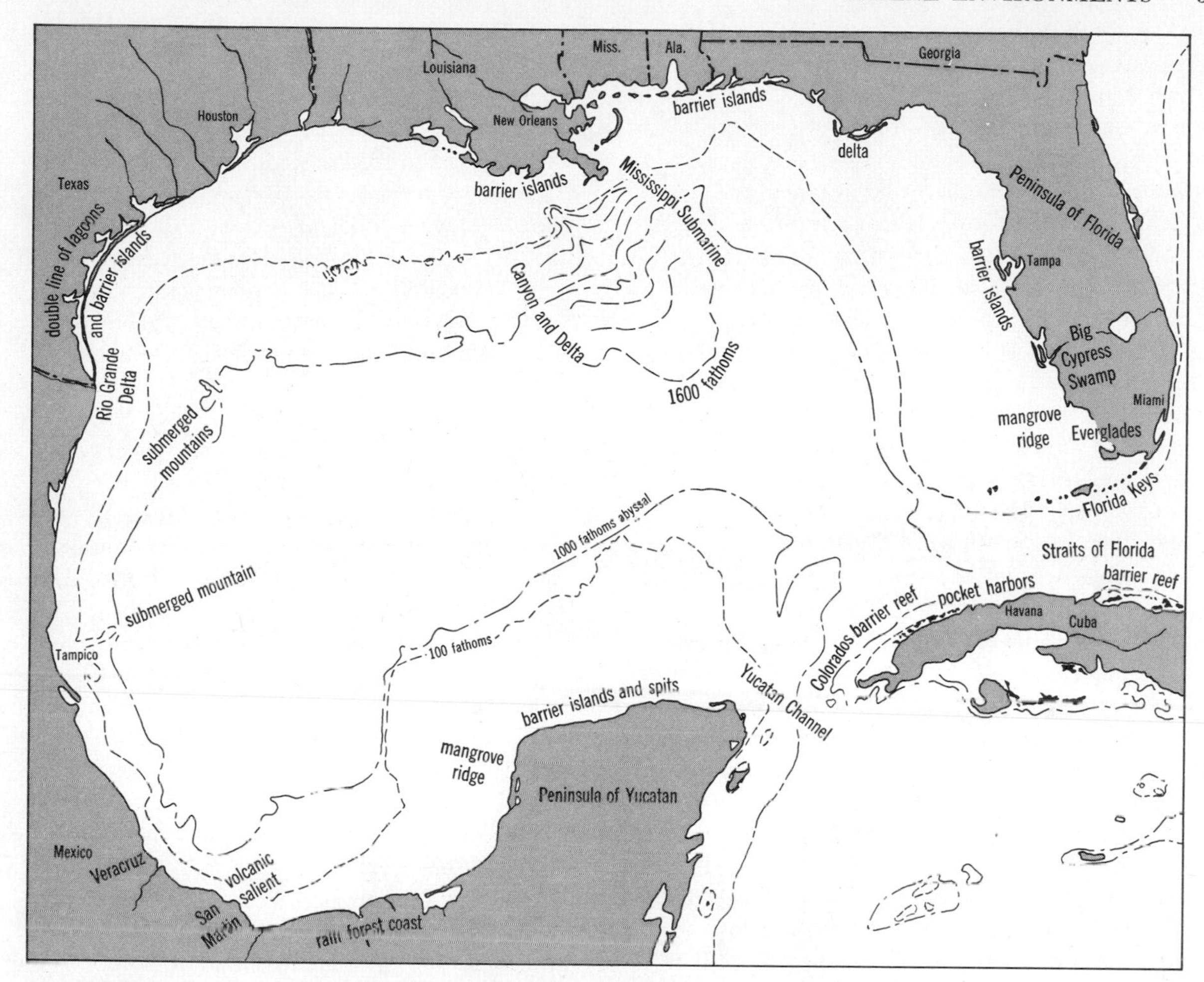

Figure 25.1. A map of the Gulf of Mexico. (*Modified from W. A. Price, "Shores and Coasts of the Gulf of Mexico." In* **Gulf of Mexico: Its Origin, Waters and Marine Life.** *Fishery Bull. 89 of the Fish and Wildlife Service, Vol. 55.*)

Marine Environments

Oceans present the most extensive of all habitats, occupying more than two thirds of the surface of the earth. As our example we have chosen the Gulf of Mexico. It represents an ocean in miniature in its features, and forms a convenient unit for the description of marine environments.

General Features. Figure 25.1 is a topological map of the Gulf of Mexico showing some of its principal features. The Gulf is essentially a large basin with an opening in one side. The periphery of the basin is formed by a *continental shelf* at depths down to 100 fathoms (600 feet). As the map shows, the 100-fathom contour line indicates a wide shelf for most of the Gulf, except along the coast of Mexico. Continental shelf areas are important in all oceans because an abundance of life inhabits them. Bottom-dwelling plants and animals, *plankton* (forms with limited swimming ability), and *nekton* (swimmers) form rich and elaborate food webs and pyramids. Many open-ocean fishes return to continental-shelf areas to spawn or to feed. The continental shelf and coastal communities of the Gulf include coral reefs (atolls, fringing reefs, and barrier reefs; see Figures 25.2 and 25.3), extensive saltwater

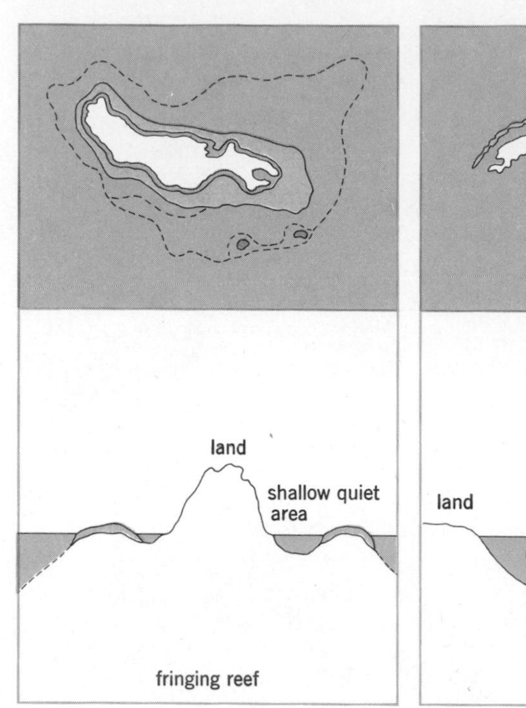

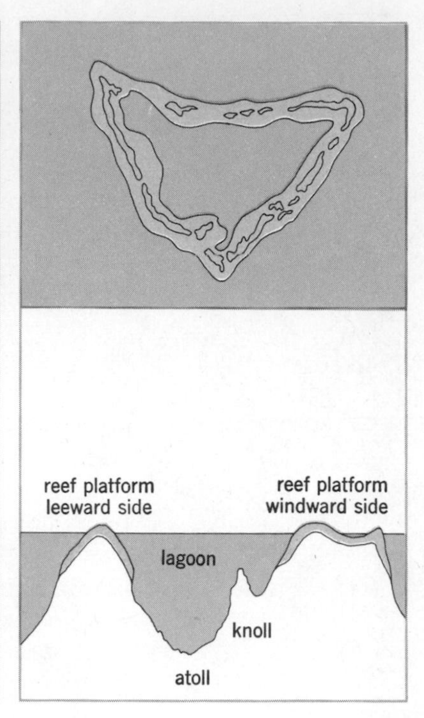

Figure 25.2. Types of coral reefs.

Figure 25.3. A massive growth of staghorn coral.

marshes and mangrove swamps, numerous barrier islands built by storm action, and extensive submarine "meadows." Such variety of community types supports an enormous assortment of producers and consumers.

Extending across a portion of the Gulf is a large triangular area with depths exceeding 2000 fathoms. At this depth is the cold, lightless, *abyssal zone* into which all energy supplies must be imported from above. A large, submerged mountain range lies off the coast of Mexico, and a deep submarine canyon is located offshore from the Mississippi coast.

Generally speaking, the Gulf is a warm body of water. Surface readings show a summer mean temperature of 84°F and a winter mean temperature of around 70°F. In the depths it is much cooler; from 900 meters down the temperature has been measured at 40°F.

The open Gulf has a salinity of about 35 ‰, but salinities along the coastline fluctuate widely, especially near the draining areas of the rivers. Hence, off the mouth of the Mississippi, salinities of less than 24 ‰ have been recorded.

Tides and Currents. Tides are caused by the gravitational pull of the moon and, to a lesser degree, the sun. When the two are pulling together, as happens twice a month (full moon and new moon), the largest extremes of high and low tides occur. When these influences are opposed, the least amount of flow and ebb takes place. Additional influences on tidal movements are those of wind and change in barometric pressure. This is especially true in the Gulf with its relatively small ranges of tides and shallow coastal waters. The mean annual tide on the Gulf coast does not normally exceed two or three feet (compared with ten feet along the Atlantic coast). This small tidal fluctuation is enough to alternately flood and drain extensive areas of salt marsh, mangrove swamp, and mud flats. One of the dangers of hurricane-type storms in the Gulf is the possibility that large amounts of water will pile up against the low-elevation coastline, resulting in flooding.

Community Structure and Function. The sea in most aspects offers a hospitable environment for living things; organisms are bathed in salts and minerals, temperature variations are moderate, dessication is not a problem, food is frequently abundant, and heavy skeletal supports are unnecessary. Generally speaking, marine communities share some features with the pond ecosystem described previously; that is, algae form the producers, zooplankton are the primary consumers, and so on. There are also major differences. Some algae (seaweeds) that are important producers are macroscopic (Figure 25.4). In many marine communities, insects are absent, their ecological equivalents being small crustaceans. There is a greater variety of life forms than in fresh water. In contrast to fresh water, the sea contains many *sessile* (attached) organisms such as sponges, corals, and sea anemones (Figure 25.5) which may be ecologically important. In fact, major groups such as the sponges, coelenterates, echinoderms (starfish and sea urchins, Figure 25.6), and annelid worms are abundant and important in the sea but play a minor role or are lacking in freshwater communities.

Plankton play a large role in the energy flow in marine communities both as producers (phytoplankton) and as consumers (zooplankton). Producer plankton are mostly *diatoms* and *dinoflagellates* (Figure 25.7). Diatoms are unicellular algae encased in two halves or valves that are impregnated with silica, an important feature (to biologists) since these forms may be represented in fossil formations. Their significance to marine and freshwater communities is that they are the most abundant microscopic plants, and hence the major component of the producer level.

Dinoflagellates, small photosynthetic forms, are also abundant plankton. Some are luminescent and give rise to spectacular displays at nights. Others like *Gymnodinium brevis* cause the death of marine creatures by the production of toxic substances. The consequences, known as red tide, may assume catastrophic proportions in some areas. A notable feature of all phytoplankton is their occasional tremendous increase in numbers, a so-called plankton "bloom." This may occur seasonally, for example, in the spring and late summer, or may result in some areas from a welling up of nutrient minerals.

The other major microscopic plankton consist of *Foraminifera* and *Radiolaria*. Members of these groups live in tiny, sometimes ornate, shells that they secrete. Their numbers are so great that portions of ocean bottoms may consist of foraminiferal or radiolarian oozes of great thickness.

The small crustacea, particularly copepods, are among the most important and abundant primary

Figure 25.4. Kelp, a large marine alga.

Figure 25.5. A sea anemone, a sessile predator.

Figure 25.6. Starfish and sea urchins.

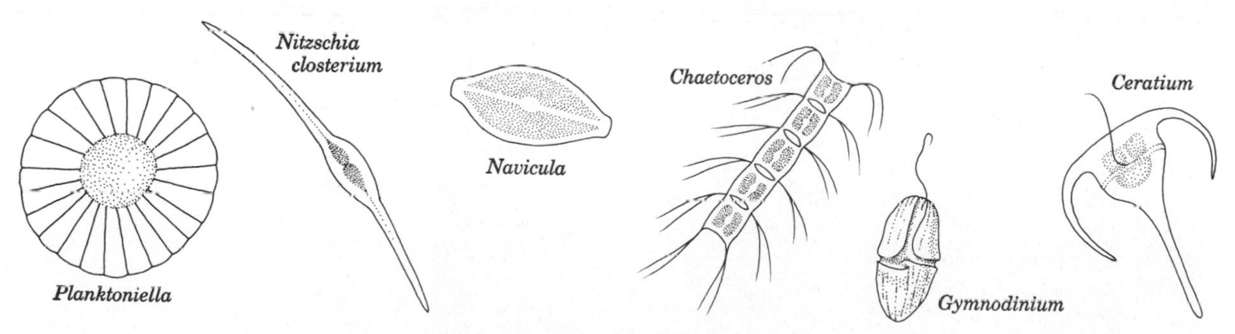

Figure 25.7. Some common kinds of phytoplankton.

consumers, which in turn are utilized by secondary consumers such as small fish. Other important crustacea are ostrocods, amphipods, and small shrimp and prawns (Figure 25.8).

The following paragraph from a paper by Gunter et al. (1948) summarizes the kind of chain reaction that plankton may exhibit.

"There was first the appearance of numbers of *Gymnodinium brevis* mixed in with other normal plankton types, mostly diatoms. . . . Locally, or over large areas there then appeared a "bloom" of *Gymnodinium*, and in these areas the mortality occurred. This was then followed by the decomposition of many dead organisms, with the consequent release into the water of much nutrient material. Bacteria and/or phytoplankton utilized this nutrient material and then were themselves utilized, especially by the *Copepoda*, which consequently increased enormously in the plankton . . . . The *Copepoda* devoured all the suitable diatoms, and left only the species of *Rhizosolenia*, which would be very difficult for the copepods to handle. . . ."

A common technique in studying plankton is to count the numbers in a small sample of water (one milliliter, for example) and then use this finding to calculate the number in larger volumes. Hence, biologically "poor" waters might have a hundred or less plankton per liter, whereas during a plankton bloom these waters might contain many thousands per liter. Quantitative studies have indicated that an increase in phytoplankton may be followed by an increase in zooplankton, a reasonable expectation. On the other hand, zooplankton have been observed to be abundant when phytoplankton were few in number, which seems contradictory to our concept of food pyramids. Recent investigations have shown that sea water frequently contains a considerable amount of organic detritus (tiny bits and even molecules of organic matter) concentrated on the interfaces of bubbles at the surface. Presumably, microcrustacea and small zooplankton feed on these.

Nekton, the large consumers, consists of fishes, marine mammals like the killer whale (Figure 25.9), sea birds (Figure 25.10), molluscs (squid), and turtles.

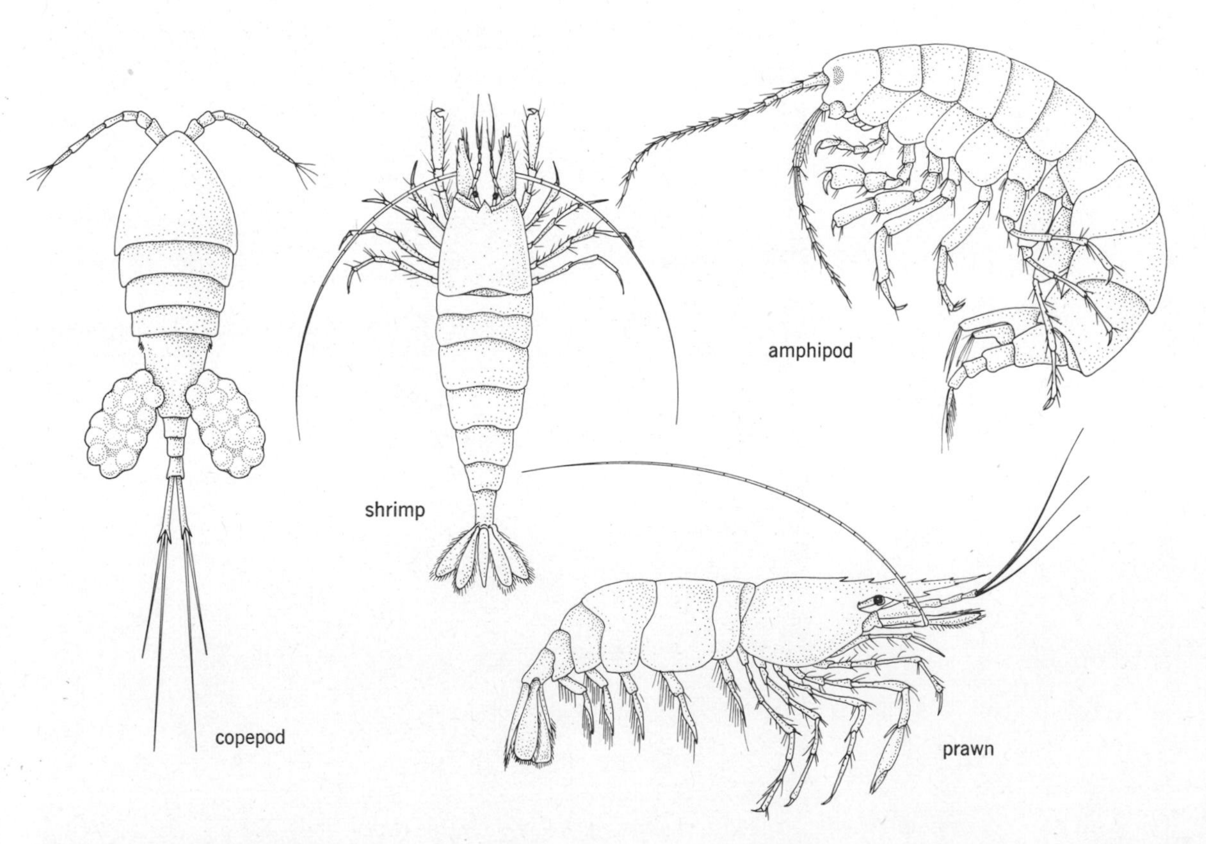

Figure 25.8. Assorted crustacea.

Although members of these groups may range the open sea, they are found in the largest concentrations near their energy sources: over continental shelves and along coastlines. These are important as well as frequently spectacular members of marine communities. Some of the fishes, mammals, and turtles are of economic value to man, a situation that means they are often unwisely utilized. But in this case, with the exception of a few of the mammals and turtles, man has not apparently damaged these marine crops.

Example of a Marine Community

Marine communities along the Gulf coast frequently form extensive underwater meadows of grasslike seed plants such as turtlegrass and manateegrass. Green algae may also contribute to this photosynthetic layer. These extensive underwater grasslands constitute a region of rich marine life, with burrowing and filter feeding perhaps the commonest adaptations.

Figure 25.9. Killer whale, one of the largest marine predators.

Figure 25.10. Brown pelicans. This species may now be endangered in California by insecticide pollution.

Many varieties of shellfish (pelecypods), burrowing worms, starfishes, marine snails, and crabs are the dominant forms of animal life. Bacteria probably reach their greatest abundance in the mud bottoms, and in turn provide a food source for many protozoans and zooplankton. In addition to filter feeding, many organisms are "deposit-feeders" in that they sift bottom sediments for bits of organic material, bacteria, and various plankton. Larger consumers such as fishes, porpoises, sea turtles, and birds feed extensively over these communities. Studies on the Gulf coastal area indicate that these shallow water areas serve as nurseries for a number of larger fish—their young dwell and feed there until they are large enough to enter deeper waters.

Extensive organismic interaction is found here, as in many marine communities. These communities change (undergo succession) as do land and freshwater ecosystems. One application of this idea is utilized along the Gulf coast to improve fishing. Debris like concrete blocks and old auto bodies are dumped offshore to create artificial reefs. Small marine life soon inhabit the reef which in turn attracts large consumers. The final consumer in this case is, hopefully, the fisherman!

Many additional communities, of course, exist in the sea, such as the organisms of the open sea, those dwelling in abyssal areas, coral reef inhabitants, and so on. (The Suggested Readings provide a detailed account that we cannot provide here.)

Concluding Remarks

In our preface we suggested that an informed citizen should be able to make an intelligent evaluation of the discoveries and issues in science. We have attempted to contribute to this objective by presenting some of the important topics in modern biology. If this helps you to read scientific articles in magazines and newspapers with greater understanding and intellectual enjoyment, or helps you to better assess the issues in local controversies involving sewage disposal, air pollution, fluoridation, and the like, this book has fulfilled a major part of its intention.

We conclude the book by proposing the question: "What are some of the major lessons or guidelines the field of biology offers mankind for tomorrow?" The most basic issue pertains to the relationship between energy and life. Energy considerations are basic to life from the level of molecules to that of biomes. This becomes particularly crucial to man since as his populations increase, his demands for carbon compounds (food) also increase. The source of these compounds are plants; photosynthesis is the vital energy-providing reaction. In the sections on ecology we noted that a community in nature is composed in such a way that its energy producers are also its energy bottleneck. In other words, there is a limit to the number or amount of photosynthesizers in an ecosystem: hence there is a limit to the amount of energy available for life. Man changes the energy relations in ecosystems. He may increase the energy output temporarily for his own use by replacing the natural community with an artificial one (crops). Eventually, though, even man encounters the inevitable photosynthetic bottleneck. What happens when all available lands on the earth are farmed efficiently?

Man often changes the energy relations in his environment unwittingly, that is, as a by-product of some other activity. Perhaps this is the greatest danger we face. Pollution, industrialization, urbanization, and farming each plays a role in disturbing or destroying natural habitats. Pollution of all kinds from cities and industries must be better controlled because it is a destroyer of producer organisms, not to mention its other undesirable powers. Smog, for example, is detrimental to all organisms that breathe it (Figure 25.11). Severe atmospheric pollution can even completely denude the landscape (see Figure 25.12). Pollution can and must be controlled, but this will require large expenditures of money and, in some instances, legal enforcement. Hopefully, we may someday see river systems like the Potomac or Hudson restored to a state where they can safely be used for recreation and water consumption.

Man's industries and dwellings often wipe out natural communities. There is little that can be done about this in most instances although there are a few encouraging signs across North America. In the pulpwood industry, for example, large acreages of forests are carefully managed so that a supply of pulpwood will always be available. This management not only protects the woodland but also provides a refuge for wildlife and, in some cases, recreation areas. This is in sharp contrast to the activities of the lumber industry a few decades ago. At that time, vast areas of forest land were completely stripped of trees and abandoned. Fires and erosion often followed.

Figure 25.11. The city of Los Angeles (a) *on a clear day,* (b) *on a day when the smog is trapped by a temperature inversion approximately 300 feet above the ground, and* (c) *on a day when the smog engulfed the Los Angeles civic center. The base of the temperature inversion is approximately 1500 feet above ground level. The inversion layer, a layer of warm air above the stratum of cool air near the ground, prohibits the natural dispersion of air contaminants to the upper atmosphere. (Los Angeles County Air Pollution Control District.)*

Figure 25.12. Areas around smelters at Copperhill, Tennessee, have been turned into wastelands by fumes from the smelters.

The development of suburbs proceeds rapidly in most parts of the United States, unfortunately for the natural communities they replace. Seldom are even small portions of the natural suburban landscape preserved as parks. This is the sort of thing that can be changed only by the action of citizens working through planning and zoning committees.

The wholesale destruction of natural areas by farming and home building has often brought on drastic consequences. A common consequence is flooding because the normal watersheds that absorb or contain excess rainfall are no longer present. Efforts are then made to retain the water artificially by dams and dikes; then the problems multiply. Thus, in southern Florida, efforts to drain swampland for use in farming were so successful that much of the Everglades were destroyed. As a by-product, the Everglades National Park is at this time in serious difficulty and quite likely doomed.

Our growing populace descends on the countryside during the summer months seeking relaxation. Most of the national parks fill to capacity with campers and visitors. It is evident that much larger recreation areas of this type must be provided soon, not only for the benefit of people but also to keep certain unusual portions of the country from being ruined. Areas like Yellowstone National Park or the Cape Cod National Seashore have educational, aesthetic, and recreational value. There appears to be considerable interest in creating more areas of this type—hopefully this is a trend for the future.

One of the key factors in man's ultimate survival or failure on this planet depends on his treatment of natural communities, biomes, and natural resources in general. Conservation once meant preservation; now it is more meaningful to think of it in the terms of *wise use*, in the sense of practices consistent with empirically derived biological concepts rather than personal, political, or economic dictates or desires. In this light, we must think in terms of the conservation of entire ecosystems, not just certain species in it or parts of it.

Principles

1. The same basic ecosystem concepts of structure and function apply to aqueous and terrestrial environments, despite their different physical and chemical features.
2. Marine environments occupy approximately two

thirds of the surface of the earth and contain the majority of the earth's producer and consumer populations.

Suggested Readings

Batdach, J. E., "Aquaculture," *Science,* Vol. 161, No. 3846 (September 13, 1968).

Bates, Marston, *The Forest and the Sea.* Vintage Books, Random House, New York, 1960.

Boyko, H., "Salt-water Agriculture," *Scientific American,* Vol. 216, No. 3 (March, 1967).

Callison, Charles H., editor, *America's Natural Resources,* revised printing. For the Natural Resources Council of America. Ronald Press Co., New York, 1967.

Carson, Rachel, *The Sea Around Us.* The New American Library of World Literature, New York, 1964.

Carson, Rachel, *Silent Spring.* Houghton Mifflin, Boston, 1962.

Clark, John R., "Thermal Pollution and Aquatic Life," *Scientific American,* Vol. 220, No. 3 (March, 1969).

Edwards, Clive A., "Soil Pollutants and Soil Animals," *Scientific American,* Vol. 220, No. 4 (April, 1969).

Egler, Frank E., "Pesticides in our Ecosystem," *American Scientist,* Vol. 52, No. 1 (March, 1964).

Galtsoff, Paul S., coordinator, "Gulf of Mexico. Its Origin, Waters, and Marine Life," *Fishery Bulletin* 89, Vol. 55, United States Department of the Interior, Fish and Wildlife Service, 1954.

Gunter, G., R. H. Williams, C. C. Davis, and F. G. W. Smith, "Catastrophic Mass Mortality of Marine Animals and Coincident Phytoplankton Bloom on the West Coast of Florida, November 1946 to August 1947," *Ecol. Monographs,* Vol. 18, 1948, pp. 309–324.

Hickman, Kenneth, "Oases for the Future," *Science,* Vol. 154, No. 3749 (November 4, 1966).

Hutner, S. N. and J. J. A. McLaughlin, "Poisonous Tides," *Scientific American,* Vol. 199 (August, 1958).

Kestwen, G. L., "A Policy for Conservationists," *Science,* Vol. 160, No. 3830 (May 24, 1968).

Long, Capt. C. John, USNR (ret.), *New Worlds of Oceanography.* Pyramids Publications, New York, 1965.

Macinko, George, "Saturation: a Problem Evaded in Planning Land Use," *Science,* Vol. 149, No. 3683 (July 30, 1965).

Murphy, Robert C., "The Oceanic Life of the Antarctic," *Scientific American,* Vol. 207 (September, 1962). Offprint No. 864, W. H. Freeman and Co., San Francisco.

Odum, Eugene P., *Ecology.* Holt, Rinehart and Winston, New York, 1963, pp. 112–123.

Plass, Gilbert N., "Carbon Dioxide and Climate," *Scientific American,* Vol. 201 (July, 1959). Offprint No. 823, W. H. Freeman and Co., San Francisco.

Williams, Carroll M., "Third Generation Pesticides," *Scientific American,* Vol. 217 (July, 1967). Offprint No. 1078, W. H. Freeman and Co., San Francisco.

Questions

1. Why do underwater movies taken at considerable depths show blue as the dominant color?
2. If 300 feet is the limit of photosynthesis, what is the energy source for animals that dwell below that depth?
3. Why is dissolved carbon dioxide important to aquatic plants?
4. What is the origin of the salts that maintain the salinity of the oceans?
5. Why, then, are lakes not also salty?
6. Describe what happens to the food pyramid in a lake when chemical fertilizers are added.
7. In what respects does the sea provide a beneficial environment for living things? In what respects is it a detrimental environment?
8. Propose a series of succession stages that could take place on a man-made reef.
9. What kinds of "crops" does man harvest from the sea at present? What additional organisms might be utilized in the future?

APPENDIX

Classification

No satisfactory classification scheme exists. Many have been suggested; each has its advantages and its disadvantages. We present here a scheme modified and simplified from R. H. Whittaker (*Science* **163**:150–160, 1969). We chose the basic scheme he presents because it emphasizes the types of nutrition in the various groups. This will be of help to the beginning student as he analyzes the system of classification.

Kingdom Monera

Cells lack nuclear membranes, true plastids, and mitochondria. The nutrition of most of these organisms is absorptive. Some carry out photosynthesis or chemosynthesis.

Phylum Cyanophyta—chlorophyll *a* and a blue pigment, phycocyanin, present; photosynthesis yields oxygen; movement by gliding (Figure 1).

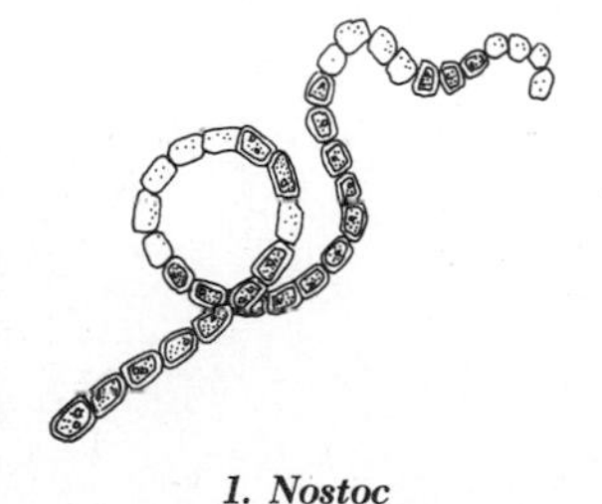

1. Nostoc

Phylum Myxobacteriae—movement by gliding, nutrition by absorption.

Phylum Eubacteriae—movement, if present, by flagella; unable to actively deform the shape of the cell wall; reproduction by transverse binary fission.

Phylum Spirochaetae—able to actively deform the shape of the cell wall; have a contractile axial filament external to the main portion of the cell.

Kingdom Protista

Cells contain true nuclear membranes and mitochondria. The nutrition is carried out by any of the three major types (absorption, ingestion, photosynthesis) or a combination of these. Most organisms are unicellular but some are colonial.

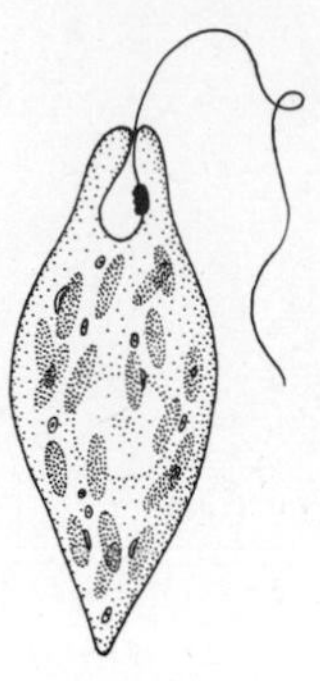

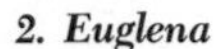

2. *Euglena*

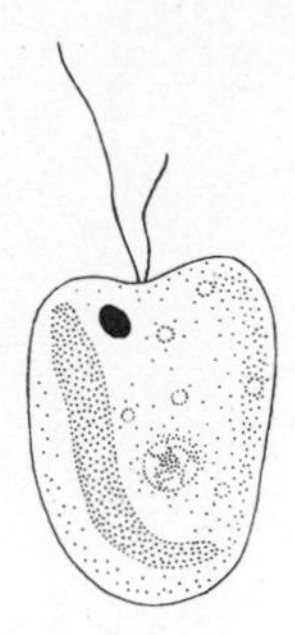

3. *Ochromonas*

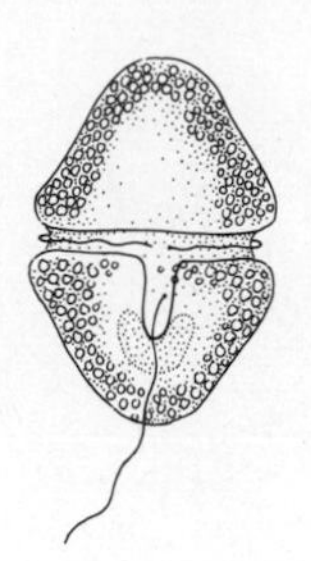

4. *Glenodinium*

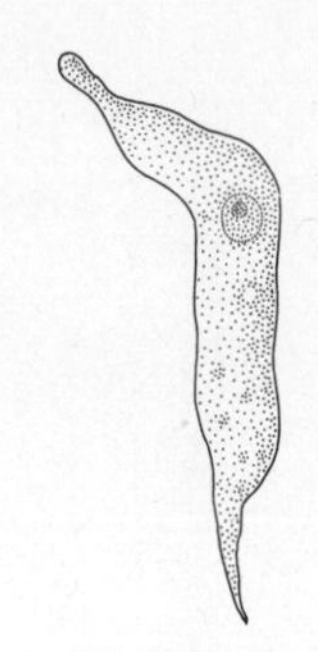

5. *Schaudinella*

Phylum Euglenophyta (euglenoids)—single-celled or colonial; contain chlorophylls *a* and *b*; reproduce by cell division; flagellum present; usually lack a cell wall (Figure 2).

Phylum Chrysophyta (diatoms)—golden to golden-brown or yellowish-green; contain chlorophyll *a*; majority are unicellular or colonial; store food as an oil or a carbohydrate, leucosin (Figure 3).

Phylum Pyrrophyta (golden-brown algae)—contain chlorophylls *a* and *c*; dinoflagellates comprise main class; heavy cell wall divided into plates with furrows, one longitudinal and one transverse (Figure 4).

Phylum Sporozoa—endoparasites; spore-forming; no special organs for locomotion; uni- and multinucleate (Figure 5).

Phylum Mastigophora—unicellular; one to many flagella for locomotion throughout life or at certain stages; generally uninucleate (Figure 6).

Phylum Sarcodina—pseudopodia for locomotion; uni- and multinucleate (Figure 7).

Phylum Ciliophora—locomotion by cilia; micro- and macronuclei (Figure 8).

Kingdom Plantae

Cells contain true nuclear membranes, plastids, mitochondria, and cell wall. Multicellular, photosynthetic organisms. Structural differentiation of cells is found in most forms.

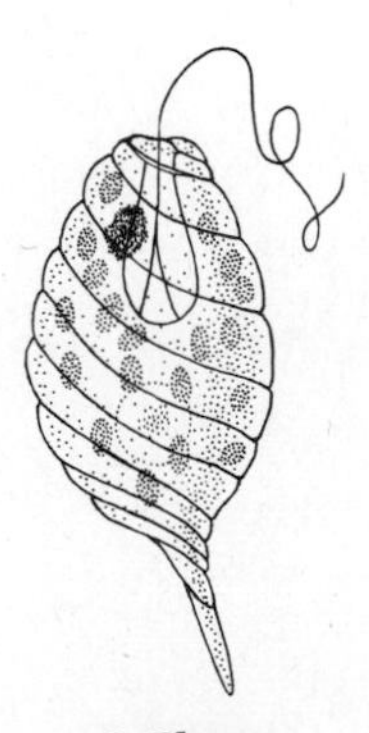

6. *Phacus*

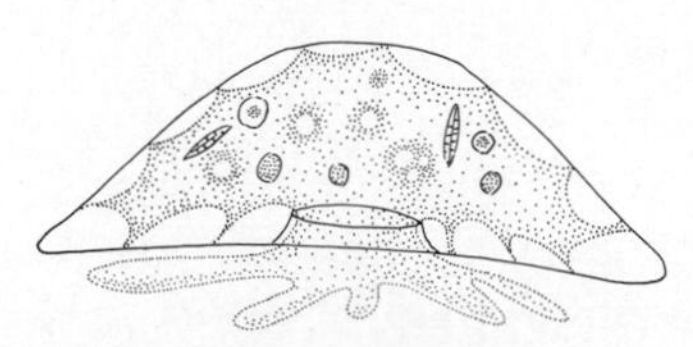

7. *Arcella*

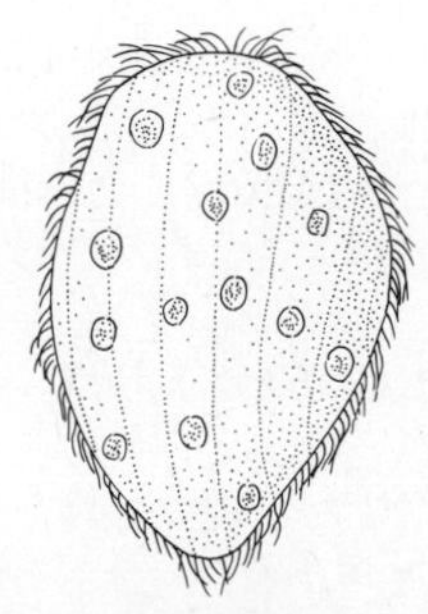

8. *Opalina*

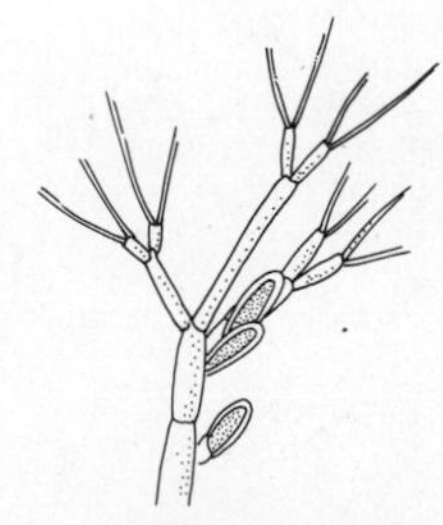

9. *Chantransia*

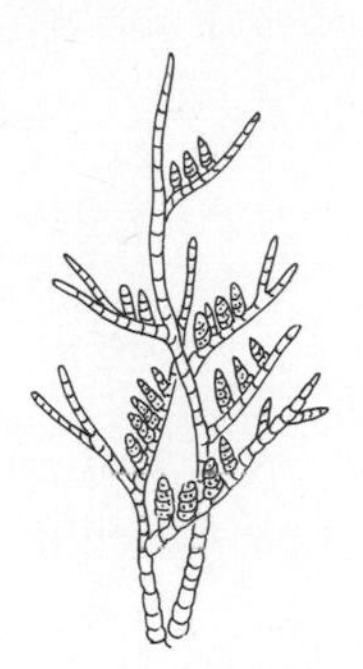

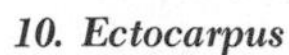

10. Ectocarpus

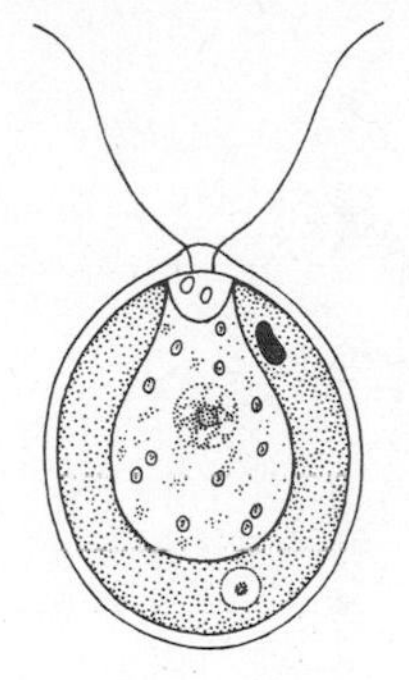

11. Chlamydomonas

12. Marchantia, a liverwort

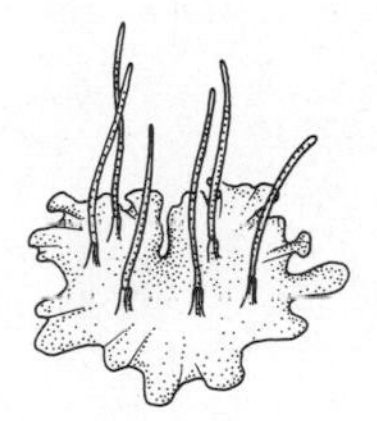

13. Anthroceros, a horned liverwort

Phylum Rhodophyta (red algae)—plastids red, owing to the pigment phycoerythrin; chlorophylls *a* and *d* present; food stored as a starchlike compound; usually conspicuous (Figure 9).

Phylum Phaeophyta (brown algae)—plastids brown, owing to the pigment fucoxanthin; chlorophylls *a* and *c* present; usually large; food stored as an oil or as carbohydrates (Figure 10).

Phylum Chlorophyta (green algae)—aquatic plants; contain chlorophylls *a* and *b*; food stored as starch in plastids; unicellular and multicellular forms; little structural cell differentiation (Figure 11).

Phylum Bryophyta (liverworts and mosses)—chlorophylls *a* and *b* are present; food stored as starch in plastids; generally small land plants found in moist habitats; gametophyte is the conspicuous plant; do not have true roots, stems, or leaves; sperm swims to egg in a layer of water. Contains liverworts, horned liverworts, and mosses, respectively (Figures 12, 13, and 14).

Phylum Tracheophyta (vascular plants)—chlorophylls *a* and *b* present; food stored as starch in plastids; primarily land plants, contain vascular tissue (xylem and phloem), sporophyte is the prominent plant.

Subphylum Psilopsida (chiefly fossil)—rare plants, usually tropical; true roots and true leaves absent (Figure 15).

Subphylum Lycopsida (clubmosses and quillworts)—true roots present, leaves small, arranged spirally; stems not jointed; spores borne on the upper surfaces of some leaves (Figures 16 and 17) (both clubmosses).

14. Polytrichum, a moss

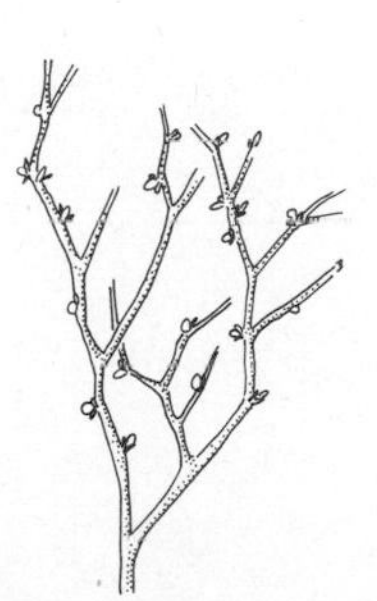

15. Psilotum

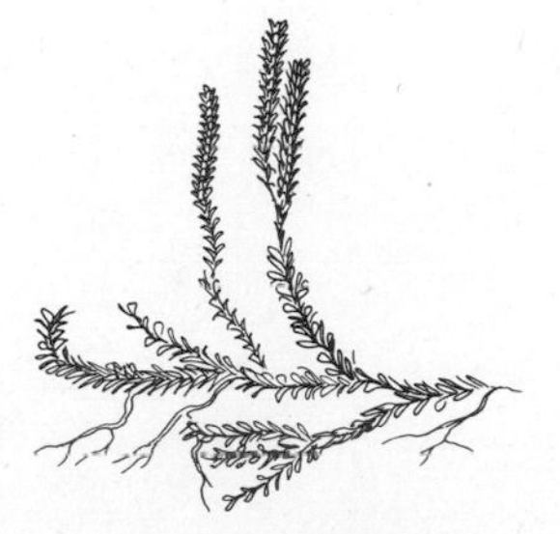

16. Lycopodium

17. Selaginella

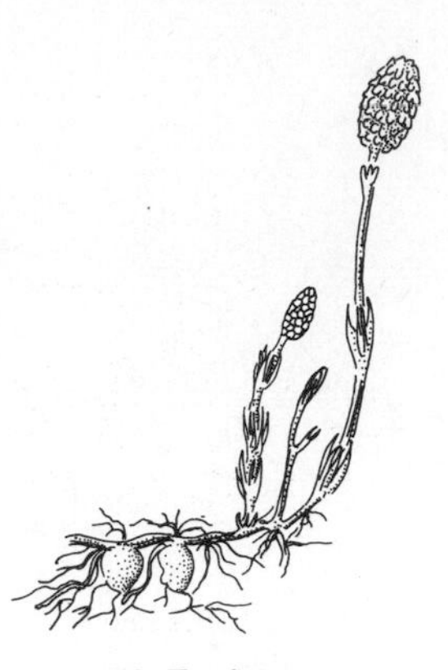

18. Equisetum

19. Dryopteris

20. Cycas

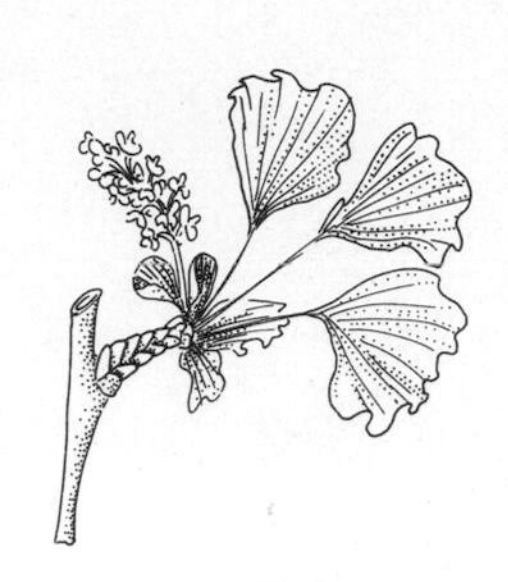

21. Ginkgo

Subphylum Sphenopsida (horsetails)—leaves small, arranged in whorls; stems jointed (Figure 18).

Subphylum Pteropsida—leaves generally large and complex, all well-known plants are in this group—ferns, conifers, and flowering plants.

Class Filicinae (ferns)—horizontal stem bearing roots and leaves; alternation of generations; sporophyte dominant generation (Figure 19).

Class Gymnospermae (cone-bearing seed plants)—bear naked seeds (embryo plus maternal tissue) (Figures 20, 21, and 22).

Class Angiospermae (flower-bearing seed plants)—bear seeds enclosed in the ovary.

Subclass Monocotyledoneae—vascular tissue of stem in scattered bundles; cambium usually absent; leaves with parallel veins; embryo with single cotyledon (Figure 23).

Subclass Dicotyledoneae—vascular tissue of stem forms a cylinder; cambium present; leaves with net venation; embryo has two cotyledons (Figure 24).

22. Pinus

23. Atropa

24. Cassia

25. *Marchella*

26. *Mycophyta*, a mushroom

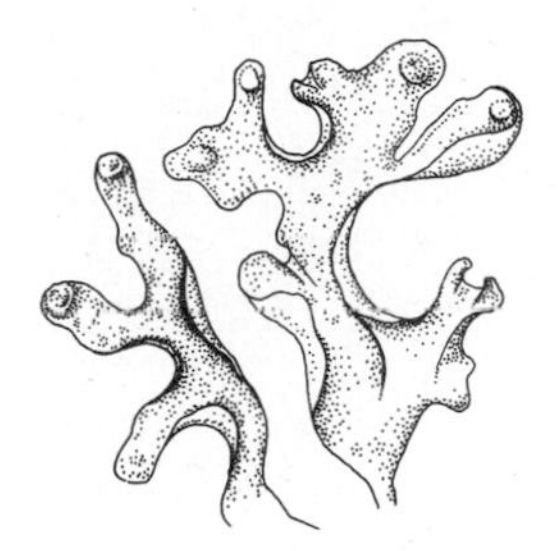
27. Lichen

Kingdom Fungi

28. *Scypha*

Cells contain true nuclear membranes, mitochondria, and a cell wall; plastids and photosynthetic pigments are absent; absorption is the main type of nutrition; little cell differentiation.

Phylum Myxomycota (plasmodial slime molds)—life cycle has stages of separate cells and aggregations of cells; in the multicellular stages, cell walls are present only in the sporangium; swarm cells have flagella.

Phylum Acrasiomycota (cellular slime molds)—life cycle has stages of separate cells and aggregations of cells; cell walls present in multicellular stages; no flagellated cells produced.

Phylum Oomycota—zoospores have two flagella; cell walls contain cellulose.

Phylum Chytridomycota—zoospores have one flagellum; cell walls lack cellulose; mainly aquatic.

Phylum Zygomycota (conjugation fungi)—no flagellated zoospores; cross walls lacking except in reproductive structures.

Phylum Ascomycota (sac fungi)—tubular filaments called hyphae have cross walls; no flagellated spores present. Spore that is produced after fertilization is developed in an ascus. (See Figure 15.8 and Figure 25.)

Phylum Basidiomycota (club fungi)—no flagellated spores present, spores are borne in club-shaped structures; mushrooms are familiar members of this group (Figure 26). Lichens are composite plants made up of algae and fungi in a symbiotic relationship (Figure 27).

29. *Euplectella*

Kingdom Animalia

Cells contain true nuclear membranes and mitochondria but lack plastids, photosynthetic pigments, and cell walls; the more common type of nutrition is ingestion; marked cellular differentiation in most groups.

Phylum Porifera (sponges)—central cavity with pores to the outside; no tissue organization; all aquatic; attached to a substrate (Figures 28, 29, and 30).

Phylum Coelenterata or Cnidaria—aquatic; tentacles surrounding mouth, the only opening to the gastrovascular cavity; single or colonial polyp, which is sessile; or a floating medusa; nematocysts (stinging cells) present.

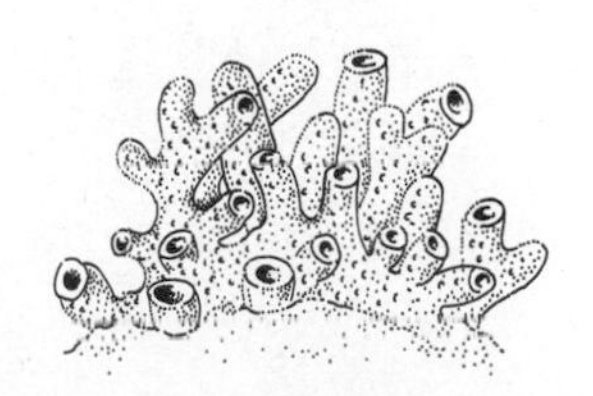
30. *Demospongia*

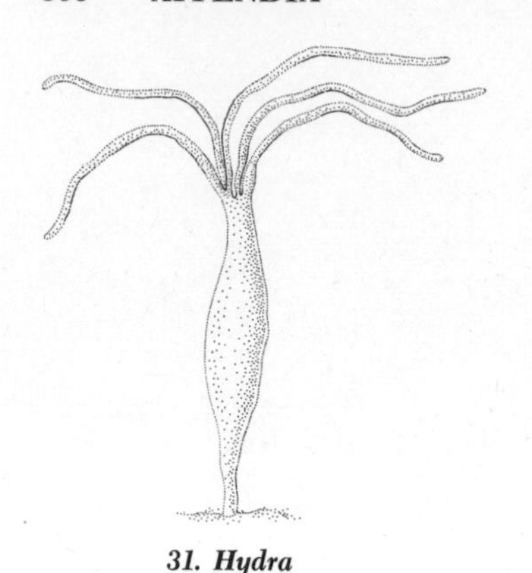

31. Hydra

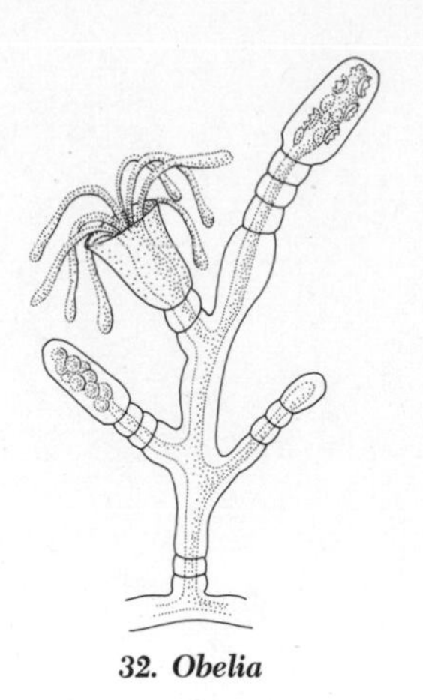

32. Obelia

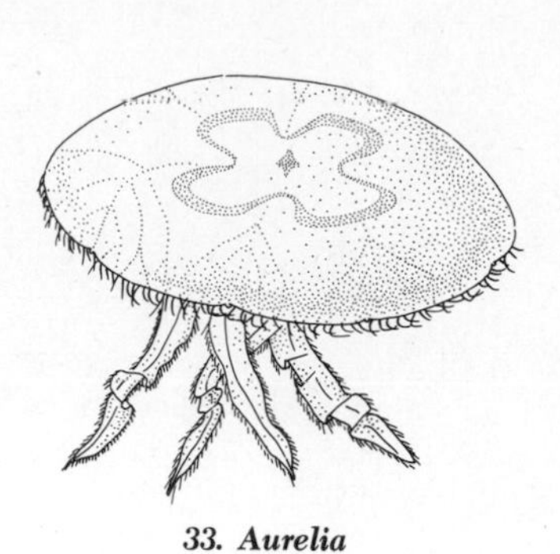

33. Aurelia

Class Hydrozoa—both medusa and polyp forms; solitary or colonial (Figures 31 and 32).

Class Scyphozoa (true jellyfishes)—medusae predominant (Figure 33).

Class Anthozoa—polyp form only; solitary or colonial (Figures 34 and 35).

Phylum Ctenophora—all marine; free swimming; with or without tentacles; locomotion by eight meridional comb plates (Figure 36).

Phylum Platyhelminthes (flatworms, flukes, and tapeworms)—body flattened dorsoventrally; digestive system without anus; absence of circulatory and respiratory systems; no body cavity.

Class Turbellaria (free-living flatworms)—epidermis without cuticle; digestive system present (Figure 37).

Class Trematoda (flukes)—digestive system present, epidermis absent; body covered with cuticle; suckers present (Figure 38).

Class Cestoda (tapeworms)—epidermis absent; digestive system absent; cuticle present (Figure 39).

Phylum Nemertinea (*Rhynchocoela*)—digestive system complete (mouth and anus present); lack a body cavity; circulatory system present; sexes separate; proboscis eversible (Figure 40).

Phylum Aschelminthes—wormlike; superficial segmentation; cuticle present; body cavity a pseudocoel; number of cells or nuclei constant in a group.

Class Rotifera—microscopic; wheel organs at anterior end; mostly fresh water (Figure 41).

***34. Metridium*, an anemone**

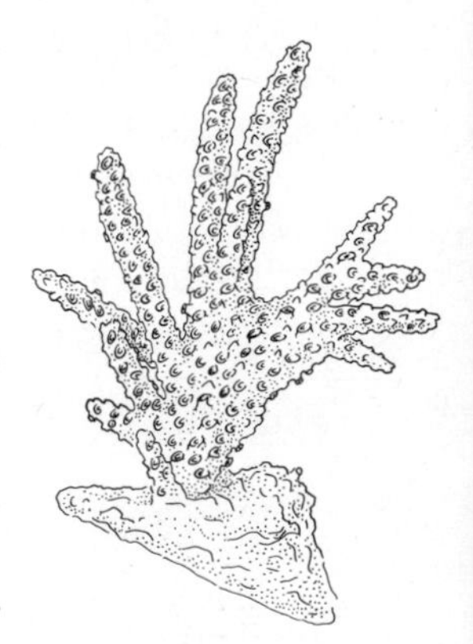

***35. Acropora*, a coral**

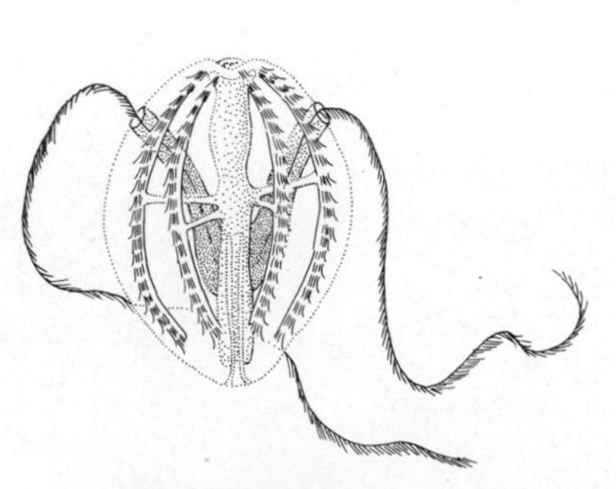

36. Pleurobrachia

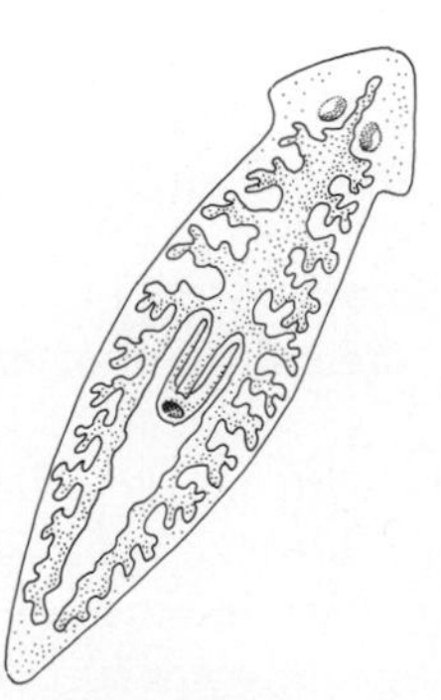

37. Planaria

38. Fasciola

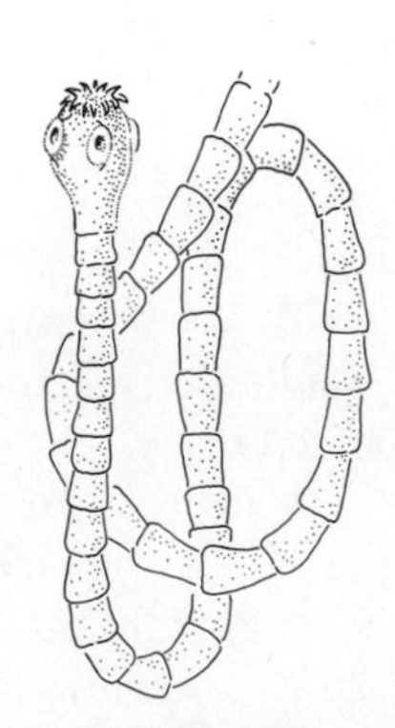

39. Taenia

40. A ribbon worm

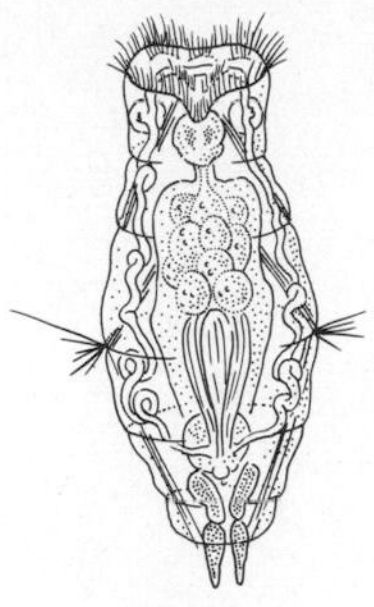

41. Rotifer

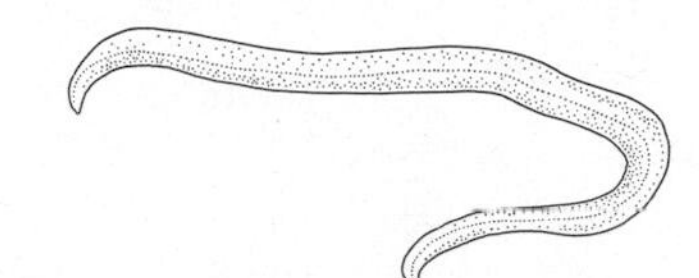

42. *Ascaris*

Class Nematoda (round worms)—body elongate and cylindrical; free-living and parasitic (Figure 42).

Phylum Echinodermata—spiny skinned; all marine; calcareous endoskeleton; pentamerous symmetry; water-vascular system.

Class Holothuroidea—body elongate, sausagelike; skeleton reduced to microscopic ossicles; secondary bilateral symmetry (Figure 43).

Class Asteroidea (sea stars)—stellate shaped; arms number 5 to 50, usually not sharply set off from disc; ambulacral grooves open (Figure 44).

Class Echinoidea (sea urchins and sand dollars)—globular or disc shaped; arms absent; endoskeleton fused; ambulacral grooves closed (Figures 45 and 46).

Class Ophiuroidea (brittle stars, serpent stars, basket stars)—arms sharply set off from disc; ambulacral grooves closed; madreporite on oral side (Figure 47).

Phylum Chordata—pharyngeal gill slits; dorsal hollow nerve cord; notochord.

Subphylum Urochordata (tunicates)—body surrounded by tunic; unsegmented; pharynx with gill slits and endostyle; nerve cord and notochord usually absent in adults; marine; sessile or pelagic (Figure 48).

Subphylum Cephalochordata (lancelets)—marine, sand-dwelling; dorsal nerve cord and notochord present throughout life; pharynx with gill slits and endostyle (Figure 49).

Subphylum Vertebrata—possess a vertebral column, internal skeleton of bone or cartilage, skeleton at least partly encloses brain.

Class Agnatha (lamprey and hagfish, etc.) jawless; no bone present, no appendages, vertebrae small, notochord persistent in adult.

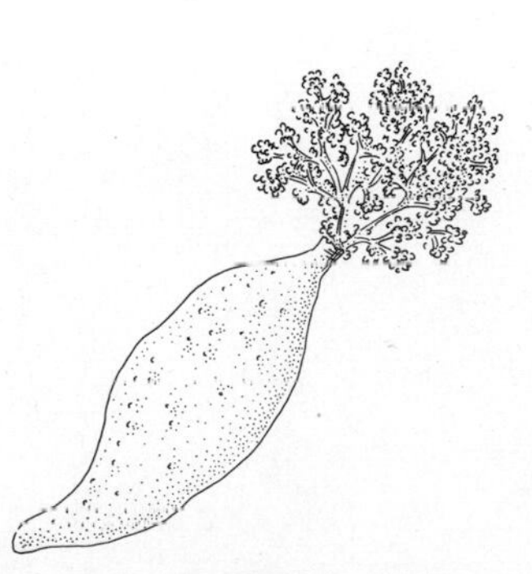

43. *Thyone*

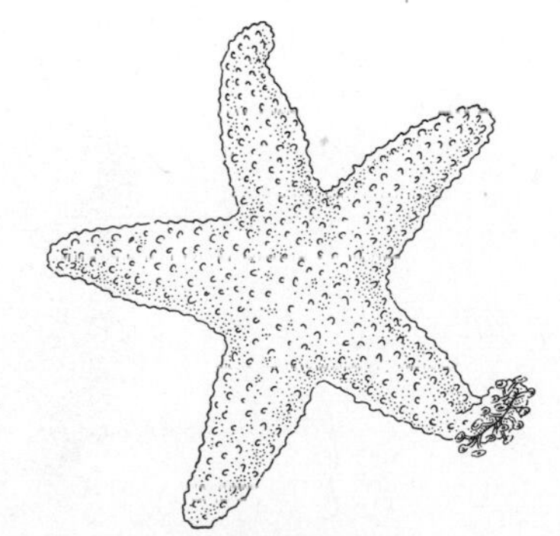

44. *Asterias*

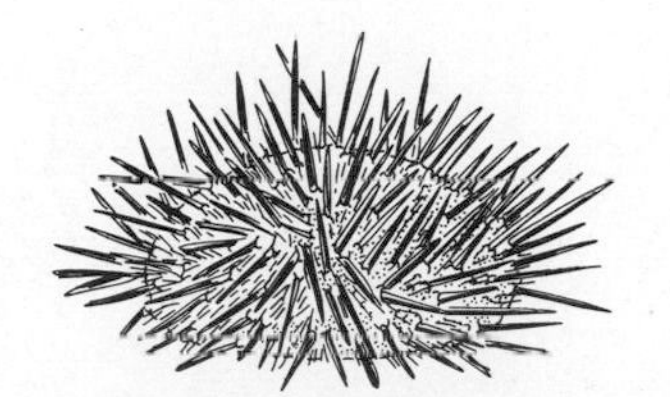

45. Sea urchin

46. Sand dollar

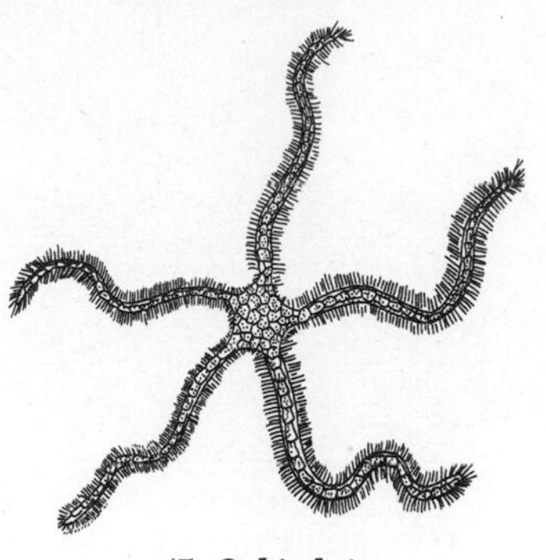

47. *Ophiothrix*

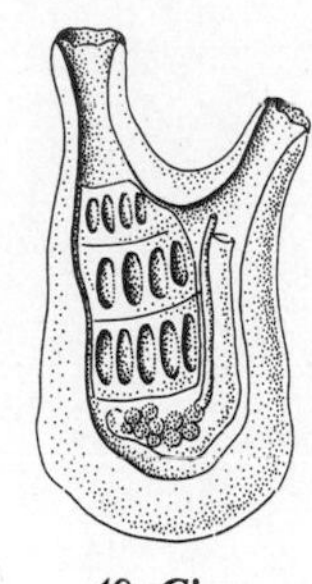

48. *Ciona*

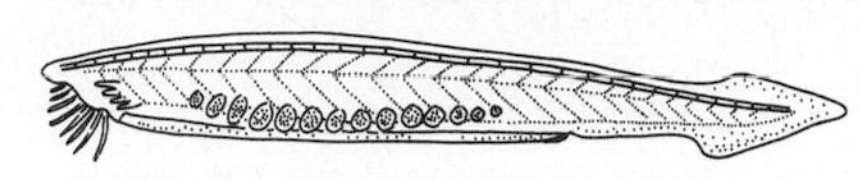

49. Amphioxus

Class Chondrichthyes (sharks, rays, skates, etc.)—skeleton is cartilaginous; possess jaws and paired appendages, fertilization is internal.

Class Osteichthyes (the bony fish)—skeleton is mostly bone; possess jaws and paired appendages.

Class Amphibia (frogs, toads, and salamanders, etc.)—most have a smooth, moist skin without scales; four limbs; usually have lungs.

Class Reptilia (snakes, lizards, turtles, crocodiles)—egg has a protective coating; skin is usually scaly; heart is three chambered.

Class Aves (birds)—feathered, homoiothermous; heart is four chambered.

Class Mammalia (mammals)—have hair; suckle the young; homoiothermous; possess a four-chambered heart.

Phylum Bryozoa (*Ectoprocta*) (moss animals)—mostly marine; colonial; sessile; liphophore; U-shaped digestive system (Figure 50).

Phylum Mollusca—soft-bodied; body composed of head, foot, and visceral hump; body enclosed in mantle which secretes an exoskeleton; digestive system with radular organ; respiration by gills.

Class Amphineura (chitons)—body elongate; shell consists of 8 dorsal plates; head reduced; tentacles and eyes absent (Figure 51).

Class Gastropoda (snails and related forms)—visceral hump spirally coiled; with or without torsion; shell generally present; head with eyes and one or two pairs of tentacles; large flat foot (Figure 52).

Class Scaphopoda (tooth shells or tusk shells)—slightly curved, tubular shell, open at both ends; gills absent; foot modified into burrowing organ (Figure 53).

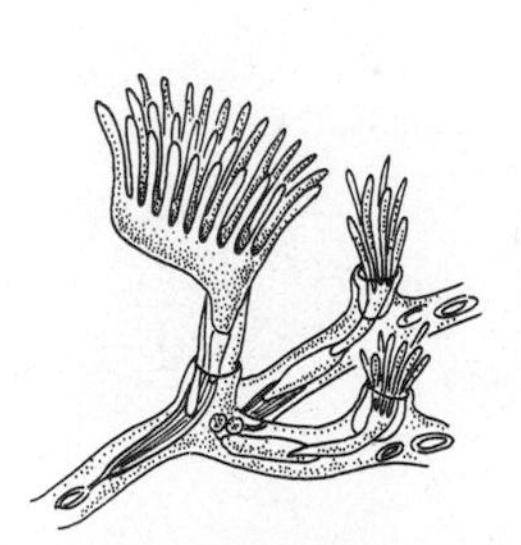

50. *Plumatella*

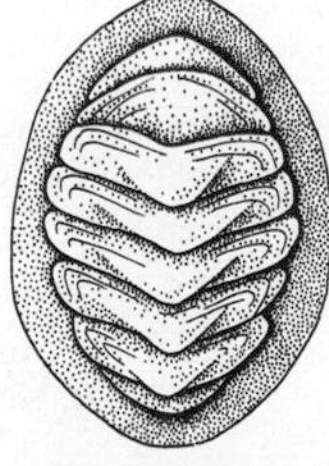

51. *Chiton*

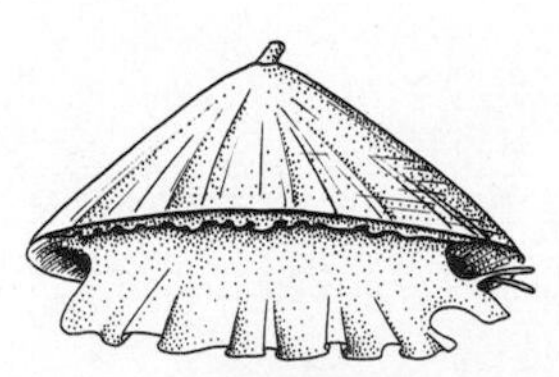

52. *Patella*

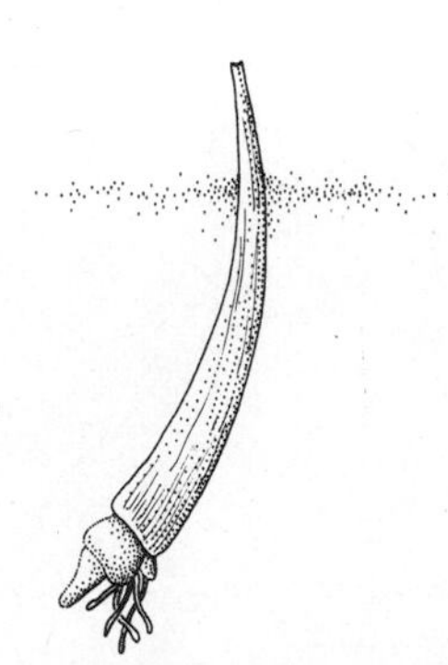

53. *Dentalium*

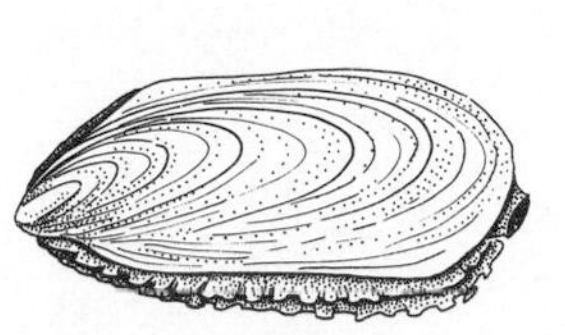

54. Mytilis

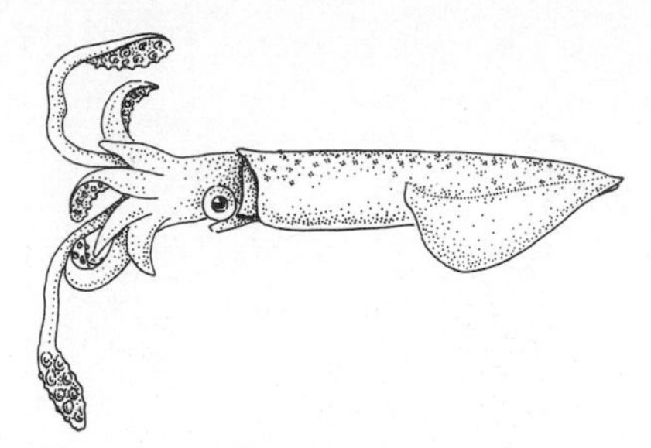

55. Loligo

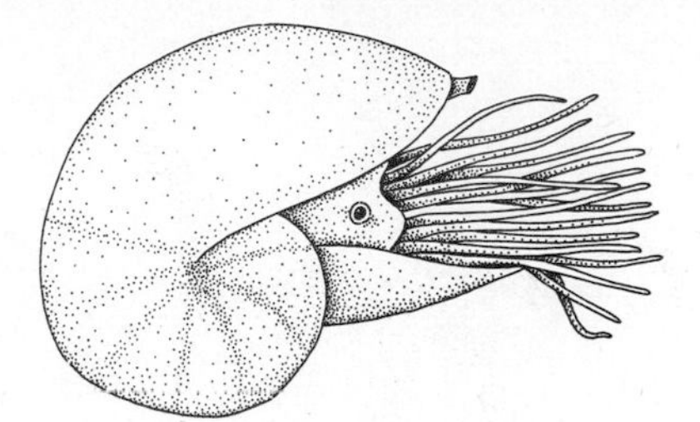

56. Nautilus

Class Pelecypoda (bivalved molluscs)—two lateral valves, hinged dorsally; head absent; radula absent; foot hatchet shaped, used for burrowing; mouth with labial palps; crystalline style in stomach in most species (Figure 54).

Class Cephalopoda (squids, octopi, nautili)—head large and with tentacles; shell external and chambered or internal and reduced or not present; mouth with horny beak and radula (Figures 55, 56, and 57).

Phylum Annelida (segmented worms)—chitinous setae; organ systems segmentally arranged; closed circulatory system; septate coelom; respiration by epidermis or gills.

Class Oligochaeta—primarily terrestrial and fresh water; parapodia absent; setae few per somite; head absent; obvious segmentation (Figure 58).

Class Polychaeta—primarily marine; numerous somites with parapodia bearing many setae; head end with tentacles (Figure 59).

Class Hirudinea (leeches)—primarily freshwater parasites; parapodia and setae lacking, tentacles lacking, large posterior sucker, occasionally smaller sucker at anterior end (Figure 60).

Phylum Arthropoda—body segmented and jointed externally; paired, segmented limbs on some or all of the segments; chitinous exoskeleton; open circulatory system.

Subphylum Chelicerata—body of cephalothorax and abdomen; six pairs of appendages found on cephalothorax; antennae absent. First pair of appendages are pincerlike (chelicera).

Class Arachnida (spiders, ticks, mites, scorpions)—simple eyes present; abdomen contains distinct segments; respiration by book lungs.

Class Merostomata (Horseshoe crabs)—compound and simple eyes present, abdominal segments fused, respiration by book gills.

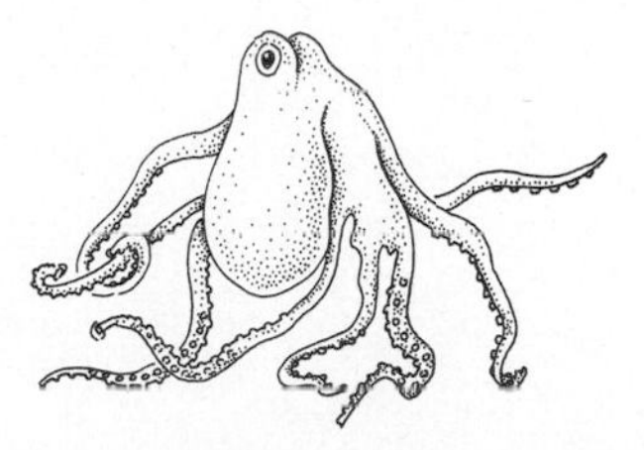

57. Octopus

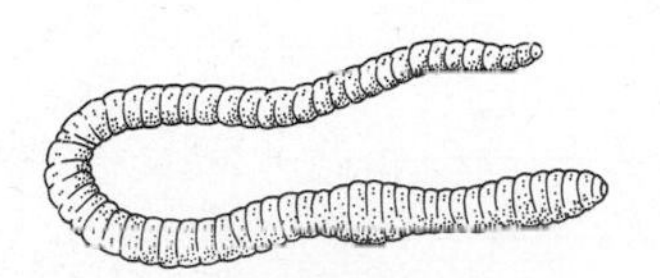

58. Lumbricus

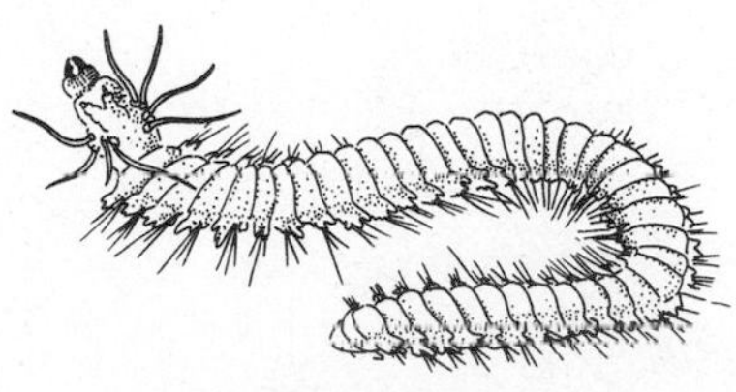

59. Nereis

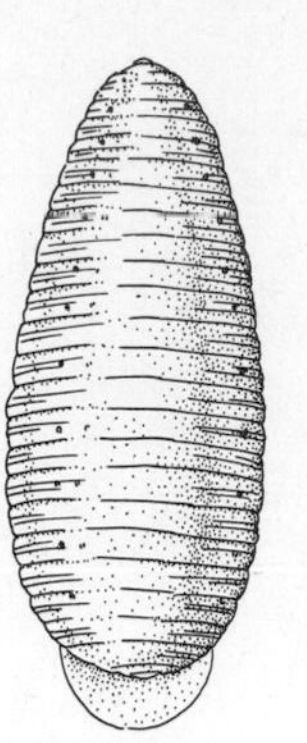
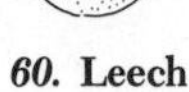

60. Leech

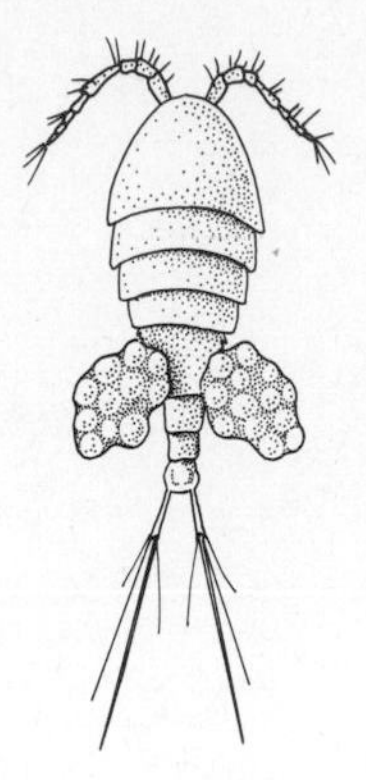

61. Cyclops

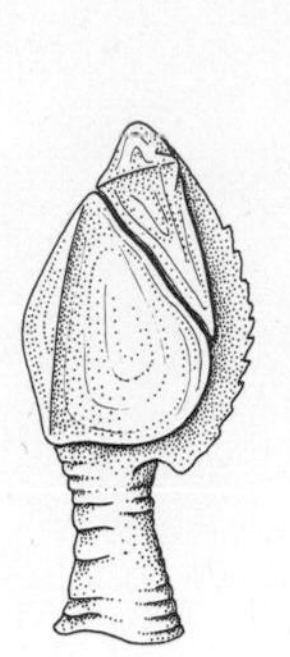

62. Lepas

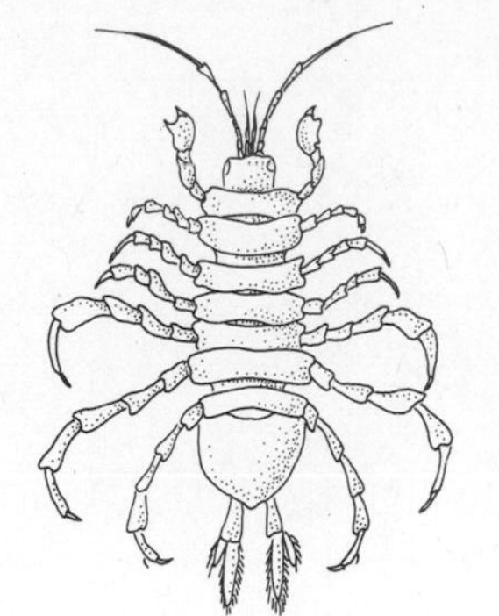

63. Asellas

64. Gammarus

Subphylum Mandibulata—body of cephalothorax and abdomen or head, thorax, and abdomen; cephalothorax or head externally unsegmented; thorax or abdomen segmented; one or two pairs of antennae on head; next pair of appendages are mandibles.

Class Crustacea—aquatic; two pair of antennae. This class is an amazingly diverse group that contains copepods (Figure 61), barnacles (Figure 62), isopods (Figure 63), amphipods (Figure 64), and crabs (Figure 65).

Class Chilopoda (centipedes)—terrestrial; respiration by tracheae; one pair of antennae; one pair of walking legs on each segment of body except the first one and the last two (Figure 66).

Class Diplopoda (millipedes)—terrestrial; respiration by tracheae; one pair of antennae; each abdominal segment bears two pair of walking legs (Figure 67).

Class Insecta (hexapoda, insects)—respiration by tracheae; three distinct body regions, head, thorax, abdomen; one pair of antennae and three pair of walking legs (Figure 68).

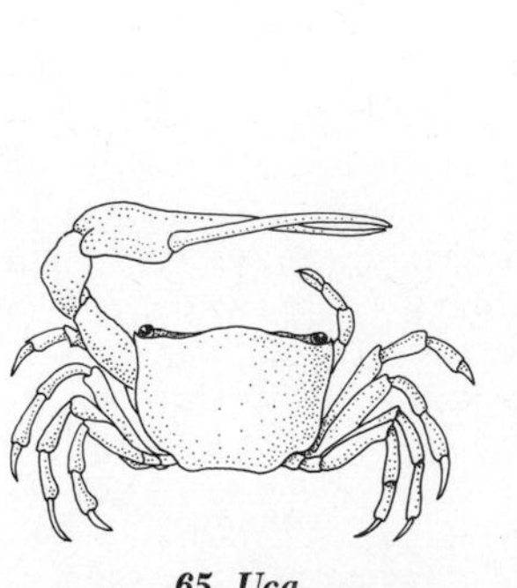

65. Uca

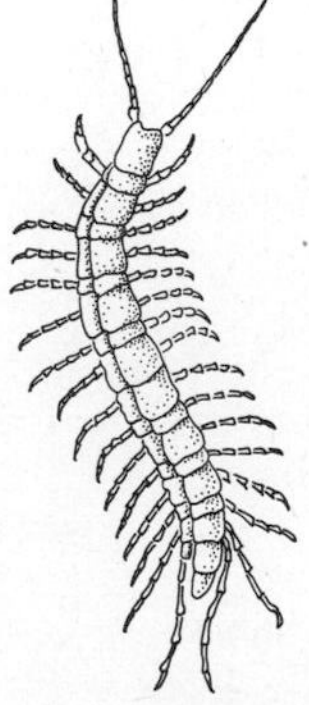

66. Lithobius

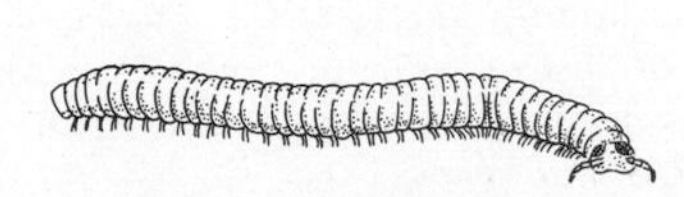

67. Iulus

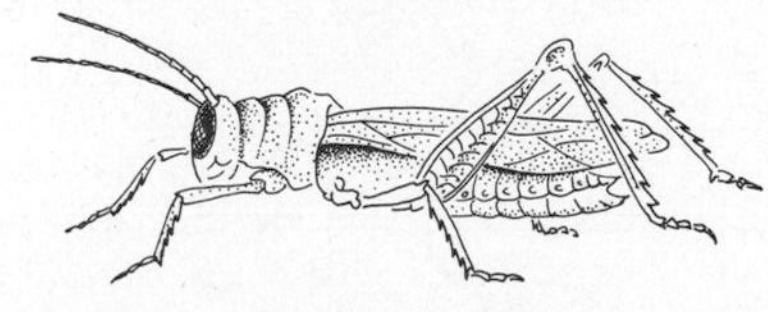

68. Grasshopper

Glossary of Important Terms from the Text

Abiotic component. A nonliving part of the environment such as minerals, sediments, etc.

Absorption spectrum. The pattern of absorption of the various wavelengths of light by a pigment.

Abyssal zone. A cold, lightless zone in the depths of the sea.

Acetabularia. A group of complex single-celled green algae.

Acetate. The salt of a two-carbon organic acid.

Acetylcholine. The chemical transmitter agent at most of the peripheral synapses in the vertebrates.

Acetylcholinesterase. An enzyme that breaks down acetylcholine.

Acid. Any substance that can act as a proton donor; it dissociates in water releasing hydrogen ions.

Activation energy. The amount of energy that must be added initially to start a reaction.

Active transport. The movement of materials across a cell membrane that requires an energy expenditure by the cell.

Adaptation. (1) The process by which an organism becomes progressively better fitted to its environment; (2) Any characteristic of an organism, structural or functional, that contributes to its fitness.

Adaptive radiation. The formation of distinct species from a common ancestor that results from the progressive adaptation of several isolated groups to different environments.

Adenine. One of the organic bases found in DNA and RNA.

ADP (adenosine diphosphate). A molecule that contains an organic base (adenine), a sugar, and two phosphate groups. The bond between the two phosphates is relatively energy rich.

Albinism. A condition in which pigment is entirely lacking from the skin, hair, and eyes.

Algae. A heterogenous assemblage of relatively simple plants containing chlorophyll.

Allele. Any one of the contrasting forms of a gene that may occupy the same site on a chromosome.

All-or-none law. A nerve carries an impulse at full strength or not at all.

Amino acid. An organic molecule that contains at least one carboxyl group and one amino group. Proteins are composed of amino acids.

Analogous structures. Structures that have similar functions but are derived from different sources in the embryo.

Anaphase. The stage of cell division during which the chromosomes move from the center of the cell to the poles.

Antibody. A protein produced by an organism in response to an antigen, inactivating the antigen.

Antigen. A substance that causes the production of an antibody when injected into an organism.

Ascus. A specialized, podlike structure that contains the diploid nucleus and, later, the eight haploid spores of the fungus, *Neurospora*.

Atom. The smallest particle of an element that can enter into a reaction.

ATP (adenosine triphosphate). A molecule that contains an organic base (adenine), a sugar, and three phosphate groups. The bonds between the first and second, and between the second and third phosphate groups are relatively energy rich.

Autosome. A chromosome other than the sex chromosomes.

Autotroph. An organism that can produce organic compounds from inorganic compounds such as water and nutrient salts.

Auxin. One group of plant growth hormones.

Bacteria. A group of single-celled organisms that lack distinct nuclei and plastids.

Bacteriophage. A virus that invades bacteria.

Barr body. A small speck of material, in the nucleus of female cells, that is thought to represent an X-chromosome.

Barriers. Features of the environment or of organisms that prevent gene exchange between two groups of organisms.

Base. Any substance that can act as a proton acceptor; it dissociates in water releasing hydroxyl ions.

Behavior. Any externally directed activity.

Biogeochemical cycles. Pathways through which elements circulate from environment to organisms and back to environment.

Biogeography. The study of the distribution of organisms.

Biome. A large basic unit of terrestrial ecosystems. All of the climax communities of a biome have the same general appearance.

Blastula. An early animal embryo stage having the form of a fluid-filled cavity surrounded by a single layer of cells.

Blastulation. The formation of the blastula.

Bond. The force of attraction binding two atoms in a molecule.

Bond, covalent. A force uniting atoms when they share electrons.

Bond, ionic. A force uniting atoms that bear opposite electrical charges.

Bud. A portion of the parent's body that remains attached to the parent during some of its growth and eventually becomes another organism.

Budding. The process of asexual reproduction by means of buds.

Cancer. A malignant tumor in which the rate of growth is uncontrolled.

Capillary. A small, thin-walled blood vessel that joins a small artery to a small vein. Capillaries exist in networks representing sites where materials enter and leave the blood stream.

Carbohydrate. One of the major classes of organic compounds in living organisms. Each carbohydrate is composed of carbon, hydrogen, and oxygen; there are twice as many atoms of hydrogen as of oxygen.

Carboxyl group. A $-C{\overset{\displaystyle O}{\diagup\!\!\diagup}}\atop{\diagdown OH}$ group that denotes most organic acids.

Carnivore. A flesh-eating animal.

Carotenoid. Yellow and orange pigments associated with chlorophyll in leaves.

Catalyst. A substance that changes the rate of a reaction without being changed by the reaction.

Cell plate. A structure that forms in plant cells to separate daughter cells during cell division.

Cellulose. Complex carbohydrate formed by plants and constituting the major component of cell walls.

Centriole. A structure within the cell composed of nine tubules in a circle. It plays a role in cell division.

Centromere. The portion of the chromosome to which the spindle fiber becomes attached.

Cephalization. An evolutionary trend resulting in the concentration of sensory and neural organs in the head end of animals.

Chalones. Chemicals, normally found in tissues, that inhibit the rate of cell division.

Chlorophyll. The green pigments in plants that captures energy from sunlight to drive photosynthesis.

Chloroplast. A chlorophyll-containing structure found in the cytoplasm of green plants.

Chromatid. One of the two strands of a chromosome during cell division.

Chromatography. A laboratory technique that uses paper strips to separate mixtures of similar materials.

Chromosome. A threadlike body in the nucleus of a cell that contains the hereditary information.

Citric acid cycle. See Krebs cycle.

Cleavage. First divisions of a zygote in which the number of cells increases but the amount of protoplasm remains the same.

Climax. The fairly stable stage of community succession.

Closed circulatory system. A vascular system in which blood is confined to vessels as it circulates through the body.

Commensalism. A symbiotic interaction in which only one member benefits and neither is harmed.

Community. The total of all organisms inhabiting any given area.

Compensation point. The point at which the plant's energy production from photosynthesis is equaled by its energy output in cellular respiration.

Compound. A substance formed by the combining of atoms of two or more different elements.

Conifer. A cone-bearing tree such as a spruce, pine, or fir.

Conjugation. The exchange of nuclear or genetic material by two single-celled organisms.

Consumer. Any organism that uses other plants or animals as an energy source.

Continental shelf. The edge of continental land masses that extends under water to a depth of 600 feet.

Convergence. Independent evolutionary development of similar characters in two groups associated with similar habits or environments.

Corridor. A means of distribution for organisms between two geographic areas.

Cotyledon. The first leaf or leaves developed by an embryo of a seed plant.

Crepuscular. Active in the twilight.

Crossing-over. The exchanging of homologous parts by homologous chromosomes during prophase of meiosis.

Cyst. A stage in which the organism is surrounded by a tough, protective coat.

Cytochrome system. A series of complex molecules involved in hydrogen and electron transport in respiration and photosynthesis.

Cytokinesis. Division of the cytoplasm occurring near the end of mitosis.

Cytokinins. Plant hormones that accelerate cell division and differentiation.

Cytology. The study of cells.

Cytoplasm. The portion of the cell outside the nucleus.

Cytosine. One of the organic bases found in DNA and RNA.

Decarboxylation. Removal of carbon dioxide from a compound.

Deciduous. Loss of leaves during certain seasons.

Decomposer. An organism, often a bacterium or a fungus, that breaks down the complex compounds of dead organisms.

Dehydration. (1) In chemistry, the reaction in which water is removed from a compound; (2) The loss of water from an organism.

Dehydrogenation. The removal of hydrogens from a compound.

Deletion. A type of mutation in which a segment of the chromosome is lost.

Depolarization. The reversal of electrical charge on the nerve cell membrane.

Detritus. A product of the disintegration of some kind of matter.

Diatoms. Planktonic algae enclosed in cases constructed of silica. A major producer organism in many instances.

Differentiation. The specialization of cells and tissues for specific functions.

Diffusion. The movement of molecules away from areas where they are most concentrated.

Digestion. The degradation of substances into smaller molecules by reactions that involve the addition of water to broken bonds.

Dihybrid cross. A genetic cross involving two pairs of genes.

Dinoflagellates. Microscopic organisms with an armorlike covering and two whiplike organelles termed flagella. Often involved in red tides.

Diploid: Two sets of chromosomes, the number present in cells prior to meiosis in most plants and animals.

Diurnal. Active during daylight hours.

Divergence. A pattern of evolution in which groups become less like each other.

DNA (deoxyribonucleic acid). The macromolecule in which all hereditary information is encoded.

Dominance. The state in which one member of a pair of allelic genes is expressed.

Drumstick. Term applied to a small speck of material in the nucleus of white blood cells of female mammals.

Duplication. The type of mutation in which a segment of the chromosome is repeated.

Ecdysis. Molt of the external skeleton of certain invertebrates.

Ecological equivalents. Organisms performing similar activities in widely separated biomes. Example, kangaroo and antelope.

Ecology. The study of the structure and function of nature; a study of the relationships between organisms and their environment.

Ecosystem. A functional unit in nature consisting of nonliving components, producers, consumers, and decomposers.

Ectoparasite. A parasite inhabiting the outside of its host.

Effector. A muscle or gland that acts in response to a nerve impulse.

Egg. The gamete from the female parent.

Electron. Light, negatively charged particles found in atoms.

Electrophoresis. A laboratory technique that utilizes an electrical current to separate mixtures of similar substances.

Element. A substance whose atoms all contain the same number of protons and electrons. One of the fundamental substances that constitute all matter.

Embryo. The organism during its early stages of development.

Embryology. The study of development.

Embryo sac. The female gametophyte of flowering plants; located in the pistil.

Endoparasite. A parasite that inhabits the interior of its host.

Endoplasmic reticulum. A system of membranes located in the cytoplasm of cells.

Endosperm. A tissue in the seed that nourishes the developing plant.

Energy. The capacity for doing work.

Enhancement. A phenomenon in which chlorophylls *a* and *b* together produce a higher photosynthetic rate than chlorophyll *a* alone.

Environment. The surroundings of an organism.

Enzyme. A biological catalyst composed of protein and produced by cells.

Epiphyte. A plant that obtains its water and minerals from the air, often dwelling on another plant.

Estivation. A state of torpor during hot seasons.

Evolution. Descent with modification or change through time.

Fat. One of the major kinds of organic substances. The molecules are composed of glycerol and three fatty acids.

Fatty acid. Organic acids that contain a carboxyl group $\left(-C\begin{matrix} /\!\!/ O \\ \backslash OH \end{matrix}\right)$ and are obtained from the digestion of fats.

Feedback. That part of a self-regulating system in which the substance being controlled regulates the activities of the controller. For example, temperature fluctuations act as feedbacks to a thermostat.

Fermentation. Anaerobic respiration of carbohydrates forming ethyl alcohol.

Fertilization. The union of sperm and egg nuclei.

First Law of Thermodynamics. Energy may not be created or destroyed.

Fission. A form of asexual reproduction in which a single-celled organism divides into two.

Flower. An organ of higher plants that contains the male and female gametophytes.

Fossil. Any evidence of ancient life.

Fruit. Modified parts of a flower that function in seed dispersal.

Fungi. A group of plants that lack chlorophyll, true nuclei, and that reproduce by means of spores.

Gamete. A sex cell.

Gametophyte. The generation of the plant life cycle that produces gametes. A microscopic stage in higher plants.

Gastrovascular cavity. A cavity having only one opening that serves for both digestion and transport.

Gastrula. The developmental stage in animals in which the first morphogenesis occurs. Typically a hollow sac with one opening.

Gastrulation. Formation of the gastrula from the blastula stage.

Gene. A hereditary factor.

Gene, operator. A gene that operates like an on-off switch in controlling one or more structural genes.

Gene, regulator. A gene that controls the production of a protein that may repress the functioning of an operator gene.

Gene, structural. A gene that controls the production of a protein that participates directly in cellular function as an enzyme, structural protein, etc.

Gene pool. The total of all genes contained in a population.

Genetic drift. Random changes in gene frequency that are found in small populations.

Genotype. The description of the genes contained by an organism for the trait or traits being considered.

Gibberellins. A group of plant hormones that promote seed germination, stimulate the synthesis of auxins, induce root growth in seedlings, and help regulate plant growth.

Gill. A respiratory structure found in organisms inhabiting water and moist environments.

Gill slit. In all chordates (at least in the embryonic stages) paired openings from the environment into the pharynx (common region of the respiratory and digestive tracts).

Glucose. The commonest of the six-carbon sugars in living systems.

Glycerol. A three-carbon molecule that forms a portion of the fat molecule.

Glycogen. A carbohydrate used for energy storage in animals.

Glycolysis. The initial stage of the respiration of carbohydrates.

Golgi body. Membraneous structures piled up in a platelike series in the cytoplasm of cells.

Gonad. An organ that produces gametes.

Granum. Closely packed chlorophyll-containing membranes in a chloroplast.

Greenhouse effect. An increase in temperature due to carbon dioxide acting like glass panes in a greenhouse to trap the sun's heat rays on the earth.

Guanine. One of the organic bases found in DNA and RNA.

Haploid. A single set of chromosomes characteristic of gametes, often called the reduced number of chromosomes.

Herbivore. An organism that eats plants.

Hermaphrodite. Containing both ovaries and testes.

Heterogametes. Gametes from the two parents differ in appearance, that is, eggs and sperm.

Heterozygous. Having contrasting genes for a trait.

Hibernation. A state of torpor that occurs with some animals in cold weather.

Homeostasis. A self-regulating or steady-state condition.

Homoiothermal. Possessing an internal temperature-regulating system.

Homologous structure. Similar in developmental details and basic structure as a consequence of a common ancestral background.

Homozygous. Having two genes of the same kind for a trait.

Hormone. A chemical produced by a cell or cells and functioning as a coordinating agent somewhere in the body.

Host. The organisms upon which parasites live.

Hydrolysis. A chemical reaction in which water is added to the broken bonds in a substance. Digestion is a hydrolysis reaction.

Hydroxyl ion. The OH ion.

Hyphae. Slender filamentlike extensions of a fungus produced asexually as the fungus grows.

Hypothalamus. An area of the floor of the brain immediately above the pituitary gland.

Hypothesis. A tentative explanation relating a series of observations.

Indicator species. A fossil found only in rocks of a particular age and hence useful for indicating other rock formations of the same age.

Interphase. Time or stage in which a cell is not in division stages.
Ion. An electrically charged atom or compound.
Isogametes. Sex cells in which eggs and sperm are alike.
Isotope. A different form of an atom owing to a different number of neutrons in its nucleus.

Krebs cycle. A portion of respiration that involves the removal of hydrogen and carbon dioxide.

Lactate. Any salt of lactic acid.
Lactic acid. A three-carbon organic acid that is a product of anaerobic respiration in muscle tissue.
Larva. A developmental stage that does not resemble the adult form.
Law of Mass Action. The rate of a chemical reaction is proportional to the product of the concentrations of the reactants.
Liana. A vine.
Linkage. The inheritance of two or more genes as a unit because they are located on the same chromosome.
Lipid. One of the major kinds of organic compounds common in living systems. Most lipids contain glycerol and at least one fatty acid.
Lymph. A watery fluid surrounding cells and transported by the lymph vessels.
Lysogenic. Strains of bacteria in which viruses live without harming the bacteria because the viruses reproduce only once in each bacterial generation.

Meiosis. A type of cell division in which the chromosome number is reduced from diploid to haploid.
Membrane, differentially permeable. A membrane through which certain kinds of particles other than the solvent molecules may pass.
Membrane, semipermeable. A membrane through which only the solvent molecules may pass.
Mendelian population. A reproductive unit composed of all individuals that can interbreed and their offspring.
Metabolism. The total of all chemical reactions in an organism.
Metaphase. The stage in cell division when chromosomes are located across the middle of the spindle.
Mimicry. Similarity in appearance between two organisms that aids the mimic species in avoiding predators.
Mitochondrion. A structure found in the cytoplasm. It is the site of most cellular respiration.
Mitosis. A type of cell division that maintains the chromosome number unchanged.
Molecule. The smallest combination of atoms that retains the properties of the substance.
Monohybrid cross. A genetic cross involving only one pair of genes.
Morphogenesis. Taking on the basic form and structure of an organism.
Mutation. A change in the genetic material.
Mutation, chromosomal. A change in genetic material caused by the loss of all or a large part of a chromosome.
Mutation, point. A change in genetic material involving minor changes in the structure of a DNA molecule that cannot be detected under a light microscope.
Mutualism. An interaction between organisms beneficial to both parties.

NADP. An organic compound that functions as a hydrogen carrier in cells.
Natality. Birth rate.
Nekton. Aquatic organisms capable of controlling their movements by swimming.
Nephron. The basic functional unit of the vertebrate kidney.
Nerve cell body. The portion of the nerve cell that contains the nucleus.
Nerve fiber. Any of the cytoplasmic extensions of the nerve cell.
Neuron. A nerve cell.
Neuron, association. A nerve cell that conducts impulses from a sensory neuron to the brain or to a motor neuron.
Neuron, motor. A nerve cell that transmits impulses to an effector organ.
Neuron, sensory. A nerve cell that receives impulses from a receptor cell.
Neurosecretion. The production of hormones by nerve cells.
Neutron. A particle within the nucleus of an atom. It carries no electrical charge.
Niche. A portion of the environment occupied by an organism.
Nocturnal. Active during the night.
Nucleic acids. One of the major kinds of organic molecules found in cells. They are composed of long chains of nucleotides.
Nucleolus (nucleoli, plural). A dense mass of granular material found in the nucleus.
Nucleotide. One of the fundamental units making up nucleic acids. Each contains a sugar, a phosphate group, and an organic base.
Nucleus. One of the two major portions of the cell. It controls cell activity.

Open circulatory system. A vascular system in which blood flows into spaces around the body organs rather than being confined to blood vessels.
Operon. A unit in a DNA molecule composed of an operator gene and the structural genes it controls.
Operon theory. A theory that accounts for the differential activity of genes in cells.
Organizer. A mass of cells in a particular region of an embryo that controls the differentiation of some other specific regions of the embryo.
Osmoregulation. The control of the salt and water content of the body.
Osmosis. The diffusion of water molecules through a semipermeable membrane.
Oxidation-reduction. One of the major kinds of chemical reactions within living cells, involving a change in the number of electrons controlled by an atom.

Pampas. A term applied to grasslands in South America.

Parasitism. A symbiotic relationship in which an organism spends part or all of its life cycle on or within its host and uses the host for food.

Parasympathetic nervous system. A portion of the autonomic nervous system, usually antagonistic to the sympathetic portion.

Parthenogenesis. The ability of an unfertilized egg to develop into an adult.

Peptide bond. A carbon-to-nitrogen bond linking two amino acids.

Peripheral nervous system. Nerves outside of the spinal cord and brain that transmit impulses to and from the central nervous system.

PGAL. Phosphoglyceraldehyde, a three-carbon relative of sugars.

pH. A symbol denoting the negative logarithm of the hydrogen ion concentration. Used in indicating acidity and alkalinity.

Phagocytosis. The ability of a cell to surround and engulf particles too large to pass into the cell by diffusion.

Phenotype. The observed characteristic produced by a set of genes.

Pheromone. A chemical released into the environment that influences the behavior of another organism of the same species.

Phloem. Living, tubular cells that transport organic materials throughout plants.

Phospholipid. An organic compound similar to a fat except that one of the fatty acids is replaced with a phosphate group and a nitrogen-containing base.

Photoperiodism. Seasonal changes exhibited by many organisms in response to the number of daylight hours in a twenty-four hour cycle.

Photosynthesis, dark reaction. The incorporation of carbon dioxide into a carbohydrate molecule and the consequent formation of PGAL.

Photosynthesis, light reaction. Light-energized chlorophyll is used to split water and to build ATP.

Phytoplankton. Microscopic plant life dwelling in fresh water or marine environments.

Phytotoxin. A material poisonous to plants.

P_i. Symbol for an inorganic phosphate group.

Pinocytosis. A process whereby portions of fluid-filled canals are pinched off in the cytoplasm.

Plankton. Water dwellers with limited swimming power.

Plastid. A cytoplasmic body composed of membranes functioning either as a pigmented body or as a bearer of chlorophyll.

Poikilothermal. Lacking an internal temperature regulating system.

Polyploid. Containing more than two haploid sets of chromosomes.

Population. A group of organisms inhabiting an area.

Predator. An animal that feeds on other animals.

Prey. An animal used by another for food.

Producer. Photosynthetic organism.

Progeny test. A mating in genetics between a homozygous recessive and an unknown phenotype.

Prophase. The initial stage of mitosis and meiosis.

Protein. An organic material composed of amino acid units.

Protists. Small organisms such as algae and protozoans.

Proton. A positively charged particle in the nucleus of an atom.

Punnett square. A set of boxes used in finding all possible gametic combinations when working with genetics problems.

Purine. A class of organic bases of the nucleic acids including adenine and guanine.

Pyrimidine. A class of organic bases of the nucleic acids including uracil, thymine, and cytosine.

Pyruvate. A three-carbon organic acid important in living systems.

Quantitative inheritance. Inheritance involving the simultaneous action of many genes in an accumulative or additive fashion. Height, weight, and body size are inherited in this way.

Radiation, ionizing. High energy transmissions, usually short wavelengths, which bring about the formation of charged atoms or molecules.

Radioautography. A laboratory technique in which a radioisotope taken in by a cell or tissue can be traced with sensitive photographic film.

Radioisotope. A radioactive isotope of an element.

Receptor. A specialized cell or body that converts an environmental change into nerve impulses.

Recessive. The member of a pair of contrasting genes, that is not expressed.

Recombination. The bringing together of chromosome sets (genes) as in the union of egg and sperm.

Reduction reactions. Removal of oxygen, addition of hydrogen, or addition of electrons.

Repression. When a protein combines with the operator gene and prevents the initiation of messenger RNA synthesis along a segment of DNA.

Respiration, anaerobic. An energy-yielding event in which a substance, other than oxygen, is used as a hydrogen acceptor.

Respiration, cellular. The sum of energy-yielding chemical events within cells.

Respiration, external. The exchange of respiratory gases between the organism and the environment.

Ribosome. Tiny granular bodies in the cytoplasm that function in the synthesis of proteins.

RNA. Ribonucleic acid; long chains of nucleotides connected by phosphate-to-sugar bonds.

RNA, messenger. A long, single-stranded chain of nucleotides that transfers a coded chemical message from DNA to the cytoplasm.

RNA, transfer. A shorter, coiled strand of nucleotides coded for transporting a specific amino acid.

Salinity. Grams of dry salts per 1000 grams of water, hence parts per thousand.

Savanna. Term applied to grasslands in South America.

Second Law of Thermodynamics. When energy is changed from one form to another, some of the energy is dispersed into unusable heat.

Secretion. The product of a cell or gland that is used by an organism.

Seed. A plant embryo surrounded by layers of food material and a tough outer coat.

Segregation. The separation of homologous chromosomes into daughter cells during meiosis.

Selection, natural. Survival of the best adapted individuals.

Sessile. Attached to a substrate.

Sex-linked inheritance. Refers to traits located on the sex chromosomes, especially the X-chromosome.

Signal. An auditory, visual, chemical, or tactile means of communication between animals.

Social aggregates. Flocks, herds, or other groupings of one kind of organism.

Speciation. The formation of two or more species from one.

Species. An interbreeding population or group of populations that is reproductively isolated from other such populations.

Spindle. A structure made of protein fibers that functions in separating chromosomes during cell division.

Spore. A reproductive cell, often surrounded by a protective coat, and serving the function of dispersal.

Sporophyte. The spore-producing portion of a plant life cycle.

Sporulation. The forming of spores.

Standing crop. The amount of abiotic and biotic materials present in an ecosystem at any given time.

Starch. A storage form of carbohydrate found in plant cells.

Steady state. A self-regulating or homeostatic system.

Steppe. Term applied to grasslands in Russia.

Stimulus. A detectable change in the environment.

Stomata. Small openings in the surface of leaves that allow gases to enter and leave.

Substrate. A compound acted upon by an enzyme.

Succession. The orderly, predictable changes a community undergoes until it attains a fairly stable climax stage.

Sugar, simple. The smallest carbohydrates. Glucose is an example.

Summation. Initiating a nerve impulse by a succession of stimuli.

Symbiosis. An intimate association between two organisms such as mutualism, commensalism, and parasitism.

Sympathetic nervous system. A portion of the autonomic nervous system, often antagonistic to the parasympathetic system.

Synapse. The tiny gap between the end of one nerve fiber (axon) and the beginning of the next (dendrite).

Synthetic reaction. Chemical reaction in which molecules are combined into larger ones.

Telophase. The final stage of mitosis or meiosis, leading to the formation of daughter cells.

Territory. An area defended by an animal.

Test cross. See Progeny test.

Tetraploid. Containing four haploid sets of chromosomes.

Thymine. One of the important organic bases found in DNA.

Tissue. A group of cells with a specialized function.

Transduction. Genetic transfer between bacteria involving the use of a virus as the transmitting agent.

Transformation. A genetic change in one strain of bacteria brought about by exposure to freshly killed bacteria of another strain.

Transpiration. The emission of water vapor from plant surfaces.

Triplet code. The coding of one amino acid by three nucleotides.

Tropism. A movement in response to a difference in stimulation of the two sides of the organism.

Unit membrane. A term applied to cell membranes which consist of two layers of proteins and two layers of lipids.

Uracil. One of the important organic bases found in RNA.

Vacuole. Fluid-filled spaces in cells with various functions such as food vacuoles and contractile vacuoles.

Vascular tissue. Specialized tissue for conducting materials in plants or animals.

Veld. Term applied to grasslands in South Africa.

Vertebrate. A fish, amphibian, reptile, bird, or mammal.

Virus. A submicroscopic organism, consisting of a DNA or RNA molecule covered with a protein coat.

X-chromosome. A chromosome concerned with sex determination.

Xerophyte. A plant that is adapted to living in an arid environment.

Xylem. Specialized tubelike cells that conduct water in plants.

Y-chromosome. A chromosome concerned with sex determination and found only in males.

Zooplankton. Small consumer organisms inhabiting fresh-water and marine environments.

Zygote. A fertilized egg.

Index

Italicized numbers indicate illustrations